ADVANCES IN HEAT TRANSFER

Volume 6

Contributors to Volume 6

A. F. CHARWAT
G. A. DREITSER
U. GRIGULL
W. HAUF
E. K. KALININ
B. S. PETUKHOV

Advances in
HEAT TRANSFER

Edited by

James P. Hartnett
Department of Energy Engineering
University of Illinois
at Chicago
Chicago, Illinois

Thomas F. Irvine, Jr.
State University of New York
at Stony Brook
Stony Brook, Long Island
New York

Volume 6

 1970

ACADEMIC PRESS · New York · London

Copyright © 1970, by Academic Press, Inc.
ALL RIGHTS RESERVED.
NO PART OF THIS BOOK MAY BE REPRODUCED IN ANY FORM,
BY PHOTOSTAT, MICROFILM, RETRIEVAL SYSTEM, OR ANY
OTHER MEANS, WITHOUT WRITTEN PERMISSION FROM
THE PUBLISHERS.

ACADEMIC PRESS, INC.
111 Fifth Avenue, New York, New York 10003

United Kingdom Edition published by
ACADEMIC PRESS, INC. (LONDON) LTD.
Berkeley Square House, London W1X 6BA

LIBRARY OF CONGRESS CATALOG CARD NUMBER: 63-22329

PRINTED IN THE UNITED STATES OF AMERICA

LIST OF CONTRIBUTORS

A. F. CHARWAT, *Department of Engineering, University of California, Los Angeles, California*

G. A. DREITSER, *Moscow Aircraft Institute, Moscow, USSR*

U. GRIGULL, *Technische Hochschule München, Institut A für Thermodynamik, Munich, Germany*

W. HAUF, *Technische Hochschule München, Institut A für Thermodynamik, Munich, Germany*

E. K. KALININ, *Moscow Aircraft Institute, Moscow, USSR*

B. S. PETUKHOV, *High Temperature Institute, Academy of Science of the USSR, Moscow, USSR*

CONTENTS

List of Contributors v
Preface ix
Contents of Previous Volumes xi

Supersonic Flows with Imbedded Separated Regions

A. F. Charwat

I. Introduction 1
II. The Chapman–Korst Model (Cavities Controlled by Recompression) 7
III. The Structure of the External and Recirculating Internal Regions 43
IV. Cross-Stream Interaction (Cavities Controlled by Separation) 83
    Symbols 121
    References 122

Optical Methods in Heat Transfer

W. Hauf and U. Grigull

I. Introduction 134
II. Principles of Geometrical Optics 135
III. Boundary-Layer Optics 147
IV. Theory of Shadowgraph and Schlieren Methods 161
V. Theory of Interference Methods 191
VI. Application Examples and Their Evaluation 312
VII. Appendix 348
    Nomenclature 360
    References 362

Unsteady Convective Heat Transfer and Hydrodynamics in Channels

E. K. Kalinin and G. A. Dreitser

Introduction	367
I. Statement of Theoretical and Experimental Investigations of Unsteady Heat Transfer and Hydrodynamics in Channels	370
II. Hydrodynamics of Unsteady Fluid Flow in Channels	386
III. Unsteady-State Heat Transfer in Channel Flow	411
IV. Some Aspects of Unsteady Heat Transfer	474
Nomenclature	484
References	486

Heat Transfer and Friction in Turbulent Pipe Flow with Variable Physical Properties

B. S. Petukhov

I. Introduction	504
II. Analytical Method	507
III. Heat Transfer with Constant Physical Properties	521
IV. Heat Transfer and Skin Friction for Liquids with Variable Viscosity	528
V. Heat Transfer and Skin Friction for Gases with Variable Physical Properties	533
VI. Heat Transfer and Skin Friction for Single-Phase Fluids at Subcritical States	543
VII. Conclusion	560
Nomenclature	561
References	561

Author Index	565
Subject Index	576

PREFACE

Research in heat transfer continues at an ever-increasing pace. This is borne out by the increased number of papers appearing in the well-established journals such as the *Journal of Heat Transfer* and the *International Journal of Heat and Mass Transfer*. The many international conferences as exemplified by the one held in Japan in 1967, those held in Yugoslavia in 1968 and 1969, and those scheduled for Yugoslavia and Paris in 1970 not only give evidence of this increased activity but also a growing need for improved communication.

"Advances in Heat Transfer" contributes to this objective of improved communication by presenting coordinated and unified monographs on important subdivisions of the general field of heat transfer, drawing on research literature published in the United States and abroad.

The favorable response to the first five volumes by the scientific and engineering community supports the view that this serial publication is playing a significant role. Against this favorable backdrop, the editors are pleased to announce Volume 6 and trust that it will join the earlier volumes in providing a unique service to scientific researchers, practicing engineers and educators.

CONTENTS OF PREVIOUS VOLUMES

Volume 1

The Interaction of Thermal Radiation with Conduction and Convection Heat Transfer
 R. D. Cess
Application of Integral Methods to Transient Nonlinear Heat Transfer
 Theodore R. Goodman
Heat and Mass Transfer in Capillary-Porous Bodies
 A. V. Luikov
Boiling
 G. Leppert and C. C. Pitts
The Influence of Electric and Magnetic Fields on Heat Transfer to Electrically Conducting Fluids
 Mary F. Romig
Fluid Mechanics and Heat Transfer of Two-Phase Annular-Dispersed Flow
 Mario Silvestri

AUTHOR INDEX—SUBJECT INDEX

Volume 2

Turbulent Boundary-Layer Heat Transfer from Rapidly Accelerating Flow of Rocket Combustion Gases and of Heated Air
 D. R. Bartz
Chemically Reacting Nonequilibrium Boundary Layers
 Paul M. Chung
Low Density Heat Transfer
 F. M. Devienne
Heat Transfer in Non-Newtonian Fluids
 A. B. Metzner
Radiation Heat Transfer between Surfaces
 E. M. Sparrow

AUTHOR INDEX—SUBJECT INDEX

Volume 3

The Effect of Free-Stream Turbulence on Heat Transfer Rates
  J. KESTIN
Heat and Mass Transfer in Turbulent Boundary Layers
  A. I. LEONT'EV
Liquid Metal Heat Transfer
  RALPH P. STEIN
Radiation Transfer and Interaction of Convection with Radiation Heat Transfer
  R. VISKANTA
A Critical Survey of the Major Methods for Measuring and Calculating Dilute Gas Transport Properties
  A. A. WESTENBERG

AUTHOR INDEX—SUBJECT INDEX

Volume 4

Advances in Free Convection
  A. J. EDE
Heat Transfer in Biotechnology
  ALICE M. STOLL
Effects of Reduced Gravity on Heat Transfer
  ROBERT SIEGEL
Advances in Plasma Heat Transfer
  E. R. G. ECKERT and E. PFENDER
Exact Similar Solution of the Laminar Boundary-Layer Equations
  C. FORBES DEWEY, JR., and JOSEPH F. GROSS

AUTHOR INDEX—SUBJECT INDEX

Volume 5

Application of Monte Carlo to Heat Transfer Problems
  JOHN R. HOWELL
Film and Transition Boiling
  DUANE P. JORDAN
Convection Heat Transfer in Rotating Systems
  FRANK KREITH
Thermal Radiation Properties of Gases
  C. L. TIEN
Cryogenic Heat Transfer
  JOHN A. CLARK

AUTHOR INDEX—SUBJECT INDEX

Supersonic Flows with Imbedded Separated Regions

A. F. Charwat

Department of Engineering, University of California, Los Angeles, California

I. Introduction	1
II. The Chapman–Korst Model (Cavities Controlled by Recompression)	7
A. Point-Recompression Assumption	9
B. The Shear Layer	12
C. Distributed Recompression	27
D. Confrontation with Experiments	33
III. The Structure of the External and Recirculating Internal Regions	43
A. The Separation at an Expansive Corner	44
B. The Internal Flow	58
C. Separated Regions in Notches	68
IV. Cross-Stream Interaction (Cavities Controlled by Separation)	83
A. Measurements in Separated Regions Ahead of Steps and Compression Ramps	84
B. Numerical Integration of the Interaction Equation.	98
C. Moment Methods	107
D. Concluding Remarks	118
Symbols	121
References	122

I. Introduction

Regions of separated flow appear when the no-slip boundary condition for an attached flow wetting the surface of an immersed body cannot be satisfied; the inviscid pressure gradient is such that the boundary layer separates from the surface. A portion of the boundary of the through-flow then appears to be replaced by a free streamline enclosing a region with closed internal flow. The theoretical problem of attached flow can be posed unequivocally, if not always solved. The problem of a flow with separated bubbles is very much more complex: it involves matching of separate domains along the dividing free streamline, the shape of which is not known *a priori*. The uniqueness of such composite flow structures is demonstrated experimentally, although a theoretical proof has yet to be given.

In slow flow (very low Reynolds numbers), there are genuine theoretical methods (asymptotic expansion techniques) which show the formation of wakelike regions as the flow develops from rest (for instance, the wake downstream of a sphere). Recent numerical investigations using high-speed computers to integrate the full Navier–Stokes equations for steady flow [for instance, Keller and Takami (*1*)], or modeling numerically the development of the flow from rest by forward integration of the unsteady problem in time [for instance, the work of Fromm (*2*) and associates], or using numerical Monte Carlo techniques (*3*) exhibit the formation of wakelike regions. Perhaps complete solutions for arbitrary Reynolds numbers may become possible in the future, at least in individual cases. For the moment, however, the practically interesting domain of Reynolds and Mach numbers is not amenable to rigorous analysis, not even with the use of computers. The problem can only be approached by formulating a physical model in which some elements of the complex interaction are assumed to dominate over others.

Historically, the first scientific paper dealing with separated flow was based on an inviscid flow model. This is the Kirchhoff (*4*) free-streamline theory (1869) for the wake behind a blunt body. It amounts to finding the shape of an isobaric surface enclosing a "bubble" filled with quiescent fluid with no dynamic coupling between them: within the framework of the inviscid theory the external flow slips over this free boundary. There is a family of bubble shapes corresponding to a (arbitrary) choice of bubble pressure which is equivalent to the choice of the point of breakaway of the free streamline from the body surface. Inviscid theory provides no condition which allows one to close the problem uniquely.

The reason for the incompleteness of Kirchoff free-streamline solution became clear 50 years later with the development of boundary-layer concepts. It was realized that a real viscous flow cannot be considered inviscid in the neighborhood of certain singular surfaces, such as the separated dividing streamline even in the limit of infinite Reynolds numbers. Viscous forces are dominant in the neighborhood of the surfaces and, although these forces act locally, they generate crucial constraints on the overall motion. Whether Kirchhoff's model is an appropriate asymptotic solution for an infinite Reynolds number is still a controversial question.

One of the consequences of the role of viscosity is that the "bubble" must be closed: the tangential viscous stresses exerted along the free boundary of the bubble accelerate the fluid contained in it downstream and there must be a reattachment or closure point at which the scavenged inner flow is returned towards separation. This remains true in a time-average sense when the wake flow is unsteady; for instance, when it sheds a double row of periodic vortices—the Karman vortex street. The closure of the wake constitutes a key condition, which was not recognized until the 1950s, and marks the beginning of a modern approach to the entire problem.

Another consequence of the viscous stress along the dividing streamline is that the fluid in the bubble cannot be truly quiescent. A circulation will build up, the time-averaged strength of which is determined by the balance between the driving shear work along the dividing streamline and the dissipation against the solid boundaries of the bubble. This may contain the key to the problem of unsteadiness and introduce an effect of the geometry of the solid boundaries of the inner flow region.

Yet another example of the importance of local viscous constraints on overall problem is this: consider a two-dimensional separation of the free streamline occurring on a continuous surface. The point of separation is usually taken as that point at which the shear at the wall becomes zero and the wall velocity reverses direction. The well-known local Taylor expansion of the Navier–Stokes equation (5) prescribes an angle relative to the surface at which the separating streamline must leave the surface. It is related to the ratio of the shear gradient to the pressure gradient along the wall, approaching separation. It is clear that such a discontinuous change in the slope of the equivalent boundary of the external flow is associated with a relatively strong pressure perturbation such that equilibrium is established by an essential interaction mechanism: the viscous conditions for separation to occur and the geometry and pressure distribution of the external flow are intercoupled and must be determined simultaneously as part of the solution.

The analysis of separated regions proceeded during the past 25 years by successively focusing on one special geometry after another in which some of these problems could be assumed to play a minor role. This period coincides with the period of rapid development of supersonic aerodynamics which provided many practical applications for this research. Moreover, an external hyperbolic flow proved to be much simpler to deal with in treating interaction problems such as this one. Highly refined experimental techniques among which visualization of the flow by schlieren optical methods provided much excellent data which helped in the formulation of simple semiempirical models and in checking assumptions with experiment. Consequently, a large body of theoretical and experimental work in this field pertains to supersonic and, more recently, to hypersonic flows. The basic structure of the theories and models of separated flows is, however, applicable to subsonic flows as well, although results in that regime are much scarcer.

Figure 1 shows schematically a sequence of increasingly complex separated regions, in the sense of an increasing role of the interaction between the "inner" and "outer" flow. Perhaps the simplest case is that of a wake occurring behind a downstep in the boundary (Fig. 1a), or the analogous case of the wake behind a blunt body (which differs from the first case by the line of symmetry replacing the solid wall). The separation

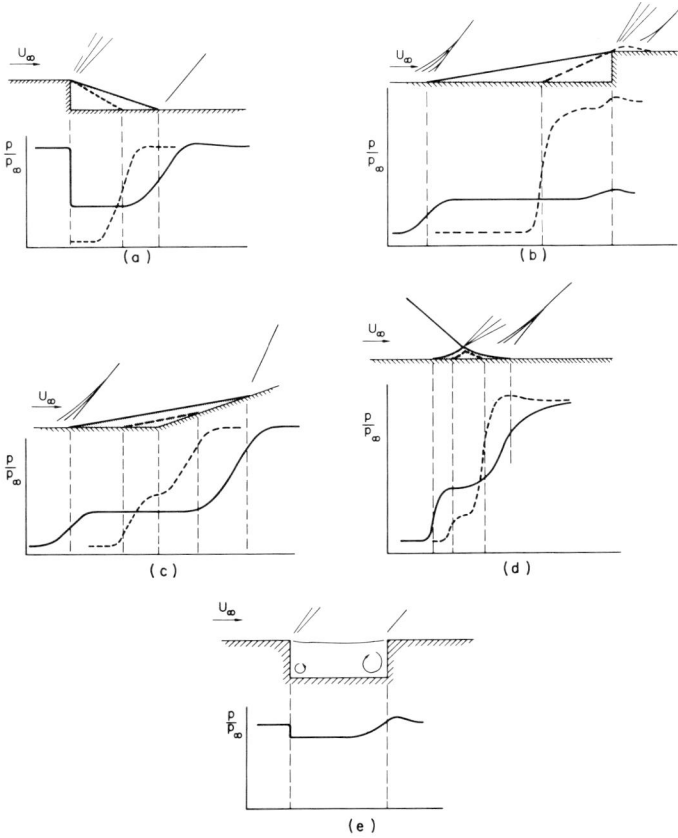

FIG. 1. Sketches of various types of flow separations and characteristic pressure distributions: (———) typical laminar flow; (– – –) typical turbulent flow. (a) Downstream-facing step; (b) upstream-facing step; (c) ramp-induced separation; (d) shock-induced separation; (e) notch (fully spanned).

point can be located a priori at the sharp corner on heuristic grounds (this does not mean that viscous constraints do not exist in the neighborhood of this singular point but, at least in gross outlines, they do not determine the point of separation). A closure condition is sought in phenomena which govern the recompression. The pioneering paper dealing with this problem is due to Crocco and Lees (6). The great bulk of research done during the 1950s was based, however, on a simpler model developed by Chapman (7) for laminar flow and, independently and simultaneously, by Korst (8) for turbulent flow. This model isolates the simplest conservation condition, which is: the mass scavenged by the thin shear layer along the dividing streamline must be returned to the wake at the point of recompression. This provides the lacking closure criterion. The model and a series of semiempirical improvements on it yielded practically useful results for supersonic flow at intermediate Reynolds number, including the effect of transition.

Figure 1b shows another typical separated region which forms ahead upstream-facing steps. In this case one can grossly locate the recompression, but not the separation point. The local conditions for the boundary layer to separate were discussed above. The argument shows that there must be precompression of the flow with an associated change in the thickness and the profile of the boundary layer. This interaction mechanism provides the criterion for the analysis of this type of cavity.

Figure 1c shows the next more complicated configuration: a cavity formed in a compressive corner. In this case both the separation and the recompression points are free and must be found as part of the problem. The region of separated flow due to the interaction of a shock wave with a boundary layer in supersonic streams is similar to that in a compression corner, but geometrically more complex (Fig. 1d). The structure of these cavities is still dominated by the interplay of constraints at the separation and the recompression points, although the matching of the external and internal flows along the entire dividing streamline must play an increasingly important role. Successful theories were developed only recently, using approximate moment methods which emphasize the viscous–inviscid flow interaction mechanism.

Figure 1e shows a cavity contained in a groove or a notch in the wall. In this case both the separation and recompression points are essentially fixed, and the only interesting questions pertain to the nature of the internal flow. The nature of the internal flow in general, and specifically in notch cavities, presents the most difficult analytical problem. Except for a model proposed by Batchelor (9) (very high Reynolds numbers, steady, laminar flow) and some numerical solutions for very low Reynolds numbers, there is no theory. Experiments are very difficult, mostly because, in the mid-regime of Reynolds numbers, the internal flow is found to be higly "turbulent" (fluctuations of the order of 100% of the mean velocity), three-dimensional, often unsteady.

The examples of Fig. 1 are typical, but not exhaustive. There are many other, more complicated situations where a "bubble" of separated flow can be observed. For instance, immediately downstream of the reattachment corner of an upstream facing step, such as shown in Fig. 1b, one will normally find another separation. The flow in the shear layer approaches the corner at an angle to the reattachment surface which it cannot negotiate (in the ideal, inviscid case, the dividing streamline would turn sharply through a centered Prandt–Mayer fan emanating from the corner). This type of separation is sometimes referred to as "breakaway." The shear layer remains "free," bends over gradually, and reattaches to the surface downstream of the corner proper. Another example of this type of "breakaway" separation occurs downstream of a sharp corner of a flat-nosed body in supersonic flow.

The subject of separated flows is very complex. There is no unifying theory and the existing models are insufficiently refined to explain many of the important observable phenomena over the full range of Mach and Reynolds numbers of practical interest, even in the particular case for which they are designed. One can distinguish two levels of information. Base pressures and the gross aspect of the flow field can be correlated to an engineering accuracy by relatively simple semiempirical theories. Any improvement of the precision of these semiempirical theories, questions pertaining to the fine details of the pressure distribution, the flow field at the closure point of the near wake (an important initial condition for the study of the far wake), etc., require that a totally different level of analysis involving very extensive numerical work be brought to bear on the problem. There seems to be no intermediate step between the elementary model and a virtually complete solution of the entire flow field over the body. A particular deficiency of existing theories is that they provide no reliable framework for dealing with the important problem of heat transfer to the walls of the cavity. Heat transfer problems are still mainly a question of judicious interpolation of experimental data.

A very large volume of descriptive literature on separated flows has been published. A number of reviews are available or in preparation. Among them, Roshko (*10*) discusses wakes and cavities, focusing mainly on incompressible flow and on the theoretical problem of the asymptotic limit of high Reynolds numbers. Lykoudis (*11*) reviews results for the hypersonic wake. Berger (*12*) is preparing a report emphasizing the mathematical theory. A two-volume proceeding of the AGARD symposium on separated regions (*13*) and a similar symposium dealing specifically with hypersonic wakes (*14*) summarize many of the recent experimental results.

In this review we selected mainly illustrative data pertaining to cavities with recompression against a solid boundary rather than blunt-base wakes, because the latter are adequately covered in the other reviews. We do not discuss special features of hypersonic flows, such as aerothermochemical effects and the far wake, for the same reason.

Section II deals with recompression-controlled cavities downstream of steps (and blunt bases) which have been most thoroughly investigated both theoretically and experimentally. This section presents the semi-empirical model of Chapman and Korst and confronts it with evidence.

In Section III more sophisticated, numerical studies, which go beyond the assumptions of the Chapman–Korst model and illustrate its insufficiencies and limitations, are reviewed. This section also gives an account of experimental data for cavities in notches.

Section IV outlines the features of the free-interaction model which isolates the dominant mechanism governing separation-controlled cavities. Illustrative data pertaining to upstream-facing steps, ramps, and shock-induced cavities are presented and correlated in terms of similarity parameters derived from this model. Numerical solutions of the interaction-dominated boundary-layer problem starting with either the differential equation or a multimoment approximation to the equations are described. These constitute the most complete and sophisticated treatment of flows with imbedded separated regions and apply in principle to all types of cavities.

II. The Chapman–Korst Model (Cavities Controlled by Recompression)

The simplest analytical description of cavities formed downstream of blunt bases and expansion steps is provided by a model developed simultaneously by Chapman (*7, 15*) for laminar and Korst (*8, 16–19*) for turbulent, supersonic flows. Figure 2 shows the flow field schematically. There are two coupled regions:

(a) The cavity proper (Region b), which is fully described by

$$p_b = \text{const} + O(u_b/u_\infty)^2; \qquad (u_b/u_\infty) \to 0 \qquad (1)$$

that is, the cavity is "dead air" at least in the sense that dynamic pressures generated by the internal motion are negligible.

(b) The inviscid external flow (Region e) is bounded by the walls and isobaric free streamline joining the point of separation (s) and the point

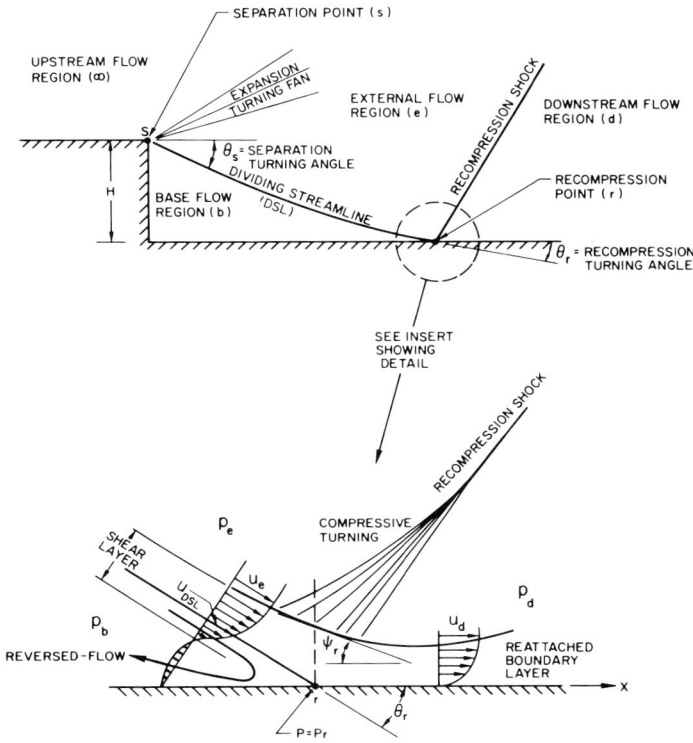

Fig. 2. Schematic of the Chapman–Korst separated flow model.

of rattachment (r). The cavity is assumed to be closed, and the separation point is fixed and known. The model is therefore directly applicable only to bodies with a sharp step at s. There is nothing fundamental, however, which would specialize it to supersonic flow; in subsonic flow Region e is governed by an elliptic equation and the shape of the section s–r of its boundary is *a priori* unknown. This makes the subsonic problem a much more difficult one than its supersonic equivalent. In supersonic flow, Region e becomes known explicitly as soon as the oncoming stream is given (for instance, data along a line perpendicular to the surface immediately upstream of separation) and one parameter describing the amount of expansion at s is specified (either θ_s or p_e/p_∞). If the oncoming free stream is uniform and two dimensional, the streamline s–r is straight. However, in axisymmetric flow or when the initial (preexpansion) flow is nonuniform, it is curved. In such cases the shape of the dividing streamline must be calculated by methods of characteristics. The expanded flow in Region e is terminated by a compressive shock rooted at the recompression point which turns it back to the direction of the recompression wall (or line of symmetry, in the case of a free wake).

The problem is now closed by providing a recompression condition which prescribes uniquely the amount of expansion at s (therefore the base pressure ratio p_b/p_∞ and the strength of the recompression shock) as follows. It is recognized that the "free streamline" s–r is in effect a shear layer in which viscous forces cannot be neglected, regardless of Reynolds number. The Chapman–Korst model assumes that the layer is thin, isobaric, and does not interact with the external flow (in analogy with first-order boundary-layer theory); nonetheless, mass is entrained from the internal region and, by continuity, must be returned to it in the recompression process. A solution for the velocity profile in this viscous layer is obtained independently; the initial conditions for it are prescribed at separation and the boundary conditions along s–r are uniform flow above and stagnant gas below. Thus the internal motion, which is now implied, is neglected both in respect to the dynamic perturbations to the cavity pressure and the internal boundary condition for the shear-layer analysis.

The model assumes that the recompression process along the dividing streamline (DSL) occurs isentropically (free of viscous effects) so that the pressure at r is simply the stagnation pressure corresponding to the velocity u_{DSL} of the flow along this streamline in the shear layer (as it approaches the recompression region). There is some direct experimental evidence [20] of this; the stagnation line r (two-dimensional step) was determined by visualizing the flow reversal on the reattachment wall, and the pressure at that point was contrasted with pitot-tube traverses of the shear layer. At usual Reynolds numbers of aerodynamic interest, the assumption of isentropic recompression along the DSL is found to hold with sufficient accuracy.

The above are the basic elements of the Chapman–Korst model: isentropic recompression, stagnant base region, inviscid external flow, and a very thin shear layer without interaction with the external inviscid flow. Additional information must be supplied and this involves additional assumptions which vary, leading to several versions of the basic theory.

A. Point-Recompression Assumption

It is necessary to relate the recompression pressure p_r at the rear stagnation point on the dividing streamline to the pressure existing in the external flow downstream of the recompression process. Here Chapman and Korst both made the simplest assumption which corresponds to the case of a shear layer of zero thickness (note: this concerns the shear layer ahead of r, not the boundary layer ahead of s) whereby the turning of the flow occurs discontinuously at a singular point r. This point is then the root of the recompression shock and the pressure at r equals the pressure p_d in the inviscid flow after it is turned parallel to the recompression surface (or line of symmetry). This is expressed by the relation

$$p_r = p_b(1 + \tfrac{1}{2}(\gamma - 1) M_{DSL}^2)^{\gamma/(\gamma-1)} = p_d \qquad (2)$$

The situation corresponds to the sketch of Fig. 1, disregarding the true detail of the recompression process as depicted on the insert in that figure.

If one assumes that the Busemann or the Crocco integral holds in the shear layer for adiabatic flow or with heat transfer, respectively, then there is a relation between the Mach number along the dividing streamline and that in the external flow. In particular, for the adabatic case

$$M_{DSL}^2 = \frac{(\phi)^2 M_e^2}{1 + [(\gamma - 1)/2] M_e^2[1 - (\phi)^2]} \ ; \quad \phi = \frac{u_{DSL}}{u_e} \qquad (3)$$

The oblique shock in the inviscid flow is often weak enough so that one can assume no total pressure loss across it and write

$$\frac{p_e}{p_d} = \left(\frac{1 + [(\gamma - 1)/2] M_d^2}{1 + [(\gamma - 1)/2] M_e^2} \right)^{\gamma/(\gamma-1)} \qquad (4)$$

Combining these equations one obtains the characteristic relation across the recompression shock

$$M_d^2 = [1 - (\phi)^2] M_e^2 \qquad (5)$$

Clearly, the key "physical" assumption is that involved in Eq. (2); the other ones are merely numerical approximations and can be removed without difficulty. For instance, it is simple to use, instead of (4), data from oblique shock tables.

The elements of the model can now be put together. If $\nu(M)$ is the local Prandtl-Meyer angle corresponding to a Mach number M, then the external flow at s turns by

$$\theta_s = \nu_{[M_\infty]} - \nu_{[M_e]} \qquad (6)$$

It turns back at the recompression point r by

$$\theta_r = \nu_{[M_d]} - \nu_{[M_e]} = \nu_{[M_e(1-(\phi)^2)^{1/2}]} - \nu_{[M_e]} \qquad (7)$$

The net turning of the external flow is known from the boundary conditions. If the wall after recompression is parallel to the wall before separation and if the flow before separation was uniform and two-dimensional, the net turning is zero:

$$\theta_s - \theta_r = 0 = \nu_{[M_\infty]} - \nu_{[(1-(\phi)^2)^{1/2}M_\eta]} \tag{8}$$

where the subscript $\eta = [\nu_{[M_\infty]} - \theta_s]$.

The ratio of the base pressure to the free stream pressure before separation can be solved for explicitly. It is given by the following simple equation:

$$\frac{p_b}{p_\infty} = \left[\frac{1 + [(\gamma-1)/2]M_\infty^2}{1 + [(\gamma-1)/2][M_\infty^2/(1-\phi^2)]}\right]^{\gamma/(\gamma-1)} = \text{funct}\{M_\infty, \phi\} \tag{9}$$

The extension of this model to cases where the recompression wall is not parallel to the initial direction of the flow before separation is obvious. When the external flow is nonuniform, in the sense that the pressure (and Mach number) vary in a plane perpendicular to the boundary immediately ahead of separation, a more complicated calculation is required (21). Such nonuniformities in the supersonic free stream are propagated along characteristics of a family opposite to the expansion fan at s. These reflect off the isobaric free streamline s–r as characteristics of the opposite family. For an expanding flow the reflection will be compressive and result in an upward turning of the free streamline. Instead of Eq. (8), the geometry of the free streamline is expressed by

$$\theta_s - \theta_r = \int_s^r \frac{d\theta}{ds} ds \tag{10}$$

where the rate of change of direction of the free streamline along the shear layer is known from, say, a characteristic solution of the external flow. It is a function of the nonuniformity, which we can express, for example, by a series expansion on the pressure,

$$\frac{p}{p_\infty} = 1 + \frac{x}{\lambda} + \cdots; \quad \lambda = \left[\frac{1}{p_\infty}\left(\frac{dp}{dx}\right)_{x=0}\right]^{-1} \tag{11}$$

and assume that, for most practical cases, the linear term is a sufficient approximation. We see occurring a length scale λ dependent only on the pressure gradient such that the geometrical characteristics of the cavity, including the base pressure, become dependent not only on the initial Mach number ahead of separation and on the recompression parameter ϕ but also on the relative height of the step H/λ:

$$p_b/p_\infty = \text{funct}(M_\infty, \phi, H/\lambda) \tag{12}$$

Steps in the surface of axisymmetric bodies, when their height is small compared to the radius of the body, can be treated as two dimensional [Eq. (7)]. Otherwise the axisymmetric compression of the external flow after the Prandtl–Mayer fan at the corner must be taken into account. In general, the base pressures in axisymmetric flow will be higher than in two-dimensional flow.

B. The Shear Layer

1. Asymptotic Solution

It is still necessary to prescribe the ratio u_{DSL}/u_e from a solution of the viscous flow in the shear layer. If one assumes that the length of the shear layer is sufficient so that the influence of the initial velocity profile of the viscous layer at its origin (immediately behind completion of the expansion process at the separation corner) has been attenuated, a self-similar, asymptotic solution of the boundary-layer equations can be obtained. The equations are (incompressible flow)

$$\frac{\partial}{\partial x}(uy^j) + \frac{\partial}{\partial y}(vy^j) = 0$$
$$u\frac{\partial u}{\partial x} + v\frac{\partial u}{\partial y} = u_e \frac{du_e}{dx} + \frac{\nu}{y^j}\frac{\partial}{\partial y}\left(y^j \frac{\partial u}{\partial y}\right) \quad (13)$$

with $j = 0, 1$ describing plane and axisymmetric flow, respectively (x, y are streamwise and cross-stream coordinates, respectively).

Defining a stream function ψ as

$$\partial\psi/\partial y = uy^j; \qquad \partial\psi/\partial x = -vy^j \quad (14)$$

and assuming the free stream velocity u_e follows the variation $u_e \sim x^m$, a single similarity equation can be derived:

$$f''' + ff'' + \beta(1 - f'^2) + j\left[2\frac{f''}{\eta} - \frac{f'}{\eta^2} + \frac{f}{\eta^3} + (2 - 3\beta)\frac{ff'}{\eta} - 2\frac{f^2}{\eta^2}\right] = 0 \quad (15)$$

where

$$\beta \equiv 2m/(m+1), \qquad \eta = \frac{y}{x}\left(\frac{u_e x}{\nu}\right)^{1/2}\left(\frac{2}{1+m}\right)^{j-1/2}$$

$$f_{(\eta)} = \psi/y^j \left[\frac{2}{1+m} u_e \nu x\right]^{1/2}$$

This equation describes a one-parameter (β) family of velocity profiles for incompressible, laminar flow. Its solutions are readily extended to compressible flows through the generalized Stewartson–Illingworth transformation.

Shear-layer solutions of the above equation have been obtained for different sets of boundary conditions. Chapman (22) studied the two-dimensional, semi-infinite jet ($j = \beta = 0$). Chapman's boundary conditions are

$$f'_{(\infty)} = 1, \quad f_{(0)} = 0, \quad f'_{(-\infty)} = 0 \tag{16}$$

Typical examples of the Chapman profile for various free stream Mach numbers are given in Fig. 3. Another analysis with application to shear-layer flows is due to Stewartson (23) who considered the family of Falkner–Skan solutions containing regions of reversed flow. Stewartson employed the same boundary conditions that are used in the familiar Blasius flat-plate boundary-layer problem: namely,

$$f'_{(\infty)} = 1, \quad f_{(0)} = f'_{(0)} = 0 \tag{17}$$

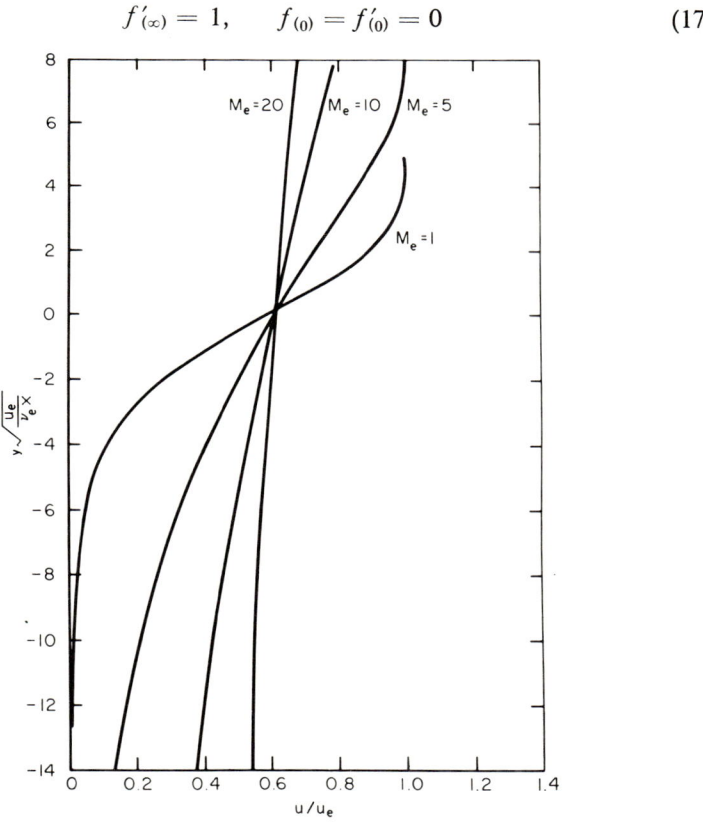

FIG. 3. The Chapman velocity profile for adiabatic flow (assuming $\mu \sim T^{0.76}$).

The third set of solutions was obtained for wake flows where the boundary conditions are

$$u = u_e \quad \text{at} \quad y \to \infty \quad \text{or} \quad \lim_{\eta > \infty}\left[f' + j\frac{f}{\eta}\right] = 1$$
$$v = \frac{\partial u}{\partial y} = 0 \quad \text{at} \quad y = 0 \quad \text{or} \quad f_{(0)} = f''_{(0)} = 0 \quad (18)$$

which specify that the flow is symmetric about the wake centerline $y = 0$. The plane flow solutions ($j = 0$) for this case were obtained by Kennedy (24) and the axisymmetric solutions ($j = 1$) were given by Kubota *et al.* (25). This family of solutions is shown in Fig. 4. Wake

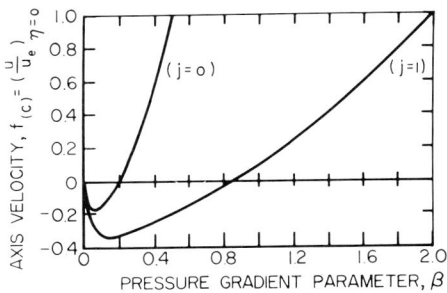

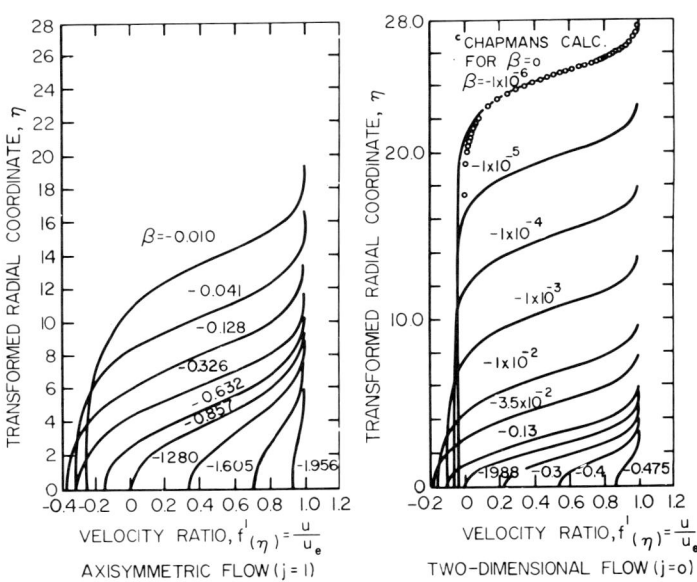

Fig. 4. Wakelike velocity profiles for plane and axisymmetric flow.

profiles are possible for [see Eq. (15)] $0 \leqslant m \leqslant -0.20$ in plane flow ($j = 0$) and for $0 \leqslant m \leqslant -0.50$ in axisymmetric flow. The "squeezing" effect along the axis in axisymmetric flow is evidenced by the range of reversed flow velocities.

The Chapman profile for incompressible flow is also included. That Kennedy's solutions should contain the Chapman solution as a special case is not immediately obvious. However, the equivalence has been demonstrated by Kennedy and can be explained as follows. In the Kennedy reversed flow profiles, the dimensionless stream function f is zero at two positions of η: once at the wake symmetry line $y = \eta = 0$ and once again at the dividing streamline (since the net mass flux in the wake recirculation region must vanish). Kennedy demonstrates that the distance between the two zeros of the stream function f increases indefinitely as $\beta \to 0$ from below. Therefore it is not incompatible to obtain the same profile with the two different boundary conditions $f'_{(-\infty)} = 0$ and $f''_{(0)} = 0$ because all higher-order derivatives must also vanish at $\eta \to -\infty$ if $f'_{(-\infty)} = 0$. Equivalence is then established between Chapman's solution and Kennedy's solution for $\beta \to 0$ if we identify Chapman's $f_{(0)} = 0$ streamline with the outer zero of the stream function f in Kennedy's solution. In other words, Chapman's origin is translated outward from the origin used by Kennedy.

The important velocity ratio ϕ of the Chapman profile varies slightly with Mach number. It is shown, as well as its equivalent for the turbulent case, in Fig. 5.

The solution for turbulent mixing was developed by Korst et al. (17), and extended to the nonisoenergetic case by Page et al. (19). In these papers the authors considered the effect of the initial boundary layer at s, which we will omit, however, and return to later in a section discussing corrections to the asymptotic theory.

The difficulty with the analysis of a turbulent shear layer lies in our persistent inability to prescribe the mixing properties of the flow with any rigor. Korst's solution is based on the equation

$$u\, \partial u/\partial x + v\, \partial u/\partial y = \epsilon\, \partial^2 u/\partial y^2 \tag{19}$$

which contains a phenomenological eddy diffusivity coefficient ϵ, assumed not to be a function of the transverse coordinate y. Furthermore, Korst linearizes the equation and writes

$$u_e\, \partial u/\partial x = \epsilon\, \partial^2 u/\partial y^2 \tag{20}$$

which is not a good approximation in the inner portions of the shear layer.

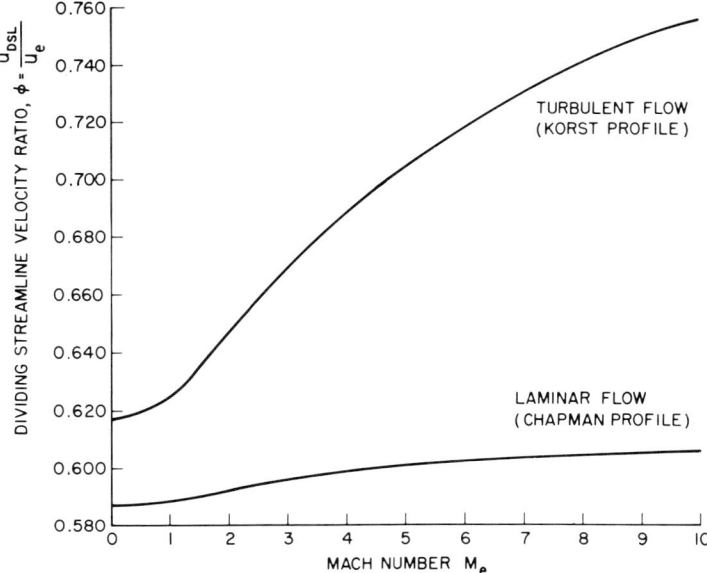

Fig. 5. The influence of Mach number on the asymptotic value of the dividing streamline velocity for laminar and turbulent flow.

The eddy viscosity coefficient is assumed to depend linearly on the streamwise distance

$$\epsilon = c u_e x \qquad (21)$$

The usual boundary conditions are that the streamwise velocity and the velocity gradient are zero at the inner boundary, and $u = u_e$ at the outer boundary of the viscous zone. The well-known solution of Goertler is then obtained (asymptotic case):

$$\frac{u}{u_e} = \tfrac{1}{2}(1 + \operatorname{erf} \eta)$$
$$\eta = \frac{1}{(2c)^{1/2}} \frac{y}{x} = \sigma \frac{y}{x} \qquad (22)$$

where $\sigma = 1/(2c)^{1/2}$ is the so-called jet spreading parameter characterizing the mixing properties of the turbulence.

When applied to compressible flow, the analysis implies an assumption that the turbulent Prandtl number is unity and thereby permits the use of a Crocco-type integral to relate the velocity and temperature profiles. As a result, the density profile for isoenergetic flow is expressed by

$$\frac{\rho}{\rho_e} = \frac{1 - (\mathrm{Cr}_e)^2}{1 - (\mathrm{Cr}_e)^2 (u/u_e)^2} \tag{23}$$

where Cr is the so-called Crocco number,

$$\mathrm{Cr} = \frac{u^2}{2c_p T_{\mathrm{stag}}} = \frac{M^2}{2/(\gamma - 1) + M^2} \tag{24}$$

The solution is written in an intrinsic coordinate system which places the ratio $u/u_e = \frac{1}{2}$ at the origin of the y coordinate. To find the dividing streamline, an integral of the momentum deficiency over the uniform free stream between positive infinity and the dividing streamline is equated to the streamwise momentum from negative infinity to the dividing streamline [using Eqs. (22) and (23)]. This determines the velocity ratio at the dividing streamline that is plotted in Fig. 5. The results depend, as indeed does the whole analysis, on the form of the mixing parameter σ. This topic has an extensive literature of its own. A recent review was given, for example, by Channapragada (26). The form used to calculate the result given in Fig. 5 is

$$\begin{aligned} \sigma &= 47.1\ \mathrm{Cr}_e{}^2 \quad \text{for} \quad \mathrm{Cr}_e{}^2 > 0.28 \quad (M_e > 1.23) \\ \sigma &= 11 \quad \text{for} \quad \mathrm{Cr}_e{}^2 \leqslant 0.23 \end{aligned} \tag{25}$$

and was suggested by Tang et al. (27). Sirieix and Solignac (28) provide another extensive correlation of data, including their own (up to Mach numbers of 4) which verifies the results of Channapragada (26) (based on data up to $M = 2$).

2. Corrections for the Effect of the Initial Boundary Layer and Base Bleed

In spite of its basic success, the simple model which assumes an asymptotic shear layer and a point recompression is unable to account for significant variations in even the simplest indices of the wake dynamics, such as base pressure. Notably, there is no Reynolds number dependence whatever in the theory, whereas such a dependence is clearly evident in experimental data, particularly at higher Mach numbers. Measurements showing base pressures lower than those predicted by the Chapman–Korst model were obtained, which is surprising, since the nature of the the theory would lead one to believe that it always gives a minimum bound.

The obvious first step in improving the formulation outlined in the preceding, which was already proposed by Korst in his initial paper, is to account for the influence of the boundary layer existing upstream of separation. This led to the publication of a number of semiempirical corrections, among which the best known are those of Kirk (29) and Nash (30).

Account must be taken of the turning of the boundary layer at the separation corner. This turning which is associated with a drop in pressure does work on the viscous layer and decreases its momentum thickness. The process is rapid and is assumed to take place without the action of viscous forces. Thus, for isoenergetic, isentropic turning of the flow in any stream tube, one can write

$$u^2 + 2c_p T = u_e^2 + 2c_p T_e = \text{const}$$

$$T_2/T_1 = \left(\frac{p_b}{p_1}\right)^{\gamma-1/\gamma} \tag{26}$$

where stations 1 and 2 are, respectively, before and after turning. Combining, one has

$$\frac{1 - u_2^{*2}}{1 - u_1^{*2}} = \frac{M_{1e}^2}{M_{2e}^2}; \qquad u^* = \frac{u}{u_e} \tag{27}$$

or

$$u_2^* = \left\{1 - \frac{M_{1e}^2}{M_{2e}^2}(1 - u_1^{*2})[2 - (1 - u_1^{*2})]\right\}^{1/2} \tag{28}$$

This can be expanded in a series and used to evaluate the integrals appearing in the definition of the momentum thickness δ^{**} yielding, to first order in $(1 - u_1^{*2})$,

$$\frac{\delta_2^{**}}{\delta_1^{**}} = \frac{\rho_{1e} u_{1e}}{\rho_{2e} u_{2e}} \frac{M_{1e}^2}{M_{2e}^2}$$

$$= \left(\frac{1 + [(\gamma-1)/2] M_{2e}^2}{1 + [(\gamma-1)/2] M_{1e}^2}\right)^{1/2(\gamma-1)} \left(\frac{M_{1e}}{M_{2e}}\right)^3 \tag{29}$$

Tang et al. (27) checked this expression against a similar analysis in which the expansion is performed streamtube by streamtube so that, after turning, a new profile is obtained. The profile is then integrated to obtain its momentum thickness. Expression (29) is found to be very good. It is worth noting that the entire treatment neglects the subsonic portion of the boundary layer.

Separated Region Supersonic Flows

The profile of the viscous layer changes during the expansion process and can be calculated as above, but this requires rather extensive numerical work (as well as prescribing the initial profile of the attached boundary layer). Kirk assumes that the profile of the viscous layer after turning can be represented approximately by the asymptotic profile (Chapman's or Korst's for laminar and turbulent flow, respectively). However, the location of the dividing streamline is then free and must be determined by a mass and momentum balance. The procedure for locating the dividing streamline is identical to that used by Korst for the turbulent boundary layer. Conservation of mass and momentum between a station (subscript 2) immediately after turning and a station (subscript 3) downstream yield, respectively,

$$\int_0^\infty \rho u \, dy \bigg]_2 = \int_{y_{DSL}}^\infty \rho u \, dy \bigg]_3$$

$$\int_0^\infty \rho u \, dy \bigg]_2 = \int_0^s \tau \, ds + \int_{y_{DSL}}^\infty \rho u^2 \, dy \quad (30)$$

where

$$\int_0^s \tau \, ds = \int_{-\infty}^{y_{DSL}} \rho u \, dy$$

which states that the action of shear along the dividing streamline is balanced by the increase of streamwise momentum imparted to the scavenged fluid. Combining these equations with the definition of the momentum thickness yields

$$\delta_2^{**} = \int_{y_{DSL}}^\infty \frac{\rho u}{(\rho u)_\infty} \, dy - \int_{-\infty}^\infty \frac{\rho u^2}{(\rho u^2)_\infty} \, dy \quad (31)$$

which determines the location y_{DSL} of the dividing streamline. For the self-preserving profiles, these integrals can be evaluated once for all. The final result has the form

$$I_{(\phi)} = -\xi^{-1/2(1+l)} \quad (32)$$

where the function I of the dividing streamline velocity ratio $\phi = u_{DSL}/u_e$ is known, and

$$\xi = \begin{cases} \dfrac{s}{\delta_2^{**} \operatorname{Re}_{\delta^{**}}} = \dfrac{s\nu}{\delta_2^{**2} u_e} & \text{and } l = 0 \text{ for laminar flow} \\[1em] \dfrac{s}{\delta_2^{**}} \left(\dfrac{1 - C r_e^2}{\sigma} \right) & \text{and } l = 1 \text{ for turbulent flow} \end{cases} \quad (33)$$

where ξ is a reduced distance measured along the shear layer. For laminar flow, the proper dimensionless grouping can be derived directly from the boundary-layer equations for isobaric flow when the initial momentum thickness δ_2^{**} is recognized as the characteristic length of the problem.

The resulting corrected ϕ is plotted in Fig. 6. Figure 7 compares this correction in laminar flow to results of a numerical integration of the preasymptotic flow in the shear layer (*31–33*) [see (23)].

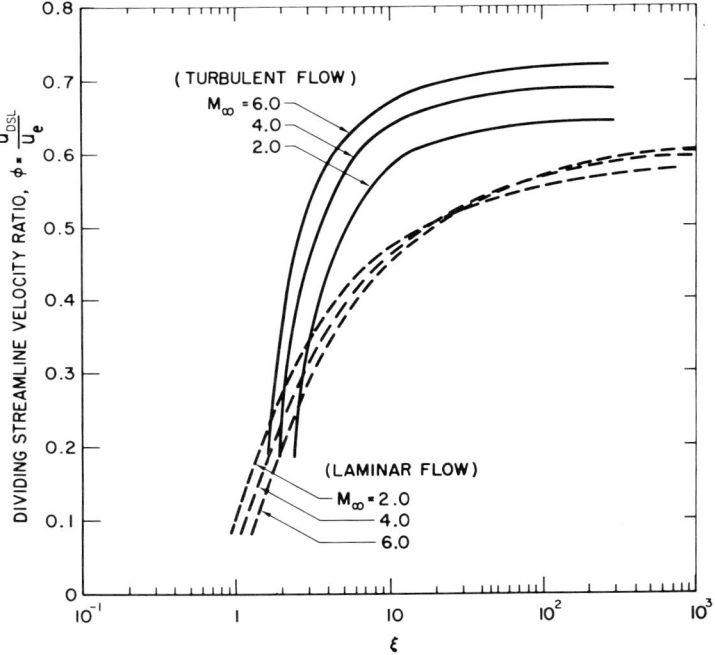

FIG. 6. The dividing streamline velocity ratio corrected for the effect of an initial boundary layer upstream of separation by the Nash–Kirk method.

The base pressure problem is now reduced to an iteration. The length of the shear layer s depends on the base pressure ratio

$$\frac{H}{s} = \cos\theta_s = \text{funct}\left(\frac{p_b}{p_\infty}, M_\infty\right) \tag{34}$$

This equation, in conjunction with Eqs. (32) and (9), yields a value of ϕ and a base pressure corrected for the effect of the initial boundary layer (within the present framework, of course).

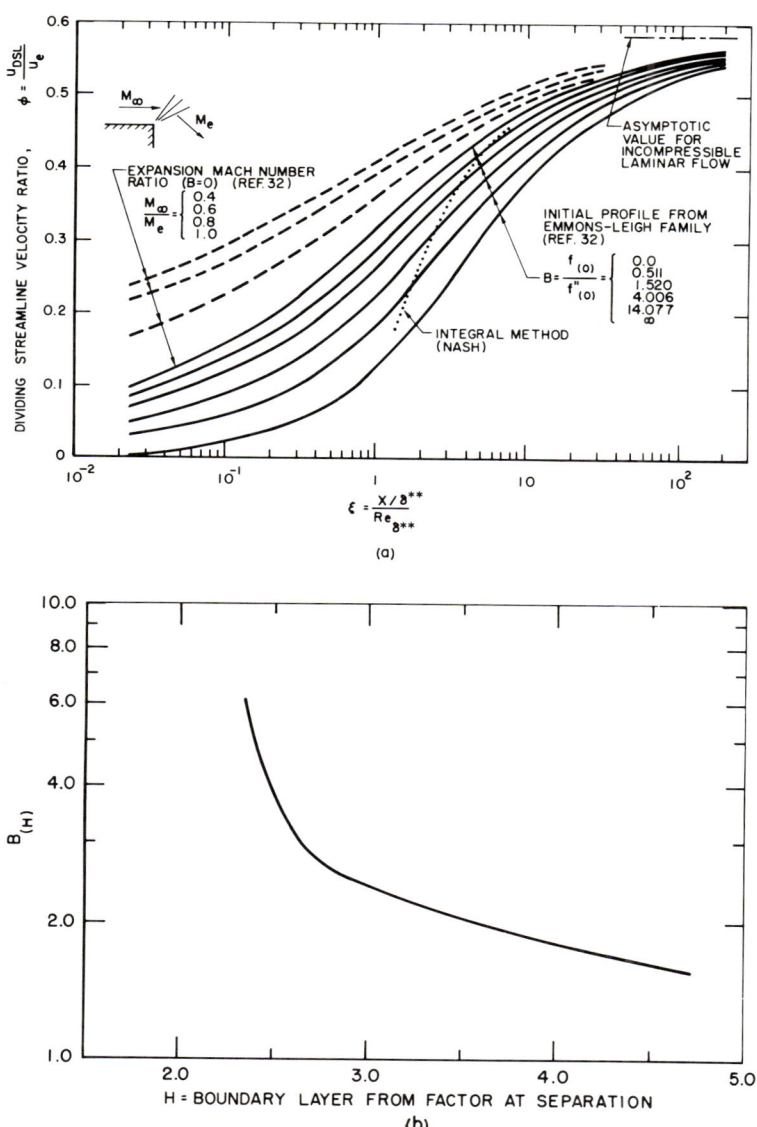

Fig. 7. The preasymptotic development of the dividing streamline velocity for various initial velocity profiles and expansion ratios in laminar flow.

Up to this point the base wake was considered closed. However, the Chapman–Korst model permits inclusion of bleed into the wake without modification of its conceptual structure.

Provided that mass is added at a sufficiently low velocity (transpiration) the assumption that the pressure in the dead-air region is constant remains true. The structure of the basic flow does not change except that the conditions of closure at recompression are modified so as to preserve continuity of mass. The stagnation point in the recompression region occurs now on a streamline originating in the inner portion of the shear layer such that the mass between that streamline and the "dividing streamline" just equals the mass added into the cavity. If that streamline is denoted by the subscript r the mass balance equation has the form

$$\frac{\dot{m}/W}{(\rho u)_e} = \int_{y_r}^{y_{DSL}} \frac{\rho u}{(\rho u)_e} dy \qquad (35)$$

where W denotes the width of the base region. This simply leads to replacing δ_2^{**} in Eq. (33) by the factor

$$\Lambda = \delta_2^{**} + \frac{\dot{m}/W}{(\rho u)_e} \qquad (36)$$

It follows that the influence of bleed on the base pressure is the same as that of a boundary layer upstream of separation. At low Mach numbers and to the order of accuracy of this overall model, the effect is verified by experiment.

3. *Numerical Analyses of the Preasymptotic Development (Laminar)*

The correction for the initial boundary layer described in the preceding section is obviously not entirely satisfactory. A number of other, more refined analyses based on momentum methods were published [e.g., Kubota and Dewey (*34*)]; however, in laminar flow it is also possible to obtain exact profiles by numerically integrating the shear-layer equations beginning with a prescribed upstream profile for the boundary layer. A series of such studies were published by Baum, King, and Dennison (Refs. *31* and *32* which are summaries of a series of reports). In their calculations they assumed $\rho\mu$ constant, Prandtl and Lewis numbers equal to unity, and preserved the Chapman–Korst boundary conditions for the viscous layer, that is, a uniform external flow of velocity u_e above and a stagnant region of constant pressure below. The Stewartson transformation is applied which uncouples the energy and momentum equation and reduces the latter to its incompressible form. Defining a streamwise scale of the type already discussed, Eq. (33), the problem (for laminar flow) becomes parameterless and dependent only on the profile prescribed at the starting station.

The boundary conditions are

$$\frac{u}{u_e} = \begin{cases} 1 & \text{at } y \to +\infty, \\ 0 & \text{at } y \to -\infty, \end{cases} \qquad \frac{\tau}{\mu} = \frac{\partial u}{\partial y} = 0 \quad \text{at } y \to \pm\infty \qquad (37)$$

Their nature favors the use of the Crocco plane (shear, velocity) where the integration domain is bounded. However, in that plane, the equations exhibit a mathematical singularity at the inner edge of the layer at separation where the shear changes discontinuously from the finite value on the wall (in the attached boundary layer) to zero. To overcome this difficulty a "starting solution" in the form of a locally valid expansion of the shear in terms of one-third powers of velocity was used (*33*), following the method of Goldstein (*35*). This provides a starting profile (the first term of the inner-layer expansion matched to the outer portion of the prescribed boundary-layer profile) "close" to the trailing edge (the closeness is coupled to the choice of mesh size of the integration) which is used to start the numerical forward-marching solution downstream. Numerical experiments on the decay of the disturbance due to the matching are presented by Charwat and Gomez (*36*).

The results show the buildup of the dividing streamline velocity along an *s*-shaped curve, examples of which are given in Fig. 7. All profiles ultimately reach the asymptotic value ($\phi = 0.587$ for laminar, incompressible flow) but the distance required to reach a given value of u_{DSL}/u_e varies considerably with the shape of the starting profile. Separation-type low shear boundary layers (positive pressure gradient or blowing over the body) lead to an initially slow and ultimately rapid increase in the dividing velocity ratio. The opposite is true for boundary layers typical of accelerated flow (high shear).

There are enough cases calculated and presented in the literature to propose an engineering correlation in terms of the simplest integral parameter which describes the initial boundary layer, the form factor H [actually, a two parameter correlation in terms of, for instance, the skin friction and the form factor is needed (*37*)]. If we use as a guide the approximate analysis of Nash [Eq. (32)] and express the function $I_{(\phi)}$ as a linear relation, a correlation equation of the following form can be suggested:

$$I_{(\phi)} = A_{(H)} + B_{(H)}\phi = -\xi^{-1/2} \quad \text{(laminar flow)} \qquad (38)$$

The functions $A_{(H)}$ and $B_{(H)}$ are determined by using the numerical results at the points $\phi = 0.40$ and $\phi = 0.50$. It is found that they are related by

$$B_{(H)} = -1.7 A_{(H)}$$

which contains the correct asymptotic limit and reduces Eq. (38) to a single function correlation:

$$\phi = 0.587 - (B_{(H)}^2 \xi)^{-1/2} \tag{39}$$

The function $B_{(H)}$ is included in Fig. 7.

The corner expansion distorts the profile of the attached boundary layer and therefore influences the growth of the dividing streamline velocity. Using the stream-tube expansion method described earlier and a Blasius profile ahead of expansion, a modified starting profile for the numerical integration was derived and the subsequent evolution of the shear layer calculated (32). The results of this calculation for several expansion ratios are also shown in Fig. 7. After expansion the profile is fuller and corresponds to a higher value of the form factor H. In accord with the general trend discussed previously, the buildup of the dividing streamline velocity then occurs at a much greater rate.

Base bleeding (assuming that the mass enters at a negligible velocity does not modify the calculated shear-layer profiles when the Chapman boundary conditions are retained. However, since now a streamline other than the dividing streamline must come to stagnation, a correction for bleed involves the use of the true nonasymptotic profile at recompression. The equivalence between the effect of the upstream boundary layer and base bleed exhibited by Nash's analysis is destroyed. Base bleed decreases the value of ϕ resulting in longer wakes and increased base pressures.

Heat transfer to the body surface leads to an enthalpy level in the inner "dead-air" region which differs from the recovery enthalpy (stagnation enthalpy when $\text{Pr} = 1$) of the external flow. This also affects the value of the dividing streamline velocity. Specifically, base cooling results in higher values of ϕ, lower base pressures, and shorter wakes.

One other interesting result of the preasymptotic analysis is that the Crocco integral relation is found to be invalid in the preasymptotic region. This is important since its use in computing mass flow and momentum quantities in the shear layer turns out to lead to fairly significant errors.

These results carry the correction to the dividing streamline velocity ratio in laminar flow as far as is reasonable. It is clear that the use of the asymptotic solution initially proposed by Chapman is not generally acceptable and can be improved. On the other hand, the dividing streamline velocity ratio at recompression is shown to be very sensitive to the nature of the initial profile, which is not normally known with sufficient precision. The problem of the corner expansion is also to be reviewed, particularly at high Mach numbers (see Section III,A). It is significant that these corrections do not introduce a dependence of the base pressure on the Reynolds number. Indeed, for geometrically similar bodies, a change of characteristic Reynolds number (based, say, on the base height) brings about simultaneously a change in thickness of the boundary layer at separation and a change in shear-layer length. These cancel approximately with regard to their effect on ϕ at recompression (we are not speaking here of laminar–turbulent transition).

4. *Remarks on the Development of a Turbulent Shear Layer*

No meaningful analysis of the development of a turbulent shear layer in the preasymptotic region downstream of the separation of a turbulent boundary layer is possible, simply because nothing is known about the evolution of the apparent turbulent transfer properties in this region. In terms of phenomenological theories, the "eddy viscosity" must change from the behavior characteristic of attached boundary layers, where it is certainly a function of distance from the boundary, to that characteristic of the fully developed shear layer, where it is usually taken to be constant across the layer (Goertler). An example of measurements very close to separation, of which very few exist, is shown in Fig. 8 (Charwat, unpublished). An insert shows the apparent evolution of the mixing parameter σ determined from its relation to the gradient at the dividing streamline as in the asymptotic solution, a definition which is not necessarily meaningful in this case. The parameter σ is not constant, as it is in the self-similar solution. However, it is important to observe how rapidly this parameter tends to the asymptotic value. See also Bauer (186 a,b).

This rapid approach toward an asymptotic solution is understandable noting that turbulent boundary layers are much fuller, i.e., characterized by a much larger value of H than laminar ones. In the spirit of the correlation found for the laminar case, the preasymptotic region should indeed be very short. Also, the large mixing (eddy viscosity) should have the same effect.

Fortunately, therefore, corrections for the nonasymptotic profile shape of the shear layer near recompression are likely not to be important in turbulent flow, and the use of the asymptotic dividing velocity ratio in conjunction with integral corrections such as that proposed by Nash should be adequate. Figure 9 shows measurements at $M = 3$ of the velocity profile for both plane and axisymmetric shear layers [no turning at the separation corner (27)]. Some 10 to 15 thicknesses of the initial

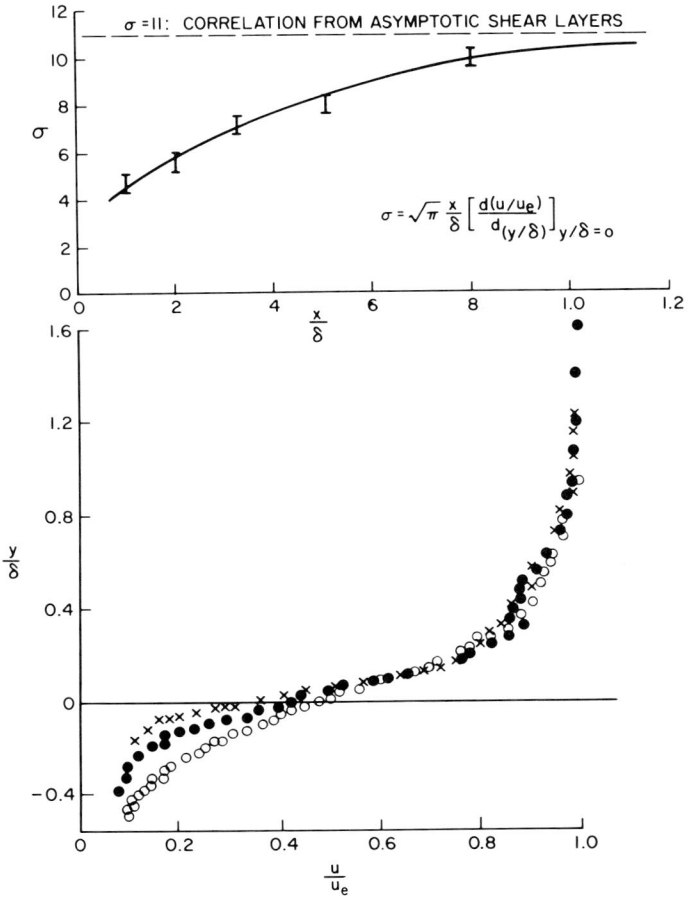

Fig. 8. The development of the turbulent mean velocity profile and the jet mixing parameter σ in the flow near separation: x = distance downstream from separation point; y = distance measured transverse from the separation point; δ = boundary-layer thickness at separation. ($\times$) $x/\delta = 1.05$; ($\bullet$) $x/\delta = 2.00$; ($\circ$) $x/\delta = 3.25$.

boundary layer downstream of separation, the profile is well represented by the error function solution [Eq. (22)]. The magnitude of the length of the preasymptotic region seems to be the same in incompressible flow (Fig. 8) and at the relatively high supersonic Mach number of 3. The development of the shear layer can then be portrayed by the asymptotic solution started from a fictitious origin—exactly the formulation underlying the Nash–Kirk correction. This also means that the

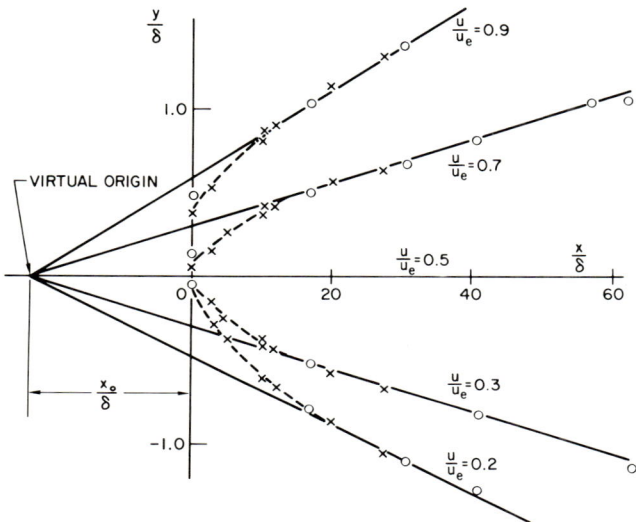

FIG. 9. Lines of constant velocity near separation for plane (×) and axisymmetric (○) shear layers at $M_\infty = 3$ (27).

equivalence of the effect of the upstream boundary layer and base blowing is preserved for turbulent flows.

Sirieix and Solignac (28) point out that a simple expression for the virtual origin of the shear layer can be given explicitly, if one approximates the velocity distribution (u/u_e) by a one-fifth power law in x/δ. One obtains then

$$x_0/\sigma\delta = 0.85$$

independent of Mach number, verifying the measurements reasonably well.

C. Distributed Recompression

It is clear that the recompression of a shear layer, which may be thin but does have a finite thickness, does not occur discontinuously. The region of recompression is schematically shown by the insert in Fig. 2. The pressure rise along the wall (or plane of symmetry) spreads out over a distance of the order of the thickness of the shear layer. The rear stagnation point pressure p_r corresponding to the reattachment of the dividing streamline is part way up the total rise in the external flow from Region e to the final pressure in Region d. Geometrically, the closure point r corresponds to some location in the middle of the compressive fan.

There are good reasons for considering this aspect of the problem in more detail: base pressure lower than that predicted by the Chapman–Korst theory and the point-recompression assumption are sometimes measured. These cannot be explained by the type of correction to $u_{\rm DSL}/u_{\rm e}$ considered in the preceding, which always leads to pressures higher than those given by the simple theory.

A simple, empirical correlation parameter was introduced by Nash in an attempt to account for these phenomena. Nash (30) defines a parameter N,

$$N = (p_{\rm r} - p_{\rm b})/(p_{\rm d} - p_{\rm b}) \tag{40}$$

which measures the fraction of the pressure rise between the base and the stagnation point on the dividing streamline to the overall pressure rise. With the introduction of this parameter the base pressure is expressed by the set of equations

$$\frac{p_\infty}{p_{\rm b}} = \Big(\frac{1 + [(\gamma-1)/2]\,M_{\rm e}^2}{1 + [(\gamma-1)/2]\,M_\infty^2}\Big)^{\gamma/(\gamma-1)}$$

$$\Big[\frac{1 + [(\gamma-1)/2]\,M_{\rm e}^2}{1 + [(\gamma-1)/2]\,M_{\rm e}^2(1-\phi^2)}\Big]^{\gamma/(\gamma-1)} - 1 \tag{41}$$

$$= N\Big[\Big(\frac{1 + [(\gamma-1)/2]\,M_{\rm e}^2}{1 + [(\gamma-1)/2]\,M_\infty^2}\Big)^{\gamma/(\gamma-1)} - 1\Big]$$

which replaces the previously given Eq. (9) and reduces to it when N is taken equal to unity.

Nash gives no conceptual framework to relate N to the physics of the recompression process and has been criticized for this. N is determined solely from experiment. In the process the value of ϕ corresponding to the measurements must be known and this introduces additional uncertainties. Using a form of the integral-momentum thickness correction (outlined before) and an aggregate of the best data available to him in the range of Mach numbers of 1.5–4, Nash finds that N lies between 0.4 (which seems best for laminar flows) and 0.35 (for turbulent flows). Using these values the base pressure is plotted against the generalized upstream boundary layer and bleed parameter in Fig. 10. It is interesting to note that according to this theory, for *turbulent* boundary layers only, the trend with Mach number is reversed when Λ is large as compared to $\Lambda = 0$ (no upstream boundary layer).

Figure 11 shows a summary of experimental measurements for turbulent boundary layers at $M = 1.5$ and 2.0. Superimposed are curves representing calculated values using the Nash correlation (with $N = 0.35$)

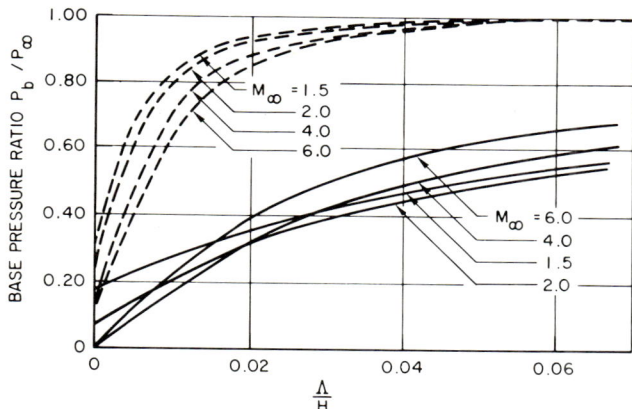

Fig. 10. Base pressure as a function of the generalized upstream boundary layer/base bleed parameter Λ/H: (- - -) laminar flow; (——) turbulent flow.

as well as results of a more sophisticated method involving the integration of moment equations describing the viscous flow with interaction (*38*), which is discussed in Section IV,C. Both analyses suggest that the proper correlation parameter is δ^{**}/H. The data and the theory concur in that the effect of the upstream boundary layer on base pressure in turbulent

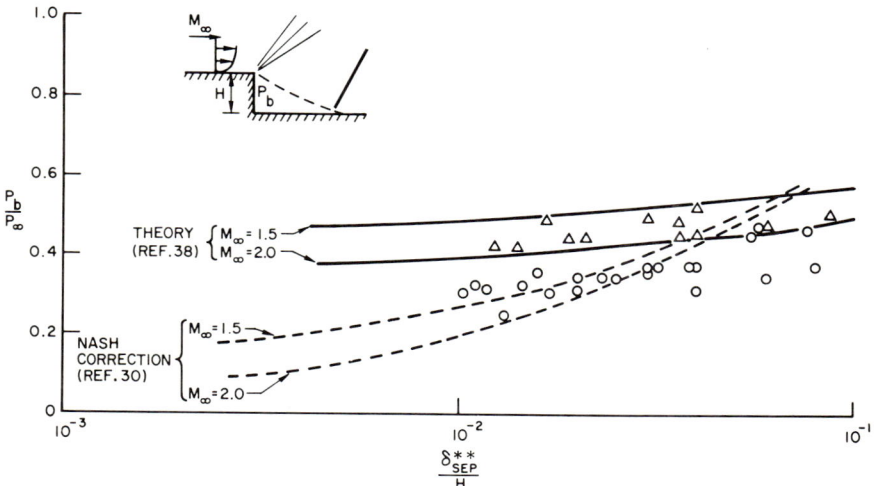

Fig. 11. The effect of an upstream boundary layer on base pressure with turbulent flow: (△) data for $M_\infty = 1.5$, (○) data for $M_\infty = 2.0$; data taken from Refs. (*20, 30, 39–49*).

flow is not strong, weaker than in laminar flow. Nash's correction appears to overestimate this effect.

The base pressure measurements in accelerated flow (21) can be used to extend the concept of N. To first approximation the pressure gradient should not strongly influence the profile of the shear layer which develops along the isobaric dividing streamline. One can thus assume that ϕ can be determined as in uniform flow and that changes due to the pressure gradient are all due to a change in the recompression process. The recompression process occurs over a distance of the order of the shear-layer thickness. Thus, a proper correlation parameter should be δ^{**}/λ (where λ is the pressure gradient parameter defined by Eq. (11). Figure 12

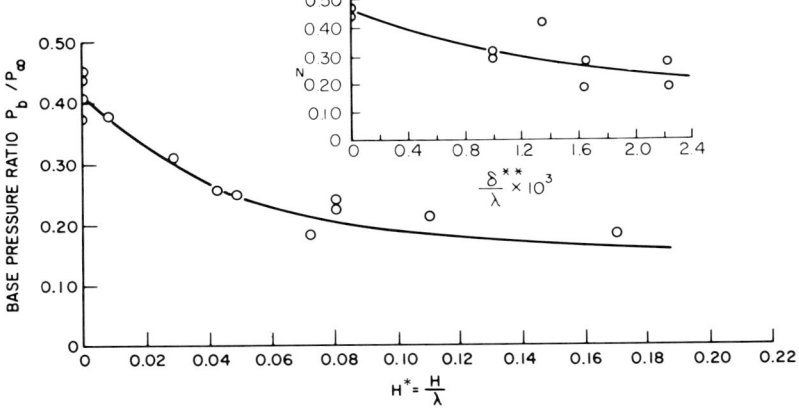

FIG. 12. The influence of an accelerated free stream on base pressure and the Nash recompression parameter N (turbulent flow; Mach number at separation is 2).

shows the results both in the form of base pressure ratios and the corresponding value of N. The approximate value of 0.4 for N at no pressure gradient is recovered and it appears that the acceleration decreases the value of N. Admittedly this data is not very conclusive but it is all that is available.

A somewhat more sophisticated method has been recently proposed (50). It is also purely empirical and, from a genuinely theoretical viewpoint, equally arbitrary. However, it does provide a physical framework for fixing the level of the pressure at r "somewhere between" p_b and p_d. The correction factor turns out to be a function of Mach number (which Nash's N is not) and also depends only on parameters pertaining to the flow before recompression (p_d need not be known), which has a certain advantage. In supersonic flow the downstream region cannot influence the flow ahead of r, that is, the base region. This holds even when the viscous layer is taken into consideration since the critical point or throat of the near-wake is somewhat downstream of r. This has been demonstrated by measurements (20, 51) and it plays a crucial role in more sophisticated analytical treatments such as the Lees–Reeves method (Section IV,3). In this sense, Nash's correction, which depends explicitly on p_d, is conceptually unsatisfactory.

The key empirical statement is that the distance (normal to the reattachment wall) to the edge of the viscous layer is minimum at the reattachment point:

$$\left.\frac{d\delta}{dx}\right]_r = 0 \qquad (42)$$

The true angle of the edge of the layer with respect to the wall, ψ_r, at the reattachment station (see insert in Fig. 1) is

$$\tan \psi_r = (v/u)_r \qquad (43)$$

and it is smaller than the angle θ_r made by the dividing streamline at r. Using an integral of the continuity equation normal to the wall with the conditions above one finds

$$\tan \psi_r = -\frac{1}{\rho_r u_r y^j} \frac{d}{dx} \int_0^\delta \rho u y^j \, dy \qquad (44)$$

where $j = 0$ and 1 for two-dimensional and axisymmetric flow, respectively. The integral is evaluated (it represents the mass flow in the viscous layer) using the turbulent profile in the mixing layer immediately before the recompression zone:

$$\frac{d}{dx}\int_0^\delta \rho u r^j \, dy \sim \frac{\rho_e u_e}{\sigma}\left[\left(\frac{s}{y}\frac{dy}{ds}\right)^j + 1\right] \qquad (45)$$

where s is the coordinate along the shear layer (its length) and $dy/ds = \sin \theta_r$ is the angle of the dividing streamline with the axis. The result is

$$\tan \psi_r \sim \frac{1}{\sigma}\frac{\rho_e u_e}{\rho_r u_r}\left[1 + \left(\frac{s}{y}\sin \theta_r\right)^j\right] \qquad (46)$$

ρ_r and u_r are the density and the velocity in the external flow after partial compression from Region e, which is associated with the turning $(\theta_r - \psi_r)$ of the external stream. These quantities are related to the pressure p_r.

One can now define a parameter

$$\chi = \tan(\psi_r - \theta_r) \Big/ \Big[1 + \Big(\frac{s}{y}\sin\theta_r\Big)^j\Big] \quad (47)$$

and evaluate it by an inverse treatment of base pressure measurements, like in the method of Nash (the base pressure with the point-recompression theory equations yields θ_r). Remarkably good correlations result which are expressible analytically by (see also Fig. 13)

$$\chi = 0.187 + 0.165 M_e - 0.021 M_e^2, \quad 1.5 < M_e < 5 \quad (48)$$

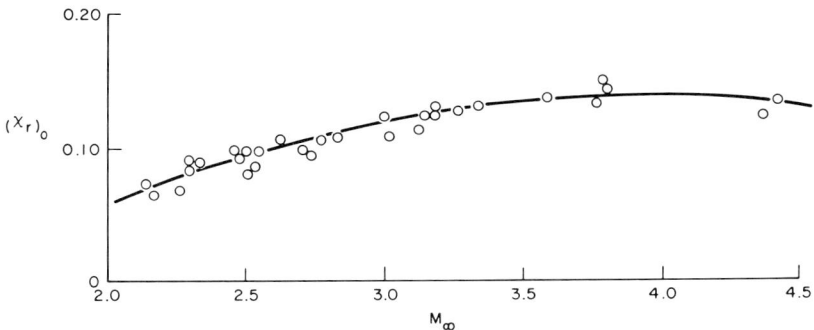

FIG. 13. The recompression flow turning parameter χ_{r_0} as a function of the free stream Mach number.

Both two-dimensional and axisymmetric configurations are included, but it should be noted that only axisymmetric cases for which y increased along the shear layer (expansive flow) are used here. It is possible that for convergent axisymmetric wakes, especially when the radius at r is very small, this correlation may have to be changed.

The manipulation of this method is somewhat more complicated than the simpler correction of Nash, but it does appear to further improve the correlation of data (only turbulent flow was analyzed). The correction for the boundary layer ahead of separation and for based bleed can be included by writing for ψ_r

$$\psi_r = \psi_{r_0} + \Lambda \, d\theta_r/d\Lambda + \cdots \quad (49)$$

where ψ_{r_0} is given by Eq. (46) or Fig. 13 for negligible upstream boundary layer and/or base bleed, and the linear correction in the parameter Λ [see Eq. (36)] is assumed to be identical to what it would be in the context of the point-recompression theory. The authors demonstrated satisfactory results applying this idea.

D. Confrontation with Experiments

1. *Pressure Distribution*

Figure 14 shows pressure distributions on the wall of a downstream-facing step taken from the careful and systematic measurements of

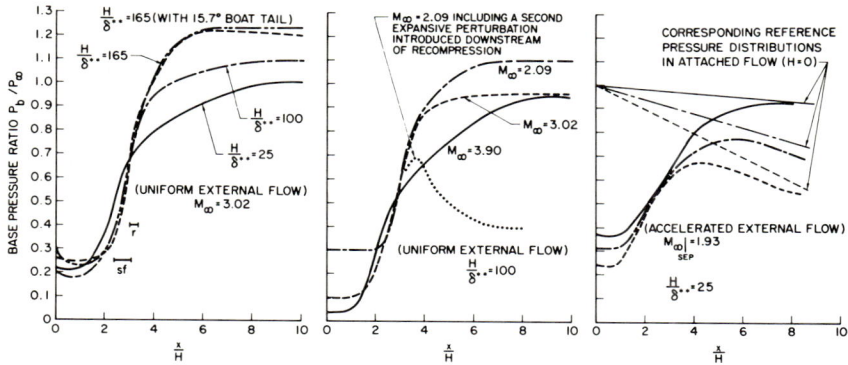

FIG. 14. Typical pressure distributions behind a downstream-facing step in turbulent flow.

Roshko and Thomke (20). The model is a step (0.25–1.68 in.) in the surface of a large (6-in. radius) axisymmetric body. The variation of the streamline radius is sufficiently small so that the flow can be considered two dimensional while being free of aspect-ratio effects. The influence of aspect ratio is now well recognized, if not understood. It is customary to use side fences on two-dimensional models (see, for instance, Lewis *et al.* (52) for work on compression corners; Charwat and Yakura (53) on blunted-wedge wakes; Ginoux (54, 55) for flow visualization). Side fences tend to lower the pressure level.

In their general features, these pressure distributions support the ideas incorporated in the Chapman–Korst model. A region of very nearly constant base pressure exists; other investigators [for instance, Chapman *et al.* (7)] verified that protrusions, fillets, and variations in the geometry of the walls inside the cavity have very little effect on the pressure in it. Consequently, to a certain level of approximation and with respect to pressure data, the concept of "dead air" appears justified. The comparison between typical data for a model with a sharp separation corner and a boat-tailed one (56) (boat-tail angle only slightly less than the total turning angle at separation) indicates no important effect due to the separation mechanism and confirms the assumption that the structure of the cavity is controlled by the recompression process.

A group of measurements involving different step heights (20), a group showing different Mach numbers (20), and a group illustrating the effect of nonuniformity (acceleration) in the oncoming external flow (21) are shown in Fig. 14. The first group indicates the position of Point r, calculated from the measured base pressure, which corresponds to the recompression point of Fig. 1. It is seen that, indeed, this point lies well before the final recovery pressure, in accord with the discussion of Section II,C. Shown also is the physical point of recompression determined from observation of flow reversal by visualization (marked sf). It lies somewhat ahead of the calculated recompression point, but close to it.

All the curves show a remarkable tendency for the pressure rise regions to become superimposed in the region of maximum slope in terms of the dimensionless distance x/H. This is also true in the case of an accelerated external flow, in spite of the large change in base pressure proper. Throughout the range of these test, the Chapman–Korst model predicts very little change in the geometrical location of Point r; thus, the data confirms the idea that the cavity is governed largely by developments along the dividing streamline. Morever, the trends in the base pressure with Mach number and with δ^{**}/H (which does not affect the geometry, but does influence the dynamics of the recompression process) are all in accord with the general predictions of the Chapman–Korst distributed-recompression model. A very interesting experiment is illustrated in the second group of curves (the curve for $M_\infty = 2.09$). It pertains to the search of what has become known as the "critical point" in the wake, from downstream of which no disturbances can propagate into the base region. The curve shown is taken from Roshko and Thomke (20); similar experiments which used different methods of disturbing the flow downstream were reported by Sirieix and Solignac (27). This critical point is found somewhat downstream of the ideal recompression point r; if the disturbance is moved upstream of that point, the cavity pressure begins to change.

In spite of the general conceptual coherence between the model and observations, attempts to correlate typical base pressure data result usually in some scatter (typically 10%). The errors are often due to poor or incomplete control over the experimental situation. For instance, many early papers emphasized good accord of measurements with the simple (point-recompression) model, which can be traced to a cancellation of two essential factors: the fact that recompression is distributed results in lowering the predicted pressure and, simultaneously, the boundary layer upstream increases the calculated pressure (see Fig. 11). There are also, however, obvious simplifications in the model; the data show, for instance, differences and (reproducible) irregularities mainly in the downstream portion of the region of pressure rise. These are due to phenomena which are also partly responsible for the scatter in base pressure data; they will be discussed further in a subsequent section of this review.

The Chapman–Korst model introduces a Reynolds number effect on the base pressure only through its influence on the key recompression parameter ϕ. The effect is felt mainly in the transition from laminar to turbulent flow. This general trend is exhibited by illustrative data in Fig. 15 (53, 57–61). Transition "travels" along the free shear layer from downstream of the critical point in the recompression region to ahead of separation as a typical Reynolds number increases causing in the base pressure to change from the higher laminar-flow level towards the lower turbulent-flow value. There are several interpretations as to exactly

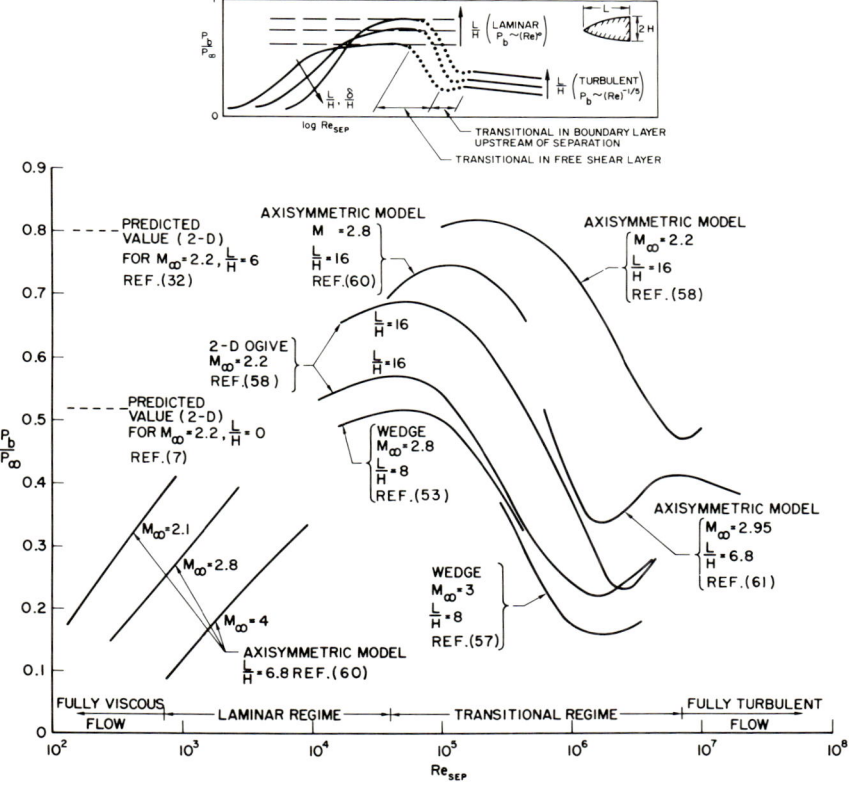

FIG. 15. Base pressure as a function of the Reynolds number at separation.

where "transition" is located within this region [see Holder and Gadd (57), Crocco and Lees (6) for most widely held interpretation, and Roshko (10) for a slightly different view]. We lack a meaningful criterion for transition and any detailed knowledge of this phenomenon even in much simpler situations. Some detailed experimental work is in progress (62–66), mainly in far wakes downstream of the neck of the cavity (blunt-base bodies). By extrapolation one can determine at least the Reynolds number at which turbulence begins to affect the recompression region. We recognize, however, that the stability and transition properties of free shear layer are dependent both on the local balance between viscous and inertia forces (a Reynolds number based on the length of the shear layer) and also on the local velocity profile, which is influenced by the initial profile of the viscous layer ahead of separation (a Reynolds number based on forebody length and its shape). It follows that at least two independent parameters, say Re_L and L/H, or Re_L and δ/H, as well as the forebody geometry, will influence the problem—as indeed is demonstrated by the data of Fig. 15. Note that transition occurs in a region of model Reynolds numbers such that it is difficult to design experiments totally free of transitional effects (unless the Mach number is high).

Figure 15 shows both wake and step data. There is no essential difference between them (at intermediate Reynolds number), although wake pressures tend to be slightly lower. Data for an axisymmetric wake is also shown. Axisymmetric step data is bracketed (other factors remaining the same) by the two-dimensional step (radius change along the shear layer is large) and the axisymmetric wake (recompression on axis).

The numerical calculations of Baum et al. (32), which are the most "sophisticated" without abandoning the basic assumptions of the Chapman–Korst model, demonstrate that the combined effect of Reynolds number on the thickness of the boundary layer ahead of separation and on the subsequent preasymptotic development of the shear layer is such that, in purely laminar flow, viscous effects drop out and only the influence of forebody and step geometry remains. The base pressure increases with increasing L/H (Fig. 15). The Chapman–Korst model does not predict the slight, but evident, base pressure variation with Reynolds number. Figure 16 shows it clearly; this is high Mach number data for the wake of two-dimensional ogival-nose bodies (66, 67) which was carefully monitored for transition so that the near wake is indeed laminar. The figure shows both a Reynolds number plot and one against δ/H. Neither single parameter results in the collapse of all curves. Note also that even at the relatively high Mach number of 6 the simple

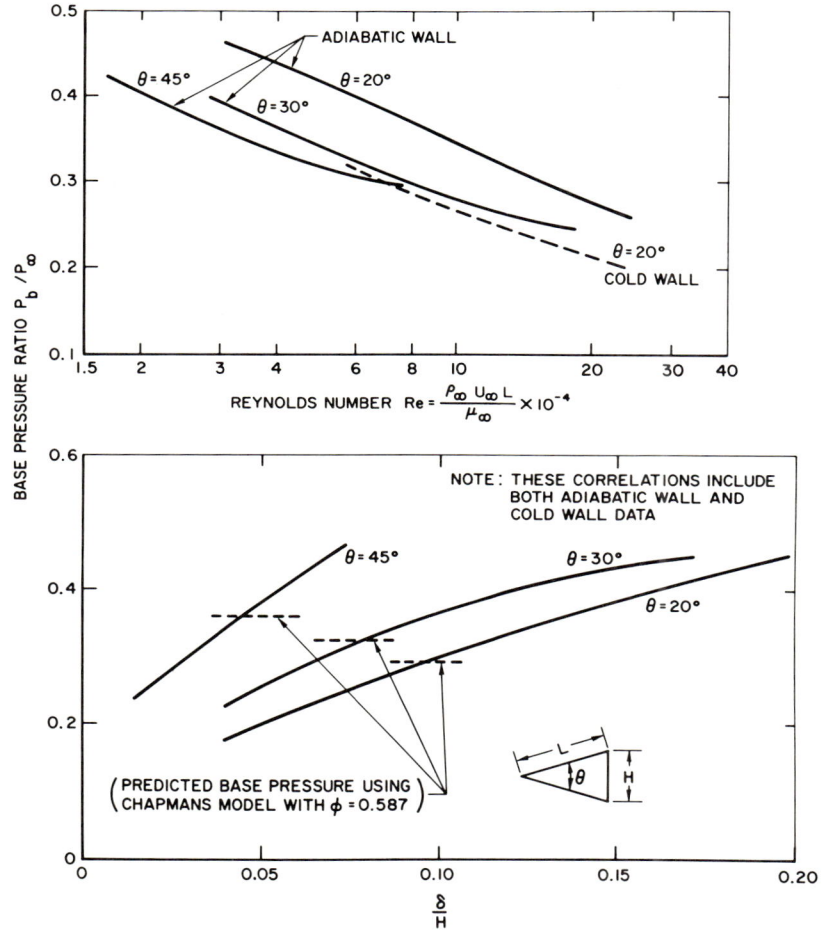

FIG. 16. Measured base pressure variations with free stream Reynolds number and boundary-layer thickness for flow over wedges at Mach 6 (66, 67).

Chapman model (which now has to be corrected for the difference between free stream pressure and body-surface pressure because the hypersonic oblique shock is too strong) gives the general level of the base pressure correctly.

In purely turbulent flow, the situation is more speculative. We have seen that the profile of a turbulent shear layer rapidly becomes undistinguishable from the asymptotic profile. Therefore the momentum correction, Figs. 10 and 11, should be good. It predicts an increase in base pressure with δ^{**}/H, which one can approximate as linear for small δ^{**}/H. The initial turbulent boundary-layer thickness to forebody length ratio δ^{**}/L varies as $Re^{-1/5}$. There is not enough good, fully turbulent data available to check this; we may refer at this point to a frankly empirical correlation of data presented by Sherberg and Smith (68) in the Mach number range of 2.5–5 and transitional to turbulent flow, which is based on their data as well as on a series of investigations of Rom (69). These authors find that the base pressure drop ($p_b - p_\infty$) is linear in p_∞ independently of Mach number with a slope dependent on L/H. The slope varies considerably: it is large for small steps and high unit Reynolds numbers and small for large steps and low unit Reynolds numbers. The nature of this correlation seems to us to be insufficiently understood and documented to allow generalization. However, the observations that $d(\Delta p)/dp_\infty$ is linear for a fixed model can be useful in extrapolating limited data.

At very low Reynolds numbers the base pressure tends to zero (free molecule flow). The combined effects of Reynolds number and L/D exhibit a reversal. This has been recently studied using Stokes and Oseen equations by Berger and Viviand (70, 71), who succeeded in explaining the seemingly anomalous behavior (for a cylinder). Some experiments in low-density flow have been published (72, 73). In this region the concept of a viscous shear layer bounded by two inviscid regions fails and a fully viscous flow about the entire forebody and wake determines the base pressure. One can hardly expect a simple correlation of data for different geometries. The trend is noted in Fig. 15, but we shall not discuss this special flow regime in any more detail.

2. Heat Transfer Measurements behind Steps and Wakes

The Chapman–Korst model admits but one assumption regarding the enthalpy (or temperature) of the "dead air": it is constant throughout the base region. One can then find a solution to the energy equation for the shear layer corresponding to the solution of the momentum equation discussed earlier. The usual assumption that $Pr = 1$ uncouples these equations. Chapman (22) provided the asymptotic self-similar solution; Baum et al. (32) included the energy equation in their numerical integration of the preasymptotic shear-layer problem. In the latter case an initial condition on the enthalpy (at separation) must be prescribed, which these authors derived by applying the Crocco integral relation to the assumed velocity profile of the initial boundary layer.

The solution yields shear-layer enthalpy distributions in terms of one parameter: the enthalpy in the cavity. Its level must be determined separately by writing an overall energy balance for the region. For this purpose a region is considered which includes the shear layer, bounded by the streamline constituting the edge of the outer inviscid region. Energy is transferred by convection into this region by the boundary-layer flow at the separation station and out of it by the shear layer at the recompression station, and by conduction to the solid boundaries of the cavity. The first two terms are integrals involving the (known) initial boundary-layer (if any) and the shear-layer enthalpy distributions (between the dividing streamline and the outer edge) which are a function of the core enthalpy.

The Chapman–Korst model leads thus to a relation between the heat flux to the solid boundaries of the base region and the core enthalpy (assumed to be uniform). It is impossible to close the problem, that is, to relate these quantities to the thermal boundary condition on the wall (the wall temperature), without making some additional assumption regarding boundary layer on these walls, that is, the nature of the internal flow.

Early papers (*74*) assumed that the thermal resistance of the wall boundary layer was very small as compared with that of the shear layer and, therefore, the cavity bulk temperature was identified with the wall temperature. Using the asymptotic solution for the shear layer, the ratio of the mean heat transfer coefficient for the cavity walls to that of an attached boundary layer in laminar flow was found to be 0.56, constant and independent of Mach or Reynolds number (*22*). In turbulent flow the same calculation yielded higher heat transfer coefficient ratios: 6.3 for incompressible flow and decreasing with an increase in Mach number (this variation depends on the assumptions regarding the turbulent mixing).

Later work proved that the thermal resistance of the wall boundary layer is not negligible, but (in the case of a wake) dominant. As a consequence, the core enthalpy is considerably higher than the wall temperature for a given heat flux, and, for a given wall temperature, the heat transfer is less than predicted by Chapman. The core-enthalpy level is very sensitive to the amount of heat removed by the walls.

In considering the overall heat transfer, one must distinguish between cavities downstream of steps and blunt-body wakes. In the latter case, the only solid boundary (the base) is away from the recompression region and exposed to well-mixed, low-velocity flow. It is reasonable to assume that the boundary layer which forms on the base wall is of the stagnation-point type. Using such an assumption a heat transfer coefficient for the inner boundary layer can be estimated. Figure 17a shows typical results obtained by such an analysis (*32*) performed in conjunction with the numerical solution of the shear layer. In turbulent flow the analysis is less

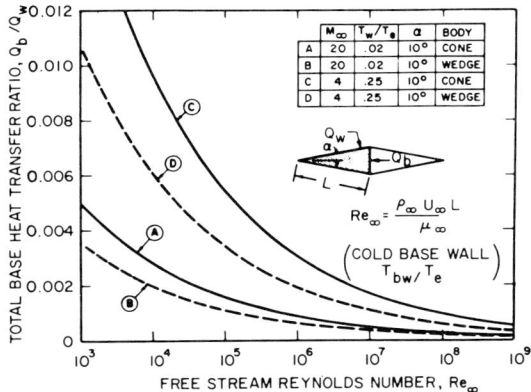

Fig. 17a. Theoretical dependence of the base heat transfer on the Reynolds number in laminar flow (32).

certain (75, 76), but good experiments have been conducted. A summary of measurements (75) showing the turbulent Stanton–Reynolds number variation for two-dimensional wakes behind ogival-profile bodies is quoted in Fig. 17b. This study, which included an independent measurement of the temperature in the bulk of the recirculating region, makes it possible to "split" the overall resistance to heat transfer into its two components: the resistance (or conversely, the conductance) of the shear layer and that of the wall boundary layer. Both are shown; the shear layer resistance by itself is not too different from that calculated

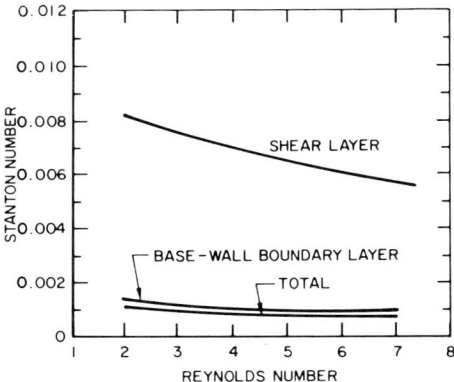

Fig. 17b. Measured variation of the shear layer, base-wall boundary layer, and total heat transfer coefficients with Reynolds number in turbulent flow (75).

as outlined above. The base boundary layer constitutes 80–90% of the total resistance.

Recent measurements of stagnation temperatures in the recirculating region in laminar and turbulent two-dimensional and axisymmetric wakes are available in Refs. *67, 75, 76, 77–82*, and others (most of these papers also give some pressure data). Muntz and Softley (*80*) measured static temperatures using an electron-beam excitation technique for a rather wide range of body temperatures. A current study conducted by Valensi (Inst. de Mecanique des Fluides, Marseille) demonstrates that the change in the bulk temperature of the fluid in the wake (fixed body temperature) can be used as a sensitive indicator of transition in the shear layer.

The problem of the downstep with reattachment to a wall is less amenable to analysis than the wake. Figure 18 shows some typical

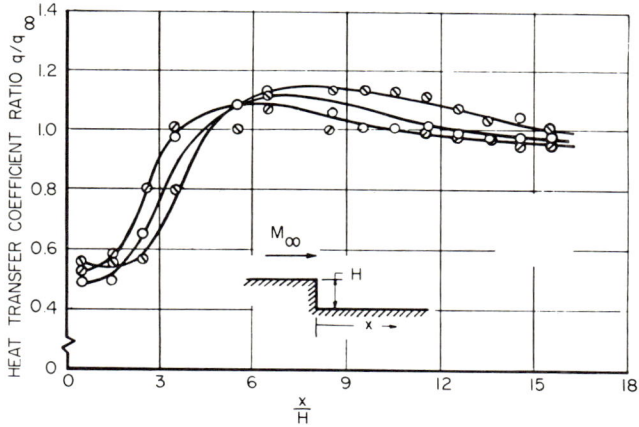

FIG. 18. Measured heat transfer distribution behind a downstream-facing step in turbulent flow (*83*): (⌀) $M_\infty = 2.1$; (○) $M_\infty = 2.9$; (⊘) $M_\infty = 3.5$.

data (*83*) (see also Refs. *84–89*). The heat transfer coefficient ratio (ratio of the base heat transfer coefficient to the one immediately ahead of separation) drops from unity to a relatively low value at separation. This minimum is generally higher than in the case of a blunt-base wake and it is rather sensitive to the ratio of the upstream boundary-layer thickness to the height of the step. Values of about 0.2 are observed for small δ/H and can increase to 0.6 for δ/H of the order of unity (*83*). Downstream, $\dot{q}/\dot{q}_\infty$ follows a recovery curve somewhat similar to that for the pressure. Unlike the pressure curve, however, the heat transfer maximum in the recompression region is extremely sensitive to transition. The maximum occurs somewhat downstream of the point of reattachment (observed, for example, visually), and somewhat downstream of the point at which the pressure has fully recovered. The value of the maximum $\dot{q}/\dot{q}_\infty$ is near unity for both fully turbulent and fully laminar flow. It is not sensitive to the free stream Mach number (see, for instance, Nestler (89) for a summary of data in the range $0 < M_\infty < 10$). When the oncoming boundary layer to step height ratio is small, $\dot{q}/\dot{q}_\infty$ tends to overshoot, but ultimately reaches a value less than unity (which is understandable because the net effect of the cavity is to thicken the boundary layer). There are cases where a maximum in $\dot{q}/\dot{q}_\infty$ as high as three or four can be observed. These can be traced to transition in the free shear layer.

Clearly the step heat transfer problem can also be schematized by considering two resistances in series: that of the shear layer, which can be calculated as outlined above, and that of the wall boundary layer, which in this case requires, however, a fairly complete statement about the inner flow. There are virtually no experiments dealing with the wall boundary layer for this configuration. Its heat transfer coefficient (average) is much higher than for the base alone because of the recompression region. The bulk temperature in the cavity is much closer to the wall temperature; i.e., the Chapman assumption is better verified than it is for wakes. The average heat transfer coefficient for a wall with a downstream-facing step separation (including the recompression region) in either pure laminar or pure turbulent flow is slightly less, typically 20%, than what it would be for an attached flow at equivalent upstream conditions. However, since steps enhance transition, the existence of a step can increase the total heat transfer to the surface considerably under certain conditions.

The behavior of the heat transfer coefficient downstream of steps in subsonic flow is qualitatively similar to that in supersonic flow [see, for instance, Seban (90, 91)].

3. Summary

There is a large volume of experimental data on separated regions published in the literature. The small selection presented in the foregoing suffices to illustrate the degree of success and the shortcomings of the Chapman–Korst model: it seems to catch correctly the dominant mechanisms which govern the flow, but it fails to portray several important details. It fails to explain correctly the Reynolds number dependence; it also fails to provide a framework for correlating heat transfer data. The heat transfer problem is intimately connected to the nature of the internal motion in the cavity, which is entirely neglected by the model.

In spite of its shortcomings, the Chapman–Korst analysis remains a useful framework for predicting base pressures to an engineering accuracy in the mid-range of Mach and Reynolds numbers. The next level of analysis leads invariably to extensive and difficult numerical calculations. There is a family of problems pertaining mainly to the far wake formed downstream of the neck of the cavity (temperature, velocity, and concentration distributions) which depend critically on the near-wake recompression zone and, therefore, require a far more precise analysis of the entire wake region, including the upstream flow field. Due to the importance of this problem in reentry aerodynamics, a large amount of highly sophisticated research became recently available. This provided considerable insight to the structure of the near wake, but the complexity of the analysis is such that little general quantitative correlation other than numerical results for specific cases can be derived from it. It seems that, at this moment, one must accept the limitations of the semiempirical analysis based on the Chapman–Korst model, judiciously monitored by comparison with specific data, for the mid-supersonic regime (Mach 1.5–6) at reasonably high Reynolds numbers ($Re > 10^4$). To improve on this model, to extend the analysis into the hypersonic regime, or to ask for details regarding the flow field in the near and far wake, it seems necessary to proceed with a rather formidable numerical program. There seems to be no intermediate stage of the analysis.

III. The Structure of the External and Recirculating Internal Regions

The Chapman–Korst model isolates three coupled but noninteracting regions: the external flow, the base region, and the shear layer. The nature of the shear layer plays a key role in the matching of the other two regions and providing the "closure condition" for the wake. All the analyses described in the preceding sections, including the numerical treatment of the shear-layer equations, were based on the assumption that both the external and the internal regions are uniform (the so-called Chapman boundary conditions). However, experiments disclose evidence of nonuniformities in the external flow as well as the fact that the internal region is not "dead air." This must influence the development of the viscous shear layer; in fact, the three regions interact so that their matching (the boundary conditions at each interface) depends on the solution.

This section presents investigations on the true nature of the external (inviscid) and the internal flows and shows thereby some aspects of the approximations involved in the Chapman–Korst model. Included in the section is a discussion of experimental data pertaining to cavities in notches and cutouts. This problem is dominated by the structure of the internal flow and hence is outside of the scope of the Chapman–Korst model entirely.

A. THE SEPARATION AT AN EXPANSIVE CORNER

Early experiments with downsteps and blunt-base wakes in supersonic flow disclosed the existence of a shock emanating from the separation point which is almost parallel to the shear layer. This shock is commonly called the lip shock. Love (92) was perhaps the first to discuss it in terms of the overturning of the boundary layer at the corner. Charwat and Yakura (53) supplied early data showing the evolution of stagnation pressure profiles in the shear layer close to the base, starting at approximately half a boundary-layer thickness from separation. Even this close to the base, the shock is already distinguishable and seems to emanate from slightly below the separation corner itself. The aforementioned investigation also showed another much fainter discontinuity which was thought to be an entropy wave (slip line). Its existence was again recently confirmed by Hama (59). It is also Hama who provided the most complete systematic study of the lip shock. The range of his experiments span Mach numbers between 2 and 4.5, and Reynolds numbers between 10^5 and 2×10^7. It follows that this is not necessarily a hypersonic phenomenon. Unfortunately, Hama's results span the region of transition from laminar to turbulent flow in the shear layer, which causes some difficulty in the interpretation of the results.

A typical example of the lip shock is shown in Fig. 19. The lip shock is sometimes directed down (relatively low Mach numbers in turbulent

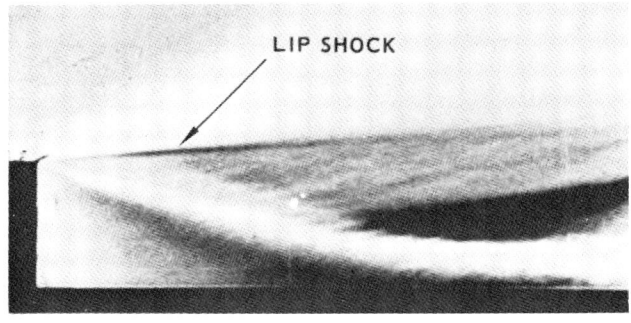

FIG. 19. A typical photograph of a lip shock.

boundary layers) but can also be directed upward (high Mach number, laminar boundary layer). Depending on its direction it can interact with the recompression shock (or, in hypersonic flow with the bow shock generated at the nose of the body) which engenders wave interactions that radiate back onto the wake. This has been identified as responsible for irregular humps appearing in the pressure distribution downstream of steps.

It is clear that the lip shock originates in the process of the turning of the boundary layer. The first adequate analysis of the problem was given by Weinbaum and Weiss (93, 94), which we shall now summarize.

1. *Rotational Characteristic Analysis*

Figure 20 is a schematic of the flow over the separation corner. There is at all times a subsonic channel connecting the base wake to the upstream boundary layer. This subsonic channel transfers the low-pressure signal upstream and causes the expansion to begin. The leading Mach wave of the expansion occurs before the corner. The primary expansion waves originate at the line at which the Mach number in the boundary layer is unity (which is not identical with the sonic streamline in the initial boundary layer because, as the flow expands, the initially subsonic portion of the boundary layer accelerates). These Mach waves traverse the rotational boundary layer, generating secondary waves of an opposite family through interaction with the entropy gradient. The secondary waves, which can be either compressive or expansive, retraverse the primary expansion and the rotational layer and reflect off the $M = 1$ line. If the pressure in the inner flow is nearly constant and the transverse gradients across the subsonic portion of the shear layer are negligible, the secondary wave reflection can be assumed to occur at an isobaric surface which replaces the thin subsonic portion of the shear layer. The impinging waves reflect as waves of opposite family off the isobaric boundary. In a practical case the secondary waves are mainly expansive and reflect as compression waves. Their coalescence forms the lip shock. Note that during the reflection process the dividing streamline curves, although the pressure in the base remains constant.

On the level of detail shown on the diagram the supersonic problem requires a simultaneous matching of the supersonic region to the flow in subsonic sublayer. As a first approximation, Weinbaum collapses the streamwise extent of the interaction zone along the sonic line to a point located at the corner, at which the pressure is singular. He replaces the distributed primary expansion by a centered fan originating at the $M = 1$ line at the corner. If one assumes, in addition, that the process is inviscid,

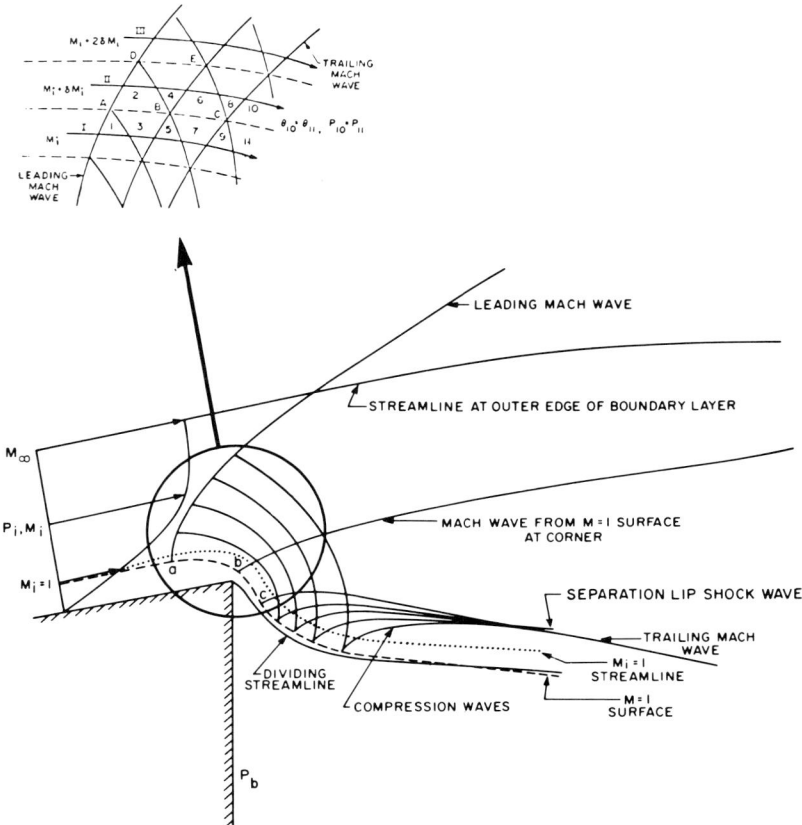

FIG. 20. Schematic of the expansion of a supersonic boundary layer at the trailing edge of a blunt-based body (93).

standard rotational characteristics methods can be used to solve the entire flow field. The solution depends on the Mach number distribution in the oncoming rotational flow (the boundary layer) and one parameter, say, the turning angle of a selected streamline, which can be conveniently taken as the dividing streamline turning angle θ_s.

An approximate treatment of this problem which is valid for small turning angles and Mach numbers not too close to unity (outer portion of the layer) gives in a concise fashion the entire analytical trends. For large turning angles, numerical methods of rotational characteristic must be used.

Consider the insert in Fig. 20 showing the wave interaction process.

Across a primary wave the pressure changes according to Busemann's second-order approximation for a weak oblique shock associated with a deflection of the flow by an angle θ_1:

$$(p_1 - p_\infty)/p_\infty = a\theta_1 + b\theta_1^2 + \cdots \qquad (50)$$

where a, b are functions only of the Mach number M_i ahead of the wave and of γ, and are given explicitly by Weinbaum (93). The primary wave bends by an angle $(\theta_2 - \theta_1)$,

$$\theta_2 = \theta_1 + \theta_1[(\theta_2 - \theta_1)/\theta_1] \qquad (51)$$

at it crosses the slipline to a region where the Mach number is $M_i + \Delta M_i$. The pressure change in the neighboring cell 2 is given by an equation analogous to (50). This equation can be worded in terms of M_i, ΔM_i, θ_1, and $(\theta_2 - \theta_1)$ by using (51) [retaining first-order terms in $(\theta_2 - \theta_1)/\theta_1$ only] and expanding the coefficients a and b pertaining to Region 2 to first-order in ΔM_i such that

$$a(M_i + \Delta M_i) = a_{(M_i)} + a'_{(M_i)} \Delta M_i + O(\Delta M_i^2) \qquad (52)$$

where new parameters a' (and b') are also functions of γ and M_i only. Retaining only terms of first order in $(\theta_2 - \theta_1)/\theta_1$, the pressure change in the neighboring cell can be expressed by an equation which is equivalent to (50):

$$(p_2 - p_\infty)/p_\infty = a\theta_1 + b\theta_1^2 + \cdots + a(\theta_2 - \theta_1) + \cdots + c\theta_1 \Delta M_i + \cdots \qquad (53)$$

The equality of flow direction and pressures is to be enforced across the slipline. In general, a wave of strength of the order of $(\theta_2 - \theta_1)$ is necessary, which leads to

$$(p_3 - p_1)/p_1 = -a(\theta_2 - \theta_1) + O(\theta_2 - \theta_1)^3 \qquad (54)$$

This equation is equivalent to (50) (to first order in $\Delta\theta$) except that Region 1 replaces Region ∞ and Region 3 replaces Region 1.

Enforcing equality of pressure between Region 2 and Region 3 one has, from (54), (53), and (50) to order $(\theta_2 - \theta_1)$, the following expression:

$$\frac{\theta_2 - \theta_1}{\theta_1} = -\left\{\frac{M_i^2 - 2}{2M_i(M_i^2 - 1)}\right\} \Delta M_i \qquad (55)$$

To write this the full expressions for the parameters a and b were used. Note that the coefficient of ΔM_i in (55) changes sign at $M_i = \sqrt{2}$. This means that the reflected wave is an expansion [$(\theta_2 - \theta_1)$ is larger than 0] when M_i is larger than $\sqrt{2}$ and a compression when M_i is smaller than $\sqrt{2}$.

The elemental interaction discussed above occurs at all interior points of the interaction net. Since the reflected waves were linearized (only the original waves were treated nonlinearly; this is what restricts the theory to small turning angles) their effects are simply additive. It follows that the change in stream tube I differs only be the contribution 9–11 (see Fig. 20) from that in stream tube II, so that

$$\theta_{11} - \theta_9 = -\frac{M_i^2 - 2}{2M_i(M_i^2 - 1)} \theta_9 \Delta M_i \tag{56}$$

is the differential change in flow angle at the rear (trailing edge) of the primary expansion fan as a function of ΔM_i and the Mach number in the stream tube under consideration immediately ahead of the leading primary expansion wave. Going from a difference to a differential form, (56) can be integrated, the result being

$$\frac{(M_i^2 - 1)^{1/4}}{M_i \theta} = \text{const} \tag{57}$$

This variation of turning angle θ of each stream tube of the incoming flow (each stream tube is characterized in this formulation by its initial Mach number M_i) is equivalent to a variation of the strength of the trailing wave of the primary expansion fan. It is substantial, especially at high Mach numbers. For example, and outer streamline along which $M_i = 20$, before the beginning of expansion, will undergo a change in pressure three times stronger than the inner streamline along which, before the interaction, the Mach number was $M_i = \sqrt{2}$. At the same time, the latter would turn three times more than the former streamline. It follows that the expansive turning of the boundary layer results in a widely divergent flow downstream and correspondingly large crossflow pressure gradients.

Equation (57) was checked against an exact characteristic calculation and shown to be quite accurate for turning angles at the dividing stream tube less than $\theta_s = 10°$ and Mach numbers more than 1.5. Weiss and Weinbaum (94) also developed an equivalent transonic and a hypersonic approximation. Using a numerical rotational characteristic program, they calculated the turning angle at the trailing wave of the primary expansion fan as a function of Mach number for a laminar and turbulent initial profile. These results are reproduced in Fig. 21 with Eq. (57) shown dotted for $\theta_s = \theta_1 = -10°$.

For small turning angles and low Mach numbers the turning of the

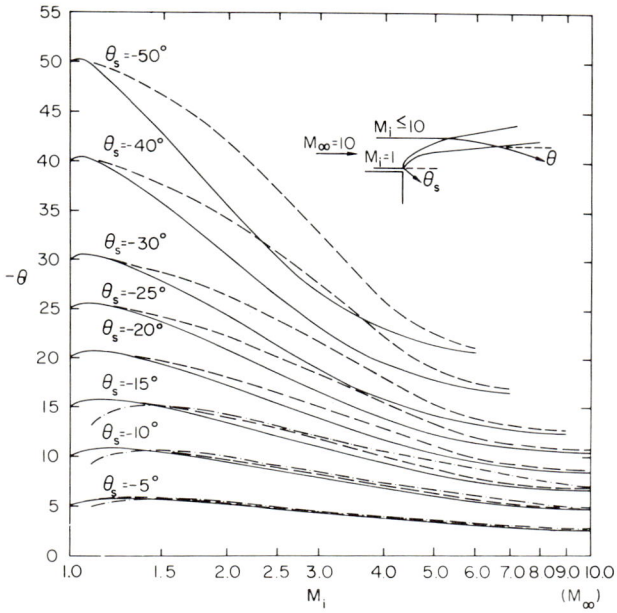

FIG. 21. Expansion of representative laminar and turbulent velocity profiles: Pr = 1; adiabatic wall conditions; $M_\infty = 10$ (93). (———) Laminar boundary layer and (– – –) turbulent boundary layer, exact numerical solution; (– · –) second-order theory, Eq. (57).

outer portion of the boundary layer is larger than that at its inner edge. This means that the waves of the primary expansion fan are weakened, compression waves are radiated back onto the dividing streamline, and, therefore, reflect as expansion waves. In such cases, no lip shock could form. However, for larger turning angles of the dividing streamline, the outer portions of the boundary layer are turned less than the inner portions, causing the divergent flow already mentioned. The primary expansion fan is strengthened (in terms of pressure). Secondary waves radiated back into the shear layer are expansion waves, and reflect as compressive waves which coalesce to form the lip shock.

The initial Mach number *distribution* in the upstream boundary layer, that is, the laminar or turbulent nature of the profile, does not appear to influence the problem very much. Weinbaum (93) discusses this unexpected results and gives some physical interpretation. In any case, the strength of the lip shock, which is proportional to the amount of differential turning across the shear layer, does not seem to be a strong function of the boundary-layer shape. It is primarily a function of the initial turning θ_s of the inner dividing streamtube and the maximum upstream Mach number at the edge of the boundary layer. In low Mach number supersonic flow the Mach number dependence is not dominant. The overturning depends primarily on θ_s, which is equivalent to the base pressure ratio. Measurements of shock strength by Hama (Fig. 22) seem

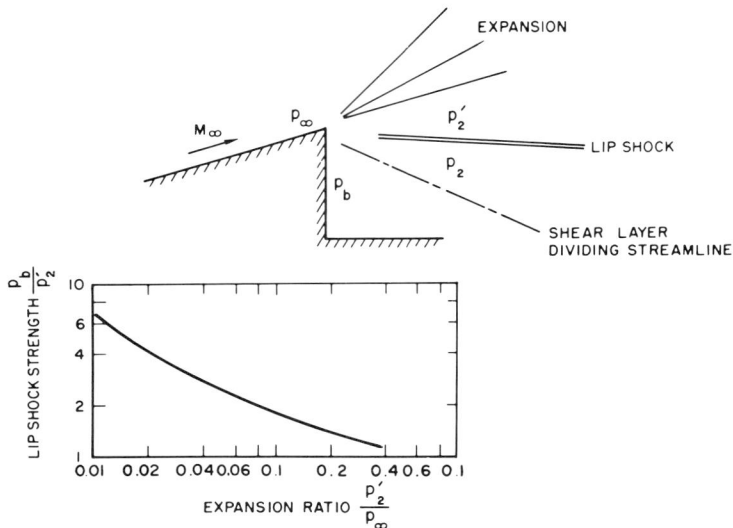

FIG. 22. Measured lip-shock strength in the Mach number range $2 < M_\infty < 4.5$ (59).

to confirm this result. His data spans the range of Mach numbers 2–4 and the laminar–turbulent transition. Still it falls fairly well on one curve when plotted in terms of base pressure ratio (which itself changes considerably in the range of this data).

The process leads to a curvature of the dividing streamline. It follows that the geometry of the wake for a given base pressure is not entirely determined by the initial expansion angle θ_s. The amount of turning of the streamline depends on the number and strength of the secondary waves which interact with it before the wake recompression point. This scales with the upstream boundary-layer thickness. Experiment verifies that the dependence of the length of the wake on Reynolds number is stronger than that of the base pressure which stems from this phenomenon. Weiss and Weinbaum (94) discuss this and give some comparisons with experiment.

The rotational characteristics solution for the corner expansion indicates the mechanism of the formation of the lip shock, but it is not entirely satisfactory. Pitot pressure profiles measured by several authors (59, 53, 66, 95, 96) indeed show the existence of a compressive disturbance at the location and with roughly the magnitude indicated by the calculations. However, the reflected compressive waves that result from this characteristic solution do not converge into a shock anywhere as near to the base as indicated by experiment. Some sources of these difficulties can be foreseen. The assumption of a centered primary expansion fan at the corner is not correct. This fan is spread over the region over which the subsonic portion of the oncoming boundary layer

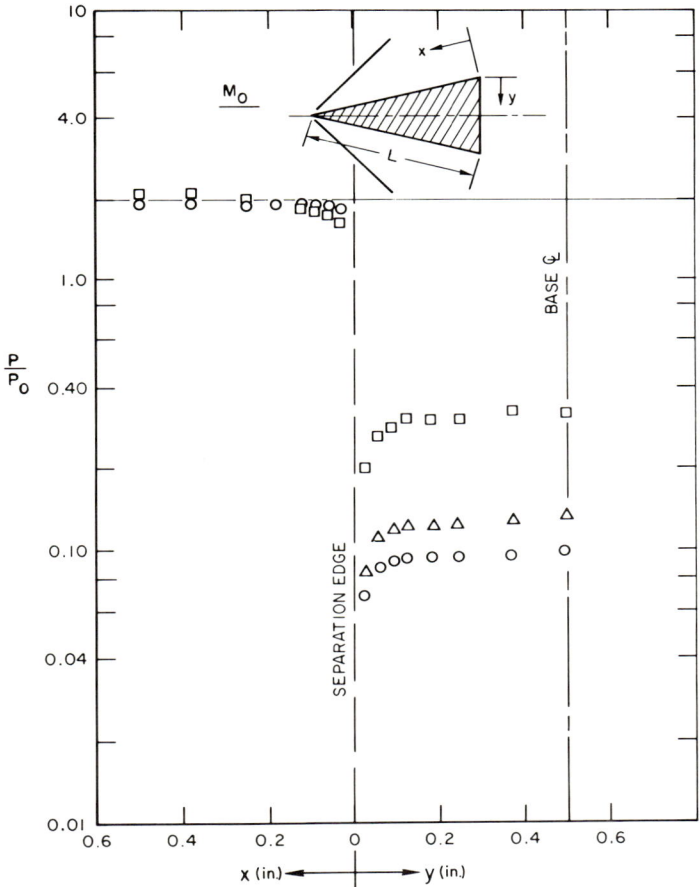

Fig. 23. The surface pressure distribution near the separation corner of a wedge (59): $M_\infty = 4.54$, $\text{Re} = \rho_0 u L / \mu_0$; (○) $\text{Re} = 2.03 \times 10^6$, (△) $\text{Re} = 1.11 \times 10$, (□) $\text{Re} = 0.22 \times 10$.

begins to thin and this has a considerable effect on the downstream interaction. The neglect of viscous forces which is justified (93) by an order of magnitude study of the terms of the Navier–Stokes equation (scaling in terms of the boundary-layer thickness) may not be entirely correct. Experimental evidence of this can be found in measurements of surface pressure immediately upstream and downstream of the corner (see Fig. 23). These show a drop in pressure ahead of the corner, as expected. They also show that the pressure immediately after turning is lower than its final value. Lip shock studies and flow mapping (59, 66, 95) suggest strongly that "separation" does not occur at the corner proper but on the base surface (very near the corner). This effect is not at all present in Weinbaum's inviscid theory.

The foregoing analysis and the corresponding detailed experimental research throw considerable light on the insufficiencies of the Chapman–Korst model. The irregularities in the pressure distribution near recompression (whether the final downstream pressure is approached from above or below) can be explained in terms of the interaction of the lip shock and the recompression shock. The need for a more refined analysis of the viscous shear layer is evident. The external flow cannot be treated as uniform. To illustrate these points, a typical complete (experimental) flow field downstream of a blunt-base wedge is shown in Fig. 24 (66). In this case the "lip shock" is seen to be imbedded in the rotational layer entirely and virtually merges with the recompression shock. The divergence of the external flow after turning is very

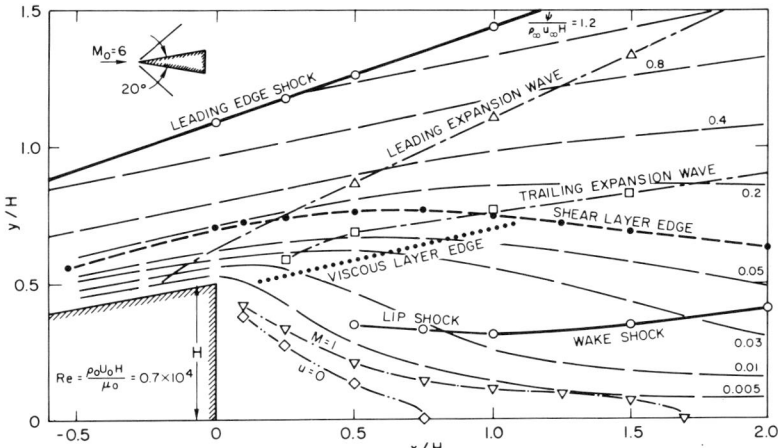

Fig. 24. A map of the flow field of the near wake behind a 20-deg adiabatic wall wedge at $M_0 = 6$ (56).

pronounced. [Similar results derived from calculations are given by Weiss and Weinbaum (*94*) for higher Mach numbers, where the flow divergence and nonuniformity are even more extreme.]

2. Stokes-Flow Region at Stagnation Points on the Dividing Streamline

The immediate neighborhood of the separation and recompression points on the dividing streamline in the shear layer were studied by Weinbaum (*97–99*). These analyses are of sufficient fundamental interest to deserve a brief review, although they cannot yet be brought to bear directly on practical aspects on the base-flow problem. They illustrate the nature of the "constraints" imposed on the overall flow field by local viscous conditions, and bear on the uniqueness of the solution and the "closure condition" which plays such an important role in the base-flow models [see also Ai (*100*)].

In the immediate neighborhood of the stagnation point the full Navier–Stokes equations can be reduced to the Stokes-flow biharmonic equation (when written in terms of the stream function). The equation is only valid within one "Stokes radius" $r = \nu/u$ from the stagnation point. Using polar coordinates, the stream function can be represented by an expansion about the origin which is located at the corner. The leading term of such an expansion must have the form

$$\psi^{(0)} = r^m f(\theta) \tag{58}$$

to satisfy a no-slip condition along the line of constant θ (the walls). After substitution into the biharmonic equation a fourth-order ordinary differential equation for f involving m as a parameter is obtained:

$$f'''' + 2[(m-2)m + 2]f'' + m^2(m-2)^2 f = 0 \tag{59}$$

m complex corresponds to viscous vortices in the corner, a class of solutions which we will not consider. The solution of this equation yields

$$f = A_1 \sin m\theta + A_2 \sin[(m-2)\theta] + A_3 \cos m\theta + A_4 \cos[(m-2)\theta] \tag{60}$$

Since the original equation is linear, combination of solutions corresponding to different m are also a solution. Now, if the flow is symmetric, $A_3 = A_4 = 0$, and, in order to have a solution which is not trivial, m must satisfy the equation

$$(m-2)\sin m\theta_0 \cos[(m-2)\theta_0] - m \sin[(m-2)\theta_0]\cos m\theta_0 = 0 \tag{61}$$

where θ_0 is the boundary at which the condition of no slip ($u = v = 0$)

$$f(\pm\theta_0) = 0, \quad f'(\pm\theta_0) = 0 \tag{62}$$

has been imposed. A corresponding equation is obtained for antisymmetric flow, where $A_1 = A_2 = 0$:

$$(m - 2) \cos m\theta_0 \sin[(m - 2) \theta_0] - m(\sin m\theta_0) \cos[(m - 2) \theta_0] = 0 \qquad (63)$$

The angle coordinate θ is defined with respect to the free stream flow at infinity. The case $\theta_0 = \pi/2$ corresponds therefore to a wall normal to the flow and a stagnation point at the origin. The case $\theta_0 = \pi$ corresponds the leading edge on an infinitely thin plate and the solution reduces to that of Carrier and Lin. The separation of a sharp downstep is represented by $\theta_0 = 3/4\pi$. In this particular case, only one solution of (61) and (63) exists:

$$m_1 = 1.544, \qquad m_2 = 1.909 \qquad (64)$$

With this result the leading term of the Stokes expansion about the corner can be written out explicitly as the sum of a symmetric and antisymmetric component. The equation for the separation streamline is

$$r = \left\{ \frac{-K_1\{\cos m_1\theta + \cot \tfrac{1}{4}(3m_1\pi) \cos[(m_1 - 2) \theta]\}}{K_2\{\sin m_2\theta - \tan \tfrac{1}{4}(3m_2\pi) \sin[(m_2 - 2) \theta]\}} \right\}^{1/(m_2-m_1)} \qquad (65)$$

The constants K_1 and K_2 appearing in this equation are related to the still undetermined coefficients A_1 through A_4 appearing in Eq. (60).

The flow is thus governed by the ratio K_1/K_2 which is related to the degree of asymmetry of the flow. For $K_1 = 0$ (purely symmetric flow) the separation streamline leaves straight along the bisector of the wall corner, as expected. As asymmetry is added, the separation streamline is depressed towards the base wall. An overall picture for $K_1/K_2 = 1$ (typical case), including also isobars of the pressure field, is shown in Fig. 25. In principle, K_1/K_2 must be determined by matching the Stokes region with the external flow. This is discussed but not resolved by the author; however, the implication that the trends exhibited by this solution are not too sensitive to the value of K_1/K_2 can be defended. The solution appears to be an equivalent to the Kutta condition which supplements the incomplete inviscid flow theory of a lifting surface. The viscous condition for separation appears thus as an essential constraint on the overall motion, although its influence on the flow field proper is very local.

This solution evokes a remarkable amount of experimental observations. It predicts a pressure which increases from the corner along the rear face of the base. It yields a finite and continuous shear at the trailing edge. Only the pressure is weakly singluar there, a problem which cannot be resolved unless one goes to a microscopic scale. The pressure

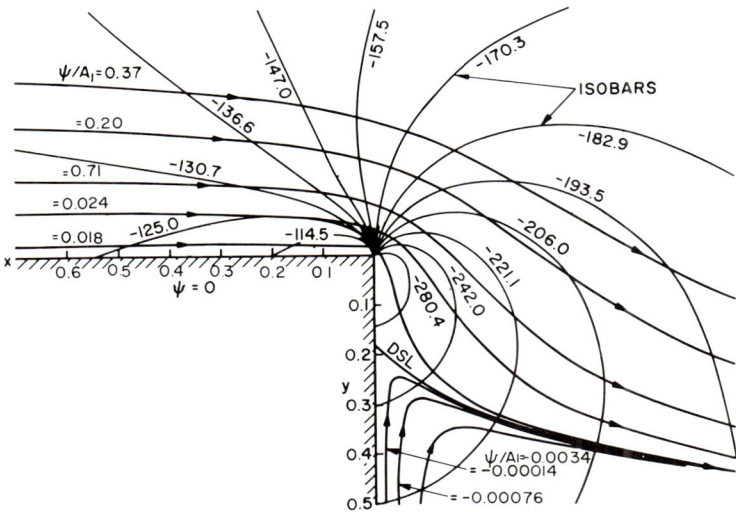

FIG. 25. Predicted streamline pattern and pressure field in the vicinity of the trailing edge obtained from a Stokes-flow analysis ($K_1/K_2 = 1$) (98).

singularity is physically more acceptable than a singularity in velocity or shear. The dividing streamline curves upwards, which means that it compresses the layers above it. Conceptually, this generates compressive waves which coalesce to generate the foot of a lip shock much ahead of where the reflected expansion waves predicted by the rotational characteristic solution of the outer flow would place it. Thus, the fact that the lip shock starts well within a distance equal to a boundary-layer thickness is verified. Also, observations that the separation process seems to emanate from below the corner now acquires some sort of framework. Hama's work (59) shows that, even for bodies with trailing edges projecting backwards from the base surface in a form of a thin knife edge, the lip shock forms immediately behind the knife edge and appears to come from underneath it.

On the other hand, the quantitative results of the Stokes analysis must not be overemphasized. The pressure on the separation surface increases towards the corner, which is contrary to observations. This is obviously because of an overriding influence of the preexpansion of the flow ahead of the corner and it could conceptually be handled by proper matching of this solution with the external flow. The problem of the mathematically sharp corner indeed requires a special treatment such as this one, but for a rounded corner, where the pressure "undershoots" just the same (see Ref. 59) and the lip shock is also formed, a solution of the ordinary boundary-layer equations matched locally to the external flow (discussed in Section IV) also reproduces the observed behavior. Therefore, while the separation problem unquestionably depends on the viscous nature of the flow near the wall, it is not certain that the analysis under discussion is essential. The most significant contribution of this theory would be an elucidation of the closure criterion and the "critical point" in the recompression region of the wake.

The same type of Stokes-flow solution applies to the immediate neighborhood of the recompression point, but the boundary conditions (equivalent to the geometry of the walls forming the separation corner) and the conditions at upstream infinity are not as clear in this case. Weinbaum discusses this problem for symmetrical flow (two-dimensional blunt-based wake). The solution can be written in terms of two parameters which Weinbaum chooses to be the angle of the dividing streamline relative to the far downstream flow and the pressure gradient in the streamwise direction. These can be related if one imposes a requirement that the pressure be continuous through reattachment. As in the case of the separation point, one obtains then a "locally valid" constraint that is independent of the external flow (analogy with the Kutta condition) involving, however, some information regarding the recirculation region in the wake proper (in analogy with the "asymmetry" ratio which governs the separation problem). For each full profile of the flow approaching the recompression region, including that portion of it containing reversed flow, there corresponds only one base pressure which allows the wake to "close" *and* the pressure to continuously approach its downstream asymptotic value along the line of symmetry of the wake (or, along the recompression wall of a step).

Weinbaum proposes two essential characteristics of a wake closure mechanism which he derives from his study of these viscous "constraints" imposed on the flow by the nature of the stagnation point. First, he suggests that the structure of the internal flow must be accounted for before a "closure condition" can be prescribed. Second, he does not admit the existence of any definite "critical throat" which cannot be crossed by signals from downstream: instead, pressure continuity is enforced. These can only be viewed as hypotheses because of the very local nature of the present solution which is incompletely matched to the outer flow. Experiments do indicate that something very close to a "sonic" throat does exist. There is no contradiction here because the base flow can well be very sensitive to disturbances occurring ahead and little sensitive to disturbances occurring behind the 'throat" region. Therefore, from the point of view of constructing a "model," it may be unnecessary to require full pressure continuity and the apparent existence of a critical point may provide a closure criterion prescribing the base flow uniquely with a satisfactory precision. We shall return to this again.

3. Weiss' Approximate Analysis of the Viscous Shear Layer (Nonuniform Adjoining Flows)

The analyses of Section III,A and analyses of the internal flow described in Section III,B show that the viscous layer developing along the dividing streamline is not isobaric nor bounded by regions of uniform inviscid flow. Consequently, the classical solution and also the numerical integration of the boundary-layer equations with a prescribed nonsimilar initial profile using the Chapman boundary conditions is not realistic. Weiss (101, 102) proposed a different approximate treatment of the problem which provides an analytical solution for the development of the viscous layer in a form suited for numerical analysis of the local matching among the regions. The solution prescribes initial conditions which (in an approximate form) portray the results of the rotational characteristic analysis of the flow over the separation corner. It prescribes the "inner" boundary condition (as a velocity variation expressible by a trigonometric series) and a variable pressure along the dividing streamline (in terms of a power series). The coefficients appearing in these boundary conditions can then be determined by matching with the inner and outer flows.

The analysis is based on an Oseen linearization of the boundary-layer momentum equation and yields a solution in closed form. Introducing the coordinates

$$\bar{y} = (u_e/\nu_e) \int_0^y (\rho/\rho_e) \, dy; \qquad \bar{x} = xu_e/\nu_e \tag{66}$$

with

$$\bar{u} = u/u_e; \qquad \bar{v} = v/u_e; \qquad \bar{p}_e = p_e/\rho_0 u_e^2 \tag{67}$$

and assuming that $\rho\mu = \rho_e\mu_e$ and $\Pr = 1$, the equation

$$c\frac{\partial \bar{u}}{\partial \bar{x}} = \frac{\partial^2 \bar{u}}{\partial \bar{y}^2} - \frac{\rho_0}{\rho_e}\frac{\partial \bar{p}_e}{\partial \bar{x}} \tag{68}$$

is obtained. The quantity ρ_0 is a reference density and c is a nondimensional, constant, linearizing velocity to be determined later.

Weiss prescribes an initial condition at $x = 0$ and a boundary condition at $y = 0$ (the dividing streamline) which are derived (approximately) from the characteristics solution for the expansion at the separation corner. The initial condition is given by the form

$$u_{(0,\bar{y})} = 1 - a \exp(-b\bar{y}), \quad a = 1 - u_s/u_e, \quad b = (a/2\delta^{**})(2-a) \quad (69)$$

where u_s is the velocity on the sonic streamline immediately after expansion and δ^{**} is the momentum thickness of the expanded vertical layer. Equivalent results can be given for the enthalpy profile. The boundary condition along the dividing streamline is specified by a Fourier series

$$\bar{u}_{(\bar{x},0)} = \left(\frac{u_{\text{DSL}}}{u_e}\right)_{\max} \sum a_n \sin \frac{n\pi\bar{x}}{l}, \quad 0 \leqslant \bar{x} \leqslant l \quad (70)$$

where l is the total length of the free shear layer. The pressure distribution is represented by a quadratic function

$$\partial \bar{p}_e/\partial \bar{x} = b_0 + b_1 \bar{x} + b_2 \bar{x}^2 \quad (71)$$

The boundary-layer equation can now be solved by Laplace transform methods, yielding an explicit solution for $\bar{u}$ (a sum of trigonometric terms and error functions). The result contains a number of parameters associated with the prescribed boundary conditions (a_n's and b's) as well as one free parameter, the "linearizing" velocity c. This is chosen so that it satisfies the boundary-layer equation on the dividing streamline in the least-square sense; i.e.,

$$\frac{\partial}{\partial c} \int_0^l (\bar{u} - c)^2 \left(\frac{\partial \bar{u}}{\partial \bar{x}}\right)^2 d\bar{x} = 0 \quad (72)$$

An equivalent solution of the energy equation is obtained by the same method. Weiss et al. (101) discusses possible extensions of the theory to the case of a streamwise varying parameter c.

B. THE INTERNAL FLOW

Questions regarding the nature of the internal flow arise if one asks for a description of the heat transfer distribution on the walls of the cavity. Moreover, the internal flow provides the internal boundary condition for the development of the shear layer. It has become obvious that for a fully satisfactory theory of base flows the coupling between the shear layer and the inner region cannot be neglected. It appears to hold the key to the problem of the Reynolds number influence on the base pressure, for instance.

1. Batchelor's Model—High Reynolds Number

The first genuine theoretical treatment of the internal flow is due to Batchelor (9, 103). He considers it to be a steady, incompressible, viscous flow with closed streamlines. The complete (Navier–Stokes) vorticity equation can be integrated along a closed streamline:

$$\oint (\mathbf{u} \times \boldsymbol{\omega}) \cdot d\mathbf{r} - \oint (\nabla P) \cdot d\mathbf{r} - \nu \oint (\nabla \times \boldsymbol{\omega}) \cdot d\mathbf{r} = 0$$

$$P = p/\rho + \tfrac{1}{2}q^2, \quad q = |\mathbf{u}|$$

(73)

The second term vanishes because P is a single-valued function of position and the first term vanishes if the integration contour is a streamline. The result, which is independent of viscosity, is

$$\oint (\nabla \times \boldsymbol{\omega}) \cdot d\mathbf{s} = 0 \tag{74}$$

where $\mathbf{s}$ is an element of streamline length. This equation is combined with the inviscid equation of motion assuming that *viscous forces are small* everywhere:

$$\mathbf{u} \times \boldsymbol{\omega} = \nabla P \tag{75}$$

Equations (74) and (75) together are sufficient to solve the problem. The solution is, for two-dimensional and axisymmetric flow (without azimuthal swirl), respectively,

$$\omega(\psi) = \text{const} = \omega_0, \quad \text{plane flow}$$
$$\omega/r = \text{const} = \alpha, \quad \text{axisymmetric flow}$$

(76)

It says that, for instance, in two-dimensional flow the vorticity in the region of closed streamlines is constant (solid-body vortex). The condition for this to be applicable is contained in the assumption of small viscous forces everywhere. It is thus necessary that none of these closed streamlines to which this solution applies enter the shear layer. Batchelor demonstrates that such flows can exist in two-dimensional and axisymmetric cases with no azimuthal swirl, but that they do not always exist in more complex geometries. For these two special cases, therefore, and if the Reynolds number is large, one can postulate that the internal flow is a solid-body vortex and attempt to join it to an inviscid external flow through the viscous shear layer. However, the internal flow need not consist of a single vortical region separated from the external flow by the shear layer. It is perfectly acceptable to have within the cavity two or more such vortical regions (for instance, the main vortex and a smaller counterrotating vortex near the corner of a step which is what is observed). The above solution would apply to each region, and they would be coupled by an internal layer of high vorticity requiring another matching procedure. Note also that the solution for the vorticity, Eq. (76), does not yet complete the problem. The distribution of velocity can be found only if the shape of the boundary (the high-vorticity region) is known. This renders a complete solution of the problem very difficult.

Using this model to represent the internal flow, Batchelor (9) proposes that a complete solution for the base bubble can be obtained in principle by a procedure which is schematically the following. For each choice of the vorticity ω_0 in the internal solid-body vortex, a requirement that the pressure be continuous along its boundary yields a boundary condition on the external, inviscid flow. The boundaries of the bubble are formed by a continuous viscous channel (vortex sheet) which follows the dividing streamtube from the separation point to the recompression point at which point the flow returns towards separation through a boundary layer along the floor and up the base wall (or, in the case of the free wake, a viscous layer along the symmetry line and up the base). The requirement that such a closed viscous channel exists and that the flow be steady closes the problem and leads to a unique choice of ω_0. The matching yields a shape for the bubble (a family of shapes for each pair of assumptions on ω_0 and the bubble pressure level, of which only one member presumably satisfies the steadiness condition). The problem is extremely complicated, primarily because the bubble boundary is free and determined only as part of the solution. Batchelor as unable to prove the uniqueness of his procedure in general. We recognize in this model, however, another interesting hypothesis regarding the "closure criterion," different from those discussed in the preceding.

An application of Batchelor's model to the study of separated flows in notches is to be found in the work of Burggraff (*104, 105*). This geometry is particularly simple inasmuch as reasonable assumptions regarding the shape and location of the free boundary can be made (a straight shear layer spanning the notch) and the difficult coupling between the inner and the outer flow can be simplified. The nature of the theory restricts it to high Reynolds numbers but laminar flows which, in practice, tend to be unrealizable. The results are applicable mainly to short cavities (length and height ratio of less than unity). Of course, as the notch becomes very short and deep (a gap) the structure of the internal flow becomes more complex, involving a series of superimposed counterrotating vertical regions; the number of vortices is approximately equal to the ratio of the depth to the length of the gap (*106*). Each is coupled to its neighbors through a viscous dissipative layer.

The flow in narrow gaps was treated by slow-flow approximation by Pan and Acrivos (*107*).

2. *Stokes-Flow Approximation*

In contrast with Batchelor's solution of the internal flow for small viscosity, Weiss (*108, 109*) developed a purely viscous solution (Stokes flow). He succeeded in matching the external (hypersonic) flow, the shear layer, and the inner flow at the cost of approximating the geometry of the bubble. Although the solution and the matching are quite approximate, this analysis demonstrated for the first time that the coupling of the shear layer with the inner flow is important and does introduce a Reynolds number dependence into the base pressure problem.

Weiss assumes that the inner flow can be represented by the biharmonic Stokes equation in the stream function (incompressible flow), with the following boundary conditions (polar coordinates):

$$(1/r)(\partial \psi/\partial \theta) = \partial \psi/\partial r = 0 \quad \text{at} \quad r = 0 \quad \text{(rear stagnation point)}$$
$$\partial \psi/\partial \theta = \partial \psi/\partial r = 0 \quad \text{at} \quad r = R \quad \text{(base)} \tag{77}$$
$$\left. \begin{array}{r} -(1/r)(\partial \psi/\partial r) = u_{\text{DSL}}(r) \\ \psi_r = 0 \end{array} \right\} \quad \text{at} \quad \theta = \pm \beta \quad \text{(symmetrical free wake)}$$

The origin is at the recompression point. The dividing streamline is straight and the base wall is distorted to correspond to a constant radius. This solution describes a base wake, but no difficulty would be encountered if the boundary condition representing a wall were imposed at $\beta = 0$ (a step). The solution is assumed to be valid up to the dividing streamline, the velocity of which is a function of r. It involves one scaling constant which can be identified with the dividing streamline velocity at some point. The variation of the velocity along the length of the dividing streamline is fixed by the nature of the solution.

The solution of the Stokes stream function equation subject to these boundary conditions can be written down without difficulty. It must now be matched to the solution for the viscous region and with the external flow. This simple version of the theory does not allow for a complete matching. For example, having assumed that the dividing streamline is straight, one cannot impose pressure continuity across the boundary (shear layer) between the internal and the external flow. Instead, assuming that the displacement thickness of the free shear layer is constant and that the velocity distribution in it is linear, Weiss fixes the magnitude of the dividing streamline velocity consistently with continuity of shear. The details of this procedure are not as interesting as the fact that it leads indeed to a Reynolds number dependence of the base pressure, which no simpler model was able to give.

3. *Numerical Solutions of the Navier–Stokes Equations for Low Reynolds Number*

A more adequate theoretical description of the recirculating region over a wider range of Reynolds numbers can be obtained only by numerical solutions of the full Navier–Stokes equation. However, numerical stability problems still impose an upper limit on the local Reynolds number. This is not objectionable in a case of hypersonic wakes where the Reynolds numbers are indeed low (*101, 102*), but continues to limit severely the theoretical analysis of cavities in general and the notch problem in particular (*104, 105*).

Weiss *et al.* (*101*) deal with the Navier–Stokes equations in the vorticity stream function form

$$\text{Re}_H \left[\frac{\partial \psi}{\partial x} \frac{\partial \zeta}{\partial y} - \frac{\partial \psi}{\partial y} \frac{\partial \zeta}{\partial x} \right] = \nabla^2 \zeta, \qquad \zeta = \nabla^2 \psi \tag{78}$$

where all quantities are rendered dimensionless by using the base height and the free stream velocity as reference quantities. The boundary conditions are

$$\begin{aligned}
\psi &= 0 & &\text{on} & x &= 0,\ y = 0,\ y = y_{\text{DSL}}(x) \\
\zeta &= 0 & &\text{on} & y &= 0 \\
\zeta &= \partial^2 \psi / \partial x^2 & &\text{on} & x &= 0 \\
\zeta &= g_{(s)} & &\text{on} & y &= y_{\text{DSL}}(x)
\end{aligned} \tag{79}$$

The first of these relations specifies the boundary of the region, avoiding, once again, the difficulty of dealing with an a priori free boundary. The third one is a condition of symmetry at the axis (the problem is then specialized to the wake) and the last one gives parametrically the distribution of vorticity on the dividing streamline. The finite-difference analog to this set of equations is written with a rectangular mesh and solved by an explicit numerical scheme. The rate of convergence of the process depends on the shape of the region, mesh-size Reynolds number, and rate of advance of the iteration. The larger the Reynolds number or the vorticity on the dividing streamline, the finer the mesh required for convergence. The critical mesh-size-based Reynolds number required for convergence is found to be about 4.5 in typical cases. Once this solution is available, the pressure field is calculated. The authors also solved simultaneously the energy equation to obtain the temperature field.

The results are expressed in terms of the boundary function $g(s)$ which represents the velocity along the dividing streamline. The authors conducted a series of parametric studies in which the shape of the dividing streamline was prescribed as a straight line at a given angle to the coordinate system and the velocity distribution on the dividing streamline was assumed to be quadratic.

Figure 26a–c shows, respectively, the stream function distribution, vorticity distribution, and temperature distribution in a typical recirculating region with a reference Reynolds number of 275. The vorticity contours and isotherms indicate the formation of the core of uniform vorticity and temperature which is in accord with Batchelor's theory. Boundary-layer regions are seen to occur on the rear wall (recall that the solution is for a free wake) and on the dividing streamline. The stream function distribution does not change character with Reynolds number. However, both the vorticity and temperature distributions depend on it strongly, and at lower Reynolds numbers the internal region of constant vorticity is lost. A number of other results are discussed in the reference. These results would vary with the assumed shape of the region and with the (assumed) distribution of the velocity along the edge of the region. For example, the effect of the wake angle on the pressure field (keeping a quadratic velocity distribution along the dividing streamline) is illustrated in Fig. 27. There is a rearward movement of the peak velocity resulting in a small pressure gradient near the body and a larger gradient near the stagnation point. These gradients are sharpened when the angle of the dividing streamline relative to the axis is increased. When this angle (recompression angle) is small—say below 30°—the pressure gradient becomes slightly favorable over the first portion of the dividing streamline. This example clearly shows the interdependence of the solution for the viscous shear layer, which is neither isobaric nor bordered by free streams of uniform velocity, and the nature of the internal flow.

4. *Discussion—Unsteadiness of the Internal Flow*

In spite of these theoretical advances, the practical problem of the internal flow remains open. In many cases (mainly in separations induced by shocks, steps, and notches) the characteristic Reynolds

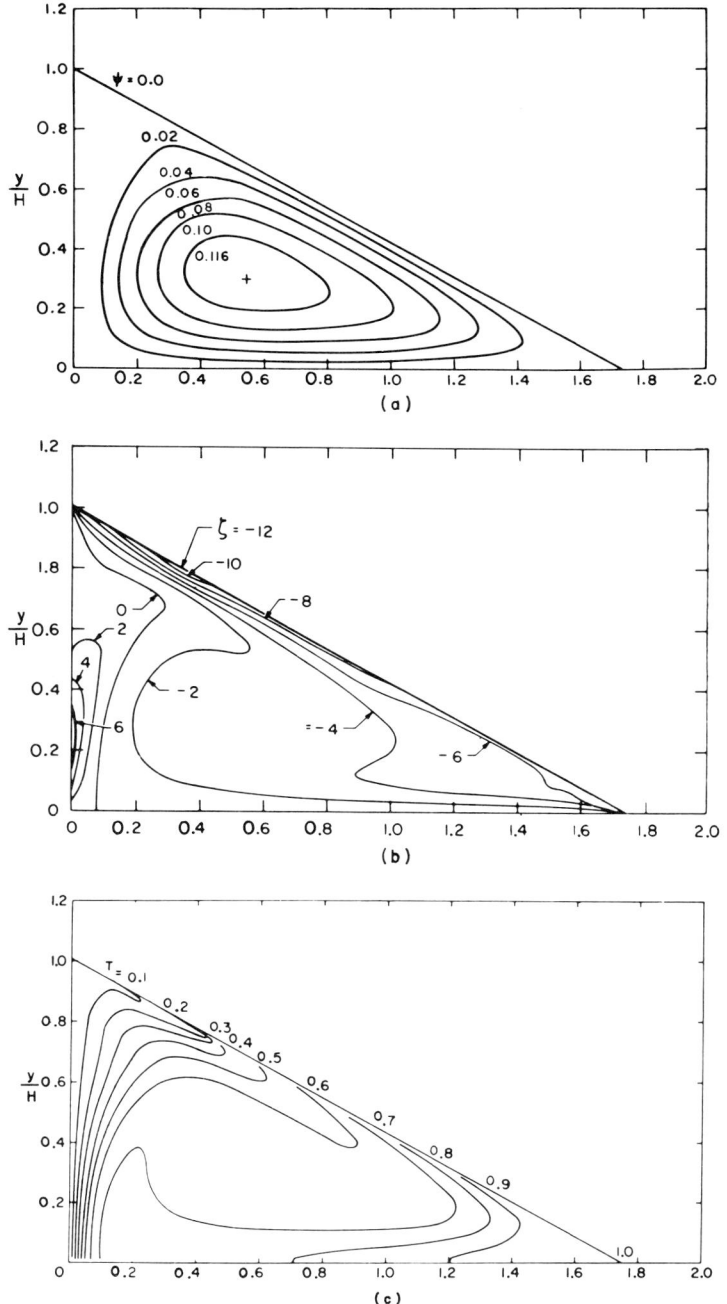

FIG. 26. Calculated recirculation region: (a) stream function contours for 30-deg recirculation region, symmetrical velocity distribution on DSL, $\mathrm{Re}(u_\infty H/\nu_\infty) = 275$; (b) vorticity contours; (c) temperature contours (*101*).

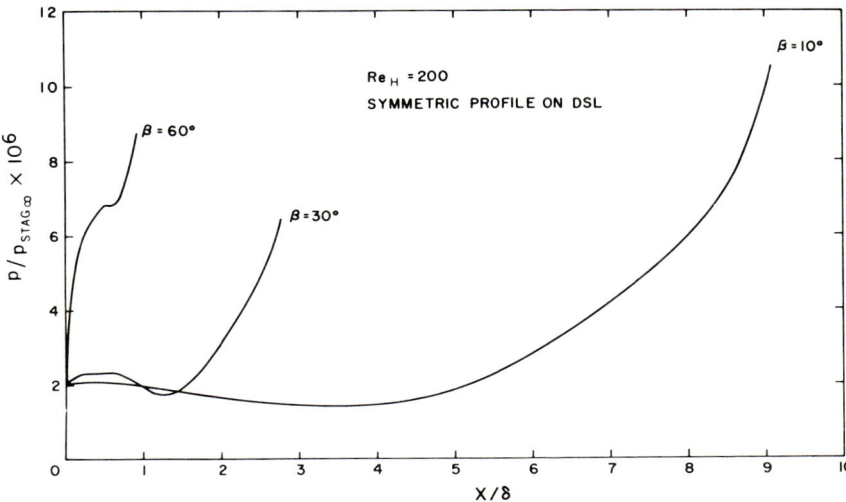

FIG. 27. The effect of the wake angle β on the pressure distribution along the dividing streamline (*101*).

number is not low as it often is in the case of hypersonic blunt-base wakes, and the geometry of the bubble is far more complicated than what can be treated analytically. Furthermore, the flow is often turbulent. It is not clear whether the internal flow is quite steady even when the shear layer is laminar (except, perhaps, at very low Reynolds numbers) but turbulent flow in the shear layer is invariably associated with an extremely high degree of fluctuations in the internal flow. Larson *et al.* (*75*), for example, reports fluctuations in hot wire output indicating over 100% "turbulence." At such high levels the concept of turbulence no longer has meaning with regard to mean heat an momentum transfer properties. Moreover, (two-dimensional blunt-wedge base wake) these fluctuations exhibit at least five narrow frequency peaks which obviously indicate a resonant unsteadiness of the flow, the nature of which is virtually unknown. Distinct pulsations are even more evident in cavities formed by notches.

In a time-average sense, a general recirculating motion exists and has been measured. Steadiest and best defined is the motion in symmetrical free wakes for which much detailed distribution data is contained in the references cited previously (*77–82*). Figure 28 shows typical integral characteristics of the velocity distribution (shear layer and returning flow) taken from Ref. (*96*) (laminar flow, this reference also presents equivalent data for turbulent flow). These are shown in a form used in connection

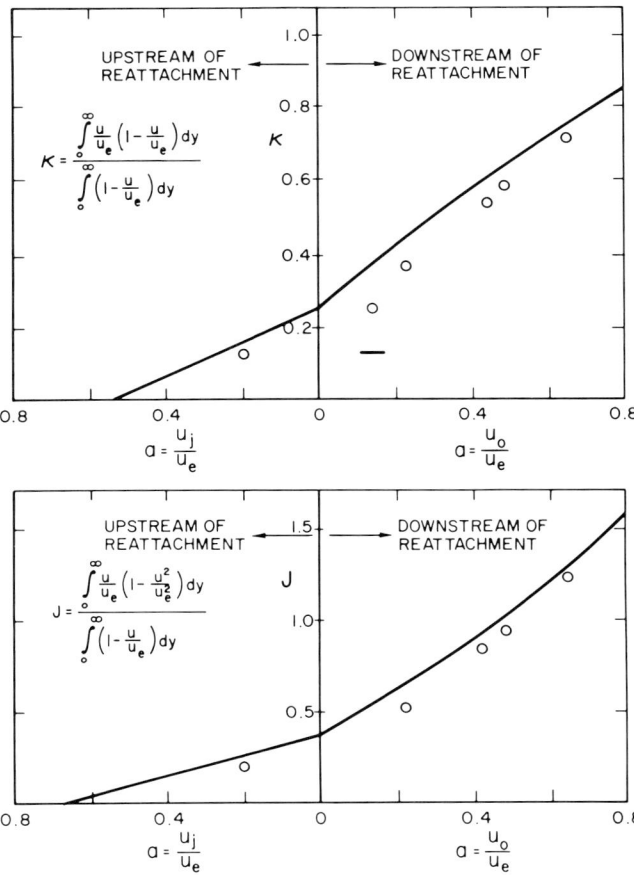

Fig. 28. Integral characteristics of the velocity profile along the wake centerline in laminar flow: u_j = centerline velocity in the reversed flow region; u_0 = centerline velocity in the streamwise flow region (96). (○) Experimental data points; (——) curve proposed by Lees and Reeves.

with the Lees–Reeves theory (Section IV,C). The boundary layer on the solid base wall seems to be of the stagnation-point type. In their treatment of it, Larson et al. (75), attempt to assess and account for the exceptionally high level of "turbulence" in the flow outside it.

The mean reverse-flow velocity profile in cavities downstream of steps is less well known. Several visualization studies have shown that the return flow which issues from the recompression zone separates before the corner, leaving in it a small vortical region with closed streamlines. The unsteadiness of the internal flow tends to be more pronounced than in free wakes.

A very interesting systematic series of experiments concerning three-dimensional perturbations shed by such flows was conducted by Ginoux (*54, 55, 110–113*). His studies include visualization, pitot pressure measurements (*110*), heat transfer studies (*112*), and cover various types of two-dimensional and axisymmetric *laminar* separations. It appears that a separating–reattaching flow is unstable in the sense that it furthers the formation of rows of counterrotating vortices. Ginoux demonstrated that the initial perturbation can come, for instance, from spanwise non-uniformities in the model leading edge or surface roughness ahead of the step, and that these perturbations are amplified and organized by the separated section of the flow and become evident downstream of its reattachment. Figure 29 shows spanwise measurements of the resulting nonuniformity in the heat transfer, which is seen to be far from small.

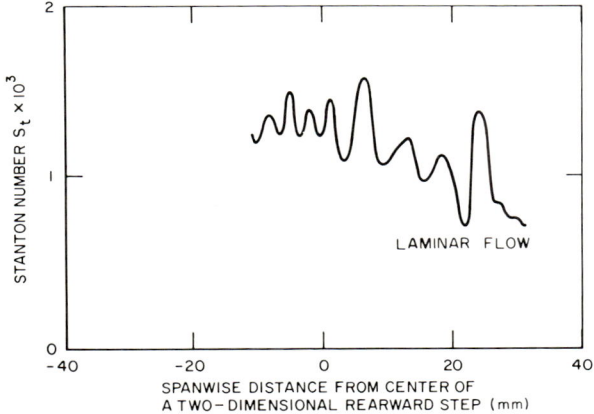

FIG. 29. Typical measured spanwise nonuniformities in the flow downstream of reattachment (*110*).

Experiments show that only on very carefully machined axisymmetric pointed bodies one can avoid this structure (transition to turbulence sets in directly). The mechanism of this phenomenon is not well understood. The vortex patterns are similar to what is observed on delta wings (instability of the crossflow), but it may also be a phenomenon of the Görtler type (in which case it would depend on the concavity of the shear-layer surface).

Cavities in which the dividing streamline recompresses at a corner (upstream-facing steps or ractangular cutouts) tend to be very unsteady. Measurements of fluctuations in the wall pressure ahead of upstream-facing steps (turbulent flow) are given by Kistler (*114*). These do exhibit a low-frequency "buffeting" which appears to be associated with changes in the geometry of the region (movement of the point of separation), and a second wide-band spectral component originating probably from standing waves on the dividing streamline. Such waves have been observed over rectangular notches. This problem is discussed in the following section entirely devoted to flow in cutouts.

In summary, pressure fluctuations originated by perturbations in the recompression process, by waves on the dividing streamline or simply by turbulence in the shear layer, are transmitted into the cavity and cause large variations in the velocity in the nearly "dead air." As a result, at large Reynolds numbers (laminar as well as turbulent flow) the flow in the boundary layers on the wall of the cavity is probably always pulsating and in some cases the vector velocity can be instantaneously reversed. Reference (*83*) discusses experiments in which the convection of heat from a source placed near the floor of a cutout was observed; they show this clearly. A time-averaged mean structure obviously exists, but it is doubtful that it can be calculated even by a numerical integration of the Navier–Stokes equation unless a theory for the apparent transfer properties of this highly disturbed flow is developed. This is also true for the properties of the boundary layers formed on the internal wall (for example, its heat transfer coefficient).

C. Separated Regions in Notches

The external flow will span a notch entirely (Fig. 30) when its length to depth ratio is roughly less than the sum of the length of the upstream and downstream step separations (*115*) (for the given free stream parameters). Otherwise the flow will reattach to the floor of the notch forming two independent cavities. The flow in the neighborhood of the critical length to height ratio is very unstable and can be observed to flip randomly from one configuration to the other. Data are available only for a turbulent and transitional boundary layer, but it may be surmised that same criterion holds when the flow is entirely laminar.

The pressure level in an "open" cavity (totally spanned) is dictated by the deflection imposed on the external flow by the geometry of the boundaries. If the internal region were truly "dead" and the dividing streamline isobaric, the pressure in the cavity would be constant and depend only on $(H_2 - H_1)/L$. Experiment shows that, indeed, this is grossly true when the cavity length to depth ratio is small (less than 4). pressure in short notches (with $H_1 = H_2$, i.e., a notch in a flat boundary) is usually slightly higher than the free stream pressure. This is because the growing shear layer and its reattachment causes a slight deflection

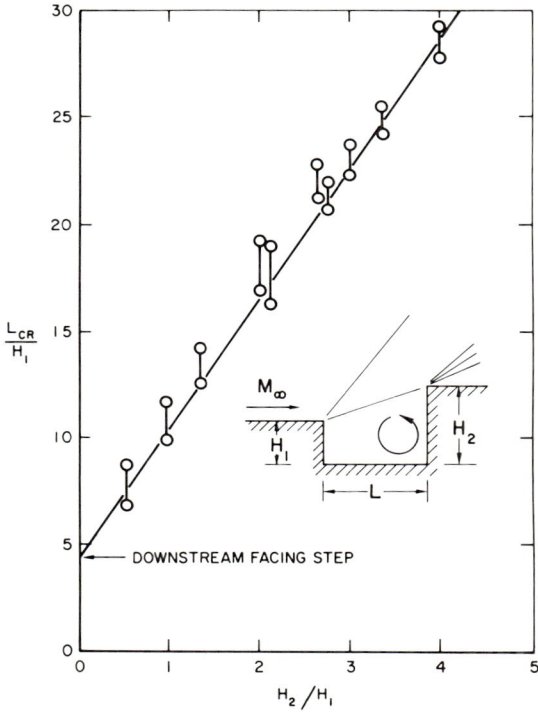

FIG. 30. The critical closure distance for notches in turbulent flow: data for $2.1 < M_\infty < 3.4$; $Re_{sep} \cong 15 \times 10^6$ (115).

of the free stream away from the wall. As the length to height ratio of the notch is increased a pressure distribution appears, showing a minimum at about one third of the length downstream of separation and a rise near the recompression step (Fig. 31) (126, 117). These variations are somewhat dependent on the Mach number (89, 115) (both the minimum and the maximum increase as M_∞ increases) but this is weak and the data is not conclusive (there are other effects such as that due to transition and a thickness of the upstream boundary layer which were not adequately controlled in the experiments).

In subsonic flow the general trends are the same but, of course, smeared out (118, 119).

A notch placed in a streamwise nonuniform supersonic flow, for instance, in the accelerating flow in an expansion nozzle, exhibits a similar overall behavior. To a first approximation (120) which is particularly true for short notches, the internal flow in the cavity is nearly

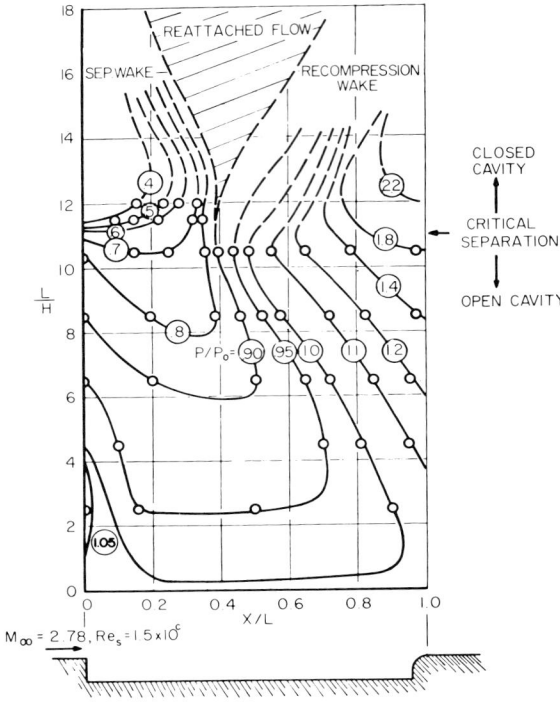

FIG. 31. An isobar map of the pressure distribution on the floor of notches: $M_\infty = 2.78$; $\text{Re}_{\text{sep}} = 1.5 \times 10^6$; $\delta/H = 0.44$, $H = 0.25$ in. (*115*).

dead and therefore the pressure in it along the dividing streamline is nearly constant. This requires a treatment similar to that discussed in Section II,A for downstep cavities in accelerated flow: waves impinging on the free streamline from the (prescribed) external flow reflect as waves of opposite family with an attendant curvature of the free boundary. The "closure" condition is supplied in this case by requiring that the curved free streamline pass through the downstream edge of the notch (to form a totally spanned cavity). The free streamline dips into the notch at the separation point through a Prandtl–Mayer fan to satisfy this closure condition. The amount of expansive turning, and thereby the mean level of the pressure in the cavity (lower than the external pressure), depends on the (prescribed) structure of the external flow and the length of the cavity. More precisely, it depends on the ratio L/λ, where λ is the length parameter characterizing the free stream pressure gradient introduced earlier [see Eq. (11)]. Results for a case in which the free stream (in the absence of the cavity) accelerates with a linearly decreasing pressure are shown in Fig. 32. These results show that the mean pressure in notches imbedded in an accelerated flow can be considerably lower than the local free stream static pressure. The cavity pressure, when referred to the pressure in it immediately downstream of separation, is quite invariant, in accord with the above simple model, except near the recompression corner. The critical "closure" length of cavities in accelerated flow seems to occur at about the sum of the lengths of a downstream-facing step separation in the given nonuniform flow, as in the case of uniform free stream. Since both these elemental separations are shorter in an accelerated stream than in a uniform stream, the critical notch length is also reduced. Data are very scarce (in spite of the frequent occurrence of such configurations) and only trends can be delineated here. In Carichner (*120*), a critical L/H of 8 was measured with the (linear) pressure gradient parameter λ of 0.173 in.$^{-1}$ ($M = 1.9$, turbulent flow). Less-well-controlled experiments can be used to supply some additional data. For instance, Wyborny et al. (*121, 122*) report measurements on

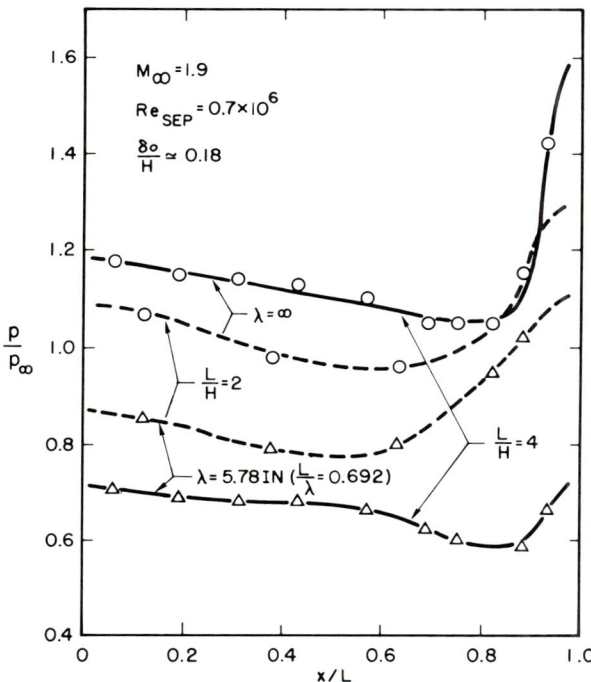

FIG. 32a. Typical pressure distributions on the floor of notches in accelerated flow.

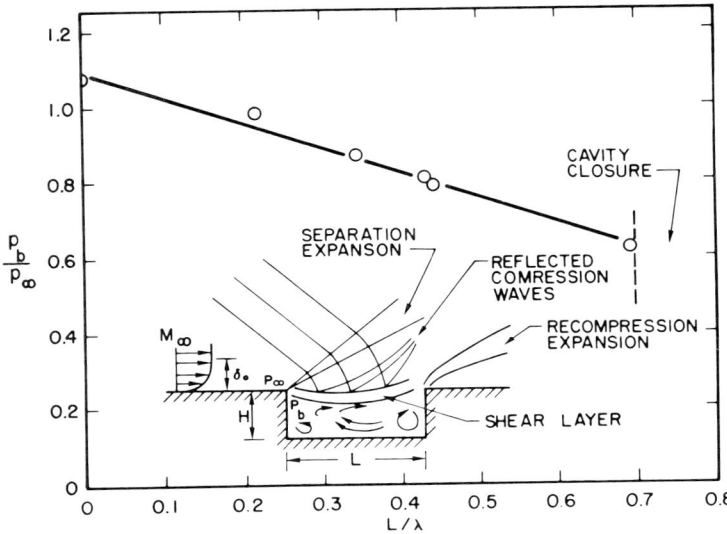

FIG. 32b. The separation pressure for notches in uniformly accelerated flow (*118*).

downstream-facing steps and rectangular cutouts in the surface of a cylinder with a spherical nose. Due to the blunt nose, the cylinder surface pressure varies from about $p/p_\infty = 3$ at the station of the separation corner to $p/p_\infty = 1$, 20 step heights downstream (this information is not included in the reference). This corresponds to a pressure gradient parameter λ of about 0.1 in.$^{-1}$. Cavity closure appears to occur at L/H between 6 and 7 ($M = 8.75$, Re $\sim 10^5$).

Variations in the pressure along the boundary of cutouts are entirely dependent on the structure of the internal flow and its coupling with the free stream through the shear layer. Some theoretical aspects of laminar, low Reynolds number flows in separated bubbles were discussed in the preceding section. Experiments in turbulent flows show that the cavity contains a strong vortex in the region of the recompression corner which has a diameter approximately equal to the step height H. There is also a weak counterrotating vortex in the separation corner. In the mid-span region one can distinguish near the wall a "boundary layer" of reverse flow. The mid-height of the center portion of the cavity is occupied by a highly unsteady "buffer layer" which is then overlayed by the shear-layer proper; see Fig. 33. The figure also shows proof that the boundary layer on the floor of the cavity suffers actual reversals in the direction of flow. This is seen from the fact that heat from a (two-dimensional) source is convected in both directions.

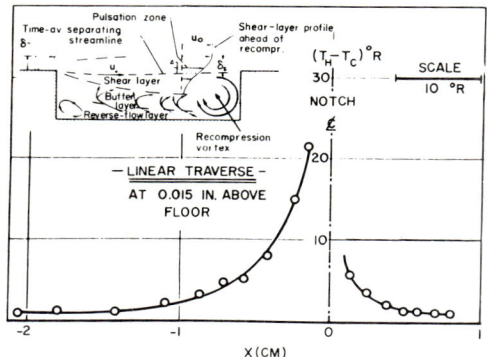

FIG. 33. Temperature distribution due to a heated wire at the cavity centerline: $M_\infty = 2.8; L/H = 9$; wire located at $y/H = 0.1$ (83).

Cutouts tend to emit strong acoustic radiation (116, 117, 123–125). A Spark–Schlieren picture of the phenomenon (in laminar flow) is shown in Fig. 34 (Charwat, unpublished). Similar behavior has also been observed in subsonic flows (126–128). The mechanism of sustenance and the acoustic modes is not well known and not necessarily always the same. Very narrow, deep cavities in low-speed flow presumably operate as Helmholtz resonators; longer and shallow cavities in high-speed flow oscillate probably due to an instability of the shear layer in the reattachment process. A number of theoretical analyes are published but the problem remains unclear. Laminar as well as turbulent shear layers exhibit acoustic radiation, although in the turbulent case one does not observe the clear two-dimensional wave (and single dominant frequency) shown in Fig. 34. From various observations which are not yet conclusive and systematic, one can suggest that the oscillations stop (or at least decrease substantially) when the notch depth is less than about three times the thickness of the oncoming boundary layer or when the length to depth ratio of the notch is less than one. The establishment of this unsteady regime depends also on the shape of the recompression face: ramp compression faces inclined at more than 45 deg to the free stream or rounded with radii of the order of H can lead to a perfectlys mooth, nonoscillating shear layer [see, for instance, experiments by Ginoux and Thiry (129)].

The oscillations seem to be associated with periodic injection and ejection of free stream mass into the cavity. Charwat and Redekopp (130) discuss experiments dealing with the penetration of small graphite particles (3μ mean diam) into the cavity. Figure 35 shows this in terms of the relative attenuation of two light beams: one traversing the free

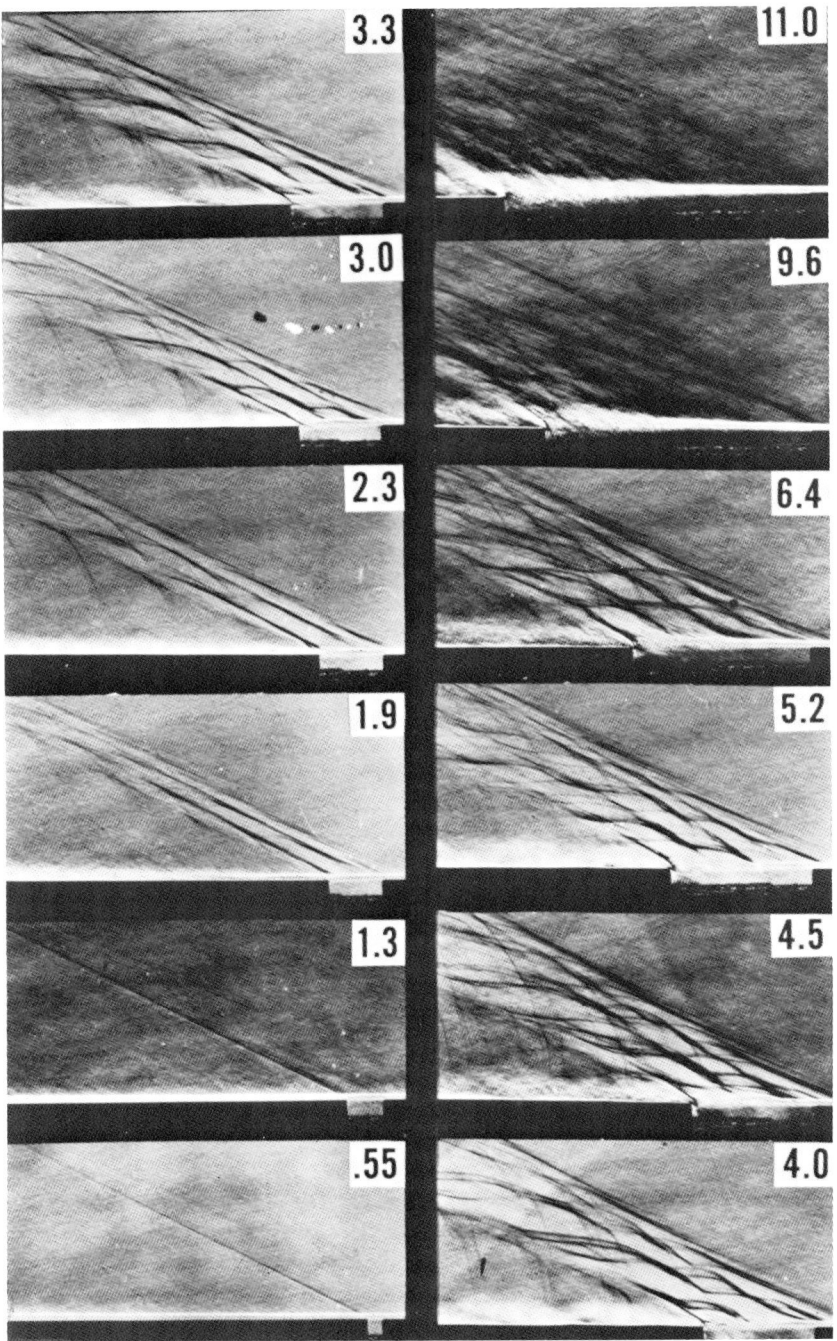

FIG. 34. A series of spark-schlieren photographs showing the acoustic radiation from a rectangular notch of variable length to depth ratio: numbers on photographs indicate the value of L/H; $M_\infty = 2.16$; $\text{Re}_{\text{sep}} \cong 6 \times 10^5$; laminar boundary-layer upstream with transition in the shear layer.

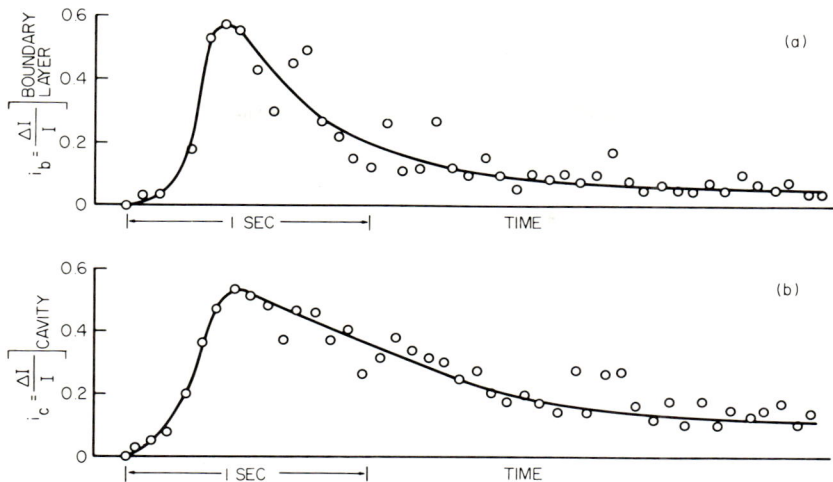

FIG. 35. Attenuation of a light beam by graphite particles in the upstream boundary layer: (a) light beam traversing the upstream boundary layer; (b) light beam traversing the cavity formed by a rectangular notch; $L/H = 2$, $M_\infty = 2.8$, $Re_{sep} = 10^6$, turbulent flow (130).

stream immediately ahead of the notch, and one traversing the notch itself, in response to a pulse of particles injected into the upstream boundary layer. The filling up of the cavity with tracer particles is rapid and follows the linear law

$$di/dt = \tau(i - i_{equ}) \qquad (80)$$

adequately well, where i is relative instantaneous attenuation of the sensing beam (proportional to the total number of particels at the test station), i_{equ} is the equilibrium value of i for a time-independent injection rate i_0, and τ is the time constant of the process (related to the rate of mass exchange). This data yields: $M_\infty = 2.8$, $Re_{sep} = 10^6$, rectangular cutout with $L/H = \frac{1}{2}$, $H = 1$ in.,

$$i_{equ} = 1.8 i_0, \qquad \tau = 0.2 \quad \text{sec} \qquad (81)$$

Measurements of heat transfer to notches in laminar and turbulent flows are illustrated in Fig. 36 (89, 130, 131). In general, there is a fairly sharp drop in the heat transfer in the separation corner (the heat transfer to the separation step is equally low) followed by a rise which starts at about the midspan of the cavity. The heat transfer (relative to that of an attached boundary layer at the same freestream conditions)

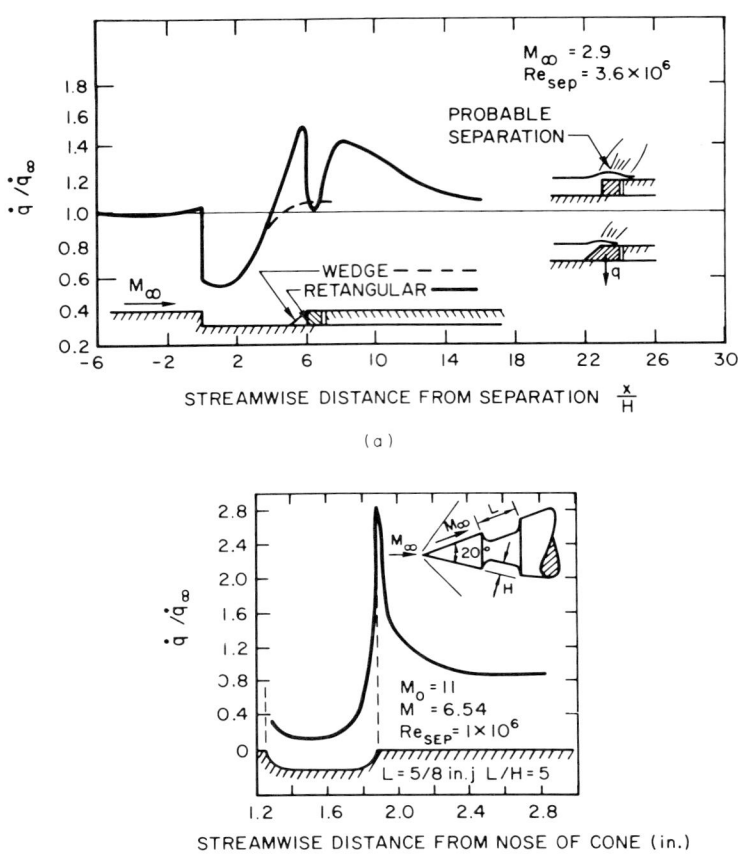

FIG. 36. Typical heat transfer distributions for flow over a notch: (a) turbulent flow over a two-dimensional notch (83); (b) laminar flow over an axisymmetric notch (131).

increases then to a maximum which is of the order of 2 near (and on) the recompression step. The general trends and even the magnitudes are the same in subsonic flow [see, for instance, Fox (132)]. When the recompression step is sharp, the peak heat transfer coefficient seems to occur somewhat downstream of the recompression corner. When the cavity is relatively long (83) ($L/H > 6$) [and if the external flow is accelerated (121, 122)] one observes a second "bubble" of separated flow on the surface downstream of the recompression step with the attendant drop and a second peak in the heat transfer coefficient ratio. This occurs because the shear layer "dips" into such cavities and approaches the corner at an angle to the final reattachment surface which it cannot negotiate (a secondary "breakaway" separation occurs).

The complexity and the unsteadiness of the internal flow in the cavity makes it very doubtful that a genuine theoretical model of heat transfer can be constructed for turbulent and even for high Reynolds number laminar flow [incidentally, the existence of cavities, and also cooling of the flow by heat transfer to a notched wall tend to promote transition (*133*)]. Conceptually, the resistance to heat transfer can be divided into that of the shear layer [which is emphasized by Chapman and Larson (*74*)] and that of the internal boundary layer as already discussed in Section II,D,2 in connection with downsteps and wakes. Recent measurements (*130*) for rectangular cavities give an empirical correlation for this (Fig. 37). It is seen that in cavities (turbulent flow) about 30% of the overall temperature drop occurs across the wall layer as

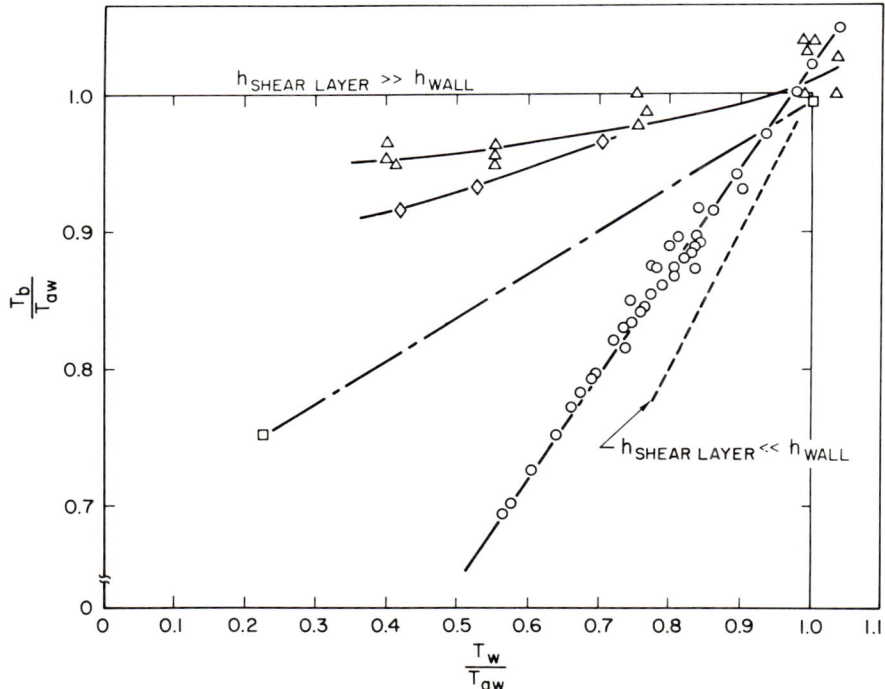

FIG. 37. The recirculation region temperature equilibrium for base wakes and notches: T_b = recirculation region temperature; T_w = base or notch wall temperature; T_{aw} = shear layer recovery temperature. (△) Near wake turbulent flow, $M_\infty = 3$ (*76*); (◇) near wake turbulent flow, $M_\infty = 3$ (*75*); (□) near wake laminar flow, $M_\infty = 6$ (*56*); (○) rectangular cutout turbulent flow, $M_\infty = 2.8$ (*130*).

contrasted with 70% in wakes. It isn't possible to conclude from this, however, that a Chapman–Larson type of model applies, because it does not account for the recompression zone. Whatever general accord between this model and experiment exists [for instance, the variation of the heat transfer coefficient with Reynolds number in laminar flow (74)] is probably coincidental. Charwat et al. (83) proposed an entirely different model based on the idea that the energy input to the cavity (and hence to the walls) is governed by the unsteady pulsatory convective mass exchange between it and the external flow. This model also predicts the trends derived from the Chapman–Larson analysis and generally observed: $\dot{q}/\dot{q}_\infty$ constant in laminar flow and decreasing somewhat with Reynolds number (as the inverse $\frac{1}{5}$ power) in turbulent flow. However, it does not constitute any more genuine a theory, if for no other reason that not all cavities pulsate.

The presence of the cavity causes a redistribution of the heat flux rather than any substantial change of it compared to an attached boundary layer on the same surface. The mean level of $\dot{q}/\dot{q}_\infty$ in the cavity seems to depend on the ratio of the thickness of the boundary layer ahead of separation to the cavity depth. Thick boundary layers lead to a smaller drop in $\dot{q}/\dot{q}_\infty$. This is shown in Fig. 38 for the *floor* of a rectangular

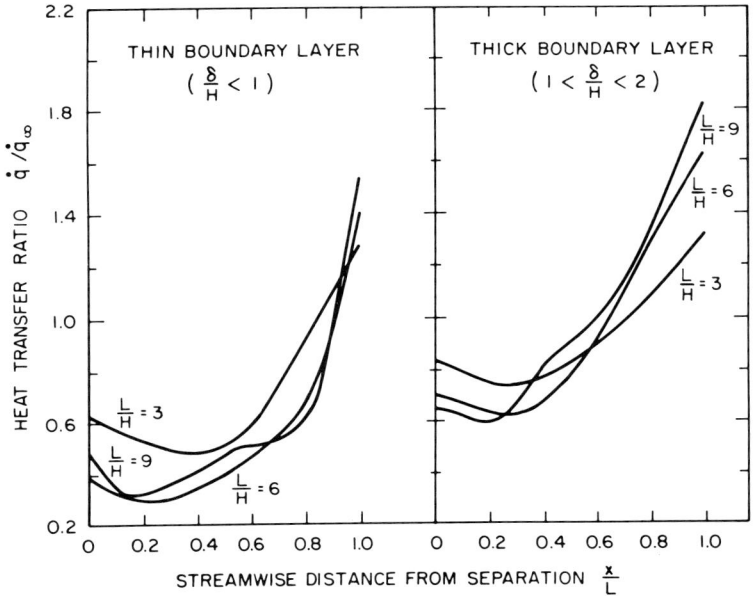

FIG. 38. Measured variation of the heat transfer distribution on the floor of a notch as a function of L/H at $M_\infty = 2.9$ (83).

cutout (*83*) (turbulent boundary layer). However, the *maximum* heat transfer coefficient ratio decreases with oncoming boundary-layer thickness. Figure 39 shows a correlation of laminar through turbulent

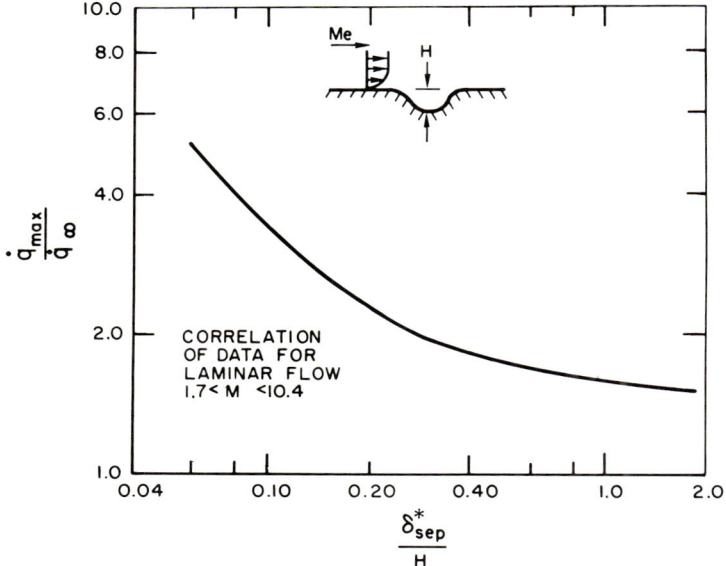

FIG. 39. Correlation of the maximum heat transfer coefficient for laminar flow over a sinusoidal cavity (*134*).

flow data (*134*). In the average (including the recompression region) the heat flux to the cavity is found to be essentially the same as that for a smooth surface of equal projected area; it is somewhat increased (10–30%) in turbulent flow and somewhat decreased in purely laminar flow, but not significantly if the cavity is long ($L/H > 4$). It is always increased if the cavity is short ($L/H < 4$) because the high heat transfer zone at recompression dominates. Further downstream of the cavity the heat transfer coefficient will usually be decreased (due to a thickening of the boundary layer by the presence of the disturbance) unless the cavity induces transition from a laminar to a turbulent boundary layer. Wyborny (*122*) shows measurements for a series of notches having L/H between 1 and 2 and spaced one cavity length apart. Such a notched surface seems to "shield" the body somewhat. The total heat flux to the notched body appears to tend asymptotically (when there are more than three successive notches) to about 70 or 80% of the heat flux to an equal length of a smooth-surfaced body ($M = 8.75$). The data of Wyborny

(*122*) were obtained in a nonuniform somewhat accelerated external flow; however, unpublished experiments by this author in uniform supersonic flow ($M = 3$) confirm these trends and magnitudes.

Several investigators beginning with Chapman drew attention to the possible combination of flow separation in cutouts and transpiration of a coolant to achieve important changes in the heat transfer to the wall. There are two aspects to this problem. The first one concerns the elimination of the high peak in $\dot{q}/\dot{q}_\infty$ occurring on and immediately downstream of the recompression point. This alone can reduce the average heat flux per unit projected area well below that of an equivalent attached boundary layer. The second problem concerns the cooling of the cavity floor as a whole.

The reduction of the peak heat transfer coefficient by blowing was recently investiaged (*129, 130, 135, 136*). Nicoll (*136*) tested steps with 90-deg sharp recompression shoulders; Ginoux (*129*) studied notches with rounded reattachment corners (radius equal to H) in the surface of a cone. Some results of this investigation are shown in Fig. 40. In these experiments, the injectant was introduced at the model surface temperature which varied depending on the length of the (transient) test run (50–100°C); the stagnation temperature was 250°C. Unfortunately, the

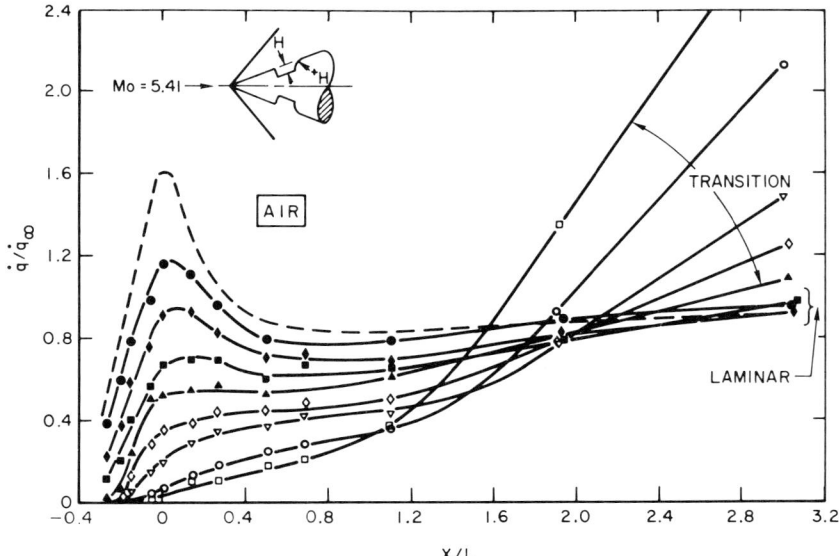

FIG. 40. The effect of injection on the heat transfer to the reattachment region and wall downstream of a groove in a conical body (*129*). $m_{inj}/m_{boundary\ layer}$: (●) 0.044, (◆) 0.088, (■) 0.135, (▲) 0.180, (◇) 0.224, (▽) 0.270, (○) 0.362, (□) 0.448.

thermal boundary condition was not controlled any closer. The pressure peak at recompression as well as the heat transfer peak are strongly decreased by injection. The heat transfer ratio for the floor only varied from 68% (no injection) down to 10%. The heat transfer coefficient ratios on the recompression surface are shown on the figure.

Cooling of a cavity as a whole by injection of secondary gas into it involves two mechanisms: (1) the absorbtion of the heat transferred through the shear layer by the coolant, and (2) changes in the film transfer resistances of both the shear layer and the wall layer due to modifications of the structure of the internal flow by the injection. The first mechanism depends on the enthalpy level of the injectant, the second depends on the injector geometry, jet velocity, direction, and so on.

Figure 41 shows typical results (*130*) of heat transfer coefficient distributions on the floor of a rectangular notch (supersonic, turbulent free stream) with coolant (in this case, air) injected through discrete orifices upstream along the floor of the cavity from the recompression step. This injector configuration was found most effective; other jet configurations (upstream blowing from the separation step, blowing at

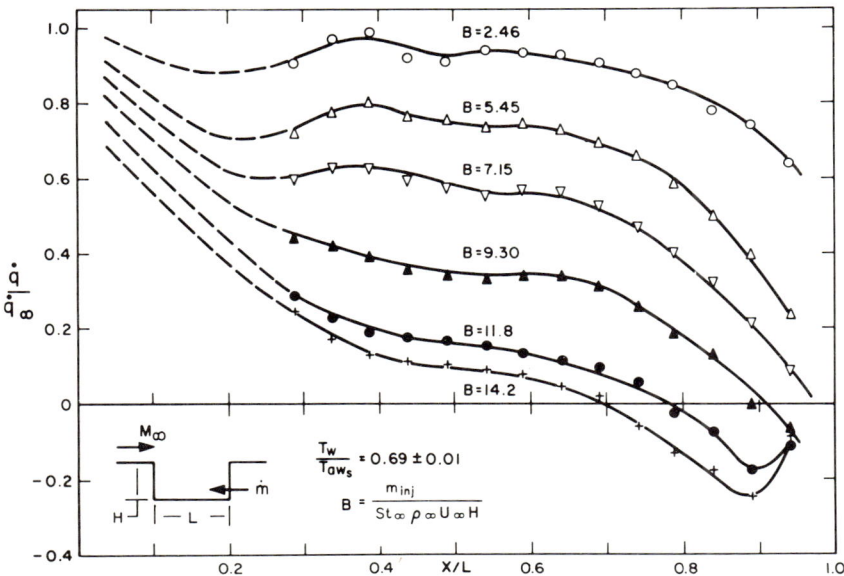

FIG. 41. Effect of injection (air) upstream from the recompression step on the heat transfer to the floor of a rectangular notch: $L/H = 2$, $M_\infty = 2.8$, $Re_{sep} \cong 10^6$, turbulent flow (*130*).

40 deg up or down from either step) always disturbed the region of the low heat transfer coefficient near the separation step so as to increase its conductance. In the case shown the secondary jets are directed tangentially to the rim of the recompression vortex, shielding the wall from its effect without disturbing the pocket of nearly quiescent gas in the separation corner. Transpiration through porous walls was not investigated. Charwat and Redekopp (*130*) propose a semiempirical generalization of these measurements.

A correlation of virtually all the available heat transfer measurements to notches in turbulent flow is presented in Ref. (*89*) and Fig. 42. This

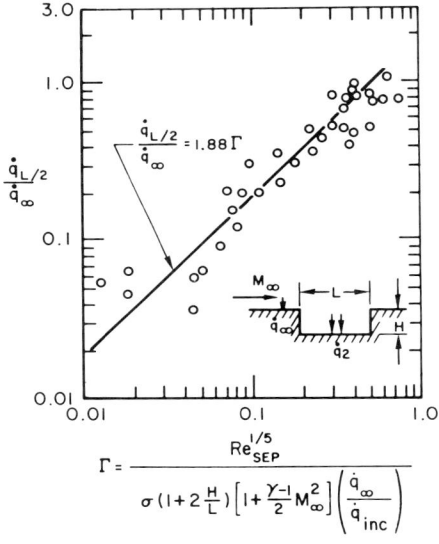

FIG. 42. Correlation of the heat transfer to the midpoint of notches in turbulent flow (*98*). Data from Refs. (*74, 83, 89, 137–139*).

gives the ratio of the heat flux to the floor at the mid-span to the heat flux ahead of the separation corner in terms of a correlation parameter derived from the Korst model. This parameter simply expresses the transfer properties of an (asymptotic) shear layer relative to those of an attached boundary layer. The explicit Mach number dependence carries from the temperature recovery properties of the shear layer, and the Reynolds number dependence from the variation of the transfer properties of an attached incompressible boundary layer. The factor $\dot{q}/\dot{q}_i$ expresses a correction for compressibility applicable to the attached, turbulent boundary layer and therefore contains the Mach number implicitly (*136*a). This correlation is strongly empirical but nonetheless appears useful, especially for short cavities, in allowing one to establish a level of $\dot{q}/\dot{q}_\infty$ to which one must add a judgment regarding the distribution of this parameter.

IV. Cross-Stream Interaction (Cavities Controlled by Separation)

Certain regions of separated flow such as those formed ahead of upstream-facing steps, in compression corners, due to the interaction of shocks with a boundary layer on a continuous surface are strongly dependent on, if not entirely controlled by, the mechanism governing separation. The upstream-facing step provides the clearest examples of this. The reattachment point is always fixed to the recompression corner and the geometry of the separated region as well as the pressure in it depend on the equilibrium position of the point of separation, which is free to "slide" on the continuous wall ahead of the step.

Boundary-layer separation received much attention, but for a long time it was discussed only within the strict framework of the first-order theory: a prediction of the point of separation for a *prescribed* pressure gradient. Since there is no theory capable of describing the shape of the free streamline enclosing the separated region, the usual iterative procedure of correcting the first-order boundary-layer solution by accounting for the induced pressure due to the deflection of the (inviscid) flow around the body-bubble combination cannot be carried out. The interaction between the displacement effect of the viscous layer and the external flow must be considered locally and the pressure field over the region where the flow separates must be determined simultaneously with the solution.

The "free" interaction model is based on the observation that a boundary layer has a short "memory" of its past history. The length of the interaction region is small. At the beginning of this region, which can be isolated, the flow is nearly that which would exist if there were no downstream separation. Its development downstream of such an initial station is governed by the local pressure gradient induced by a cross-stream displacement–compression coupling between the boundary layer and the external flow. The viscous flow in the interaction region continues to be adequately described by the boundary-layer approximation to the Navier–Stokes equation. It remains a parabolic problem in which the "initial" conditions determine the evolution of the solution.

Having neglected the terms in the equations of motion which describe the upstream propagation of disturbances from downstream, the model can give no information on the strength and nature of the perturbation propagated to the "initial" station, which starts the interaction process. To each such perturbation corresponds a possible flow field, that is, a different "body." It is then necessary to identify some distinguishable downstream characteristic of a particular case of interest and proceed by an iterative search for the desired solution by adjusting the "perturbation" at the initial station.

The following section begins by describing experimental data for regions of separated flows, the structure of which is dominated by the interaction process. A semiempirical formulation of the ideas above provides a framework for correlating the data. A rigorous formulation of the model involving a numerical integration of the differential boundary-layer equations or a multimoment approximation to these equations is discussed in Sections IV,B and IV,C, respectively. The results of these analyses indicate that the model is remarkably good. The cross-stream interaction theory is capable of describing laminar and, with the usual difficulties, turbulent flow with all types of imbedded separations. It also provides much more satisfactory "closure" conditions than the semiempirical criteria discussed in the preceding sections. The cost is a very large amount of numerical effort.

A. Measurements in Separated Regions Ahead of Steps and Compression Ramps

1. *Free Interaction Similarity Parameters*

The mechanism of the separation of a supersonic boundary layer was described approximately by Chapman (7) who termed it "free interaction." He assumed that over the (relatively) short length of the separation region only the self-induced pressure gradients are important and the change in the profile of the boundary layer under the influence of the gradual compressions caused by its thickening is self-similar and dependent only on the initial conditions at the beginning of the interaction. The initial "signal" is propagated to the boundary layer from downstream by a mechanism which need not be specified; once started, the interaction will lead to separation and to a final pressure which is uniquely determined by the free interaction mechanism itself. For example, consider the case of the upstream-facing step. The free interaction mechanism establishes the pressure at the point of separation which is the same as the pressure in the "dead-air" region beneath the separated shear layer. This corresponds to an overall (oblique shock) deflection of the external flow. The separation region will now shift along the surface ahead of the step, maintaining a constant angle of deflection of the free stream until the dividing streamline passes through the recompression point at the corner of the step.

Chapman's "free interaction" theory provides valuable similarity parameters which correlate data for a variety of separated regions with remarkable success (although they are strictly applicable only to the "free interaction" portion of the flow ahead of the separation point). This is not only true for the flow about upstream-facing steps, where the separation mechanism clearly dominates, but also about ramps and shock boundary-layer separations, where the recompression point is free.

Consider a boundary layer having a displacement thickness δ^* on a continuous surface. A disturbance originating at some point downstream propagates upstream and causes the boundary layer to thicken, inducing a deflection and a corresponding compression in the free stream. This deflection angle θ can be obtained from a complete solution for the viscous layer profiles as

$$\tan \theta = (v/u)_e \tag{82}$$

where v and u are the cross-stream and streamwise velocities, evaluated at the edge of the layer. Alternately, this angle is estimated from an approximate solution of the mass conservation equation, as follows (6). The mass contained in the viscous layer at any streamwise station is

$$\bar{m} = \int_0^\delta \rho u \, dy \tag{83}$$

and the rate at which this mass increases is

$$d\bar{m}/dx = \rho_e u_e [d\delta/dx - \tan \theta] \tag{84}$$

where the term in brackets is the tangent of the angle formed by the edge of the boundary layer and the local streamline penetrating it. The classical definition of a displacement thickness is

$$\delta^* = \int_0^\delta (1 - \{\rho u/\rho_e u_e\}) \, dy = \delta - \{\bar{m}/\rho_e u_e\} \tag{85}$$

Differentiating (85) and using (82)–(84) one obtains the expression

$$\tan \theta = d\delta^*/dx - d \ln(\rho_e u_e)/dx \int_0^\delta (\rho u/\rho_e u_e) \, dy \tag{86}$$

Clearly, when the interaction is weak the approximation can be used:

$$\tan \theta \simeq \theta \simeq d\delta^*/dx \tag{87}$$

The compression generated by this displacement effect is usually sufficiently gradual to admit the use of the isentropic Prandtl–Meyer relation

$$\theta = \nu_{(M_e)} - \nu_{(M_\infty)} \tag{88}$$

where ν is the Prandtl–Meyer angle and M_∞ is the (reference) undistributed free stream Mach number corresponding to $\theta = 0$. This expression can be put into several forms and approximated further, depending on the requirements of the problem; in particular, the differential Prandtl–Meyer pressure deflection relation is useful:

$$dp/p = (\gamma M^2/(M^2 - 1)^{1/2})\, d\theta \tag{89}$$

Conversely, it is possible to envisage an exact numerical solution of the inviscid flow over the edge of the viscous layer using the method of characteristics; for instance, accounting for wave reflections when the inviscid flow is strongly rotational, but this has not been done to this date.

For the purpose of deriving a gross similarity parameter for the free interaction, it is sufficient to use the simplest form of the pressure deflection relation:

$$\theta \simeq d\delta^*/dx = \int_{p_\infty}^{p_e} -((M^2-1)^{1/2}/\gamma M^2)(dp/p) \tag{90}$$

If the length of the curved portion of the separating viscous layer is l and the subscripts zero indicates properties at a station immediately before interaction begins, a linearized form of (90) combined with Eq. (87) yields (88)

$$\frac{p_{(x/l)} - p_\infty}{(\gamma/2)\,p_\infty M_\infty^2} = \frac{2}{(M_0^2 - 1)^{1/2}} \frac{\delta_0^*}{l}\left[\frac{d(\delta^*/\delta_0^*)}{d(x/l)}\right] = \frac{2}{(M_\infty^2 - 1)^{1/2}} \frac{\delta_0^*}{l}\left[f_1\!\left(\frac{x}{l}\right)\right] \tag{91}$$

The bracket term $[f_1(x/l)]$ describes the evolution of the thickness of the viscous layer which is assumed to be self-similar. In other words, this term is a universal function of x/l.

On the other hand, the boundary-layer equations yield at the wall ($u = v = 0$)

$$dp/dx = d\tau/dy)_w \tag{92}$$

Integrating and introducing appropriate reference quantities,

$$\frac{p_{(x/l)} - p_\infty}{(\gamma/2)\,p_\infty M_\infty^2} = \frac{l}{\delta_0^*} \frac{\tau_{w_0}}{(\gamma/2)\,p_\infty M_\infty^2} \int_0^{x/l} \left(\frac{\partial \tau/\tau_{w_0}}{\partial y/\delta_0^*}\right)_w d(x/l) \tag{93}$$

By virtue of the similarity hypothesis, the integral must be a function of (x/l) only, so that

$$\frac{p_{(x/l)} - p_\infty}{(\gamma/2) p_\infty M_\infty^2} = \frac{l}{\delta_0^*} C_{f_0} f_2\left(\frac{x}{l}\right) \tag{94}$$

(l/δ_0^*) can be eliminated between the above expressions for the pressure rise. In particular, evaluating the expression at $x/l = 1$ (the end of the interaction zone), one obtains for the total pressure increase by this "free interaction" mechanism

$$\frac{p_p - p_\infty}{(\gamma/2) p_\infty M_\infty^2} = K_1 \left[\frac{2 C_{f_0}}{(M_\infty^2 - 1)^{1/2}}\right]^{1/2} \tag{95}$$

and for the distance over which it acts

$$l/\delta_0^* = K_2 [(M_\infty^2 - 1)^{1/2} C_{f_0}]^{-1/2} \tag{96}$$

The magnitude of the constants must be determined from experiments, namely from measurements of the "plateau" pressure (p_p) in the separation region. There is no difficulty in laminar and transitional flows[1] where a plateau is well defined. The situation is more ambiguous for turbulent flow where often no plateau can be clearly identified. Typically K_1 has the value 1.47 in laminar flow and $K_1 = 6$ in turbulent flow. The correlation of the pressure plateau data is shown in Fig. 43.

Various investigators contributed extensions to this basic theory. Erdos and Pallone (140), who provide plots of various intermediate functions appearing in the analysis and discuss its concordance with experiment, proposed a somewhat different development in which they use a second universal function representing the (self-similar) gradient of pressure across the separating viscous layer. However, this factor becomes ultimately included in a single constant (K_1) which must be evaluated from experiment, so that their "theory" is not essentially different. Carriere (141) extended the analysis to a nonuniform free stream by accounting, in an approximate manner, for the interference between the compression waves generated in the interaction process and (expansion) characteristics impinging on the interaction zone. The analysis brings another factor into the basic formula above—a function of the pressure gradient that would exist over the region of the interaction if there were no separation (which is evaluated at the start of the

---

[1] Reference (185) presents data which suggests that the plateau is most pronounced in transitional flow and less clearly evident in laminar flow. One must remember, however, that the results are also influenced by a geometric scale of the model relative to the thickness of the oncoming boundary layer, not only its nature.

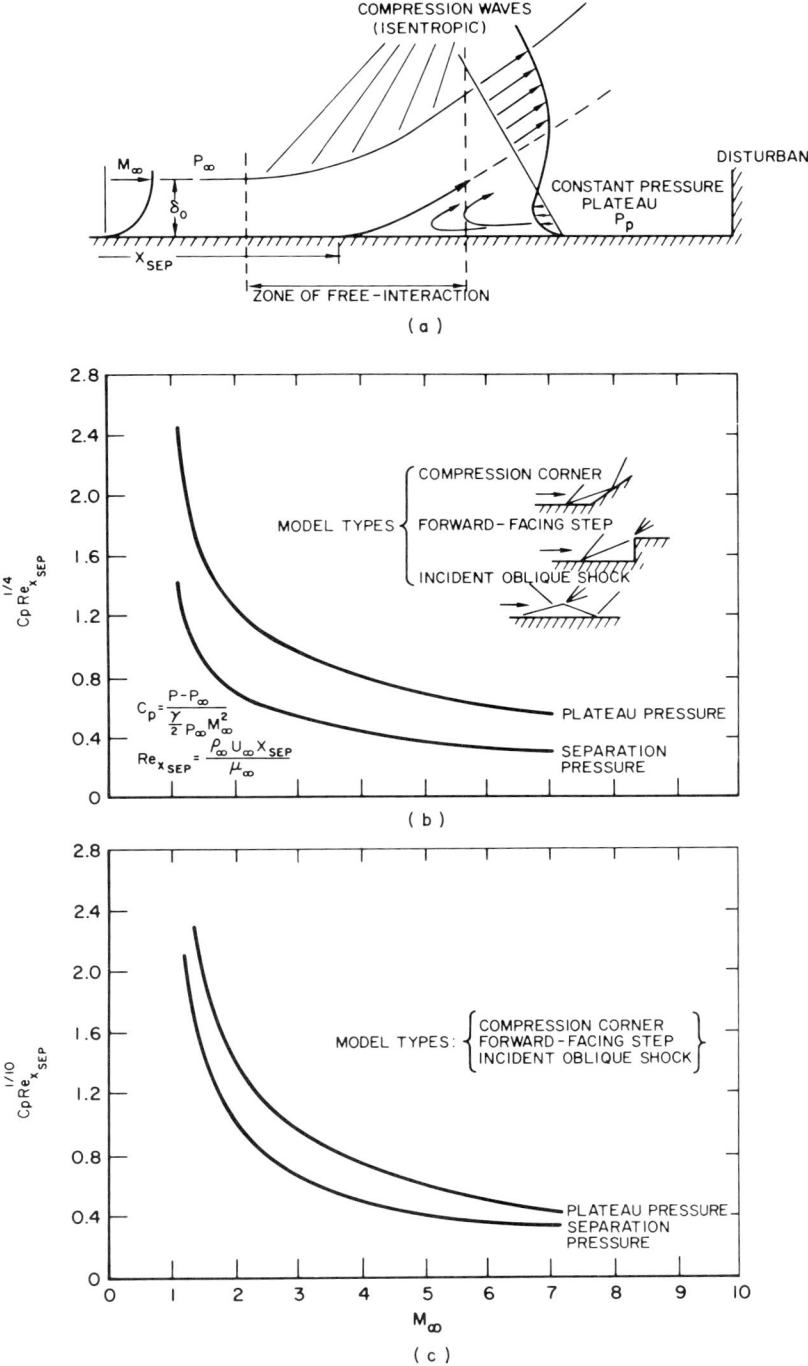

interaction). The correction seems to yield good results and it is needed in situations such as the separation of the boundary layer inside of an overexpanded nozzle where the free stream is strongly accelerated. Lewis et al. (52) [see also Lewis (142)] includes one more term in the expansion for the pressure coefficient to extend the analysis to hypersonic flows. The similarity variables become then

$$\pi = \left(\frac{p - p_\infty}{p_\infty}\right)\chi_\infty^{-1/2}; \quad X = \frac{x - x_0}{x_0} \frac{M_\infty^2 C_\infty^{1/2}}{[Re_x(\delta^*/x)]^{1/2}} \chi_\infty^{-1/2} \quad (97)$$

where χ_∞ is the hypersonic interaction parameter and C_∞ the viscosity–temperature proportionality factor.

The effect of heat transfer to the wall ahead of separation enters the problem through the reference parameters. Using an approximate integral of the energy equation, on a Polhausen-type analysis, Curle (143) finds that the pressure variable is unchanged but the distance variable in the scaling becomes

$$X = [(x - x_0)/x_0] M_\infty^2 (T_w/T_\infty)^{-1} \chi_\infty^{-1/2} \quad (98)$$

2. *Geometrical Characteristics of Separations Induced by Steps, Compression Corners, and Shock Waves*

The preceding section outlined the mechanism of free interaction which determines the "plateau" pressure in the cavity. If the separation is caused by a sharp (90°) upstream-facing step, the entire geometry of the cavity is thereby roughly determined because the reattachment point is known. The deflection angle of the dividing streamline is calculated (oblique shock deflection of the free stream up to plateau pressure) using an appropriate theory for the boundary-layer parameters upstream of the free interaction region (C_{f_0}, M_∞). Such a calculation is shown in Fig. 44. The shape of the cavity is approximately determined by drawing a straight, free streamline inclined at the calculated angle to the free stream and passing through the recompression corner. The sharp rise in pressure at the foot of the step which is due to the recompression of the shear layer against its face seems to be a "local" phenomenon: it does not extend more than one step height upstream of recompression, and does not affect the calculation outlined above.

The situation is more complicated in the case of compression ramps because the reattachment point is now free to move along the continuous downstream surface. One needs a "closure condition" similar to that

---

FIG. 43. Free interaction mechanism: correlation of the pressure plateau (7, 140). (a) Schematic of free interaction; (b) pressure correlations for laminar, adiabatic flow; (c) pressure correlations for turbulent, adiabatic flow.

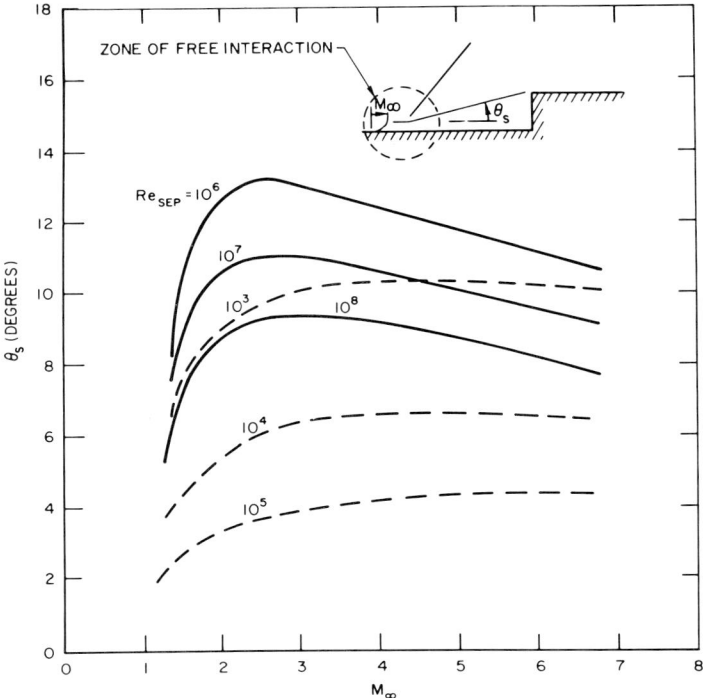

Fig. 44. Deflection of the free shear layer over separation ahead of upstream facing step (*140*): (———) turbulent flow; (– – –) laminar flow.

which determines the geometry of a wake behind a downstream-facing step (see Section IV,C,3). In principle, concepts of mass conservation in the cavity such as those underlying the Chapman criterion are applicable, but their elaboration would be much more uncertain in this case. Erdos and Pallone (*140*) propose a frankly empirical correlation for the length of the (straight) dividing streamline in terms of the overall pressure rise determined by the total turning angle imposed on the inviscid flow by the boundary of the ramp. Their curves are given in Fig. 45. This correlation, toghether with the angle of the dividing streamline, which is uniquely determined by the separation mechanism and given by Fig. 44, solves the problem for compression ramps.

The geometry of regions of separated flow induced by the interaction of shocks with a boundary layer established on a flat surface can also be roughed out to a surprisingly good accuracy by using the present framework. The known strength of the impinging shock provides the

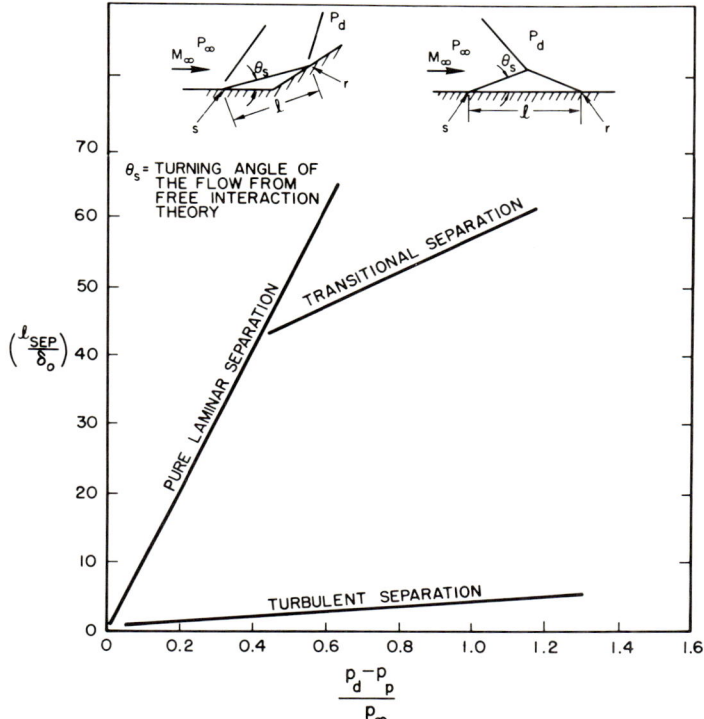

FIG. 45. Correlation of the length of separation in compression corners and shock-induced separations (*140*).

overall pressure rise suffered by the external flow. The free interaction calculation provides the angle at separation (angle θ_s; see sketch of Fig. 45) and the plateau pressure p_p in the cavity. With these data and the correlation of Fig. 45, the total length (l in the sketch, Fig. 45) of the region as well as points s and r can be located.

3. *Typical Pressure and Heat Transfer Distributions*

The free interaction similarity parameters are basically applicable only to the "free interaction region" preceding separation, but it turns out that they correlate the pressure distribution for the entire separated region. Figure 46 shows, for example, measurements (*52*) of pressure in a compression corner with heat transfer in laminar and transitional flow (flow transition occurring along the shear layer) plotted in terms of Curle's correlation, which is seen to unify the measurements very well.

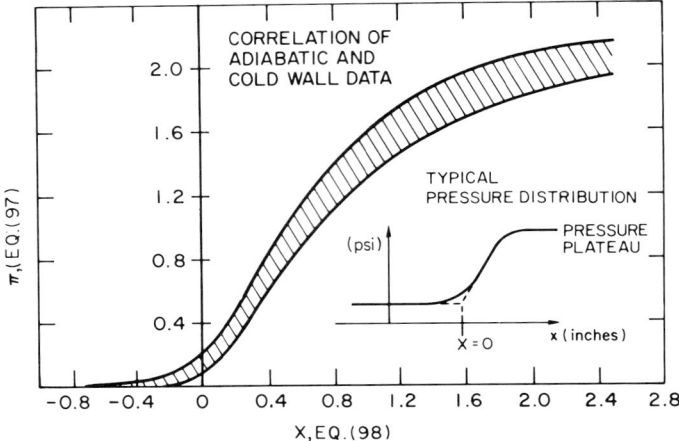

FIG. 46. Correlation of ramp pressure distribution data in the free interaction region for nonadiabatic laminar and transitional flow (52).

The form of these similarity parameters is good for both laminar and turbulent flow. In turbulent flow, however, the pressure plateau which represents the end of the free interaction region is less clearly distinguishable, and there is considerably more uncertainty about the empirical constants to be used in the analysis.

In turbulent flow the length of the free interaction region is much shorter and the pressure gradient is steeper than in laminar or transitional flow. The pressure tends to overshoot the final value. These aspects of the data are exemplified by Fig. 47, which also gives measurements (144) of heat transfer in the region of the separation (on a 15-deg ramp at $M_\infty = 2.02$). It should be noted that the turbulent data shown involve artificial transition.

Transition in the shear layer does not appear to influence the free interaction region strongly, until the transition point moves upstream into the attached boundary layer. Thus transitional data continue to exhibit a fairly clear pressure plateau. As transition moves upstream of the reattachment point the separation point moves slowly towards the disturbance, the length of the free streamline decreases, but important changes only appear when the entire flow becomes turbulent.

The shape of the pressure distribution ahead of steps is about the same as that in compression corners except that the peak pressure immediately ahead of the step often "overshoots" the final value on the recompression surface. A summary of turbulent-flow data was recently published by Zukoski (148). It shows that in the range $2 < M < 4$ the

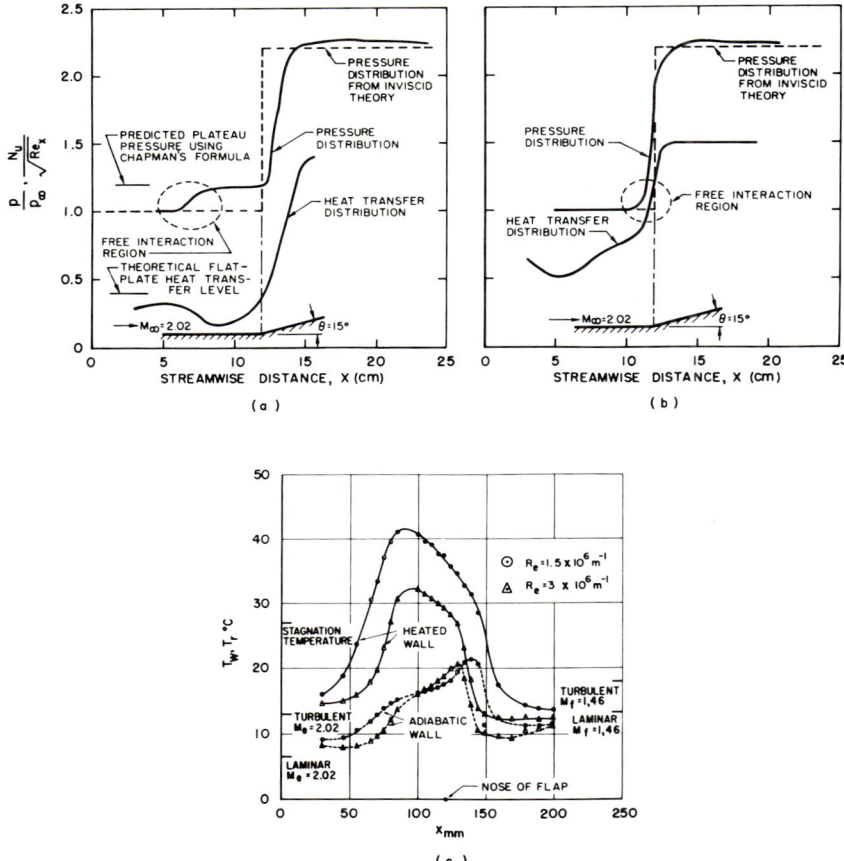

FIG. 47. Typical pressure, heat transfer, and recovery temperature measurements in laminar and turbulent compression corners (*144*): (a) laminar separation data, Re = 1.5 × 10⁶/meter; (b) turbulent separation data, Re = 3 × 10⁶/meter, boundary layer tripped; (c) temperature distributions for a flap angle of 15 deg.

"triangular" wedge model of the shape of the separated region is quite good. Moreover it suggests a useful approximation which is

$$\theta_s \simeq 13°, \quad \Delta p_p/p_\infty \simeq \tfrac{1}{2} M_\infty \qquad (99)$$

($2 < M_\infty < 4$; $3 \times 10^4 < \mathrm{Re}_{\mathrm{sep}} < 10^6$; turbulent).

The heat transfer coefficient distribution in a compression ramp was illustrated in Fig. 47. The Nusselt number is seen to decrease beginning in the "free" interaction zone, reaches a minimum in the fully separated flow (the "dead-air" region, corresponding to the region of the pressure plateau), and then increases rather rapidly in the zone of recompression. It usually "overshoots" its final value on the reattachment surface of the ramp.

Note that the data presented here are reduced, using the difference between the *local recovery* and wall temperature as the driving potential for heat transfer:

$$h = q/(T_w - T_{aw}); \quad \text{Nu} = hx/k_\infty \tag{100}$$

(k_∞ is the thermal conducitivity in the undisturbed flow). The recovery temperature was measured in this experiment and it is shown in the figure. It is seen to increase by more then a factor of two in the region of separation (up to 50% of the difference $T_w - T_{stag}$), a variation which is more pronounced than in other types of cavities (cutouts and downsteps). The heat transfer coefficient defined by (100) is conceptually the correct one, and the trend exhibited by this data is logical. However, there are very few measurements of the recovery temperature in regions of separated flow. The majority of heat transfer data for ramp and step separations are reported as, approximately, the heat flux ratio $\dot{q}/\dot{q}_s$ (even though authors most often write it as a ratio of coefficients h defined using the difference of the *stagnation and the wall temperature*). The aspect of the curves is then quite different (see Fig. 48). The ratio $\dot{q}/\dot{q}_s$ begins by increasing in the free interaction region, then decreases to a minimum in the "dead-air" region (which is, however, higher than unity), and finally rises near recompression to a value about twice that of the plateau.

The available data is rather incomplete and covers different, limited ranges of free stream conditions (virtually only turbulent flow) which are not always stated in detail. There are also variations in the definition of the reference parameters. Only a rough empirical correlation can be suggested, which implies that the transfer properties of the shear layer and the internal flow (the wall boundary layer) are insensitive to the free stream condition (turbulent flow) so that the heat flux to the walls depends primarily on the pressure (density) level in the cavity. Indeed, from measurements (89, 144–148) one finds that the average heat flux ratio $\bar{\dot{q}}/\dot{q}_s$ over the region of the separation is approximately one half the plateau pressure rise:

$$\bar{\dot{q}}/\dot{q}_s \simeq \tfrac{1}{2} p_p/p_\infty \tag{101}$$

which combined with Zukoski's approximation for the plateau pressure (99) gives

$$\Delta \bar{\dot{q}}/\dot{q}_s \simeq \tfrac{1}{4} M_\infty^2 \tag{102}$$

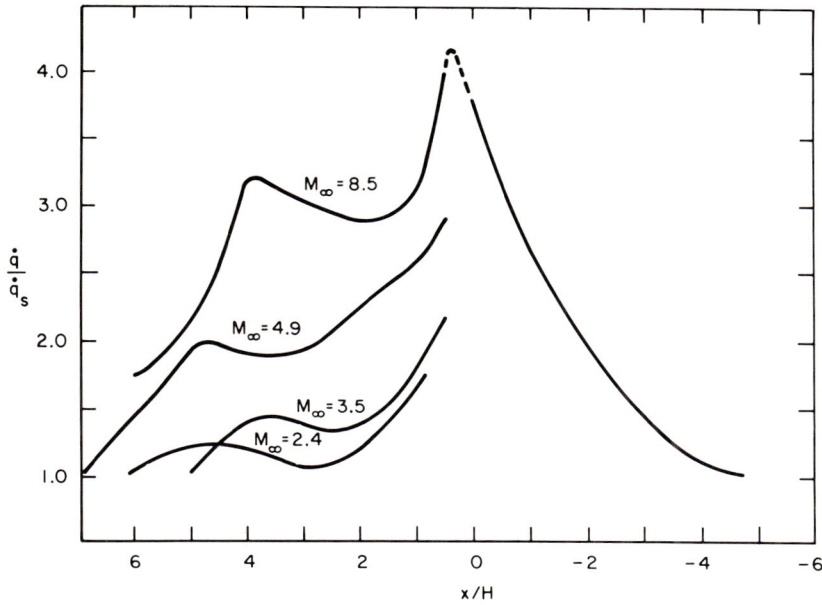

FIG. 48. Approximate correlation of turbulent heat transfer data for forward-facing steps. Data obtained from Refs. (89, 156–158).

In addition to references already quoted in the text, Refs. (149–157) present studies of various types of separation, but usually with a limited range of free stream parameters. Note also the extensive article by Pearcey in Lachman's book on shock-induced separation and the design problems associated with it (158).

4. Incipient Separation

Separation occurs when the attached boundary layer cannot negotiate compression imposed on it by the turning of the inviscid flow. It was originally thought that it occurred when the overall pressure rise ($p_d - p_\infty$) as determined by the geometry of the ramp or the strength of the impinging shock exceeded the plateau pressure rise determined by the free interaction theory. However, recent data show that the pressure rise at which separation just occurs (incipient separation) is considerably higher (the phenomenon exhibits a sort of hysterisis loop). This is exemplified by the data in Fig. 49 (144). The 5-deg corner in this example exhibits a gradual, monotone rise in pressure. The pressure profile for the 7-deg ramp shows, instead, the characteristic plateau of a separated region. Evidently these tests span the condition at which the

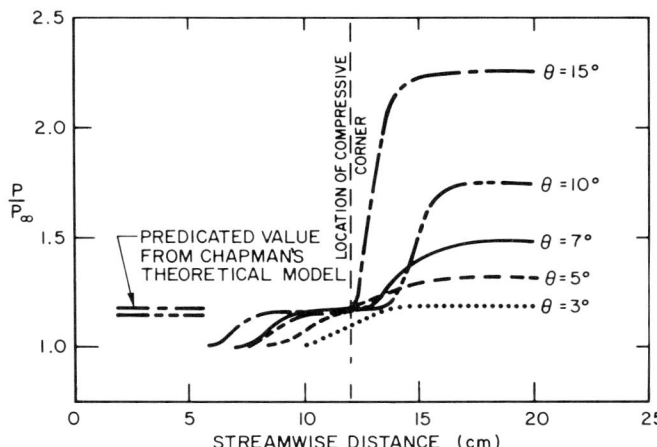

FIG. 49. Effect of ramp angle on the pressure distribution (incipient separation): $M_\infty = 2$, $Re = 3 \times 10^6$/meter (144).

boundary layer actually separates. The total compression in the first case is larger than the plateau pressure in the second case.

Roshko and Thomke (184) analyzed this problem by a method similar to that used by Chapman to define the free interaction correlation parameter. The difference is that now the pressure rise across the region under consideration is fixed by the geometry of the boundary and not by the oncoming flow. The result suggests that the pressure coefficient for incipient separation should depend linearly on the skin friction (not on the square root of it, as in the free interaction region). Using data for corners and transitional shock-induced separation measurements on airfoils, the authors prepared a correlation curve which is reproduced in Fig. 50. At the same time they pointed out, however, that there is a fundamental inconsistency between the incipient separation and the free interaction theory which becomes evident as the Reynolds number increases: The correlation of Fig. 50 should not be extended past $Re = 10^7$. They undertook further experiments, which are still in progress [see Roshko and Thomke (184b)]. These experiments indicate that the incipient pressure ratios are higher than predicted at high Reynolds numbers and that the Reynolds number trend is reversed (personnel communications). Figure 50 should be viewed only as an empirical correlation of data over a limited range. The subject remains open.

Heat transfer appears to have a considerable influence on the problem of incipient separation. Cooling decreases the length of the detached flow (see, for instance, Curle's similarity parameter given in Section IV,A,1). An analysis based on a solution of the boundary-layer

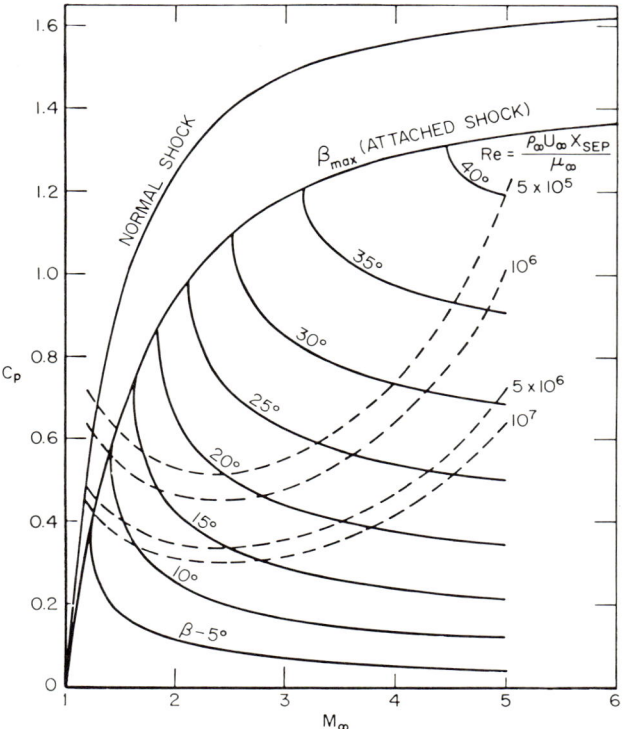

FIG. 50. Correlation of the incipient separation point (A. Roshko, unpublished): (——) compression corner; (– – –) incipient separation.

equations by a moment method which is outlined in Section IV,C,4 indicates that there exists a critical ratio of the wall temperature to the stagnation temperature at which the length of the free streamline goes to zero. This critical ratio depends only on Mach number (independent of Reynolds number). The analysis involves a number of approximations (such as, for instance, that the temperature in the separated region equals the wall temperature) which may affect the quantitative results; there is no data to check them. The trends should be valid, however. The following table gives some of the calculated points ($\gamma = 1.4$, 10° compressive corners, Pr $= 1$).

M_∞	2	4	6	8	10
$\left(\dfrac{T_w}{T_{stag}}\right)_{crit}$	0.09	0.21	0.25	0.27	0.28

One can also expect a limit of incipient separation in the domain of low Reynolds numbers, especially when the compressive region is finite in extent (for example, an upstream-facing step between two parallel walls). There exist basic experiments and analysis of base wakes in very low Reynolds number flow; these are summarized very well in Ref. (*12*). No analytical work and few experiments deal with ramps or steps. One experimental investigation on upstream-facing steps in supersonic rarefied gas flow is reported by Rogers (*159*). This experiment shows that as the Reynolds number decreases (blow 300) the pressure rise associated with recompression of a detached flow begins to diffuse upstream; the boundary-layer thickens ahead of the step and eventually exceeds the step height; the plateau in the pressure distribution gradually disappears and the flow becomes one of a very thick boundary layer in which is inbedded a relatively small protrusion.

However, the interplay among the different parameters such as the unit Reynolds number (importance of the viscous forces) and the ratio of the step height to the thickness of the boundary layer is not yet sufficiently clear.

B. Numerical Integration of the Interaction Equation

The system of equations representing a boundary layer coupled to the external flow by a displacement effect is parabolic and lends itself to numerical integration by a forward-marching scheme, starting from an initially prescribed condition. The nature of the method and the interaction mechanism itself are illustrated in what follows by two analyses. The first one, Section IV,B,1, is a highly simplified analysis of the preexpansion of the boundary layer upstream of and up to separation at a (expansive) sharp corner. It provides a simple and clear illustration of the procedure. Section IV,B,2 discusses a series of papers where the same basic method was applied to the numerical integration of the full differential boundary-layer equations; the same problem of expansive separation is treated. This work represents the ultimate in rigor and completeness within the framework of the cross-stream interaction model.

1. *Model of the Interaction Mechanism; Preexpansion Ahead of a Corner*

The separation of the boundary layer at the corner of a downstream-facing step was discussed in Section III,A. Here we are concerned with the upstream influence of the low pressure in the cavity which was disregarded in that discussion (*160*).

Consider the *subsonic* inner portion of an established boundary layer upstream of the corner. Neglect transverse pressure gradients and replace the distribution of velocity in this channel by a mean velocity and a mean Mach number. If the pressure drops in the streamwise direction, the thickness of the subsonic channel decreases. This is compatible with an inward, expansive turning of the adjacent supersonic flow, and the pressure can remain matched in the streamwise direction across the interface separating them. However, a point will be reached when the mean Mach number in the initially subsonic channel becomes sonic. From this point on the thickness of channel has to increase in order to allow the pressure to drop further. This is associated with an outward compressive deflection of the supersonic flow. Clearly the sonic throat in the inner channel must occur exactly at the corner, at which the wall bounding the flow is as if "removed" and the sublayer can expand into the cavity while the supersonic layer continues to turn inward.

The subsonic portion of the initial boundary layer is not actually a stream tube because the outer, high Mach number portion of the profile accelerates past sonic velocity first. This can be accounted for by considering the subsonic flow to be a series of stream tubes of varying (initial) Mach number and calculating the shape of the sonic line as part of the analysis. The corner is then defined by the point of intersection of the sonic line with the boundary. This is depicted in Fig. 51.

In the immediate neighborhood of the singular corner, the model and the boundary-layer description of the oncoming flow are no longer valid. However, this occurs within a radius of the order of δ_0 from the corner while the extent of the zone of upstream influence is of the order $\delta_0 M_\infty$.

Let the subsonic layer be represented by a series of stream tubes

$$\varDelta = \sum \varDelta_j \tag{103}$$

The thickness of each stream tube (cross-sectional area per unit breath) is related to its Mach number and the pressure by the usual quasi-one-dimensional flow equation:

$$\frac{\varDelta_j}{\varDelta_{j0}} = \frac{M_{j0}}{M_j} \left[\frac{1 + [(\gamma-1)/2] M_j^2}{1 + [(\gamma-1)/2] M_{j0}^2} \right]^{(\gamma+1)/2(\gamma-1)}$$

$$M_j = \left(\frac{2}{\gamma-1}\right)^{1/2} \left[\left(\frac{p_j}{p_{\text{stag}_j}}\right)^{-(\gamma-1)/\gamma} - 1 \right]^{1/2} \tag{104}$$

The subscript zero denotes conditions at the initial station at which interaction begins. The process is assumed to be isentropic. Viscosity

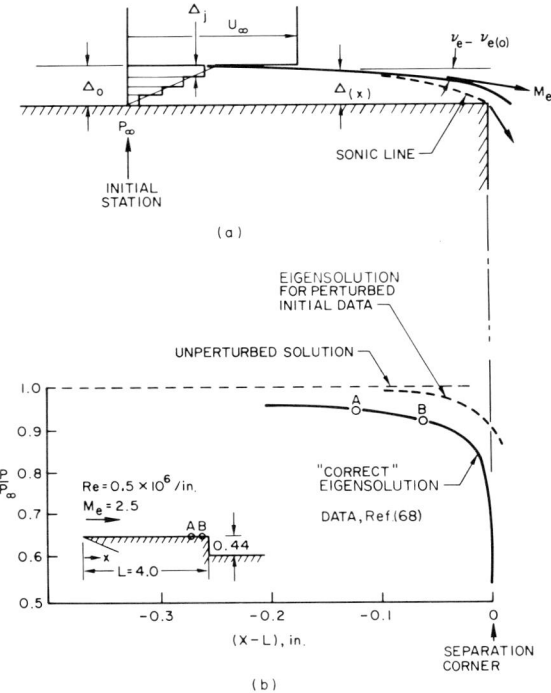

Fig. 51. Preexpansion ahead of a sharp corner (160): (a) sketch of model; (b) calculated pressure.

enters only in determining the initial profile. Normal pressure gradients are neglected so that p is constant from stream tube to stream tube at a given streamwise station.

The coupling between the subsonic sublayer and the supersonic external flow is expressed by continuity in the flow direction at the interface (88):

$$-d\Delta/dx = [\nu - \nu_{(0)}]_e \qquad (105)$$

where ν is the Prandtl–Meyer angle which can be written as a function of the pressure ratio $(p_s/p_{\text{stag}_j})_e$. The subscript e denotes conditions at the edge of the inner layer.

The initial Mach number profile M_{j0} and the free stream conditions are given (the initial conditions are of course a solution of the equations describing this model flow). The simple set of equations can be integrated (numerically) and solved for any variable, in particular, for the local pressure, as a function of streamwise distance x from the initial station. There is an infinite number of integral solutions. For instance, the solution $v_e = v_{0_e}$ ($\Delta = \Delta_0$ = const) corresponds to uniform flow with a constant thickness of the subsonic sublayer (viscosity was neglected so there is no mechanism to disturb this equilibrium). In order to generate other solutions an initial perturbation at the starting station is needed. For example, the pressure at the first streamwise mesh-point is lowered (typically, by 1%). Perturbations which lead to a streamwise decreasing pressure describe the expansive flow under consideration. To each perturbation corresponds a different evolution of flow downstream of the "initial" station and a different point of intersection of the sonic line with the wall, the index chosen for locating the corner. A truly unique solution cannot be determined. However, the "catastrophic" nature of the change in the flow properties which is found to occur some distance downstream of the starting point consitutes a feature of the method which makes it usable. The Mach line becomes virtually perpendicular to the initial flow direction. The pressure drops extremely rapidly (see Fig. 51), etc. This identifies the "corner" to a very good accuracy, or, conversely, locates the starting station.

Weiss and Nelson (*160*) present further results and discuss them in detail. This analysis, although very simplified, appears to yield realistic results. Moreover, it illustrates the nature of the interaction without requiring much mathematical effort.

2. *Numerical Integration of the Differential Boundary-Layer Equations with Interaction*

The concepts of the interaction theory can be formulated in terms of the full differential boundary-layer equations, avoiding the schematization of the flow and most of the *ad hoc* assumptions involved in the simplified example described in the preceding section. Such an approach was developed by Baum and co-workers and used to study several problems: the same problem of the preexpansion to separation off a sharp corner which was treated approximately in the preceding (*161, 162*), the development of the flow behind the downstream stagnation point of a blunt-base wake (*163*) and the separation off a rounded corner (*162, 164*).

Baum retains the inviscid transverse momentum equation, instead of making the classical boundary-layer statement $\partial p/\partial y = 0$. The need for including the effect of transverse pressure gradients, albeit approximately, is demonstrated a posteriori, especially for treating the hypersonic case.

The flow field is thus described by the following set of equations (*164*) (assuming Le = Pr = 1):

$$\rho u \frac{\partial u}{\partial x} + \rho v(1 + \kappa y) \frac{\partial u}{\partial y} + \kappa \rho u v = -\frac{\partial p}{\partial x} + \frac{\partial}{\partial y}\left[\mu(1 + \kappa y) \frac{\partial u}{\partial y}\right]$$

$$\rho u \frac{\partial v}{\partial x} + \rho v(1 + \kappa y) \frac{\partial v}{\partial y} - \kappa \rho u^2 = -(1 + \kappa y) \frac{\partial p}{\partial y}$$

$$\rho u \frac{\partial I}{\partial x} + \rho v(1 + \kappa y) \frac{\partial I}{\partial y} = \frac{\partial}{\partial y}\left[\mu(1 + \kappa y) \frac{\partial I}{\partial y}\right]$$

$$\frac{\partial(\rho u)}{\partial x} + \frac{\partial}{\partial y}[\rho v(1 + \kappa y)] = 0$$

(106)

where κ is the local wall curvature. [Actually Baum finds that his numerical scheme becomes unstable in the subsonic region of the boundary layer because of the character of the transverse momentum equation. To avoid this he uses the boundary-layer assumption $(\partial p/\partial y = 0)$ below the sonic line in the boundary layer but not above it; however, this is only a numerical approximation and could be removed if necessary.]

The boundary conditions at the inner edge are, for the attached flow ahead of separation and for the flow downstream of the wake closure point, respectively,

$$u(x, 0) = 0; \quad v(x, 0) = 0; \quad I_{(x,0)} = I_\text{w} : \text{ wall condition}$$
$$(\partial u/\partial y)_{y=0} = 0; \quad v(x, 0) = 0; \quad (\partial I/\partial y)_{y=0} = 0 : \text{ symmetry condition}$$

(107)

The outer-edge boundary conditions (only adiabatic flow was treated) are

$$dp = -\rho q \, dq; \quad q = (u^2 + v^2)^{1/2}, \quad dI = 0 \quad (108)$$

The essential coupling between the external flow and the boundary layer is expressed by the Prandtl–Meyer pressure deflection relation (89)

$$dp = (\gamma p M^2/(M^2 - 1)^{1/2})[d\theta/(1 + \theta^2) - \kappa \, dx] \quad (109)$$

where $\tan \theta = (v/u)_e$ is the flow angle at the edge of the boundary layer, relative to the undisturbed free stream found as part of the solution of the boundary-layer equations.

Adding an equation of state and a law for the viscosity–temperature variation (ideal gas, viscosity proportional to temperature, and Lewis and Prandtl numbers of unity, which are already implied by the form of the equations, are used in these calculations), the statement of the problem is complete and, except for the use of the transverse momentum equation in the supersonic portion of the boundary layer, corresponds to a standard boundary-layer formulation.

The problem is transformed into a Von Mises (streamwise distance and stream function) coordinate system and use is made of the Stewartson transformation. The dependent variables are expressed as dimensionless ratios scaled by parameters representing the undisturbed free stream. The finite-difference analog to the equations is discussed in the references. The numerical work requires fair sophistication in ordering of the equation, without which the problem becomes formidable. The calculation requires high-precision arithmetic to reduce roundoff errors. Stringent requirements of precision are characteristic of this type of analysis in which the instability of the set is an essential feature of the result.

The calculation is now started at an arbitrary station by prescribing there the initial profiles in the form of dimensionless quantities (velocity ratios, density ratios, etc.) across the boundary layer, scaled by the value of the pertinent parameters at the edge of the boundary layer. These scaling parameters differ from the undisturbed free stream parameters used to scale the equations themselves due to the induced displacement effect caused by the growth rate of the boundary layer.

The problem contains two separate scales: the set of undisturbed free stream parameters (M_∞, Re_∞, and I_∞) used to reduce the differential equations, and the set of quantities describing the edge of the boundary layer at the starting station (M_e, Re_e, θ_e, and I_e) which appear in the prescribed initial profile at the starting station. The parameters in the second set are interrelated and must be prescribed in terms of a single parameter from knowledge of the history of the flow up to the starting station.

For example, in analyzing the case of an attached boundary layer approaching an expansion corner the authors assume at the starting station a fully developed laminar, adiabatic, Blasius boundary layer (Pr = 1) with "weak" interaction. The classical solution of this case gives for the boundary-layer edge Reynolds number and the local inclination of the edge to the wall expressions of the form

$$(v/u)_e = \tan(\nu_\infty - \nu_e),$$

(110)

$$Re_e = Re_\infty \frac{M_e}{M_\infty} \left[\frac{1 + [(\gamma - 1)/2] M_\infty^2}{1 + [(\gamma - 1)/2] M_e^2} \right]^{(3-\gamma)/2(\gamma-1)}, \qquad I_e = I_\infty$$

With these auxiliary relations the description of the initial profile is given by a single parameter; in the form shown above it is M_e (other choices are possible, of course).

The problem of selecting an appropriate initial profile for the calculation is not always backed by as satisfactory a theoretical framework as in the example above. This depends on how well one knows the flow up to the starting station; in any case, the flow up to the starting station of this calculation must be known or assumed, which becomes translated into a set of dimensionless profiles involving *a single* scale parameter. Baum and Denison (*163*) study the development of a symmetrical free wake, a model of the flow downstream of the closure point of the near wake of a blunt-base body. In that case the true profile at the initial station should be derived from a solution of the problem of the shear layer and recompression process, which is not available. The authors select, instead, purely arbitrarily initial profiles (Stewartson's similar wake stagnation point profile normalized across the viscous layer). The normalization factors are chosen so that all parameters (thickness of the layer, its angle to the axis at the edge, etc.) are expressible in terms of a single eigenvalue, chosen to be, once again, the edge Mach number at the starting station.

For a prescribed free stream, the eigenvalue M_e determines one member of a family of solutions of the parabolic set of equations downstream of the starting station. A value of M_e/M_∞ corresponds to an inclination of the boundary-layer edge and thereby to a streamwise pressure gradient at the starting station. Suppose M_e is chosen to be sufficiently smaller than M_∞. The pressure gradient at the starting station is then sufficiently high to cause the boundary layer to thicken more rapidly than in the equilibrium case (the flow over a flat plate). This thickening displaces the free stream and further increases the local streamwise pressure gradient at points downstream of the starting station. It is found that the process becomes "catastrophic" at some point at which the pressure increases extremely rapidly and apparently without bounds. Conversely, if M_e is chosen sufficiently smaller than M_∞, the corresponding acceleration at the starting station thins the boundary layer, allows the inviscid flow above it to expand, and ultimately leads to a "catastrophic" drop in the pressure. The pressure is only an index: other properties of the boundary layer behave in a similar fashion.

The magnitude of the perturbation M_e/M_∞ determines the point at which the eigensolutions diverge. The two families of solutions (positive and negative "catastrophic" change in the pressure) bracket a "neutrally stable" integral curve. Baum verified (*164*) that if $\kappa = 0$ and a Blassius initial profile is assumed, the value of M_e/M_∞ characterizing this neutrally stable curve as well as all other properties of the flow check the theoretical result for a flat-plate boundary layer with weak interaction (*165*). The nature of this calculation is illustrated schematically in Fig. 52: we are now discussing the upper "tree" of that schematic, corresponding to $\kappa = 0$ everywhere.

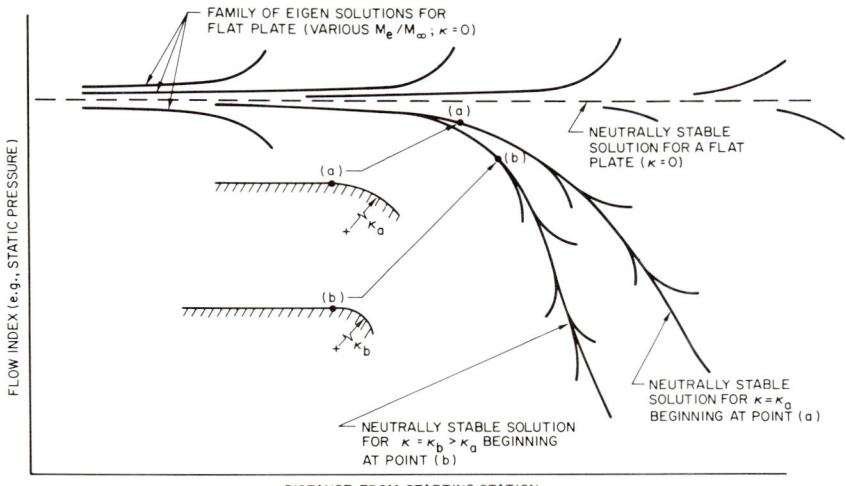

FIG. 52. Development of the iterative solution for the cross-stream interaction model.

If, beginning with the solution for attached flow over a flat plate, one of the family of integral curves exhibiting the fall in pressure is chosen $[(M_e/M_\infty)$ slightly larger than the value corresponding to the neutral solution with $\kappa = 0$], and at some point $x = x_a$ downstream of the initial station κ is made different than zero, that is, a rounded corner is introduced at that point, another neutrally stable solution can be found. It occurs for one unique value $\kappa = \kappa_a$. Such a solution is shown schematically in Fig. 52. The farther downstream one introduces the curvature, the larger must be the value of κ (the sharper the corner) to obtain a neutrally stable solution.

Figure 53 shows a specific calculation (*164*) [geometry and free stream chosen so as to correspond to an experiment (*59*)]. Members of both "families" of solutions peel off from the neutrally stable solution. If one calculation is continued along the curves which peel-off the neutrally stable solution with an increase in pressure, it is found that they exhibit a minimum and then level out to a virtually constant pressure. The corresponding velocity gradient at the wall, which is being obtained as part of the integration, is found to decrease rapidly at about

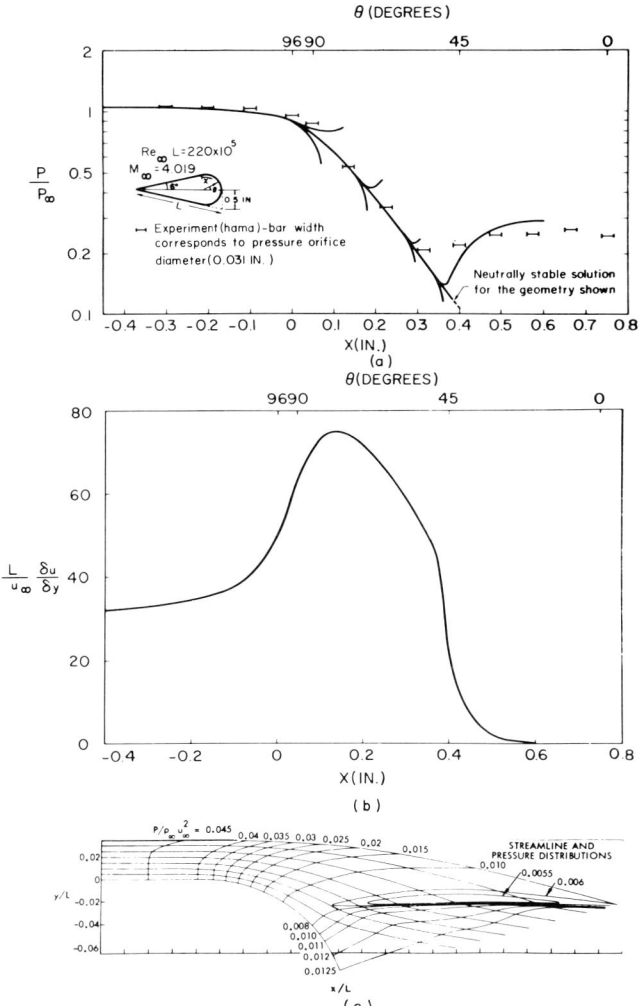

Fig. 53. Pressure, shear, and flow field map derived from Baum's solution for separation of a boat-tailed corner (*164*).

the peel-off point and then tend smoothly to zero in the nearly constant pressure section. This point is identifiable as the "separation" point in the usual sense in boundary-layer theory; it occurs in a region of slightly increasing pressure (an analytically necessary condition for separation off a smooth wall) and remarkably far downstream from the beginning of the boattail.

The iterative process thus leads to an integral curve which exhibits the following characteristic: the wall shear vanishes as the pressure approaches a value which checks measurements. This characteristic can now be used to select the desirable solution; only one integral curve possesses them. The example shown in Fig. 53 demonstrates the remarkable match between the calculated curve and pressure measurements all along the curved wall. The results are even more remarkable when one considers the map of the flow field. It exhibits clearly a region of high pressure gradients deeply imbedded in the viscous flow. The location of this "shock" (the root of the "lip" shock which the rotational characteristic analysis could not supply) agrees very well with Schlieren observations. All other available checks between the calculated flow and the detailed measurements of this control experiment are equally satisfactory.

The solution does not behave singularly at any point and there is no difficulty with continuing it downstream of the separation point, using the profile at the separation point for a "starting" profile in another calculation, with boundary conditions representing the free shear layer adjoining a recirculating flow. Work on such a streamwise patching of individual sections of the complete flow field about a separated region is being carried out by the developers of the present method. The portion up to separation (outlined above) and the section downstream of the stagnation point in the recompression region (*163*) present no difficulty. The central section, i.e., the free shear layer, demands an inner boundary condition derived from matching the flow in the shear layer with a solution of the Navier–Stokes equations in the recirculating inner flow; the difficulties and the limitations of such a solution were amply discussed elsewhere in this text. This constitutes the most formidable and perhaps the last important hurdle before it becomes possible to solve the entire (laminar) flow over a separating–reattaching region with rigor and in full detail. In parallel with the efforts of Baum and his group, Weiss *et al* (*101*) are attempting it also, using their more approximate formulation (see Section III,A,3). See also Tyson *et al.* (*187*).

C. Moment Methods

The well-known moment approximation to the boundary-layer equations consists of integrating these equations formally across the viscous layer after multiplying them by powers of velocity (momentum equation) and enthalpy (energy equation). This procedure leads to a series of first-order, nonlinear differential equations in the streamwise coordinate for certain integral properties of the profile ("thicknesses," such as the displacement, momentum, energy thickness, etc.). Information regarding the shape of the profile is lost in the process and must be resupplied by an *ad hoc* assumption (on the shape of the profile itself, as in the Pohlausen method, or by postulating a relation among the integral properties of the profile, such as in the Twaites approach). Such a statement involves an arbitrary number of "free" profile parameters which are functions of the streamwise coordinate. This determines the number of "moment" equations that are needed to close the problem. The streamwise pressure gradient must be prescribed, as always, in the framework of the boundary-layer approximation: the interaction between the boundary layer and the inviscid flow is included by prescribing not the pressure gradient directly but the displacement–compression relation (86) and (88) at the edge of the viscous layer.

The solution of the moment equations describes the flow approximately in the sense that only certain integral properties are conserved (depending on the number and the choice of moment equations), and, realistically, only if the supplementary assumption on the boundary-layer profiles is "proper"; this cannot be determined but from comparison with experiment.

In the classical Karman–Pohlausen method which proved successful in treating the attached boundary layer, the assumed velocity profiles and hence the integral properties of the flow in the boundary layer are uniquely determined by the local streamwise pressure gradient. In this case only one moment equation (conservation of mean streamwise momentum) is needed to close the problem (adiabatic flow). The formulation was extended to flows with separation (and interaction) by various authors (*143, 148, 154, 166, 167*) using different secondary approximations, different choices of the profile shape, and different methods of integration, but retaining the same baisc feature of tying the free profile parameter to the streamwise pressure gradient. The results of these analyses are not entirely satisfactory inasmuch as the solutions do not consistently reproduce some of the most characteristic features of the flow; for instance, the occurrence of a pressure "plateau."

A new approach was pioneered by Crocco and Lees (*6*) who suggested that the profile of the velocity in a separated region is not determined uniquely by the local pressure gradient. The shape of the profile continues to be described by a chosen function (polynomial, Falkner–Skan–Stewardson family, etc.) which is characterized by one "free" parameter (for instance, the velocity gradient at the wall). However, this parameter, that is, the member of the family of admissible profiles, is determined by the solution itself and not be the pressure boundary condition.

Clearly, this introduces one additional unknown for which an additional relation must be supplied. Crocco and Lees (6) and subsequently Glick (*168*) and others [for instance, Ref. (*69*)] do this by prescribing a semiempirical mixing function which specifies the rate of mass entrainment from the external flow. Solutions based on this approach are capable of portraying the experimental observations in much finer detail; however, the semiempirical nature of the mixing function constitutes a weakness. Furthermore, if more than one "free" parameter were to be used in characterizing the velocity profile, more additional semiempirical relations would be necessary to close the problem, which limits the scope of the method.

These weaknesses of the Crocco–Lees theory were removed and a self-consistent, general formulation of the multimoment technique was developed through the work of Tani (*169*), Honda (*170*) (noninteracting, attached incompressible and supersonic, adiabatic boundary layer, respectively), Abbott *et al.* (*171*), and finally (the most successful application to interacting separated flows) Lees and Reeves (*172, 173*) and their associates (*174–176*). The additional equations required to close a problem in which the velocity (and enthalpy) profiles are prescribed in terms of a (now arbitrary) number of "free" parameters are obtained by adding additional moments to the set, rather than by supplying semiempirical relations from without this established analytical framework. In its simplest form, in which the description of the profiles involves only one parameter, both the momentum (zero-order moment) and the moment of momentum equations (first-order moment) are used, in addition to the relation which determines the local pressure gradient.

The success of this multimoment theory continues to depend on the choice of the profile shape. Abbott *et al.* (*171*) applied the basic method to a boundary-layer shock-wave interaction using a quartic polynomial description of the velocity profile. This proved unsatisfactory. Lees–Reeves and co-workers also experimented with a polynomial representation but ultimately selected, instead, a different family of reversed-flow profiles (these profiles will be discussed in more detail later) and showed that the results are consistently good when the method is applied to the whole range of separated-flow configurations. Confidence in the results of moment methods can only be developed by trials and comparisons with experimental data.

In what follows we shall outline the moment method of Lees and Reeves using the simplest (two-moment) formulation originally presented by these authors. An approximate solution, which yields closed-form results and avoids the necessity of elaborate numerical work, is included. In a further section, a different approach which is based on the method of integral relations is presented. The first method is elaborated in much more detail in literature; one of its most attractive features is that it provides a clearly defined, nonarbitrary closure condition.

1. Lees–Reeves Multimoment Theory

The simplest closed set of moment equations allowing for one free parameter in the prescription of the velocity and enthalpy profiles is formed by the momentum and moment of momentum (mechanical energy) equations:

$$M_e \frac{d\delta_i^{**2}}{dX} + 2[2 + H^{-1}]\,\delta_i^{**2} \frac{dM_e}{dX} = 2\,\frac{\nu_\infty}{a_\infty}\,\frac{H\delta_i^{*}}{U_e}\left(\frac{\partial U}{\partial Y}\right)_{Y=0}$$

$$M_e \frac{d\theta_i^{*2}}{dX} + 2\left\{3 + \frac{2}{\theta_i^{*}}\left[\int_0^{\delta_i} \frac{U}{U_e}\left(\frac{I}{I_e} - 1\right)dY\right]\right\}\theta_i^{*2}\frac{dM_e}{dX} \qquad (111)$$

$$= \frac{4\nu_\infty}{a_\infty}\frac{\theta_i^{*}}{U_e^{2}}\int_0^{\delta_i}\left(\frac{\partial U}{\partial Y}\right)^2 dY$$

The equations above are derived from the usual boundary-layer approximation; that is, $\kappa = 0$ and $dp/dy = 0$ in Eqs. (106). They are written in the transformed Stewartson plane:

$$dX = (p_e a_e / p_\infty a_\infty)\,dx, \qquad dY = (\rho a_e / \rho_\infty a_\infty)\,dy, \qquad U = (a_\infty/a_e)\,u \qquad (112)$$

where a denotes the speed of sound. The streamwise pressure gradient term is replaced by the Mach number gradient at the edge of the boundary layer using isentropic flow relations. The subscript e refers to the edge of the boundary layer. The subscript i indicates that the quantity is defined in the transformed plane. The nomenclature is that of Lees and Reeves (172), where

$$\delta_i^{**} = \int_0^{\delta_i} \frac{U}{U_e}\left(1 - \frac{U}{U_e}\right)dY; \qquad \theta_i^{*} = \int_0^{\delta_i}\frac{U}{U_e}\left(1 - \frac{U^2}{U_e^2}\right)dY$$

$$\delta_t^{*} = \int_0^{\delta_i}\left(1 - \frac{U}{U_e}\right)dY + \int_0^{\delta_i}\left(\frac{I}{I_e} - 1\right)dY; \qquad H = \frac{\delta_i^{**}}{\delta_t^{*}}$$

(113)

Higher-order approximations involving more moments can be generated without difficulty. For instance, Webb (174) uses a set consisting of the zeroth, first, and second moment of momentum and the zeroth and first moment of energy, allowing for five "free" parameters in the representation of the velocity and enthalpy profiles.

The integral functions appearing in (111) and (113) are determined by prescribing the shape of the velocity distribution. Lees and Reeves chose the profiles corresponding to the reversed-flow solutions along the lower branch of the self-similar Falkner–Skan family [tabulated by Cohen and Reshotko (*177*)]. These are not viewed, however, as solutions of the boundary-layer equations, but simply as convenient functions (instead of, for instance, polynomials) which have the qualitatively expected behavior. They "unhook" the profile, in their terminology, from the boundary conditions which it implies by noting that each pair of enthalpy and velocity profiles is determined completely by a single parameter selected as follows:

When the flow is attached:

$$\alpha = [\partial(U/U_e)/\partial(Y/\delta_i)]_{Y=0}, \quad 0 \leqslant \alpha < 1.58 \quad \text{(Blasius)} \tag{114}$$

When the flow is separated:

$$\alpha = (Y/\delta_i)_{U=0}, \quad 0 \leqslant \alpha \leqslant 1 \tag{115}$$

and disregarding completely the fact that each member of this family of flows, i.e., each characteristic parameter α, corresponds (as a solution of the boundary-layer equations) to a prescribed external pressure gradient and wall to free stream enthalpy ratio.[2]

All the integral functions appearing in the moment equation can be evaluated in terms of the parameter α. Alternately, all can be expressed in terms of one of them. Lees and Reeves use the form factor H as the "free" parameter. Equations (111) can be transformed then into a set containing three independent parameters: δ_i^*, H, and M_e. The set is completed by the relation prescribing the displacement-induced pressure gradient (or Mach number gradient) which is derived from mass conservation (86), and the isentropic pressure deflection law (88). These equations are transformed into the Stewartson plane. After rearrangement, this and the two moment equations can be solved algebraically for

---

[2] This formulation "ties" the velocity to the enthalpy profiles and is appropriate only for near-adiabatic flows. In flows with heat transfer, the enthalpy profiles can be expected to have an independent "memory." To treat these cases, the velocity and enthalpy profiles must be specified in terms of independent "free parameters" and the additional moment equations (integrals of the energy equation) are needed.

the derivatives of the independent variables and written in the following form:

$$\frac{\delta_t^*}{M_e}\frac{dM_e}{dX} = \frac{1}{\tilde{R}_e}\frac{N_{1(H,M_e,h)}}{D_{(H,M_e)}}$$

$$\delta_t^*\frac{dH}{dX} = \frac{1}{\tilde{R}_e}\frac{N_{2(H,M_e,h)}}{D_{(H,M_e)}} \qquad (116)$$

$$\frac{d\delta_t^*}{dX} = \frac{1}{\tilde{R}_e}\frac{N_{3(H,M_e,h)}}{D_{(H,M_e)}}$$

where

$$\tilde{R}_e = a_\infty M_e\, \delta_t^*/\nu_\infty$$

$$h = h_{(\tilde{R}_e, M_e, M_\infty)}$$

Complete expressions for the functions D, N, and h are given in Ref. (*172*).

2. *The Critical-Point Closure Condition*

Equations (116) represent the formal (algebraic) solution for the streamwise derivatives dM_e/dX, $d\delta_t^*/dX$, and dH/dX. The variables themselves and all derived flow parameters can then be calculated by forward integration from a prescribed initial solution. For instance,

$$M_{e_{(X)}} = M_{e_{(X=0)}} + \int_0^X \frac{dM_e}{dX} dX \qquad (117)$$

When this is done, for each initial condition

$$M_{e_{(X=0)}}, \quad H_{(X=0)}, \quad \delta_{t_{(X=0)}}^*$$

the integral curves fall into one of two "families" in the same way as the solutions to the full differential problem outlined in Section IV,B,2 (see Fig. 52). Some distance downstream of the starting station the flow properties "blow up" either positively or negatively. Operating on the initial data, the solution tends to a "neutral equilibrium" curve which proceeds ever farther downstream.

However, the behavior of the present formulation can be clearly associated with the existence of a singularity, which occurs when the determinant D of the set of Eqs. (116) goes through zero. The nature of this singularity is discussed by Crocco and Lees (*6*) and Webb (*174*).

Webb *et al.* (*174*) use a simpler version of the present formulation to study it in detail. They prove (in a special case) that the singularity is of the "saddle-point" type which allows one and only one integral curve to pass smoothly through the singular point. This condition is stated by

$$N_1 = N_2 = N_3 = 0 \quad \text{when} \quad D = 0 \tag{118}$$

The character of the singular point provides a powerful and clear "closure cirterion" and determines uniquely the entire separated flow. The fact that the singularity must be of such a nature as to admit an integral curve to pass though it is evident on physical grounds. The system of Eqs. (116) represent the conservation of certain moments (momentum, energy, etc.) in the viscous layer, and, if this were not possible, the formulation would have to be rejected entirely. That it allows only one curve to pass is less obvious. In fact, D has more than one zero but (in the examples studied analytically) only one is a saddle point. The others are nodal points which admit an infinite family of integral curves. (A complete analysis of the system (116) has not been made and the generality of this characteristic is only inferred at this time.)

An evocative physical interpretation of the saddle-point singularity (*172*) is found in the analogy to quasi-one-dimensional channel flows. It represents a "throat" joining a subcritical flow in the viscous layer (in analogy to subsonic) to a "supercritical" one (supersonic). In subcritical flow the pressure rise generated downstream propagates smoothly upstream (through the process of thickening of the viscous layer). A pressure signal in supercritical flow cannot propagate upstream except through a discontinuous "jump." The singular critical point appearing in this method is analogous to the throat of a convergent-divergent nozzle; some appropriate average Mach number in the viscous channel goes through unity at that point. Sample calculations (*174*) do show that, indeed, the real average Mach number at the "throat" is near unity [while it is subsonic in the neighborhood of the other zeros of the discriminant D of (116)]. Experiment [Ref. (*51*), for instance] also suggests such a "throat" a short distance downstream of the stagnation point of the wake, in the sense that disturbances from downstream do not influence noticeably the flow upstream of it. On the other hand, the analogy should not be overemphasized. This critical singularity is a property of an approximate set of equations and not necessarily of the real flow. The effect of the remaining moment equations not included in this set may well be one of "diffusing" the critical point; the more complete solution of the differential equations proper (Section IV,B) indicates that this is the case. That solution exhibits the same general behavior. In certain regions the upstream influence of perturbations is extremely short. The moment approximation condenses this sensitive region to a single singular point.

The Lees–Reeves critical closure condition brought out a certain amount of argument which is probably not justified. It is certainly entirely correct within its own framework, that of the moment approximation. It is the most elegant and self-consistent statement of uniqueness among all the approximate and semiempirical "closure conditions" suggested so far. It is well substantiated by experiment; in fact, the arbitraries in the choice of profiles inherent in the moment method and their uniqueness seems to be a more serious criticism of this technique than the critical-point closure criterion.

3. *Approximate Solution of the Lees-Reeves Problem*

Hankey and Cross (*179*) published an approximate solution of the Lees–Reeves problem involving a number of numerical approximations. The solution illustrates the method without requiring numerical calculations and seems to yield very satisfactory results for the ramp separation as well as for downstream-facing steps.

The authors note that the quantity

$$\theta_i^*/\delta_i^{**} = (\theta_i^*/\delta_t^*)\, H^{-1} \tag{119}$$

is very nearly constant for attached flow in the range of the parameter α ($0 < \alpha < 1.44$) and also, though less precisely, for separated flow when $\alpha < 0.3$. Using this fact one can eliminate $d\delta_i^{**2}/dX$ from the set of Eqs. (111) to obtain a single equation

$$\frac{a_\infty \delta_i^{**2}}{\nu_\infty} \frac{dM_e}{dX} = \frac{\dfrac{\theta_i^*}{U_e}\left[\dfrac{\partial U}{\partial Y}\right]_{Y=0} - \dfrac{2\delta_t^* H}{U_e^2}\left[\int_0^{\delta_i}\left(\dfrac{\partial U}{\partial Y}\right)^2 dY\right]}{\dfrac{\theta_i^*}{\delta_i^{**}}\left\{H^{-1} - 1 - \dfrac{2}{\theta_i^*}\left[\int_0^{\delta_i}\dfrac{U}{U_e}\left(\dfrac{I}{I_e} - 1\right)dY\right]\right\}} \tag{120}$$

Hankey and Cross now use an approximate form of the cross-stream interaction coupling:

$$\tan\theta = d\delta^*/dx \simeq (\overline{dv/dM})(M_\infty - M_e) \tag{121}$$

where the physical displacement thickness δ^* is related to the transformed form factor by

$$\delta^*/\delta^{**} = H^{-1} + [(\gamma - 1)/2]\, M_e^2(H^{-1} + 1) \tag{122}$$

Assuming that the variation of the form factor in the interaction region is much greater than that of δ^{**} or M_e,

$$d\delta^*/dx \gg d\delta^{**}/dx, \qquad d\delta^*/dx \gg \delta^{**} M_e \, dM_e/dx \qquad (123)$$

Differentiation yields

$$d\delta^*/dx \cong \overline{\delta^{**}}(1 + [(\gamma - 1)/2] \, \overline{M_e^2}) \, dH^{-1}/dx \qquad (124)$$

where overbars denote quantities averaged over the interaction zone. Equating (121) and (124) and using

$$\frac{dH^{-1}}{dx} = \frac{dH^{-1}}{dK} \frac{dK}{dx} \cong \frac{dH^{-1}}{dK} \frac{d^2 M_e}{dx^2} \overline{\left(\frac{a_\infty \delta_i^{**2}}{\nu_\infty} \frac{p_\infty a_\infty}{p_e a_e}\right)} \qquad (125)$$

one obtains a single relation

$$L^2 \, d^2 M_e/dx^2 - (M_e - M_\infty) = 0$$

where

$$L^2 = -\frac{dH^{-1}}{dK} \overline{\left(\frac{a_\infty \delta_i^{**2}}{\nu_\infty} \frac{p_\infty a_\infty}{p_e a_e}\right)} \overline{\delta^{**}} \sqrt{1 + \frac{\gamma - 1}{2} \overline{M_e}^2} \frac{dM}{dv} \qquad (126)$$

$$K = \left(\frac{a_\infty \delta_i^{**2}}{\nu_\infty} \frac{p_\infty a_\infty}{p_e a_e}\right) \frac{dM_e}{dx} = \frac{a_\infty^4 \, p_e \delta^{**2}}{a_e^3 \, p_\infty \nu_\infty}$$

The form factor is a unique function of K determined from the family of Stewartson profiles, as in the Lees–Reeves method. This is a double-valued function with a sharp maximum at the junction between the separated flow branch and the attached flow branch. It can be approximated by two straight lines such that in either attached or separated flows the gradient dH^{-1}/dK is negative or positive, respectively, and L is constant but of opposite sign in each regime. Corresponding to these cases, (126) has two explicit solutions which are

attached flow: $\quad M_e - M_\infty = A_1 e^{x/L} + A_2 e^{-x/L}; \qquad dH^{-1}/dK < 0 \qquad$ (127a)

separated flow: $\quad M_e - M_\infty = B_1 \cos(x/iL) + B_2 \sin(x/iL); \qquad dH^{-1}/dK > 0$
$\hfill$ (127b)

Equations (127) are a closed-form solution to the interaction problem containing four constants plus the as yet unspecified value of the streamwise coordinate x corresponding to the separation and the reattachment points. The authors construct a sufficient number of boundary and matching conditions (assuming that the value of M_e is continuous at the junction of the various regions involved) to determine everything explicitly in terms of free stream parameters and the average boundary-layer thickness ahead of the interaction. They obtain a surprisingly good check with experimental results.

4. *Method of Integral Relations*

The method of integral relations is a variant of the moment method developed by Dorodnitsyn (*178*) for incompressible boundary layers and Pavlovskii (*180*) for the compressible case. It was extended to the case of separated flow by Holt (*181*), and Nielsen and co-workers (*182, 183*).

The boundary-layer equations are transformed by the Stewartson transformation (112). Next the Dorodnitsyn transformation is applied:

$$\xi = \int_0^x \frac{U_e}{U_\infty} \frac{dX}{l}; \qquad \eta = \frac{U_e}{U_\infty l} \frac{(U_\infty l)^{1/2}}{\nu} Y$$
$$u = U/U_e; \qquad v = (V/U_e)(U_\infty l/\nu_\infty)^{1/2} \tag{128}$$

The variables are related to their value at the outer edge of the boundary layer (subscript e). The undisturbed free stream is denoted by u_∞ and the characteristic length l is to be determined later. Instead of the cross-stream velocity, a new variable is defined:

$$w = v + u\eta(1/U_e)\, dU_e/d\xi \tag{129}$$

This leads to a set of equations which are

$$u\frac{\partial u}{\partial \xi} + w\frac{\partial u}{\partial \eta} = \frac{1}{U_e}\frac{dU_e}{d\xi}(1 - u^2) + \frac{\partial^2 u}{\partial \eta^2}, \qquad \frac{\partial u}{\partial \xi} + \frac{\partial w}{\partial \eta} = 0 \tag{130}$$

The method now calls for the introduction of a weighting function $f_{(u)}$. Multiplying the first equation by f' and the second by f, adding, integrating across the boundary layer with respect to η, and changing the variable of integration from η to u one obtains

$$\frac{d}{d\xi}\int_0^1 \frac{u f_{(u)}}{z}\, du = \frac{1}{U_e}\frac{dU_e}{d\xi}\int_0^1 \frac{(1-u^2)f'_{(u)}}{z_1}\, du - z_0 f'_{(0)} - \int_0^1 z f''_{(u)}\, du$$
$$z = \partial u/\partial \eta, \qquad z_0 = z_{(\eta=0)} \tag{131}$$

One now represents z as a function of u containing an arbitrary number of parameters, functions of ξ. The representation must be such that z is of order $(1 - u)$ as u goes to 1. One assigns to $f_{(u)}$ successive forms $(1 - u)$, $(1 - u)^2$, $(1 - u)^n$, where n equals the number of free parameters used in describing z. Integrals indicated in (131) can then be performed in each case. There results a set of n ordinary differential equations from which the parameters can be determined.

The simplest prescription of z which is adequate for separated flow is *(181)*

$$z = b_0(u + \alpha)^{1/2} (1 - u) \tag{132}$$

The first factor is introduced here to allow z to go to zero as α goes to zero (separation). In the post separation region, (132) must be further modified to allow for the double valuedness of z when u is negative (reverse flow). The following form is adequate:

$$\begin{aligned} z &= b_0(u + \alpha)^{1/2} (1 - u), & z &> 0 \\ z &= -b_0(1 + \alpha)(u + \alpha)^{1/2}, & z &< 0 \end{aligned} \right\} \text{ separated flow} \tag{133}$$

which matches z and $\partial u/\partial z$ at $u = 0$, continuously. This is a two-parameter formulation such that one needs to generate two moment equations by taking $f_{(u)}$ equal to $(1 - u)$ and $(1 - u)^2$, successively. The problem is completed by prescribing the usual Prandtl–Meyer interaction between the displacement effect of the viscous layer and the pressure gradient in the external stream; see Section IV,A,1.

The Reynolds and Mach numbers in the undisturbed free stream are assigned at a certain initial station ξ_0 where the parameters α and b_0 are given values corresponding to a prescribed upstream flow (for example, the Blasius boundary-layer profile for which $\alpha = 0.18704$, $1/b_0 = 1.2401 \xi^{1/2}$). In addition, one free eigenparameter which is the ratio (u_e/u_∞) at the starting station remains and must be prescribed. The forward-marching integration can then be started. For each eigenvalue (u_e/u_∞) selected "indices" such as the parameter α vary along the integration path; a decrease of α to zero signals the point at which the boundary layer separates. The solution at that point is used as initial data to continue the procedure downstream into the zone of detatched flow using now the "separated" set of equations for z (133).

The strength of the perturbation (the eigenvalue u_e/u_∞) determines the point of separation relative to the initial station as well as the subsequent development of the solution downstream of that point. A "distinguishable" characteristic is needed to select the appropriate integral curve.

In their analysis of separation in compressive corners Nielsen *et al.* *(183)* calculate the local angle of the dividing streamline relative to the undisturbed flow and select the point at which it equals the ramp angle (point of "tangency" with the reattachment surface) to locate reattachment. To each eigenvalue u_e/u_∞ correspond a point of separation and a point of reattachment (the "corner" shifts in relation to the initial station). The pressure at the point of reattachment is different in each case. One can make an assumption that it equals the pressure in the free stream after turning through an oblique shock at the corner. This sort of closure condition is approximate and must be justified empirically. Nielson et al. (183) demonstrate that the interesting properties of the flow (pressure plateau, extent of separation) do not change much with this "shift" in the location of the separation region. In a recent paper (181b) Holt and Meng show that the integral-relation system of equations involves a saddle point singularity analogous to that occurring in the Lees–Reeves method and study it in connection with the free wake problem. The nature of the solution and the "closure" condition is thus the same for both methods. However, the method of integral relation is still less thoroughly documented by comparison with experimental data than the multimoment scheme of Lees.

The results discussed in (183) were obtained, incidentally, using a four-parameter formulation instead of the two-parameter scheme outlined in the preceding. The simplest form of an energy equation was used. The temperature in the detached region below the dividing streamline was taken to be constant and equal to the wall temperature. In the attached boundary layer and in the shear layer above the DSL, the Crocco integral was assumed to hold. These are strong assumptions affecting the quantitative validity of the results for nonadiabatic flows. Nonetheless, the trends and the discovery of the effect of heat transfer on "incipient separation" are interesting: this is mentioned in Section IV,A,4. The results of this analysis indicate a power law dependence of the length of the free interaction region on the wall temperature and a relative insensitivity of the pressure level on that parameter in accord with the form of the "similarity" parameters of Curle (see Section IV,A,1).

D. Concluding Remarks

The theories outlined in this section assign a dominant role to crossstream interaction between the viscous layer and the inviscid region. The streamwise (upstream) coupling is neglected and replaced by an iterative search for a solution which describes the evolution of a prescribed initial flow satisfying certain distinguishable "closure" criteria. The success of the technique lies in that the cross-stream interaction process is unstable and exhibits regions of "catastrophic" change where indices of "closure" can be identified with good accuracy.

The attached boundary layers upstream of the separation point and downstream of reattachment on the one hand and the shear layer on the other hand can both be described by the boundary-layer approximation to the equations of motion but with different boundary conditions. The solution for the detached shear layer requires knowledge of the circulating flow in the cavity proper which in turn requires treament by some form of the full Navier–Stokes equations. The moment methods absorb this inner flow into the concept of a "viscous flow channel," making some proper (but *ad hoc*) assumption on the overall shape of the velocity profile which includes the recirculating region. This is an advantage in a sense, especially for dealing with certain types of cavities at high Reynolds numbers in which the complexity of the internal flow is such that a more precise description of the motion would necessarily require other *ad hoc* assumptions, if one can be given at all. The more elegant treatment of the differential equations coupled with a simultaneous solution of the recirculating flow seems possible only for laminar, low Reynolds number flows in the cavity (geometries where the cavity is steady, such as base wakes). When possible, the differential formulation has the advantage of supplying a *complete* solution which moment methods cannot provide.

The moment methods and also the differential equations yield individual solutions for streamwise sections of the flow about an immersed body: starting, say, at the stagnation point, the integration proceeds to the separation point; from the separation point to reattachment; from reattachment to the downstream infinity. The individual sections of the solution are joined at the "critical" points: separation and recompression. Such a treatment of a wake behind a cylinder is given by Grange *et al.* (*175*) using the Lees–Reeves moment method. The uniqueness of the solution is determined by passage through the critical "wake throat" and locates both the separation point (on the smooth cylindrical afterbody) and the recompression point. The results are in remarkably good agreement with experimental data. The same streamwise patching technique in principle can be applied to the differential formulation, resulting in a unique solution with a prescribed uniform flow at infinity downstream. A complete test case has not yet been published and therefore the results cannot yet be compared to experiment; however, the results up to separation (see Fig. 53) are extremely encouraging.

In a recent paper (*38*), Alber and Lees extended the Lees–Reeves moment method to turbulent flow. The extension is conceptually simple.

The boundary-layer equations with an "eddy viscosity" of the form

$$\epsilon = (\rho_{ref}/\rho_e)^2 \, K_{(\delta^{**})} \, U_e \delta^{**} \tag{134}$$

are transformed to the Stewartson plane and integrated across the boundary layer to yield, as in the laminar case, a (closed) set of moment equations: continuity (prescribing the interaction), momentum, and moment of momentum. The variables are the same as in laminar flow: the edge Mach number, a single profile parameter, and the transformed displacement thickness. Key additional assumptions relate to the eddy viscosity given by (134). The authors set the (incompressible) eddy viscosity proportional to the momentum thickness δ^{**}. They further assume that the ratio of the compressible to incompressible eddy viscosity is like the inverse square of the density, with the reference density ρ_{ref} identified with that at the wake centerline:

$$\rho_{ref}/\rho_e = \{1 + [(\gamma - 1)/2] \, M_e^2 1 - u_{cl}^2/u_e^2\}^{-1}$$

This is the only form of the eddy viscosity which admits a Stewartson-type transformation (the turbulent transformed streamwise coordinate differs slightly from the laminar one, $dX = a_e \rho_e / a_\infty \rho_\infty \, dx$). This treatment of the turbulent boundary-layer problem is well known.

The authors use the same Stewartson–Falkner–Skan family of velocity profiles to evaluate the integral properties of the viscous layer for turbulent and for laminar flow. The solution, including the passage through a critical "throat" somewhat downstream of the reattachment point, follows the same procedure as in laminar flow. The calculations result in fair agreement with experimental measurement; some results are included in Fig. 51.

Both the integration of the differential equations and the moment methods require extremely large computer capabilities in all respects: high numerical accuracy imposed by the essentially "unstable" nature of the integration, a large memory, and a staggering amount of computer time to complete the multiple iterations required to converge on the final solution. Not so many years ago the method would have been considered impractical; today opinions vary. The moment approach as used by Lees and Reeves is probably simpler and cheaper from this standpoint. Whether this justifies the approximation depends on the objective at hand. Note that in all the published work based on moment methods the boundary-layer assumption $dp/dy = 0$ is made; i.e., transverse gradients across the viscous layer are neglected. Baum's set of equations includes an inviscid form of the transverse momentum equation. A transverse momentum equation can also be added to the set of moments, as already suggested by Lees and Reeves in their initial paper, but it appears that the numerical work that would be necessary to solve such a more complete problem would be enormously increased and the moment method may no longer present much advantage. Transverse pressure gradients do not seem important at relatively low Mach numbers, typically below $M = 3$, but they are important in hypersonic flow.

Acknowledgment

Mr. L. Redekopp contributed substantially to this review, both by assisting in the preparation of the manuscript and curves and by critical discussions of the text. His valuable help is gratefully acknowledged.

The work was performed under the sponsorship of the National Science Foundation in connection with grant GK1244 for study of shear layers and separated flows.

Symbols

Cr	Crocco number [Eq. (24)]	β	boundary-layer pressure gradient parameter [Eq. (15)]
h	heat transfer coefficient	θ	flow turning angle; also denotes angular distance in polar coordinates
H	base step height (equal to one half the base diameter of a blunt body): also denotes the boundary-layer profile form factor		
		θ^*	energy thickness
I	total enthalpy	σ	jet spreading parameter [Eq. (22)]
L	body length upstream of separation	Λ	generalized upstream boundary-layer–base bleed parameter [Eq. (36)]
$\dot{m}$	mass injection (bleed) rate		
M	Mach number	δ	boundary-layer thickness
N	Nash recompression pressure rise parameter [Eq. (40)]	δ^*	displacement thickness
		δ^{**}	momentum thickness
p	pressure	λ	external flow pressure gradient parameter [Eq. (11)]
$\dot{q}$	heat transfer rate per unit area		
s	streamwise distance along the dividing streamline	ρ	mass density
		μ	dynamic viscosity
T	temperature		
u	along x or s direction	ν	kinematic viscosity: also the Prandtl–Meyer function
v	along y direction		
V	vector velocity	γ	ratio of specific heats (C_P/C_V)
x	physical coordinate (transverse)	τ	shear stress
Nu	Nusselt number	ϵ	eddy viscosity coefficient (turbulent flow)
Re	Reynolds number		
Pr	Prandtl number	η	similarity variable [Eqs. (15) and (22)]; also, dimensionless transverse coordinate
St	Stanton number		
ϕ	dividing streamline velocity ratio (u_{DSL}/u_e)		
		ξ	reduced shear-layer length

ψ	[Eq. (33)]; also, dimensionless streamwise coordinate stream function; also, recompression turning angle (see Fig. 2)	e	pertaining to the external flow region above the shear layer
		r	pertaining to conditions at reattachment
ω	vorticity vector $\nabla \times \mathbf{V}$	s	pertaining to conditions at separation
ζ	component of the vorticity in the x–y plane in two-dimensional flow ($k \cdot (\nabla \times \mathbf{V})$	∞	pertaining to the region upstream of separation
		0	denotes a constant reference quantity (usually denotes the value upstream of the body)

SUBSCRIPTS

b	pertaining to the base or cavity region	sep	denotes conditions immediately ahead of separation
d	pertaining to the region downstream of shear-layer reattachment	stag	denotes conditions obtained by isentropic compression to stagnation conditions
DSL	pertaining to conditions along the dividing streamline	w	wall
		aw	adiabatic wall

REFERENCES

1. H. B. Keller and H. Takami, Numerical studies of viscous flow about cylinders, *in* "Numerical Solutions of Nonlinear Differential Equations" (D. Greenspan, ed.), pp. 115–140. Wiley, New York, 1966.
2. J. Fromm, The Time Dependent Flow of an Incompressible Viscous Fluid, Univ. of California, Los Alamos Scientific Laboratory (1963).
3. G. A. Bird, Aerodynamic properties of some simple bodies in the hypersonic transition regime, *AIAA J.* **4**, 55–60 (1966).
4. G. Kirchhoff, Zur theorie freier flüssigkeitsstrahlen, *J. Angew. Math.* **70**, 289–298 (1869).
5. K. Oswatitch, Die ablösungsbedingung von grenzschichten, "Symposium on Boundary Layer Research" (H. Goertler, ed.), pp. 357–367. Springer-Verlag, Berlin, 1958.
6. L. Crocco and L. Lees, A mixing theory for the interaction between dissipative flows and nearly isentropic streams, *J. Aeron. Sci.* **19**, 649–676 (1952).
7. D. R. Chapman, D. M. Kuehn, and H. K. Larson, Investigation of Separated Flows in Supersonic and Subsonic Streams with Emphasis on the Effect of Transition, NACA TN 3869 (March 1957).
8. H. H. Korst, A theory of base pressure in transonic and supersonic flow, *J. Appl. Mech.* **23**, 593–600 (1956).
9. G. K. Batchelor, A proposal concerning laminar wakes behind bluff bodies at large reynolds numbers, *J. Fluid Mech.* **1**, 388–398 (1956).
10. A. Roshko, A review of concepts in separated flow, *Proc. Can. Congr. Appl. Mech.* Université Laval, Quebec, (May 1967).
11. P. S. Lykoudis, A review of hypersonic wake studies, *AIAA J.* **4**, 577–590 (1966).
12. S. Berger, A Review of Laminar Wakes, RAND RM-5756-ARPA (1968) (to be published).
13. Separated flows, *AGARD Conf. Proc.*, No. 4, **1** and **2** (1966).

14. Fluid physics of hypersonic wakes, *AGARD Conf. Proc.* No. 19, 1 and 2 (1967).
15. D. R. Chapman, Laminar Mixing of a Compressible Fluid, NACA Rept. 958 (1949).
16. H. H. Korst, Comments on the effect of boundary layer on sonic flow through an abrupt cross-sectional area change, *J. Aeron. Sci.* **21**, 568 (1954).
17. H. H. Korst, R. H. Page, and M. E. Childs, A Theory for Base Pressures in Transonic and Supersonic Flow, Univ. of Illinois (Engr. Exp. Station), ME-TN-392-2, OSR TN 55-89 (1955).
18. H. H. Korst, R. H. Page, and M. E. Childs, Compressible Two-Dimensional Jet Mixing at Constant Pressure, Univ. of Illinois (Engr. Exp. Station) ME-TN-392-1 (1955).
19. R. H. Page and H. H. Korst, Non-Isoenergetic turbulent compressible jet mixing with consideration of its influence on the base pressure problem, *Proc. Midwestern Conf. Fluid Mech.* 4th Purdue Univ., pp. 45–68. (September 1955).
20. A. Roshko and G. J. Thomke, Observations of turbulent reattachment behind an axisymmetric downstream—facing step in supersonic flow, *AIAA J.* **4**, 975–980 (1966).
21. A. F. Charwat, G. H. Burghart, and W. H. Nurick, Base wakes in accelerated supersonic free strams, *Proc. 1911 Heat Transfer Fluid Mech. Inst.* pp. 383–393, La Jolla, California (June 1967).
22. D. R. Chapman, A Theoretical Analysis of Heat Transfer in Regions of Separated Flow, NACA TN 3793 (October 1956).
23. K. Stewartson, Further solutions of the Falkner-Skan equation, *Proc. Cambridge Phil. Soc.* **50**, 454–465 (1954).
24. Kennedy, E. D., Wake-like solutions of the laminar boundary layer equations, *AIAA J.* **2**, 225–231 (1964).
25. T. Kubota, B. L. Reeves, and H. Buss, A family of similar solutions for axisymmetric incompressible wakes, *AIAA J.* **2**, 1493–1495 (1964).
26. R. S. Channapragada, Compressible jet spread parameter for mixing zone analysis, *AIAA J.* **1**, 2188–2190 (1963).
27. H. H. Tang, C. P. Gardiner, and J. W. Barnes. Jet Mixing Theory, Extensions, and Applications in Separated Flow Problems, Douglas Aircraft, Santa Monica, California Rept. DAC-59181 (Feb. 1967).
28. M. Sirieix and J. L. Solignac, Contribution a l'étude expérimentale de la couche de mélange turbulent isobare d'un écoulement supersonique, ONERA (France), TP No. 327 (1966) (Also published in Ref. 13).
29. F. N. Kirk, An Approximate Theory of Base Pressure in Two-Dimensional Flow at Supersonic Speeds, RAE TN Aero 2377 (AD-233-651) (Dec. 1959).
30. J. F. Nash, An Analysis of Two-Dimensional Base Flow Including the Effect of the Approaching Boundary Layer, Aero Res. Council (U.K.) R & M No. 3344 (1963).
31. M. R. Denison and E. Baum, Compressible free shear layer with finite initial thickness, *AIAA J.* **1**, 342–360 (1963).
32. E. Baum, H. H. King, and M. R. Denison, Recent studies of the laminar base flow region, *AIAA J.* **2**, 1527–1534 (1964).
33. E. Baum, Initial development of the laminar separated shear layer, *AIAA J.* **2**, 128–131 (1964).
34. T. Kubota and C. F. Dewey, Jr., Momentum integral methods for the laminar free shear layer, *AIAA J.* **2**, 625–629 (1964).
35. S. Goldstein, Concerning some solutions of the boundary layer equations in hydrodynamics. *Proc. Cambridge Phil. Soc.* **26**, Part 1, 1–30 (1930).
36. A. F. Charwat and A. V. Gomez, The Development of a Laminar Shear Layer with

Various Initial Profiles at Separation, Univ. of California, Los Angeles, Engr. Rept. No. 67-40 (August 1967).
37. A. F. Charwat and L. Schneider, The effect of the boundary layer profile at separation on the evolution of the wake, *AIAA J.* **5**, 1188-1190 (1967).
38. I. E. Alber and L. Lees, Integral theory for supersonic turbulent base flows, *AIAA J.* **6**, 1343-1351 (1968).
39. G. E. Gadd, D. W. Holder, and J. D. Regan, Base Presures in supersonic flow, ARC 17, 490. Aeronautical Research Council (March 1955).
40. M. Sirieix, Pression de culot et processus de mélange turbulent en écoulement supersonique plan, *La Recherche Aeron.* **13**, No. 78, 13-20 (1960).
41. R. C. Hastings, Turbulent Flow Past Two-Dimensional Bases in Supersonic Streams, TN Aero. 2931, Royal Aeronautical Establishment Dec. 1963.
42. D. Beastall and H. Eggink, Some Experiments on Breakaway in Supersonic Flow, Part II, TN Aero. 2061, Royal Aeronautical Establishment (Nov. 1951).
43. W. R. Winbrow, Effects of Base Bleed on the Base Pressure of Blunt Trailing Edge Airfoils at Supersonic Speeds, RM A54 A07, NACA (1954).
44. J. D. Morrow and E. Katz, Flight Investigation at Mach Numbers from 0.5 to 1.7 to Determine Drag and Base Pressures on a Blunt Trailing Edge Airfoil and Drag of Diamond and Circular Aro Airfoils at Zero Lift, TN 3548, NACA (1955).
45. K. L. Goin, Effects of Plan Form, Airfoil Section and Angle of Attack on the Pressurese Along the Base of Blunt Training Edge Wings of Mach Numbers of 1.41, 1.62 and 1.96, RM L52D21 (M. O. A. TIB 3324), NACA (1952).
46. E. J. Saltzman, Preliminary Base Pressures Obtained from the X-15 Airplane at Mach Numbers from 1.1 to 3.2, TN D-1056, NACA (Aug. 1961).
47. R. A. White, Turbulent Boundary Layer Separation from Smooth Convex Surfaces in Supersonic Two-Dimensional Flow, Ph.D., thesis, Univ. of Illinois (1963).
48. M. A. Badrinarayan, An experimental investigation of base flows at supersonic speeds, *J. Roy. Aeron. Soc.* **65**, 475-481 (1961).
49. R. Rebuffet, Effects de Supports sur l'Écoulement a l'Arrière d'un Corps, Rept. 302, AGARD (March 1959).
50. M. Sirieix, J. Delery, and J. Mirande, Recherches expérimentales fondamentales sur les écoulements séparés et applications, ONERA (France) TP 520 (1967) [Also presented at the VIII Polish Symposium on Fluid Dynamics, Torda, Poland (Sept. 1967)].
51. M. Sirieix, J. Mirande, and J. Delery, Expériences fondamentales sur le récollement turbulent d'un jet supersonique, ONERA (France) TD 326 (1966) (Also published In Ref. 13, Vol. 1, p. 353).
52. J. E. Lewis, T. Kubota, and L. Lees, Experimental investigation of supersonic laminar two-dimensional boundary Layer separation in a compression corner with and without cooling, *AIAA J.* **6**, 7-14 (1968).
53. A. F. Charwat and J. K. Yakura, An investigation of two-dimensional supersonic base pressure, *J. Aeron. Sci.* **25**, 122-128 (1958).
54. J. J. Ginoux, Experimental Evidence of Three-Dimensional Perturbations in the Reattachment Region of a Two-Dimensional Laminar Boundary Layer at M = 2.05, TCEA Tech. Note 1 (Nov. 1958).
55. J. J. Ginoux, On the Existence of Cross-flows in Separated Supersonic Streams, TCEA, Tech. Note 6 (Feb. 1962).
56. A. Roshko and G. J. Thomke, Effect of shoulder modification on turbulent supersonic base flow, *AIAA J.* **5**, 827-829 (1967).
57. D. W. Holder and G. E. Gadd, Interactions between shock waves and boundary

layers and its relation to base pressure in supersonic flow, *Symp. Boundary Layers Wakes, Boundary Layer Effect Aerodynamics, Nat. Phys. Lab. (Teddington, England), 1964*, pp. 1–65. Philosophical Library Inc., New York (1955).
58. V. Van Hise, Investigation of Variation in Base Pressure Over the Reynolds Number Range in which Wake Transition Occurs for Two-Dimensional Bodies at Mach Numbers from 1.9 to 2.92, NASA TN D-167 (1959).
59a. F. R. Hama, Estimation of the strength of lip shock, *AIAA J.* **4**, 166–167 (1966).
59b. F. R. Hama, Experimental investigations of wedge base pressure and lip shock, JPL Tech. Rept. 32-1033 (Dec. 1966).
59c. F. R. Hama, Experimental studies on the lip shock, *AIAA J.* **6**, 212–219 (1968).
60. L. L. Kavanau, Base pressure studies in rarefied supersonic flow, *J. Aeron. Sci.* **23**, 193–207 (1956).
61. S. M. Bogdonoff, A preliminary study of reynolds number effects on base pressure at M = 2.95, *J. Aeron. Sci.* **19**, 201 (1952).
62. A. Favre, J. Gaviglio, and H. Bornage, Observation sur la Transition dans un sillage transversal en écoulement supersonique, *Recherche Aerospatiale*, No. 119, pp. 22–37 (July-August 1967).
63. R. E. Slattery and W. G. Clay, Measurements of turbulence transition statistics and gross radial growth behind hypervelocity objects, *Phys. Fluids* **5**, (1962).
64. Demetriades, A., Some hot-wire anemometer measurements in a hypersonic wake, *Proc. Heat Transfer Fluid Mech. Inst. 1961*. Stanford Univ. Press, Stanford, California, 1962.
65. A. Demetriades, Hot-wire measurements in the hypersonic wakes of slender bodies, *AIAA J.* **2**, 245–250 (1964).
66. R. G. Batt, Experimental Investigation of Wakes Behind Two-Dimensional Slender Bodies at Mach Number Six, Ph.D. Thesis, California Inst. of Tech., Pasadena (June 1967).
67. C. F. Dewey, Jr., Near wake of a blunt body at hypersonic speeds, *AIAA J.* **3**, 1001–1010 (1965).
68. M. G. Sherberg and M. E. Smith, An experimental study of supersonic flow over a rearward-facing step, *AIAA J.* **5**, 51–56 (1967).
69. J. Rom, Theory for supersonic two-dimensional laminar base–type flows using Crocco-Lees mixing concepts, *J. Aeron. Sci.* **29**, 963 (1962).
70a. H. Viviand and S. A. Berger, The base flow and near wake problem at very low Reynolds numbers; Part 1: The Stokes approximation, *J. Fluid Mech.* **23**, 417–438 (1965).
70b. H. Viviand and S. A. Berger, The base flow and near wake problem at very low Reynolds numbers; Part 2: The Oseen approximation, *J. Fluid Mech.* **23**, 439–458 (1965).
71. H. Viviand and S. A. Berger, L'Écoulement de culot aux faibles nombres de Reynolds *AGARD Conf. Proc.*, No. 4, 1, 1–15 (1966).
72. P. Trepaud, R. Pery, J. P. Boehler, H. Viviand, and E. A. Brun, Étude de Sillages de cylindres et de diédres en écoulement de Gas Rarefié *AGARD Conf. Proc.*, No. 19, **1** and **2** (1967).
73. R. Cheng, S. A. Schaaf, and F. C. Hurlbut, The Measurement of Base Pressure on Wedges in Supersonic Low Density Flow, Inst. of Engr. Res., Univ. of California, Berkeley, Rept. AS-64-17 (Nov. 1964).
74. H. K. Larson, Heat transfer in separated flows, *J. Aeron. Sci.* **26**, 731–737 (1959).
75. R. E. Larson, A. R. Hanson, F. R. Krause, and W. K. Dahm, Heat Transfer Below

Reattaching Turbulent Flows, *AIAA Aerothermochem. of Turbulent Wakes Conf.*, *San Diego, California, 1965*. AIAA Paper 65–825, 1966.
76. C. J. Scott and E. R. G. Eckert, Heat and mass exchange in the supersonic base region *AGARD Conf. Proc.*, *4th* 1 and 2, 429 (1966).
77. A. Todisco and A. Pallone, Near wake flow field measurements, *AIAA J.* 3, 2075–2080 (1965).
78. A. Todisco and A. Pallone, Measurements in Laminar Near Wakes, Aerospace Sci. Meeting, New York, *1967*. AIAA Paper 67–30 (1968).
79. V. Zakkay and R. J. Cresci, An experimental investigation of the near wake of a slender cone at $M = 8$ and 12, *AIAA J.* 4, 41–46 (1966).
80. E. P. Muntz and E. J. Softley, A Study of Laminar Near Wakes, *AIAA J.* 4, 961–968 (1966).
81. A. Martellucci, A. Trucco, and A. Agnone, Measurements of the turbulent near wake of a cone at Mach 6, *AIAA J.* 4, 385–391 (1966).
82. E. M. Schmidt and R. J. Cresci, Near wake of a Slender Cone in Hypersonic Flow, *AGARD Conf. Proc.*, No. 19, 1 and 2 (1967).
83. A. F. Charwat, C. F. Dewey, J. N. Roos, and J. A. Hitz, An investigation of separated flows: Part II. Flow in the cavity and heat transfer, *J. Aeron. Sci.* 28, 514–527 (1961).
84. G. C. Lorenz and S. L. Strack, Heat Transfer at Reattachment of a Turbulent Boundary Layer at $M = 6$, Boeing Airplane Co., Document No. D2-23058 (March 1964).
85. S. L. Strack, Heat Transfer at Reattachment of a Turbulent Boundary Layer, Boeing Airplane Co., Document No. D2-22430 (April 1963).
86. A. Naysmith, Measurements of heat transfer in bubbles of separated flow in supersonic air streams, *Int. Develop. Heat Transfer* p. 378, Am. Inst. of Chem. Engrs. (1961).
87. A. Naysmith, Heat Transfer and Boundary Layer Measurements in a Region of Supersonic Flow Separation and Reattachment, RAE (Britian) TN Aero 2558 (May 1958).
88. J. Rom and A. Seginer, Laminar heat transfer to a Two–dimensional backward-facing step from the high enthalpy Supersonic flow in the shock tube, *AIAA J.* 2, 251–255 (1964).
89. D. E. Nestler, A. R. Saydah, and W. L. Auxer, Heat transfer to steps and cavities in hypersonic turbulent flow, *AIAA* paper 68–673, AIAA Fluid and Plasma Dynamics Conf., Los Angeles (June 1968).
90. R. A. Seban, A. Emery, and A. Levy, Heat transfer to separated and reattached subsonic turbulent flows obtained downstream of a surface step, *J. Aeron. Sci.* 26, 809–814 (1959).
91. R. A. Seban, Heat transfer to a turbulent separated flow of air downstream of a step in the surface plate, *J. Heat Transfer (Trans. ASME Ser. C.)* 86, 259–264 (1964).
92. E. S. Love, Base Pressure at Supersonic Speeds on Two-Dimensional Airfoils and on Bodies of Revolution With and Without Fins Having Turbulent Boundary Layers, NACA TN 3819 (1957).
93. S. Weinbaum, Rapid expansion of a supersonic boundary layer and its application to the near wake, *AIAA J.* 4, 217–226 (1966).
94. R. F. Weiss and S. Weinbaum, Hypersonic boundary layer Separation and the base-flow problem, *AIAA J.* 4, 1321–1330, (1966).
95. I. S. Donaldson, On the separation of a supersonic flow at a sharp corner, *AIAJ.* 5, 1086–1087 (1967).

96. M. Sirieix and J. Delery, Analyse Experimentale du Proche Sillage d'un Corps Elance Libre de Tout Support Lateral *AGARD Conf. Proc.*, No. 19, 1 and 2 (1967).
97. S. Weinbaum, On the singular points in the laminar two-dimensional near wake, General Electric Co. Space Sci. Lab., Tech. Memo No. 36 (Jan. 1967).
98. Weinbaum, S., Laminar Leading and Trailing Edge Flows and the Near-Wake Rear Stagnation Point [Published in Reference 14 (AGARD) 1966].
99. S. Weinbaum, Near-wake uniqueness and re-examination of the throat concept in laminar mixing theory, *Aerospace Sci. Meeting, 5th, New York, 1967*. AIAA Paper 67–65 (1968).
100. D. K. Ai, On the hypersonic laminar near-wake critical point of the crocco-lees mixing theory, *Aerospace Sci. Meeting, 5th, New York, 1967* (AIAA Paper 67–60) (1968).
101. R. F. Weiss, R. A. Greenberg, and P. P. Biondo, A New Theoretical Solution of the Laminar Hypersonic Near Wake; Part I, AVCO Everett Res. Lab., Res. Rept. 256 (Aug. 1966). Part II: Numerical Results and Comparison to Experiment, AVCO Everett Res. Lab., Res. Rept. 279 (Aug. 1967).
102. R. F. Weiss, A new theoretical solution of the laminar hypersonic near wake, *AIAA J.* **5**, 2142–2149 (1967).
103. G. K. Batchelor, On steady laminar flow with closed streamlines at large reynolds number, *J. Fluid Mech.* **1**, 177–190 (1956).
104. O. R. Burggraf, A model of steady separated flow in rectangular cavities at high Reynolds number, *Proc. Heat Transfer Fluid Mech. Inst.* (A. Charwat, ed.), p. 191–229. Stanford Univ. Press, 1963.
105. O. R. Burggraf, Analytical and numerical studies of the structure of steady separated flows, *J. Fluid Mech.* **24**, 113–151 (1966).
106. R. L. Haugen and A. M. Dhanak, Momentum Transfer in Turbulent Separated Flow Past a Rectangular Cavity, *ASME* Paper No. 66-APM-Q (1966).
107. F. Pan and A. Acrivos, Steady flows in rectangular cavities, *J. Fluid Mech.* **28**, 643–655 (1967).
108. R. F. Weiss, The base pressure of slender bodies in laminar hypersonic flow, *AIAA J.* **4**, 1557–1559 (1966).
109. R. F. Weiss, The Near Wake of a Wedge, AVCO-Everett Res. Lab. Rept. 197 (Dec. 1964).
110. J. J. Ginoux, The Existence of Three-Dimensional Perturbations in the Reattachment of a Two-Dimensional Supersonic Boundary Layer After Separation, AGARD Rept. No. 272 (April 1960).
111. J. J. Ginoux, Leading Edge Effect on Separated Supersonic Flows, *Proc. III ICAS Conf. Stockholm, 1962*, pp. 421–428.
112. J. J. Ginoux, Streamwise Vortices in Laminar Flow, AGARDograph 97 [Recent Develop. Boundary Layer Res. Part I, p. 395–422] (May 1965).
113. J. J. Ginoux, Investigation of flow separation over ramps, AEDC-TDR 65-273 (1965).
114. A. L. Kistler, Fluctuating wall pressure under a separated Supersonic flow, *J. Acoust. Soc. Am.* **35**, 543–550 (1954).
115. A. F. Charwat, C. F. Dewey, J. A. Hitz, and J. N. Roos, An Investigation of separated flows, I: The pressure field, *J. Aero. Sci.* **28**, 457–470 (1961).
116. D. J. Maull and L. F. East, Three-dimensional flow in cavities, *J. Fluid Mech.* **16**, 620–632 (1963).
117. M. G. Mozorov, acoustic emission from cavities in a supersonic air flow, *Izv. AN SSSR OTN Mekhn. i Mashinostroenie* **2**, 42–46 (1960).

118. A. Roshko, Some Measurements of Flow in a Rectangular Cutout, NACA TN 3488 (1955).
119. I. Tani, M. Iuchi, and H. Komoda, Experimental Investigation of Flow Separation Associated with a Step or a Groove, Aeronautical Research Inst., Univ. of Tokyo, Rept. 364 (April 1961).
120. G. E. Carichner, Accelerated Supersonic Flow over Notches, M. S. Thesis, Dept. of Engr., Univ. of California, Los Angeles (1968).
121. W. Wyborny, H. P. Kabelitz, and H. J. Schepers, Hypersonic Investigation of the local and average heat transfer in cavities and after steps of bodies of revolution. *AGARD Conf. Proc. 19th* **1** and **2** (1967).
122. W. Wyborny, H. P. Kabelitz, and H. J. Schepers, Wiederstands und Wärmeübergangsmessungen an Zylindrischen Körpern mit Hubräumen bei Hyperschallmachzahlen, *WGLR-DGRR-Jahrestagung Kartsruhe, 1966* Vortrag No. 75. Deutsche Versuchsanstalt för Luft und Raumfahrt, 1967.
123. H. E. Plumblee, T. S. Gibson, and T. W. Lassiter, A Theoretical and Experimental Investigation of the Acoustic Response of Cavities in an Aerodynamic Flow, Wright-Patterson AFB, WADD TR 61–75 (1962).
124. J. E. Rossiter, Wind Tunnel Experiments on the Flow over Rectangular Cavities at Subsonic and Transonic Speeds, RAE (Britian) Tech. Rept. 64037 (1964).
125. W. H. Dunham, Flow induced cavity resonance in viscous compressible and incompressible fluids, *Symp. Naval Hydrodynamics Ship Propulsion Hydroelasticity 4th Washington (1962.* ACR-73, Vol. 3, p. 939 (1962).
126. K. Krishnamurty, Acoustic Radiation from Two-Dimensional Rectangular Cutouts in Aerodynamic Surfaces, NACA TN 3487 (Aug. 1956).
127. B. M. Spee, Wind Tunnel Experiments on Unsteady Cavity Flow at High Subsonic Speeds, *AGARD Conf. Proc.*, No. 4, **2**, 941–974 (1966).
128. L. F. East, Aerodynamically Induced Resonance in Rectangular Cavities, Aero. Res. Council (U. K.) A.R.C. 27–131 (August 1965).
129. J. J. Ginoux and F. Thiry, Cone Cavity Flow at M = 5.3 with Injection ot Light, Medium and Havy Gases, Rept. VKI-TIV-35, Von Karman Inst. for Fluid Dynamics, Rhode-Saint-Genise, Belgium (November 1968).
130. A. F. Charwat and L. G. Redekopp, Effect of mass addition on the heat transfer to a cavity in supersonic flow, *Intern. J. Heat Mass Transfer* **11**, 787–804 (1968).
131. K. M. Nicoll, A study of laminar hypersonic cavity flows, *AIAA J.* **2**, 1535–1541 (1964).
132. J. Fox, Heat transfer and air flow in a transverse rectangular notch, *Intern. J. Heat Mass Transfer* **8**, 269–279 (1965).
133. F. J. Centolanzi, Heat Transfer to Blunt Conical Bodies Having Cavities to Promote Separation, NASA TN D-1975 (July 1963).
134. M. H. Bertram and M. M. Wiggs, Effects of Surface Distortion on the Heat Transfer to a Wing at Hypersonic Speeds, *Summer Meeting, Los Angeles 1962.* IAS Paper 62–127 (1962).
135. K. M. Nicoll and S. M. Bogdonoff, Experimental studies of a specific cavity configuration in laminar hypersonic flow, *Fluid Dynamics Trans.* **2** (1965).
136. K. M. Nicoll, Mass Injection in a Hypersonic Cavity Flow, ARL Rept. 65–90 (May 1965).
136a. Eckert, E. R. G., Engineering relations for heat transfer and friction in high velocity laminar and turbulent boundary layer flow over surfaces with constant pressure and temperature, *Trans. ASME* **78**, 1273–1283 (1956).
137. H. Thomann, Measurements of Heat Transfer and Recovery Temperature in

Regions of Separated Flow at a Mach Number of 1.8, FFA Rept. 82, Stokholm, Sweden (AD 216947) (1959).
138. T. Mix, Hypervelocity Kill Mechanisms Program, Progr. Rept. No. 5, NRL Rept. 1261, Sect. K (Dec. 1961).
139. E. E. Vanden Eykel and D. E. Nestler, The Effects of High Speed Particle Impact on ICBM Re-entry Vehicle Structures — A Two Year Technical Summary Report, Aerothermal Studies Section, GE MSVD TIS 61 SD 115 (Dec. 1961).
140. J. Erdos and A. Pallone, Shock-Boundary Layer Interaction and Flow Separations, AVCO Corp. Tech. Rept. RAD TR-61-23 (Aug. 1963).
141. P. Carriere, Recherches sur les Décollements dans les Tuyères Propulsires, ONERA (France) TP No. 506 (1967) [Presented at the Symposium on Coanda Effects, Bucarest, June 1967].
142. J. E. Lewis, Experimental Investigation of Supersonic Laminar, Two-Dimensional Boundary Layer Separation in a Compression Corner With and Without Cooling, Ph.D. Thesis, California Inst. of Tech. (1966).
143. N. Curle, The Effects of Heat Transfer on Laminar Boundary Layer Separation in Supersonic Flow, *Aero. Quart.* **12**, (1961).
144. J. J. Ginoux, Supersonic Flow over Flaps with Uniform Heat Transfer, Von Karman Inst. for Fluid Dynamics, Rhode St. Genese, Belgium, Rept. VKI TN 30 (Sept 1966) [See also *Proc. Symp. Advan. Problems Methods Fluid Mech. 8th, 1967*. Polish Academy of Sciences, Warsaw (1967).
145. G. E. Gadd, W. F. Cope, and J. L. Attridge, Heat Transfer and Skin-Friction Measurements at a Mach Number of 2.44 for a Turbulent Boundary Layer on a Flat Surface and in Regions of Separated Flow, ARC R & M No. 3148 (1960).
146. P. F. Holloway, J. R. Sterrett, and H. S. Creekmore, An Investigation of Heat Transfer within Regions of Separated Flow at a Mach Number of 6.0, NASA TN D-3074 (November 1965).
147. P. D. Burbank, R. A. Newlander, and I. K. Collins, Heat Transfer and Pressure Measurements on a Flat-Plate Surface and Heat Transfer Measurements on Attached Protuberances in a Supersonic Turbulent Boundary Layer at Mach Numbers of 2.65, 3.51, 4.44, NASA TN D-1372 (Dec. 1962).
148. E. E. Zukoski, Turbulent boundary-layer separation in front of a forward-facing step, *AIAA J.* **5**, 1746–1753 (1967).
149. J. R. Sterrett and J. C. Emery, Extension of Boundary Layer Separation Criteria to a Mach Number of 6.5 by Utilizing Flat Plates with Forward-Facing Steps, NASA TN D-618 (Dec. 1960).
150. G. Drougge, An Experimental Investigation of the Influence of Strong Adverse Pressure Gradients on the Turbulent Boundary Layer at Supersonic Speeds, Aeron. Res. Inst. of Sweden, Rept. 46 (1953).
151. G. E. Gadd, A theoretical investigation of laminar separation in supersonic flow, *J. Aeron. Sci.* **24**, 759–771 (1957).
152. J. D. Grey, Investigation of the Effect of Flow and Ramp Angle on the Upstream Influence of Laminar and Transitional Reattaching Flows From Mach 3 to 7, AEDC TR 66–190 (Jan. 1967).
153. S. A. Hertofilis, Pressure and Heat Transfer Measurements at Mach 13 and 19 of Flows Ahead of Ramps, Over Expansion Corners, and Past Fin-Plate Combinations, FDL TDR 66-144 (Sept. 1964).
154. G. E. Gadd, An experimental investigation of heat transfer effects on boundary layer separation in supersonic flow, *J. Fluid Mech.* **2**, 105 (1957).

155. S. M. Bogdonoff and I. E. Vas, Some experiments on hypersonic separated flows, *ARS J.* **32**, 1564–1572 (1962).
156. R. J. Hakkinen, I. Greber, L. Trilling, and S. S. Abarbanel, The Interaction of an Oblique Shock Wave with a Laminar Boundary Layer, NASA Memo 2-18-59W (1959).
157. K. N. C. Brag, G. E. Gadd, and A. Woodgen, Some Calculations by the Crocco-Lees and other Methods of Interactions Between Shock Waves and Laminar Boundary Layers, Including the Effects of Heat Transfer and Suction, Aero. Res. Council, Rept. 21, 834, FM 2937 (April 1960).
158. H. H. Pearcey, Shock-induced separation and its prevention by design and boundary layer control, *in* "Boundary Layers and Their Control" (G. V. Lachmann, ed.), Vol. 2, p. 1170–1344. Pergamon Press, Oxford, 1961.
159. E. W. E. Rogers, C. J. Berry, and B. M. Davis, An Experimental Investigation of the Interaction Between a Forward Facing Step and a Laminar Boundary Layer in Supersonic Low Density Flow, Nat. Phys. Lab. NPL Aero. No. 1139, Aeron. Res. Council (U.K.) (Jan. 1965).
160. R. F. Weiss and W. Nelson, Upstream influence of the base pressure, *AIAA J.* **6**, 466–471 (1968).
161. E. Baum, An Interaction Model of a Supersonic Laminar Boundary Layer Near a Sharp Backward Facing Step, TRW Systems Rept. BSD-TR-67-81 (Dec. 1966).
162. E. Baum, An interaction model of a supersonic laminar boundary layer on sharp and rounded backward facing steps, *AIAA J.* **6**, 440–447 (1968).
163. E. Baum and M. R. Denison, Interacting supersonic laminar wake calculations by a finite difference method, *AIAA J.* **5**, 1224–1230 (1967).
164. E. Baum, An Interaction Model of a Supersonic Laminar Boundary Layer on Sharp and Rounded Backward Facing Steps, TRW Systems Rept. BSD-TR-67-181 (July 1967).
165. W. D. Hayes and R. F. Probstein, "Hypersonic Flow Theory," Section 9.2. Academic Press, New York, 1959.
166. A. Martellucci and P. A. Libby, Heat Transfer Due to the Interaction Between a Swept Planar Shock Wave and a Laminar Boundary Layer, Proc. of the USAF-ASD Symposium on Aeroelasticity, Dayton, Ohio (Oct. 1960). Also, Tech. Rept. ASD-TR-61-727, Vol. II (Feb. 1962).
167. G. E. Gadd, Boundary Layer Separation in the Presence of Heat Transfer, *AGARD* Rept. 280 (April 1960).
168. H. S. Glick, Modified crocco-lees mixing theory for supersonic separated and reattaching flows, *J. Aeron. Sci.* **29**, 1238–1244 (1962).
169. I. Tani, On the approximate solution of the laminar boundary–layer equations, *J. Aeron. Sci.* **21**, 487–504 (1954).
170. M. Honda, A theoretical investigation of the interaction between shock waves and boundary layers, *J. Aeron. Sci.* **25**, 667–678 (1958).
171. D. E. Abbott, M. Holt, and J. N. Nielsen, Investigation of Hypersonic Flow Separation and Its Effects on Aerodynamic Control Characteristics, ASD-TDR-62-963 (Nov. 1962). (See also, Studies of Separated Laminar Boundary Layers at Hypersonic Speed with Some Low Reynolds Number Data *AIAA* Paper 63-172, AIAA Summer Meeting, Los Angeles (June 1963).
172. L. Lees and B. L. Reeves, Supersonic separated and reattaching laminar flows: I. General theory and application to adiabatic boundary-layer/shock-wave interactions, *AIAA J.* **2**, 1907–1920 (1964).

173. B. L. Reeves and L. Lees, Theory of laminar near wake of blunt bodies in hypersonic flow, *AIAA J.* **3**, 2061–2074 (1965).
174. W. H. Webb, R. J. Golik, F. W. Vogenitz, and L. Lees, A Multimoment Integral Theory for the Laminar Supersonic Near Wake, *Proc. Heat Transfer Fluid Mech. Inst., 1965*, p. 168–189 (June 1965).
175. J. M. Grange, J. M. Klineberg, and L. Lees, Laminar boundary–layer separation and near-wake flow for a smooth blunt body at supersonic and hypersonic speeds, *AIAA J.* **5**, 1089–1096 (1967).
176. R. J. Golik, W. H. Webb, and L. Lees, Further Results of Viscous Interaction Theory for the Laminar Supersonic Near Wake, TRW Systems Rept. BSD-TR-66-242 (March 1966).
177. C. B. Cohen and E. Reshotko, Similar Solutions for the Compressible Laminar Boundary Layer with Heat Transfer and Pressure Gradient, NACA Report 1293 (1956).
178. A. A. Dorodnitsyn, General method of integral relations and its application to boundary-layer theory, *Advan. Aeron. Sci.* **3** (1962).
179. W. L. Hankey and E. J. Cross, Approximate closed–form solutions for supersonic laminar separated flows, *AIAA J.* **5**, 651–654, (1967).
180. Iu. N. Pavlovskii, Numerical computation of the laminar boundary layer in a compressible gas, *Zh. Vychisl. Mate. i Mate Fiz.* **2**, no. 5, 884–901 (1962).
181a. M. Holt, Separation of laminar boundary-layer flow past a concave corner (Published in Ref. 13, AGARD CP No. 4, vol. 1, 69–87).
181b. M. Holt and J. C. S. Meng, The calculation of base flow and near wake properties by the method of integral relations, *19th Congress of the Int. Astronautical Fed., New York, Oct., 1968.* IAF Paper RE 65 (1968).
181c. D. R. Crawford and M. Holt, Method of integral relations as applied to the problem of laminar free mixing, *AIAA J.* **6**, 372–374, 1968.
181d. D. R. Crawford, Supersonic separated flow downstream of a backward facing step, Univ. of Calif., Berkeley, College of Engr. Rept. AS-67-9 (May, 1967).
182. J. N. Nielsen et al. Calculation of Laminar Separation with Free Interaction by the Method of Integral Relations, AFFDL-TR-65-107, Part I, May 1964; AFFDL-TR-65-107, Part II (Oct. 1965); AIAA Paper 65-30, AIAA 2nd Aerospace Sci. Meeting, New York (Jan. 1965).
183. J. N. Nielsen, L. L. Lynes, and F. K. Goodwin, Theory of Laminar Separated Flows on Flared Surfaces Including Supersonic Flow with Heating and Cooling (Published in Ref. 13, AGARD CP No. 4, Part I, 31–68).
184a. A. Roshko and G. J. Thomke, Correlations for incipient separation Pressure Douglas Rept. DAC — 59800, Missile & Space Systems Div., Douglas Aircraft Company, Santa Monica, California (May 1966).
184b. A. Roshko and G. J. Thomke, Supersonic, turbulent boundary layer interaction with a compression corner at very high Reynolds number, *Symp. Viscous Interaction Phenomena in Supersonic and Hypersonic Flow, Wright-Patterson AFB, Ohio, 1969,* Douglas Paper 10163 (May, 1969).
184c. G. J. Thomke and A. Roshko, Incipient separation of a turbulent boundary layer at high Reynolds number in two-dimensional supersonic flow over a compression corner, NASA-Ames Research Center, Report DAC 59819 (Jan., 1969).
185. J. J. Ginoux, Supersonic Separated Flows Over Wedges and Flares with Emphasis on a Method of Detecting Transition. Von Karman Inst. for Fluid Dynamics, Rhode-Saint-Genise, Belgium VKI TN 47 (August 1968).
186a. R. C. Bauer, An analysis of two-dimensional laminar and turbulent mixing, *AIAA J.* **4**, 392–395 (1966).

186b. R. C. Bauer, Another estimate of the similarity parameter for turbulent mixing, *AIAA J.* **6**, 925–927 (1968).
187. T. J. Tyson, J. E. Alber, and D. E. Coats, Finite difference solution for the laminar near-wake recompression behavior, *AIAA Fluid and Plasma Dynamics Conf., San Francisco, June, 1969*, AIAA Paper 69-713 (1969).

Optical Methods in Heat Transfer

W. HAUF

Technische Hochschule München, Institut A für Thermodynamik, Munich, Germany

U. GRIGULL

Technische Hochschule München, Institut A für Thermodynamik, Munich, Germany

    I. Introduction 134
   II. Principles of Geometrical Optics 135
       A. The Eikonal Equation 136
       B. Direction of Light Radiation and Wave Fronts 138
       C. The Differential Equation for a Light Ray in an Inhomogeneous Medium 140
       D. Huygens' Principle 146
 III. Boundary-Layer Optics 147
       A. Light Deflection in a One-Dimensional Refractive Index Field 147
       B. The Boundary Layer as a "Schlieren Lens" 150
       C. Optical Measurements of Boundary Layers 158
 IV. Theory of Shadowgraph and Schlieren Methods 161
       A. Wiener's Diffusion Investigation (1893) 162
       B. Iterative Calculation of an Arbitrary Refractive Index Field from the Light Deflection ϵ 166
       C. Diffraction by a Parallel Slit and at the Model Boundary, $\lambda/2$ Phase Plate 169
       D. Schlieren Methods without Image Formation (Wiener, 1893) 176
       E. Schlieren Methods with Image Formation (Toepler, 1864) 185
       F. Shadow Method (Dvořac, 1880) 189
  V. Theory of Interference Methods 191
       A. Methods of Two-Beam Interference 191
       B. The Mach–Zehnder Interferometer 202
       C. Two-Dimensional Phase Object in a Mach–Zehnder Interferometer 230
       D. Cylindrical and Irregular Phase Objects 267
       E. The Interferogram; Test Media 278
 VI. Application Examples and Their Evaluation 312

    A. Models in Gaseous Media. 312
    B. Models in Liquid Media 330
VII. Appendix 348
    A. Mathematical Formulation of the Conditions for a Basic Position of the Mach–Zehnder Interferometer (SectionV, B) 348
    B. Calculation of the Ray Path $\eta = f(\bar{z})$ of the Displacement Error $\Delta\eta$ and of the Phase Difference $S \cdot \lambda$ in Model Boundary Layers 353
    C. Calculation of the Ray Path $\eta = f[\bar{z}, (d\eta/d\bar{z})_0]$ and of the Path Difference $S \cdot \bar{\lambda}$, When the Slope of the Light Ray at the Wall ($\eta_0 = 0$) Is Not Equal to Zero at the Beginning of the Test Section 358

    Nomenclature 360
    References 362

I. Introduction

In this chapter, we are considering only those "optical methods" in which the temperature dependence of the refractive index is used to make the temperature field visible. This leads to a natural limitation of the topic by which, for example, pyrometric measurements, although doubtless belonging to the field of optical methods, are not included.

Compared with other measurement methods in the area of heat transfer, optical methods possess considerable advantages. First, the measurements do not disturb the temperature field since in most cases the energy absorbed by the medium is small compared to energy exchange by heat transfer. There are also practically no inertial errors in the optical methods so that rapidly changing processes can be accurately followed. This advantage arises from the possibility of recording the entire temperature field on a single photograph. The information usually obtained from point-by-point measurements is instead deduced from an evaluation of the photograph. These measurements are frequently of greater sensitivity and accuracy than those using other methods, e.g., calorimetric measurements or measurements of the temperature field using thermocouples.

There are also disadvantages in the optical methods. The media under consideration must be transparent to radiation. In order to obtain photographs suitable for accurate evaluation, the physical dimensions of the system must be relatively small. For media other than ambient air, the system must be enclosed, with two sides of the enclosure being glass plates of high optical quality. The optical methods basically yield a refractive index field which requires subsequent calculations for interpretation as a temperature field. Thus it is seen that, like all other methods of measurement, optical methods have a confined rather than universal area of application.

The optical methods considered are divided into two groups: the shadow and schlieren techniques, utilizing the deflection of light in the measurement media, and the interference methods based on differences in lengths of the optical paths. There are numerous references in the literature describing applications of these methods. Due to space limitations, only a selection of those which have been proved in practice is presented here. A complete treatment of methods already developed, or which are in principle possible, was neither intended for nor included within the scope of this chapter. For further information, the reader is referred to the literature ($1-7$).

In order to fully demonstrate how the results are evaluated, some selected examples will be presented. Important data for the evaluation of measurements have been presented in tables.

If, in referring to the literature, preference has been given to the German literature, the only reason is that this is more familiar to the authors. Wherever possible, internationally standard symbols (8) have been used.

There are two justifications for including a discussion of optical methods in these serial publications dedicated to advances in heat transfer. First, although the described methods have been known in principle for around a hundred years, they are being constantly developed and improved. In addition, the optical methods themselves are instrumental in the progress of knowledge of heat transfer since some problems are only tractable using optical methods.

II. Principles of Geometrical Optics

In this section, the laws governing the propagation of light through a medium with locally varying refractive index are considered; in this, preference has been given to the theory of geometrical optics. An optical inhomogeneity is designated as a "schliere," an expression which originated in glass technology. Because of the dependence of the refractive index on temperature, a thermal bounday layer e.g., is a schliere. The temperature distribution, and therefore the refractive index distribution for a laminar thermal boundary layer is governed by the known laws of boundary-layer physics. In a schliere consisting of a turbulent eddy in a rising column of flue gas, for example, this distribution is almost completely irregular. Both of these problems may be quantitatively investigated using optical methods. In the first case, naturally, more detailed information, such as the temperature distribution, is obtainable; in the second case only integrated values, e.g., of the enthalpy content of such an eddy, are available. Our treatment of thermal boundary layers will be more extensive than for the larger fields encountered in gas dynamics and ballistics.

A. THE EIKONAL EQUATION

Geometrical optics, which provides the basis for design of optical instruments, may be considered as the theory described by Maxwell's electrodynamic equations of the wave theory of light with the restriction that the wave length is vanishingly small ($\lambda \to 0$).

The physical model for describing light propagation is a three-dimensional vector field, the streamlines of which are the light rays. The following development was given by Sommerfeld (9, 10).

The three-dimensional wave equation written in cartesian coordinates, widely known as the Helmholtz equation, is the starting point:

$$\partial^2 u/\partial x^2 + \partial^2 u/\partial y^2 + \partial^2 u/\partial z^2 + \bar{k}^2 u = 0 \tag{1}$$

The symbols are defined as follows: u is the wave disturbance; $\bar{k} = (\epsilon\mu)^{1/2}\omega = 2\pi/\lambda$ wave number (this number depends on properties of the medium and light wave); ϵ is the dielectric constant, either a smoothly or irregularly changing function of position (e.g., as in a schliere or lens); μ is the permeability; $\omega = 2\pi\nu$ is the angular frequency; ν is the frequency of the light oscillation (strictly monochromatic); λ is the local wavelength; and λ_0, ϵ_0, and μ_0 are corresponding quantities in a vacuum.

The condition for geometrical optics, $\lambda \to 0$ or $\bar{k} \to \infty$, results in a degeneration of the wave equation. Using the following trial expression proposed by P. Debye, an approximate solution may be obtained:

$$u = A e^{i\bar{k}_0 E}; \qquad \bar{k}_0 = (\epsilon_0 \mu_0)^{1/2}\omega = 2\pi/\lambda_0 \tag{2}$$

where $A = A(x, y, z)$ is the amplitude factor; and $E = E(x, y, z)$ is the eikonal (the designation for the wave front in geometrical optics proposed by H. Bruns). $A(x, y, z)$ and $E(x, y, z)$ are functions which may change only moderately with position and remain finite as $\bar{k}_0$ approaches infinity. The wave disturbance u changes very rapidly with position as $\bar{k}_0 \to \infty$ (very small λ_0).

Substituting Eq. (2) into Eq. (1) results in the eikonal (geometrical wave front) differential equation. First, differentiating Eq. (2),

$$\frac{\partial u}{\partial x} = A i \bar{k}_0 \cdot e^{i\bar{k}_0 E} \cdot \frac{\partial E}{\partial x} + \frac{\partial A}{\partial x} e^{i\bar{k}_0 E} \cdot \frac{1}{A} \cdot A$$

$$\frac{\partial u}{\partial x} = i\bar{k}_0 \cdot u \cdot \frac{\partial E}{\partial x} + u \frac{\partial \ln A}{\partial x}$$

$$\frac{\partial^2 u}{\partial x^2} = i\bar{k}_0 \cdot u \frac{\partial^2 E}{\partial x^2} + i\bar{k}_0 \frac{\partial E}{\partial x} \cdot \frac{\partial u}{\partial x} + \frac{\partial^2 \ln A}{\partial x^2} u + \frac{\partial \ln A}{\partial x} \cdot \frac{\partial u}{\partial x}$$

$$\frac{\partial^2 u}{\partial x^2} = -\bar{k}_0^2 u \left(\frac{\partial E}{\partial x}\right)^2 + 2i\bar{k}_0 u \left(\frac{1}{2}\frac{\partial^2 E}{\partial x^2} + \frac{\partial \ln A}{\partial x}\frac{\partial E}{\partial x}\right) + u\left[\left(\frac{\partial \ln A}{\partial x}\right)^2 + \frac{\partial^2 \ln A}{\partial x^2}\right]$$

The partial derivatives with respect to y and z are obtained in an analogous way. Substituting these into Eq. (1),

$$\frac{\partial^2 u}{\partial x^2} + \frac{\partial^2 u}{\partial y^2} + \frac{\partial^2 u}{\partial z^2} + \bar{k}^2 u = 0$$

$$= -\left[\left(\frac{\partial E}{\partial x}\right)^2 + \left(\frac{\partial E}{\partial y}\right)^2 + \left(\frac{\partial E}{\partial z}\right)^2 - \left(\frac{\bar{k}}{\bar{k}_0}\right)^2\right]\bar{k}_0^2 u$$

$$+ 2i\bar{k}_0 u\left[\frac{1}{2}\left(\frac{\partial^2 E}{\partial x^2} + \frac{\partial^2 E}{\partial y^2} + \frac{\partial^2 E}{\partial z^2}\right)\right.$$

$$\left. + \frac{\partial \ln A}{\partial x} \cdot \frac{\partial E}{\partial x} + \frac{\partial \ln A}{\partial y} \cdot \frac{\partial E}{\partial y} + \frac{\partial \ln A}{\partial z} \cdot \frac{\partial E}{\partial z}\right]$$

$$+ u\left[\left(\frac{\partial \ln A}{\partial x}\right)^2 + \left(\frac{\partial \ln A}{\partial y}\right)^2 + \left(\frac{\partial \ln A}{\partial z}\right)^2\right.$$

$$\left. + \frac{\partial^2 \ln A}{\partial x^2} + \frac{\partial^2 \ln A}{\partial y^2} + \frac{\partial^2 \ln A}{\partial z^2}\right]$$

Using the vector operator

$$D = (\partial/\partial x)^2 + (\partial/\partial y)^2 + (\partial/\partial z)^2$$

this may be rewritten

$$\Delta u + \bar{k}^2 u = 0$$

$$= -\bar{k}_0^2 u[D(E) - \bar{k}^2/\bar{k}_0^2]$$

$$+ 2i\bar{k}_0 u[\tfrac{1}{2}\Delta E + \text{grad} \ln A \cdot \text{grad } E]$$

$$+ u[D(\ln A) + \Delta \ln A]$$

The wave equation, Eq. (1), is approximately satisfied if E and A are determined from the differential equations obtained by setting the expressions in brackets equal to zero:

$$D(E) = (\partial E/\partial x)^2 + (\partial E/\partial y)^2 + (\partial E/\partial z)^2 = (\bar{k}/\bar{k}_0)^2 = n^2 \qquad (3)$$

$$\text{grad} \ln A \cdot \text{grad } E = -\tfrac{1}{2}\Delta E \qquad (4)$$

Equation (3) is the eikonal differential equation defining a family of nonintersecting surfaces $E = $ const. According to Eq. (2), these surfaces have the same phase and are therefore wave fronts.

The known relationships between wave number and wavelength, $\bar{k}/\bar{k}_0 = \lambda_0/\lambda$, and wavelength and light velocity, $\lambda_0/\lambda = c_0/c$, lead to $\bar{k}/\bar{k}_0 = \lambda_0/\lambda = c_0/c = n(x, y, z)$; the quantities with subscript zero refer to values in a vacuum while those without are local values in the medium. The ratio $\bar{k}/\bar{k}_0$ is, therefore, the usual refractive index n, defined as the ratio of the light velocity in a vacuum to the local light velocity.

By using the integrated eikonal equation, Eq. (3), we see that Eq. (4) gives the gradient of ln A in the direction of the light ray; it says nothing about the gradients of ln A perpendicular to the gradient of E.

It is therefore possible for inconsistencies to arise, e.g., on the boundary of an opaque screen. One could, for example, calculate the distribution of intensity in the projection of a schliere, where, however, the boundaries, with diffraction effects and high values of grad ln A (focal line), must be excluded.

B. Direction of Light Radiation and Wave Fronts

The eikonal equation, Eq. (3), for $E = $ const. describes a family of nonintersecting surfaces in space (wave fronts) lying everywhere normal to the directions of the light rays. For the case of a locally changing refractive index $n(x, y, z)$ the wave fronts are not planar.

Figure 1 shows the deformation of an originally ($z \leqslant 0$) plane wave with circular boundaries after passage through a refractive index field. This field was produced by the instantaneous temperature field encountered during a nonstationary heat transfer process. A more detailed consideration of this example is presented in Section VI, B.

The wave front is a continuous surface in space graphically portrayed by its contour lines. Its topographical projection in the x, y plane is shown in the interferogram in the right half of the figure. The direction of propagation of the original plane wave front was along the z axis of the coordinate system and has been designated as "direction of light beam" in Fig. 1.

The depicted wave front is greatly amplified in the z direction (10,000 times); the distances between the contours (the closed interference lines) correspond to the wavelength of the light used, in this case 0.546×10^{-6} m. A more complete discussion of interferograms may be found in Section V.

The instantaneous picture of a wave surface at a fixed position of the

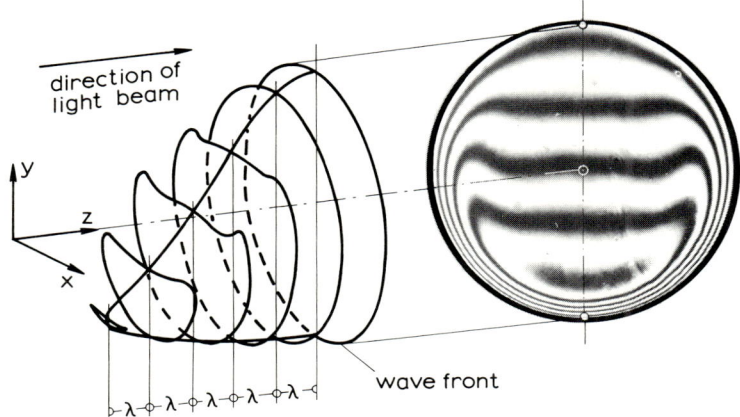

FIG. 1. Portrayal of a wave front as a surface in space. (The interferogram at the right comes from the application given in Section VI, B.)

light path in space is shown in Fig. 1. If the progressing wave front were to continue, especially traveling through other regions whose refractive index varied, a further deformation would result. The refractive index, $n = \bar{k}/\bar{k}_0 = \lambda_0/\lambda$, has been assumed in Eq. (2) to be a moderately changing function of position and time compared to the local and time-dependent events occurring in the wave disturbance. This moderately changeable function $n(x, y, z, t)$ describes the process in the schliere (test section) to be investigated.

The wave front in Fig. 1 has, however, already passed through the schliere and is continuing its motion through surrounding homogeneous air ($n = $ const) in a direction normal to its surface. The direction of propagation of a surface element is always in the direction normal to the surface. This is, at the same time, the direction of the local light ray which, in a homogeneous medium ($n = $ const), does not change. If the vectors, each of which shows the direction of the light ray for an element of the wave surface, are summed at various times as the wave progresses, they define a curve depicting the path of the entire light ray. This path is a straight line for a homogeneous medium and a curve in space for an inhomogeneous medium.

For the plane of symmetry of the wave surface shown in Fig. 1, some radiation directions and angles ϵ_z from the z axis, along which the original plane wave propagated, are shown in Fig. 2. Because of symmetry, $\epsilon_x = \epsilon_y = 0$.

The central cross section shown in the diagram is amplified 2500 times

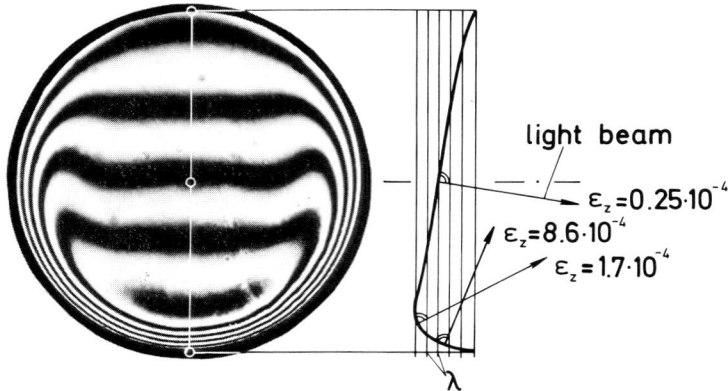

FIG. 2. Portrayal of the light beam direction as the surface normal of the wave front.

in the z direction. Deflections of the light beam by the temperature fields encountered in heat transfer processes are generally of the order $\epsilon < 10^{-2}$.

C. THE DIFFERENTIAL EQUATION FOR A LIGHT RAY IN AN INHOMOGENEOUS MEDIUM

1. *The Vector Field Describing the Ray Directions*

From the scalar field of the wave surfaces, $E = $ const, the corresponding vector field depicting the normal directions, and therefore the ray directions, may be obtained by taking gradients. The normal unity vector $\mathbf{s}(s_x, s_y, s_z)$ of the eikonal lies tangent to the curve indicating the path of the light ray and is obtained by dividing the grad E vector by its magnitude. Using the eikonal differential equation, Eq. (3),

$$D(E) = \left[\left(\frac{\partial E}{\partial x}\right)^2 + \left(\frac{\partial E}{\partial y}\right)^2 + \left(\frac{\partial E}{\partial z}\right)^2\right] = \text{grad } E \cdot \text{grad } E = n^2 \quad (5)$$

$$\mathbf{s} = \text{grad } E/[D(E)]^{1/2} = (1/n)\,\text{grad } E$$

Using a well-known relation from vector analysis, Eq. (5), written in the form $n\mathbf{s} = \text{grad } E$, leads to the relation

$$\text{rot}(n\mathbf{s}) = \text{rot} \cdot \text{grad } E = 0 \quad (6)$$

From this, the propagation of light, as described by geometrical optics, may be recognized as analogous to a spatial potential current or to thermal conduction in a solid body. The light rays correspond to the streamlines.

2. *The Spatial Path of the Ray*

In order to set up the differential equation describing the path of the light ray in space (Fig. 3), two adjacent wave surfaces E and E' are

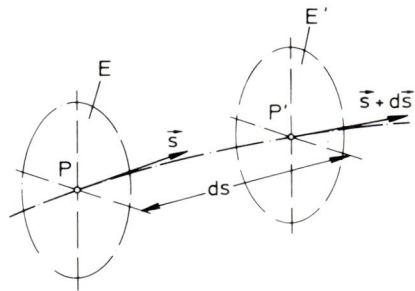

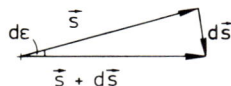

$|\mathbf{s}| = 1 \; ; \; d\mathbf{s} = d\epsilon$

FIG. 3. Spatial path of light ray. (Letters with arrows correspond to boldface in text.)

considered; P and P' are the respective points of intersection with such a path; ds is the differential displacement along the path. The two unit vectors lying tangent to the spatial path (and showing the directions of the ray) are **s** and **s** + *d***s**.

The vectors **s** and **s** + *d***s** define a plane (osculation plane) containing the line element ds and the circle of curvature with radius R. The difference of the "ray vectors" gives the change in direction $d\epsilon$ of the light ray (Fig. 3):

$$\mathbf{s} + d\mathbf{s} - \mathbf{s} = d\mathbf{s}; \qquad d\epsilon = |\,d\mathbf{s}\,|/|\,\mathbf{s}\,|$$

Since $|\,\mathbf{s}\,| = 1$, $|\,d\mathbf{s}\,|$ is equal to $d\epsilon$. The relation of the radius of curvature

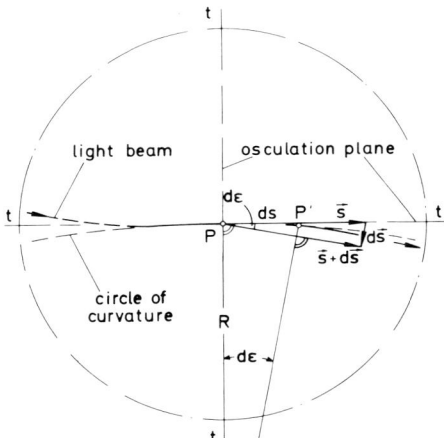

FIG. 4. Representation of the spatial path of the light ray in its osculation plane. (Letters with arrows correspond to boldface in text.)

R to the line element ds and to the change in direction $d\mathbf{s}$ may be obtained from Fig. 4, which has been drawn in the osculation plane:

$$d\epsilon = |\, d\mathbf{s}\,| = ds/R \tag{7}$$

The curvature $\mathbf{K}$ is defined as the change in direction $d\mathbf{s}$ per unit line element ds. Using Eq. (7),

$$\mathbf{K} = d\mathbf{s}/ds, \qquad |\mathbf{K}| = |\,d\mathbf{s}\,|/ds = d\epsilon/ds = 1/R \tag{8}$$

The direction of the vector $\mathbf{K}$ corresponds to the change in direction of the ray $d\mathbf{s}$; the radius of curvature R is inversely proportional to the magnitude $|\mathbf{K}|$.

Representing $\mathbf{K}$ by its components, using Eq. (8),

$$\mathbf{K} = \frac{d\mathbf{s}}{ds} = \frac{\partial \mathbf{s}}{\partial x}\frac{dx}{ds} + \frac{\partial \mathbf{s}}{\partial y}\frac{dy}{ds} + \frac{\partial \mathbf{s}}{\partial z}\frac{dz}{ds} \tag{8a}$$

where the factors dx/ds, dy/ds, dz/ds are the components of the unit vector $\mathbf{s}(s_x, s_y, s_z)$. The geometrical relationship is illustrated in Fig. 5.

In addition, by taking the gradient of $|\mathbf{s}|^2$, where $|\mathbf{s}|^2 = 1$,

$$0 = \mathrm{grad}|\mathbf{s}|^2 = 2[s_x\,\mathrm{grad}\,s_x + s_y\,\mathrm{grad}\,s_y + s_z\,\mathrm{grad}\,s_z] \tag{8b}$$

Subtracting Eq. (8b) from (8a),

$$\mathbf{K} = \frac{d\mathbf{s}}{ds} = s_x\left(\frac{\partial \mathbf{s}}{\partial x} - \mathrm{grad}\,s_x\right) + s_y\left(\frac{\partial \mathbf{s}}{\partial y} - \mathrm{grad}\,s_y\right) + s_z\left(\frac{\partial \mathbf{s}}{\partial z} - \mathrm{grad}\,s_z\right)$$

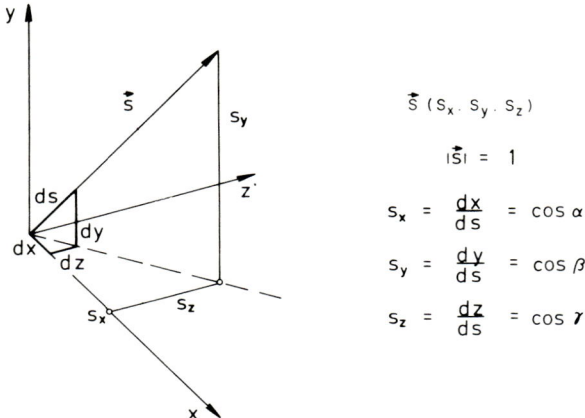

Fig. 5. Relation between the vector s, defining the ray direction, and the path element ds. (Letters with arrows correspond to boldface in text.)

Therefore, the x component of this vector equation is

$$ds_x/ds = s_y(\partial s_x/\partial y - \partial s_y/\partial x) + s_z(\partial s_x/\partial z - \partial s_z/\partial x)$$
$$= -s_y \operatorname{rot}_z \mathbf{s} + s_z \operatorname{rot}_y \mathbf{s}$$
$$ds_x/ds = -(\mathbf{s} \times \operatorname{rot} \mathbf{s})_x$$

Forming the other components, ds_y/ds and ds_z/ds, in an analogous way, the following vector equation may be written:

$$\mathbf{K} = d\mathbf{s}/ds = -[\mathbf{s} \times \operatorname{rot} \mathbf{s}] \tag{9}$$

Observing that $\mathbf{s}$ and $\operatorname{rot} \mathbf{s}$ are mutually perpendicular, this vector product may be written as the product of magnitudes as follows:

$$|\mathbf{K}| = |d\mathbf{s}/ds| = |\mathbf{s}| \cdot |\operatorname{rot} \mathbf{s}| \sin(\mathbf{s}; \operatorname{rot} \mathbf{s})$$
$$= |\mathbf{s}| \cdot |\operatorname{rot} \mathbf{s}| \tag{9a}$$

Equation (6), $\operatorname{rot}(n\mathbf{s}) = 0$, expresses the special property of the ray field that the rays are normal to the eikonal surfaces. Rewriting this equation,

$$n \cdot \operatorname{rot} \mathbf{s} - \mathbf{s} \times \operatorname{grad} n = 0$$

or

$$\operatorname{rot} \mathbf{s} = (1/n)[\mathbf{s} \times \operatorname{grad} n]$$
$$|\operatorname{rot} \mathbf{s}| = (1/n)|\mathbf{s}| \cdot |\operatorname{grad} n| \sin \alpha, \qquad \alpha = \sphericalangle(\mathbf{s}; \operatorname{grad} n) \tag{9b}$$

Substitution into Eq. (9a) yields

$$|\mathbf{K}| = |d\mathbf{s}|/ds = (1/n)|\mathbf{s}| \cdot |\mathbf{s}| \cdot |\operatorname{grad} n| \cdot \sin \alpha$$

$$|\mathbf{K}| = \frac{1}{R} = \left|\frac{d\mathbf{s}}{ds}\right| = \frac{|\operatorname{grad} n|}{n} \cdot \sin \alpha \tag{10}$$

Equation (10) is a differential equation establishing the connection between the curvature of the light ray path and the gradient of the refractive index. Integration leads to a relation for determining the desired distribution of the refractive index. This equation forms the basis for the shadow and schlieren techniques where the measured quantity is the angle of deflection of the light ray emerging from the schliere (see Section IV, B).

In practice, the gradient of the refractive index is most frequently one dimensional, so that $n = n(y)$ and $\operatorname{grad} n = dn/dy$. For this case, the differential equation for the light ray path becomes

$$|\mathbf{K}| = \frac{1}{R} = \frac{d^2y/dz^2}{[1 + (dy/dz)^2]^{3/2}} = \frac{1}{n} \cdot \frac{dn}{dy} \cdot \sin \alpha \tag{10a}$$

$$\frac{d^2y}{dz^2} = \frac{1}{n} \cdot \frac{dn}{dy} \cdot \sin \alpha \tag{10b}$$

In Eq. (10b), $(dy/dz)^2$ is negligible compared to unity when the incoming light is parallel to the z axis. In this special case the osculation plane corresponds to the y, z plane.

In the general case of a spatially curved light ray, the radius of curvature, which is perpendicular to the light ray vector $\mathbf{s}$, is given by the following equation obtained from Eqs. (9) and (9a):

$$\mathbf{K} = (1/n)[[\mathbf{s} \times \operatorname{grad} n] \times \mathbf{s}] \tag{11}$$

This double vector product may be rewritten as

$$n\mathbf{K} = \operatorname{grad} n - \mathbf{s}(\mathbf{s} \cdot \operatorname{grad} n)$$

This means that the vectors $\mathbf{K}$ (in the direction normal to the spatial path), $\operatorname{grad} n$, and $\mathbf{s}$ lie in a plane (osculation plane). According to Eq. (9b), the vector rot $\mathbf{s}$ is normal to both $\operatorname{grad} n$ and $\mathbf{s}$.

3. Snell's Law of Refraction

In the special case of a medium comprised of parallel layers lying perpendicular to the y axis, e.g.,

$$n \cdot \sin \alpha = \text{const} \tag{12}$$

Taking logarithms in Eq. (12) and differentiating,

$$dn/n + d(\sin \alpha)/\sin \alpha = 0; \qquad d(\sin \alpha) = -(dn/n) \cdot \sin \alpha$$

In addition, from the definition of the curvature $d\epsilon/ds$, where ϵ is the angle between the z axis and the path,

$$|\mathbf{K}| = \frac{d\epsilon}{ds} = \frac{1}{\cos \alpha} \cdot \frac{d(\sin \alpha)}{ds} = \frac{1}{n}\frac{dn}{dy} \sin \alpha \qquad (\alpha = \pi/2 - \epsilon)$$

where

$$d \sin \alpha/ds = \cos \alpha \, d\alpha/ds = -\cos \alpha \, d\epsilon/ds$$

Using these two equations, Eq. (12) is transformed into Eq. (10a), thereby demonstrating its validity. Equation (12) contains, in this form, Snell's law of refraction; the light ray is deflected towards the optically denser medium, i.e., in the direction of increasing refractive index gradient.

4. *The Optical Path (Fermat's Principle)*

The differential equation rot($n\mathbf{s}$) = 0, Eq. (6), may be transformed into an integral equation taken over a closed surface F:

$$\int_F \text{rot}(n\mathbf{s}) \, d\mathbf{f} = 0$$

Using Stokes' law one obtains

$$\int_F \text{rot}(n\mathbf{s}) \, d\mathbf{f} = \oint_C n\mathbf{s} \, d\mathbf{r} = 0$$

The second integral is taken along an arbitrary curve C on this surface. With the assistance of Eq. (5) this yields

$$\oint_C \text{grad } E \, d\mathbf{r} = 0$$

This means, as previously discussed, that E is a function of position only, or

$$\int_1^2 n\mathbf{s} \, d\mathbf{r} = E_2 - E_1 = \Delta E = S \cdot \lambda \qquad (13)$$

The line integral between Points 1 and 2 in a ray field is independent of the integration path and is equal to the difference in the eikonal at the two points. This property of the wave field is known as Fermat's principle. The integral value is designated as the optical path from Point 1 to Point 2. Equation (13) is the basis for interference techniques in which the phase differences $E_2 - E_1 = \Delta E$ can be measured.

In those frequently encountered cases where the refractive index distribution is one dimensional, $n = n(y)$, Eq. (13) may be simplified to

$$\int_1^2 n(y)\, ds = E_2 - E_1 = \Delta E = S \cdot \lambda \tag{13a}$$

These relations may be demonstrated utilizing the simple case of an idealized lens. The media are considered homogeneous, and the respective refractive indices of the glass and surroundings are n_{g1} and n_0, respectively; see Fig. 6.

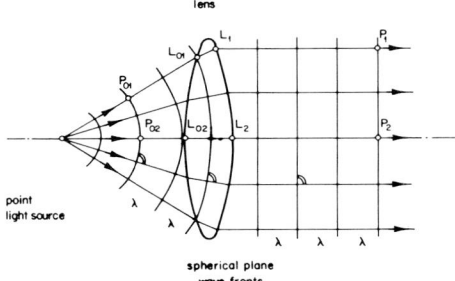

FIG. 6. Lengths of optical paths for the case of an idealized lens.

The spherical-shaped wave surfaces of constant phase, emanating from a point light source, are transformed into plane wave surfaces by the shape of the glass body. Therefore, the optical path is constant, or

$$n_0 \cdot \overline{P_{01}L_{01}} + n_{g1} \cdot \overline{L_{01}L_1} + n_0 \cdot \overline{L_1P_1} = n_0 \cdot \overline{P_{02}L_{02}} + n_{g1} \cdot \overline{L_{02}L_2} + n_0 \cdot \overline{L_2P_2}$$
$$= \text{const}$$

The corresponding geometrical paths through the media are different but the optical paths $n \cdot l$ are equal.

D. HUYGENS' PRINCIPLE

As Kirchhoff showed, this principle follows directly from the optical differential equation. Accordingly, it contains the geometrical optics description of a ray field and, in addition, describes the classical diffraction theory. Put in a way which can be visualized, Huygens' principle states that, from a given wave surface, the immediately adjacent wave surface can be constructed by considering the given surface as consisting of an infinite number of elementary point light sources. The surface tangent to the elementary spherically shaped wave fronts gives the new wave surface. Considering light rays entering a thermal boundary layer, for example, such a sequence of wave surfaces has been constructed in Fig. 7. Parallel layers of fluid are considered to lie over the heated wall.

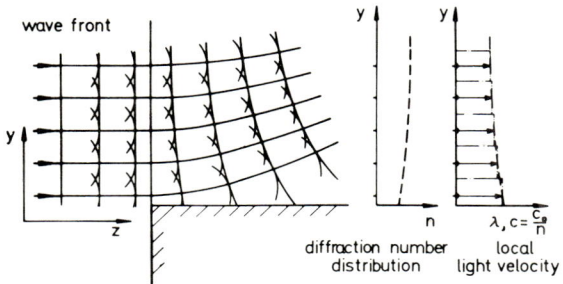

Fig. 7. Ray deflection in a boundary layer (wave surfaces constructed using Huygens' principle).

The refractive index distribution $n(y)$ is obtained from the proportionality between n and the density.

A plane wave front is considered to move in the z direction. Choosing the period duration $\tau = 1/\nu$ as the time interval between the constructed wave surfaces, the traversed light path λ is the product of the local light velocity and the period duration, i.e., $\lambda = c \cdot \tau = c_0/n \cdot \tau$. Each successively more deformed wave front, developed from the previous wave front (originally from the plane wave front entering the boundary layer), is that surface tangent to the arcs whose centers lie on the previous wave front and whose radii correspond to the local wavelength. The arcs are cross sections, in the plane of the drawing, of the spherically shaped elementary waves whose centers lie in the same plane. The traversed light path λ is larger in the vicinity of the wall as a result of the smaller refractive index; this results in a deflection of rays toward the denser surrounding medium.

III. Boundary-Layer Optics

A. Light Deflection in a One-Dimensional Refractive Index Field

A light ray, which passes through a refractive index field with locally varying values n and grad n, is now considered. The term R is the local radius of curvature and α is the angle between the light ray and grad n. According to Eq. (10),

$$1/R = (\text{grad } n/n) \sin \alpha$$

Restricting ourselves to a one-dimensional refractive index field, introducing the y, z normal coordinates shown in Fig. 8, and utilizing grad $n = dn/dy$ and $\epsilon = (\pi/2) - \alpha$, this equation becomes

$$\frac{1}{R} = \frac{y''}{(1+y'^2)^{3/2}} = \frac{dn/dy}{n} \cos \epsilon \qquad (14)$$

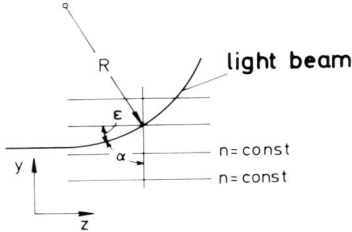

FIG. 8. Passage of light through a one-dimensional refractive index field $n(y)$.

Using the relation $\cos \epsilon = 1/[(1 + \text{tg}^2\epsilon)]^{1/2} = 1/[(1 + y'^2)]^{1/2}$, Eq. (14) yields

$$\frac{y''}{1+y'^2} = \frac{dn/dy}{n} = \frac{d \ln n}{dy} \qquad (15)$$

In addition, since $y'' \, dy = y' \, dy'$, Eq. (15) becomes

$$\frac{y' \, dy'}{1+y'^2} = d \ln n \qquad (16)$$

Integrating, and using the boundary condition $n = n_0$ for $y' = y_0'$,

$$\frac{1+y'^2}{1+y_0'^2} = \left(\frac{n}{n_0}\right)^2 = \frac{\cos^2 \epsilon_0}{\cos^2 \epsilon} = \frac{\sin^2 \alpha_0}{\sin^2 \alpha} \qquad (17)$$

where y_0', n_0, ϵ_0, and α_0 are given values for one position y_0, z_0 of the light ray under consideration.

Equation (17) states Snell's law of refraction in the form $n \cdot \sin \alpha = n_0 \cdot \sin \alpha_0$, which could also have been directly derived. Further integration of Eq. (17), and thereby the calculation of the light ray path $y(z)$, is only possible if the function $n(y)$ is known. Inversely, by measuring the two inclinations of the light ray y' and y_0' only the ratio n/n_0 can be calculated. In this case, however, it is not possible to obtain further information on the distribution of n.

In practical applications, y' is of the order $1/100$ so that approximations for the light ray path may be used. In the following development, it is assumed that every light ray in the test section passes through a region where $n' = dn/dy = $ const. Therefore

$$n = n_0 + n'(y - y_0) \tag{18}$$

This approximation is assumed valid only for a single light ray and does not require that the refractive index gradient in the entire field be constant. Since $n'(y - y_0)/n_0 \ll 1$, the approximation

$$(n/n_0)^2 \approx 1 + (2n'/n_0)(y - y_0) \tag{19}$$

may be used in Eq. (17). For simplification, it is assumed that $y_0' = 0$ for the entering light beam so that, from Eqs. (17) and (18), the path of the light ray is given by the relation

$$z = \int_{y_0}^{y} \frac{dy}{[(2n'/n_0)(y - y_0)]^{1/2}} \tag{20}$$

Integrating between the positions of entrance and exit of the light ray, $z = 0$ and $z = l$, respectively,

$$y_l - y_0 = n'l^2/2n_0 \tag{21}$$

The slope of the ray y_l' at $z = l$ is

$$y_l' = \frac{n'l}{n_0} = \frac{y_l - y_0}{l/2} \tag{22}$$

In the medium, the light ray follows a parabolic path which, according to Eq. (22), may be replaced by a path comprised of two straight lines having a slope discontinuity at the center (Fig. 9). The case of a nonzero entrance

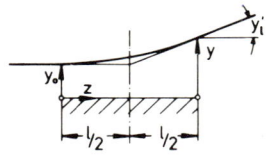

FIG. 9. Parabolic approximation of the light ray path.

angle ($y_0' \neq 0$), such as for divergent or convergent emerging light bundles, can also be treated in a similar way. It is also possible to deal with a strongly changing distribution of the refractive index $n(y)$ by including higher-order terms in Eq. (18) (see Section IV,B).

As a consequence of the curvature, the light ray not only has a longer geometrical path s (compared to a ray passing parallel to, and at a distance y_0 from, the wall) but also experiences a phase change. This difference in the optical path $\int n \, ds - n_0 l$ may be calculated using the above approximations. Using Eq. (19), (21), and (22), and neglecting higher-order terms, one obtains

$$\int_0^{s_l} n \, ds - n_0 l = y_l'^2 l/3 \cdot n_0 \tag{23}$$

where s_l is the length of the geometrical light ray path to $z = l$, i.e., at the position of emergence of the light ray.

B. The Boundary Layer as a "Schlieren Lens"

1. Imaging Law of the Boundary Layer

The relations derived in the preceding section are now used to investigate the projected image (shadowgraph) of a schliere (*11*). It is assumed that parallel light passes through the thermal boundary layer of a flat plate and falls on a projection screen (Fig. 10). The functional

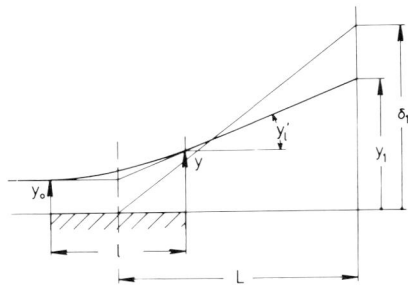

FIG. 10. Derivation of the imaging function of the boundary layer.

relation describing the path of the light ray through the region of interest (in this case, the thermal boundary layer) and the coordinate y_1 on the screen is desired. The distance between the center of the model and the screen is defined as L. Using the parabola approximation shown in Fig. 9, one obtains the following relation from Fig. 10:

$$(y_1 - y_0)/L = y_1' \tag{24}$$

where y_0 is the distance between the incoming ray and the wall. Substitution of y_1' from Eq. (22) leads to the equation

$$y_1 - y_0 = Lln'/n_0 \tag{25}$$

The deflection of the light ray $y_1 - y_0$ is now referred to the deflection δ_1 of the ray in the immediate vicinity of the wall. From Eq. (25),

$$\delta_1 = Lln_w'/n_w \tag{26}$$

where the subscript w refers to values at the wall. Using the boundary-layer thickness δ, the normalized coordinates $\eta_0 = y_0/\delta$ and $\eta_1 = y_1/\delta_1$ are introduced. From Eqs. (25) and (26) one obtains

$$\eta_1 = n'n_w/n_0n'_w + (\delta/\delta_1)\,\eta_0 \tag{27}$$

Equation (27) is the desired "imaging law of the boundary layer," i.e., the relation between η_1 and η_0.

The specific radiation impinging on the screen E^* (illumination intensity or simply illumination) is changed by the deflection of the light. If L^* designates the ray density (luminous intensity) of the entering light, then the following relation is valid:

$$L^* \, dy_0 = E^* \, dy_1 \tag{27a}$$

Using Eq. (27), one obtains the relation

$$\frac{E^*}{L^*} = \frac{dy_0}{dy_1} = \frac{\delta}{\delta_1} \cdot \frac{d\eta_0}{d\eta_1} = [1 + Ll(n_0n'' - n'^2)/n_0^2]^{-1} \tag{28}$$

where $n'' = d^2n/dy^2$ is the second derivative of the refractive index distribution $n(y)$ in the test section.

The specific radiation becomes infinite, according to Eq. (28), when the condition

$$n'^2/n_0^2 - n''/n_0 = 1/(L \cdot l) \tag{29}$$

is satisfied. In this case focal lines appear on the screen, e.g., as often arises from inhomogeneous glass plates. Naturally there are many functions $n(y)$ which are capable of satisfying Eq. (29). In addition, the loca-

2. Model for a Boundary Layer

As a concrete example consider a laminar thermal boundary layer. The temperature profile is represented by a polynomial of the 4th degree:

$$\vartheta_w = (T - T_\infty)/(T_w - T_\infty) = a + b\eta + c\eta^2 + d\eta^3 + e\eta^4 \tag{30}$$

where T_∞ and T_w are the free stream and wall temperature, respectively, and $\eta = y/\delta$ is the distance from the wall relative to the boundary-layer thickness. The following five boundary conditions may be satisfied:

$$\eta = 0: \quad \vartheta = 1; \quad \vartheta'' = 0$$

$$\eta = 1: \quad \vartheta = 0; \quad \vartheta' = 0; \quad \vartheta'' = 0$$

Then follow the relations

$$\vartheta = 1 - 2\eta + 2\eta^3 - \eta^4 \tag{31a}$$

$$\vartheta' = -2 + 6\eta^2 - 4\eta^3 \tag{31b}$$

$$\vartheta'' = 12\eta - 12\eta^2 \tag{31c}$$

The condition $\vartheta'' = 0$ for $\eta = 0$ has special implications for optical methods. This will be considered later in Section III,D.

The differential equation for a laminar thermal boundary layer having no heat production and where the wall is stationary and nonporous is

$$w_x \, \partial T/\partial x + w_y \, \partial T/\partial y = a \, \partial^2 T/\partial y^2$$

For $y = 0$, $w_x = 0$, and $w_y = 0$, so that $\partial^2 T/\partial y^2 = 0$, and, therefore, $\vartheta'' = 0$.

In order to calculate the refractive index profile from the temperature profile using Eq. (31), a relative refractive index is defined:

$$N = (n - n_w)/(n_\infty - n_w) = (n - n_w)/\Delta n$$

For simplification, it is assumed that the quotient dn/dT is constant in the range $0 < \vartheta < 1$ so that the following refractive index distributions in the boundary layer may be obtained from Eq. (31):

$$n/\Delta n = n_w/\Delta n + 2\eta - 2\eta^3 + \eta^4 \tag{32a}$$

$$n'/\Delta n = (1/\delta)(2 - 6\eta^2 + 4\eta^3) \tag{32b}$$

$$n''/\Delta n = (1/\delta^2)(-12\eta + 12\eta^2) \tag{32c}$$

In addition $n_w' = 2\Delta n/\delta$.

Substituting Eqs. (32a–c) into Eq. (27), the imaging law for a laminar thermal boundary layer may be written in the following form:

$$\eta_1 = 1 + (\delta/\delta_1 - 2b)\eta_0 - 3\eta_0^2 + (2 + 8b)\eta_0^3 - 5b\eta_0^4$$
$$-b(6\eta_0^5 - 7\eta_0^6 + 2\eta_0^7) \tag{33}$$

where, for convenience, $b = \Delta n/n_w$. Higher-order b terms have been neglected. It is often possible to set $b \approx 0$ so that calculations may be carried out using the simpler function

$$\eta_1 = 1 + (\delta/\delta_1)\eta_0 - 3\eta_0^2 + 2\eta_0^3 \tag{34}$$

The parameter δ/δ_1 in Eqs. (33) and (34) depends, for given experimental conditions, on the distance L to the screen according to the following relation obtained using Eq. (26):

$$\delta/\delta_1 = \delta n_w/(L \cdot l \cdot n_w') \tag{35}$$

From this it is seen that δ/δ_1 is inversely proportional to L.

The function $\eta_1(\eta_0, \delta/\delta_1)$, according to Eq. (34), is shown in Fig. 11

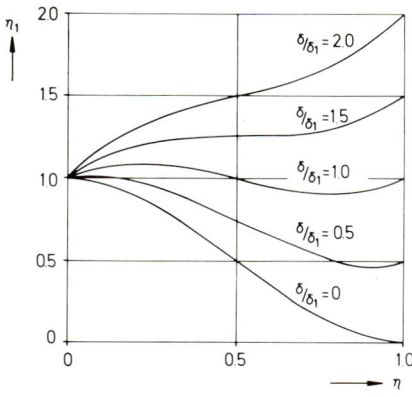

FIG. 11. Imaging law of the boundary layer.

with δ/δ_1 as parameter. Using nonzero values of b in Eq. (33) would result in only a minor shift of the curves in Fig. 11 for the values of b encountered in practice. Figure 11 presents the relation between the boundary-layer coordinate $\eta_0 = y_0/\delta$ and the projection screen coordinate $\eta_1 = y_1/\delta_1$, as indicated by the arrows. According to Eq. (28), the specific impinging radiation E^* (illumination intensity) upon the screen depends on the slope $\delta\, d\eta_0/\delta_1\, d\eta_1$ and becomes infinite for $d\eta_1/d\eta_0 = 0$, for arbitrary values of δ/δ_1. This means that a focal line develops for those screen coordinates η_1 for which the $\delta/\delta_1 = $ const curve in Fig. 11 has a horizontal tangent.

3. Envelope as a Focal Line

Writing Eq. (34) in the form

$$y_1/\delta = (\delta_1/\delta)(1 - 3\eta_0^2 + 2\eta_0^3) + \eta_0 \tag{36}$$

one obtains the equation for a family of lines having the coordinates y_1/δ and δ_1/δ and with η_0 as parameter. A family of lines may describe envelopes which satisfy the equation $F(y_1/\delta, \delta_1/\delta, \eta_0) = 0$ if the parameter η_0 is eliminated by using the condition $\partial F/\partial \eta_0 = 0$. From this, one obtains the following two distances from the wall as the solution to the quadratic equation:

$$\eta_{a,b} = \tfrac{1}{2}[1 \pm (1 - 2\delta/3\delta_1)^{1/2}] \tag{37}$$

from which the condition $\delta_1/\delta \geq \tfrac{2}{3}$ follows.

Substituting the η values from Eq. (37) into Eq. (36), the envelopes shown in Fig. 12 are obtained. The relation, between the distance to the screen L and the ratio δ_1/δ is

$$\delta_1/\delta = L l n_w'/\delta n_w \tag{38}$$

If one imagines that the screen in Fig. 12 is set up at different locations, characterized by specified values of δ_1/δ, then the light ray coming from the wall ($\eta_0 = 0$), i.e., the straight line through the zero point in Fig. 12, may be traced. This is, according to definition, the screen coordinate $\eta_1 = 1$. All of the light rays entering the boundary layer ($1 > \eta_0 > 0$) fall on the screen within the two envelopes. These envelopes are cross sections of two focal surfaces whose intersections with the screen denote the focal lines. For large values of δ_1/δ, the ray from the wall ($\eta_0 = 0$) and that tangential to the boundary layer ($\eta_0 = 1$) form the two asymptotes of the envelopes. The ray from $\eta_0 = 0.5$ forms the bisecting angle between the two envelopes, which intersect at the coordinates $\delta_1/\delta = \tfrac{2}{3}$ and $y_1/\delta = \tfrac{5}{6}$.

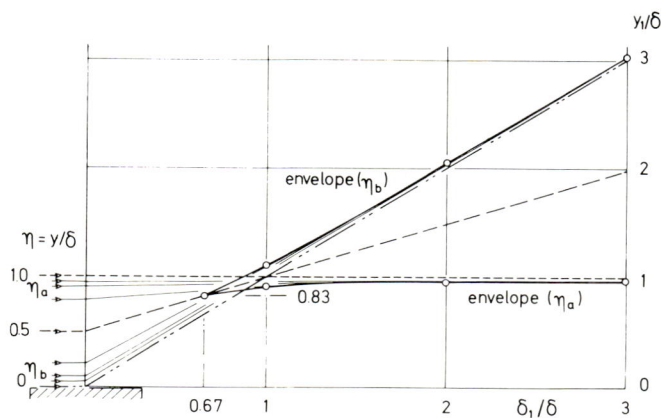

FIG. 12. Envelopes of a light ray bundle.

Some related values giving the coordinates of the envelopes, the screen coordinates, and the boundary-layer coordinates η_a and η_b, are given in Table I.

TABLE I

PROPERTIES OF SCHLIEREN LENS

δ_1/δ	y_1/δ	η_1	η_a	η_b
0.67	0.83	1.25	0.5	0.5
1.0	0.908	0.908	0.788	—
1.0	1.096	1.096	—	0.211
2.0	0.956	0.478	0.908	—
2.0	2.045	1.022	—	0.092
3.0	0.972	0.324	0.941	—
3.0	3.027	1.009	—	0.059

The light rays forming the focal lines on the screen come, for every screen placement $L \sim \delta_1/\delta$, from different locations in the boundary layer. It is therefore possible to consider the boundary layer as a "Schlieren lens" whose focal length is dependent on the distance from the wall. The imaging law of this lens may be determined equally well from either Fig. 12 or Fig. 11, where the same physical relation is shown in two different ways. For screen distances L, corresponding to $\delta_1/\delta < \frac{2}{3}$, no focal lines appear. However, substantial changes in the specific radiation (intensity of illumination) falling on the screen would, in this case also, be expected as a result of the deflection of the light rays as they pass through the boundary layer.

4. *Experimental Confirmation*

The relations derived above are now confirmed by an example illustrated in Fig. 13. The light bundle entering the thermal boundary layer

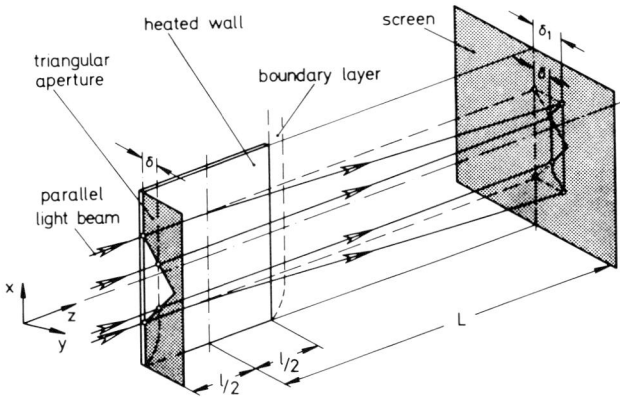

FIG. 13. Schlieren method using a mask.

on a vertical flat plate is outlined by using a mask with a triangular aperture. The distortion of the projected aperture on the screen is used to confirm the imaging law of the boundary layer. The term L is the distance to the projection screen, δ the thickness of the thermal boundary layer, and δ_1 the coordinate on the screen of the light ray closest to the wall.

Figure 14 shows the shadowgraph for a screen distance $L = 1$ m which, for the given conditions, corresponds to $\delta_1/\delta = 0.25$. From Fig. 12, no focal line would be expected for this value. The brightening of the projection as a result of the reduction in size of the original triangle, illustrated with broken lines in the figure, may be seen clearly. The coordinate δ_1 in the figure shows the projected position of the wall; the projection of the region close to the wall is displaced but not greatly distorted. This results from the "wandbindung" of the laminar thermal boundary layer where, according to Eq. (31c), $\vartheta'' = 0$, and therefore $n'' = 0$ for $\eta = 0$. The rays close to the wall proceed approximately parallel and the parabola approximation in Eq. (22) is satisfied here in a finite range of wall distances.

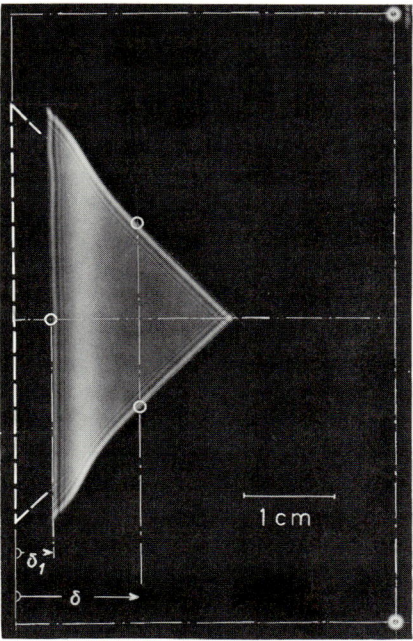

Fig. 14. Shadowgraph of a laminar thermal boundary layer using a triangular aperture: $\delta_1/\delta = 0.25, L = 1$ m.

For a screen distance $L = 2$ m, corresponding to $\delta_1/\delta \approx 0.75$, the projection of the triangular aperture is reproduced in Fig. 15. According to Fig. 12, a focal line appears for the first time from Table I, $\delta_1/\delta = 0.67$, $\eta_1 = 1.25$). This line does not run exactly parallel to the wall because the thickness of a free-convection boundary layer on a vertical flat plate is dependent upon the height.

The projection for screen distances L of 5 and 7 m are shown in Figs. 16 and 17, respectively; the chosen screen distances correspond to $\delta_1/\delta = 1.25$ and $\delta_1/\delta = 1.75$. According to Fig. 12, two focal lines are to be expected in these regions. For much greater distances, the focal lines stem from those light rays which form the asymptotes to the envelopes in Fig. 12.

The contours in Figs. 15–17 display considerable fuzziness, accompanied by interference lines arising from diffraction phenomena. The quantitative evaluation of these shadowgraphs becomes considerably more difficult (see also Section IV,D). However, the photographs qualitatively confirm the theoretical considerations in the previous section.

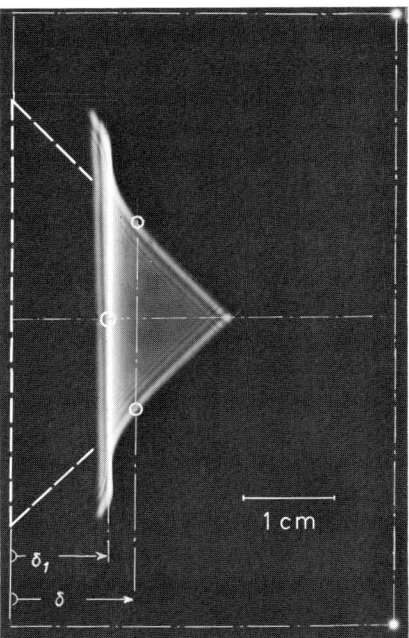

Fig. 15. Shadowgraph of a laminar thermal boundary layer using a triangular aperture: $\delta_1/\delta = 0.75$, $L = 2$ m.

C. Optical Measurements of Boundary Layers

Let us consider again an optical arrangement such as that shown in Fig. 10. Using shawdowgraph methods, the only obtainable information is the change in direction of the light rays; using interference methods, the information concerns the change of phase of the light rays during their passage through the test section. According to Fig. 2, both phenomena are inseparably connected. It is not possible to obtain local measurements along the paths of the light rays inside the model since methods which give the local direction and phase of these rays are not as yet known. It would be possible to use models of various lengths, but the usable range is quite limited due to the decrease in sensitivity with decreasing model lengths and the increase of the deflection, or density of the interference lines, with increasing lengths.

In any case, the curvature of the light rays precludes obtaining information on the layers closest to the wall. This is true for a boundary layer either cooled or heated by the wall. However, the heated wall, as shown in Fig. 10, offers various practical advantages. Therefore, in evaluating

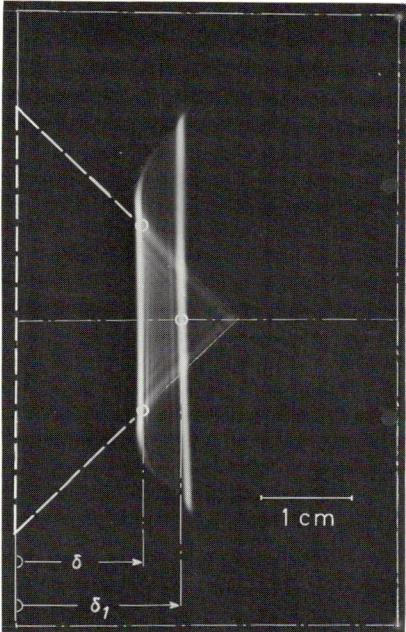

Fig. 16. Shadowgraph of a laminar thermal boundary layer using a triangular aperture: $\delta_1/\delta = 1.25, L = 5$ m.

the data from the photographs, it is necessary to extrapolate to the position of the wall. Mostly it becomes a question of determining the local refractive index gradients and from these the local temperature gradients.

In this connection the wandbindung referred to in Section III,B,2, plays a decisive role. The temperature gradient in a boundary layer without internal heat generation is constant in the immediate vicinity of a stationary nonporous wall. This results from the following equation, already given in Section III,B,2, where $w_x = w_y = 0$ at the wall:

$$w_x \, \partial \vartheta/\partial x + w_y \, \partial \vartheta/\partial y = a \, \partial^2 \vartheta/\partial y^2 \tag{39}$$

This wandbindung considerably simplifies the application of optical methods in heat transfer.

In other cases, the gradient at the wall is no longer constant; for example, at a moving wall such as the surface of a rotating cylinder, where w_x has a predetermined nonzero value, or at a porous wall, where the nonzero velocity w_y is given. As a result, the temperature profile in the vicinity of the wall changes in a characteristic way. It is usually possible,

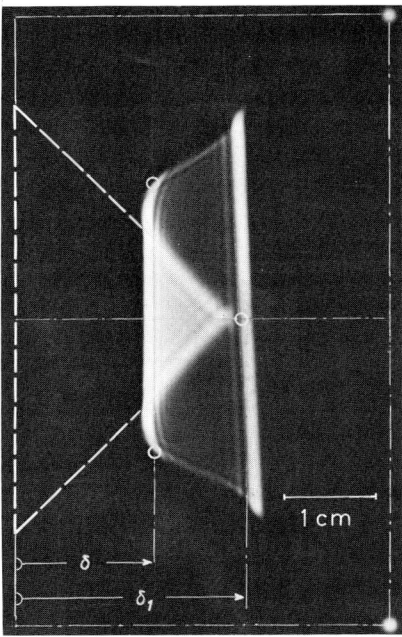

FIG. 17. Shadowgraph of a laminar thermal boundary layer using a triangular aperture: $\delta_1/\delta = 1.75, L = 7$ m.

however, to predict the shape of the gradient in the vicinity of the wall either in an approximate or in an exact way by utilizing the established mathematical methods of handling the boundary layer. Nonstationary heat conduction problems, e.g., those which arise when a wall is suddenly heated, may be treated in an analogous way. In this case also, the shape of the gradient at the wall can usually be calculated from measured quantities.

In coupled heat and mass transfer processes, according to the following relation, the change in the refractive index is dependent on two quantities:

$$dn = (\partial n/\partial T)_C \, dT + (\partial n/\partial C)_T \, dC \qquad (40)$$

where C is the local concentration. If both partial derivatives in Eq. (40) were known, it would still be difficult in general to obtain both the temperature and concentration fields from the measured refractive index field. It is often attempted to obtain one of the two fields, either by calculation or by other nonoptical methods of measurement. One may also attempt to satisfy the following relation by choosing appropriate experimental materials:

$$(\partial n/\partial T)_C \, \Delta T \gg (\partial n/\partial C)_T \, \Delta C$$

here ΔT and ΔC mean the greatest encountered difference in temperature and concentration occurring, e.g., between the wall and the free stream. The condition is satisfied for combinations such as water vapor and air. In nonstationary processes it is possible to use the fact that the two fields develop at different rates since for liquid media the relation $a \gg D$ is mostly valid (here, D is the coefficient of diffusion). This facilitates studying the effect of, e.g., thermal diffusion.

Almost exclusive consideration has been given in this discussion to plane two-dimensional fields which do not change, or do not change strongly, in the direction of propagation of the light ray. For cylindrical or spherical symmetry, the temperature field or concentration field may be determined from the measured refractive index field using known calculation procedures; in this case, the measurement accuracy is usually considerably smaller. For completely irregular three-dimensional fields, average values can be determined using the fact that the resultant influence on the light ray is the integrated influence of the local refractive indices along the light ray path through the test section. The average enthalpy of a volume may be determined, in this way, for example. This integration effect may be utilized also for the measurement of flow processes such as tube flow. When light is shone through the tube in the flow direction, values of the gradient at the wall, averaged over the length of the tube, may be obtained.

It should always be observed that in all optical methods a refractive index field is obtained initially. Its conversion to a temperature or concentration field is a problem in its own right. Many optical methods which are in principle possible break down in practice since the conversion cannot be effected with the required accuracy.

IV. Theory of Shadowgraph and Schlieren Methods

The total information on the region of optical inhomogeneity being investigated comes from the deformation of a wave front, whose shape was originally known, as it passes through the region (see Section II,B). The form of the deformed wave front at every point in the field of view is to be determined using optical methods. These methods fall naturally into two groups:

(1) shadowgraph and schlieren methods (methods utilizing intensity, and light ray identification);

(2) interference methods (recording the phase differences).

As already shown in Section II,C, the deflection of the light ray ϵ and the phase difference S are mutually dependent properties describing the deformation of the wave front. The shadowgraph and schlieren methods record the deflection; the interference methods transform the invisible phase differences into differences in intensity allowing these local phase differences to be determined.

In the following description of the methods and their uses, the following Cartesian coordinate system is defined: The coordinates x, y define planes in the region of interest which are perpendicular to the direction of the entering light ray bundle. It is assumed here that the rays comprising this bundle are always parallel. The coordinate z therefore lies in the direction of the light rays. Since the methods are mostly applied to boundary-layer investigations, the usual definitions of boundary-layer physics are used, i.e., y is taken normal to, and away from, the wall.

A large number of optical arrangements are known which could be included under the heading "shadowgraph and schlieren methods." These are used to show, either quantitatively or qualitatively, the light ray deflection ϵ in the region of interest. Schardin (*2*), Weinberg (*3*), and Wolter (*4*), for example, have given comprehensive reviews of these methods, to which the interested reader is referred. In the following, only a few particularily useful methods, in connection with measurements in the area of heat and mass transfer, are presented. As an introduction, the principle of operation of a schlieren apparatus, used to measure diffusion coefficients of liquids, will be described. This apparatus was originated by Wiener (*12*) in 1893; an improved apparatus developed by Philpot and Svensson (*13, 14*) in which the recording process was partially automated, is more widely used. It will be shown that the essential elements of Wiener's schlieren apparatus may also be used in the area of heat transfer.

A. WIENER'S DIFFUSION INVESTIGATION (1893)

1. *Review of Wiener's Methods*

Using a slit diaphragm set at a 45° angle, a narrow light bundle was produced and allowed to pass through a glass cuvette (see Fig. 18). Two liquids of different specific weights, with refractive indices of n_1 and n_2,

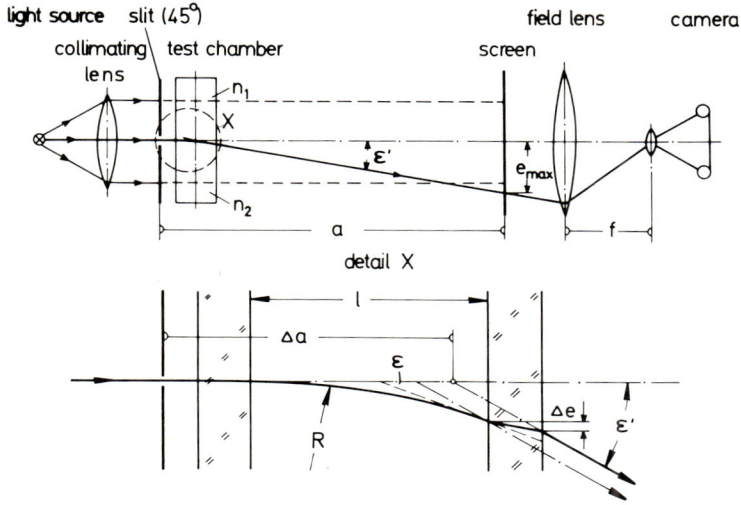

FIG. 18. Wiener's schlieren apparatus for measuring diffusion coefficients.

respectively, are allowed to lie in layers in the container. The mixing of the fluids through diffusion is to be observed. The lighter liquid is assumed to lie over the denser so that there is no movement due to convection. A concentration gradient develops in the y direction, being always greatest at the interface, which produces the observed light deflection. By using the obliquely placed slit, it is possible to identify each coordinate in the y direction (height) with a point in the slit. The deflection, through the angle ϵ, of the light bundle passing through such a point of the slit is proportional to the local refractive index gradient, dn/dy, which here, for simplification, will be assumed proportional to the concentration gradient dC/dy. The deflection e is observed on a projection screen.

If this pattern is to be reduced in size and photographically recorded in order, for example, to obtain a slow-motion cine of the nonstationary process, it is recommended that a so-called field lens be used. This lens, or a concave mirror, of large diameter and focal length is set a short distance behind the original plane of the screen. The screen is then removed. The divergent bundle of rays is deflected by the optical element so that it passes into the camera aperture. The lens of the camera is located approximately at the focal point of the field lens. The combination of the field lens and the camera lens system projects the original plane of the screen on to the film.

2. Calculation of the Deflection ϵ'

Because of the shortness of the glass cuvette, it will be assumed for simplification that the light path lies in a region of locally constant refractive index gradient dn/dy. In addition, because the light deflection angle ϵ is expected to be small, the ray path is replaced by its circle of curvature having a horizontal tangent at the entrance. According to Eq. (10a), $1/R = 1/n \cdot |\, dn/dy\,|$. Also, from the light ray equation, $\epsilon = l/R$. The light ray undergoes an additional deflection as it enters the air so that the final angle of deflection ϵ' is, according to Snell's law of refraction ($n_{air} \approx 1$),

$$\sin \epsilon' / \sin \epsilon = \epsilon'/\epsilon = n$$

or

$$\epsilon' = n\epsilon = l \cdot (dn/dy) \tag{41}$$

The effect (displacement) on the light ray due to its passage through the glass wall of the cuvette (see Fig. 18) for small angles ϵ remains unaccounted for. It is clear from the figure that one would have to calculate ϵ' from the equation $\epsilon' = e/(a - \Delta a)$; however, Δa may be neglected, since it is small compared to the distance a between the screen and the slit diaphragm.

Another reason for neglecting these corrections is that the diffraction disturbances coming from the boundaries of the slit make it impossible to take measurements, especially of the position of maximum deflection e_{max}, with a sufficient accuracy from the screen. This is true when light having a wave length λ is used; for white light, consisting of a mixture of all wave lengths (e.g., corresponding to the continuum from a carbon arc), the refractive index gradient in the diffusion zone behaves like a dispersion prism. Therefore, at the location of maximum deflection, especially at the beginning of the experiment, the light falling on the screen is decomposed into its spectral colors.

3. Equation for Determination of the Diffusion Coefficient

This one-dimensional diffusion process is described by Fick's equation in the following form:

$$\partial C/\partial t = D \cdot \partial^2 C/\partial y^2$$

where, as in Fig. 19, C is the concentration, y the coordinate in the direction of diffusion, and t the time from the start of the experiment; D is the desired diffusion coefficient. In solving Fick's equation only small

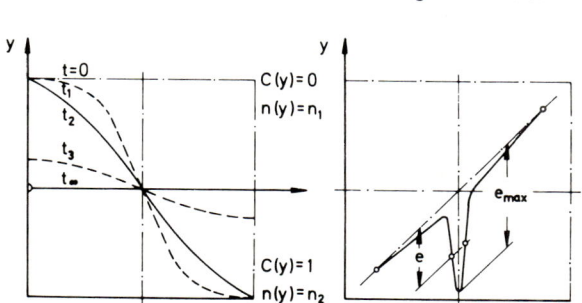

FIG. 19. Schematic of the screen image and the distribution of refractive index from Wiener's diffusion experiment.

time intervals shall be considered. Furthermore, it will be assumed that the concentrations at the boundaries of the fluid, which are parallel to the original phase boundary, are practically constant. The solution is thereby simplified, since the fluids can then be considered to be semi-infinite. If the refractive index n is taken to be proportional to the concentration C, the following solution of Fick's equation results:

$$n = [(n_1 - n_2)/2] \, \Phi(y/2(Dt)^{1/2}) + (n_1 + n_2)/2 \qquad (42)$$

where

$$\Phi = \frac{2}{\sqrt{\pi}} \int_0^{y/2(Dt)^{1/2}} e^{-z^2} \, dz$$

is the Gaussian error function and n_1 and n_2 are the refractive indices, respectively, of the fluids one and two. For $y = 0$ and arbitrary values of $t \neq 0$ it then follows that

$$n(0) = (n_1 + n_2)/2 \qquad (43)$$

This may also be seen from the $n = f(y, t)$ diagram in Fig. 19. The maximum refractive index gradient occurs at the position $y = 0$, independent of time, and is given by

$$\partial n/\partial y = (n_1 - n_2)/2(\pi Dt)^{1/2} \qquad (44)$$

The maximum deflection angle ϵ' is given by Eq. (41), $\epsilon' = l \cdot dn/dy$, whereby

$$\epsilon' = [(n_1 - n_2)/2(\pi Dt)^{1/2}] \, l \qquad (45)$$

Wiener used this equation to determine the diffusion coefficient D. The time interval close to $t = 0$ must be excluded since the deflection becomes infinitely large as t approaches zero; therefore, for convenience, one obtains relative values at various times $t > 0$.

This problem is analogous to that of nonstationary heat conduction in a semi-infinite body having constant wall temperature for $t \geqslant 0$ as the boundary condition.

The same method will be described in Section VI,B for determining the thermal conductivity k_c, which is analogous to the coefficient of diffusion in this problem. In this case, however, the boundary condition of constant heat flux for $t \geqslant 0$ can be more accurately realized in practice.

B. Iterative Calculation of an Arbitrary Refractive Index Field from the Light Deflection ϵ

Excluding disturbances from the boundary, it is presumed here that the refractive index fields consist of plane parallel layers which lie parallel to the incoming light bundle and perpendicular to the refractive index gradients.

For the case of a boundary layer, as already demonstrated for a thermal boundary layer in Section III, the form of the refractive index profile, either totally or in part, is calculable. This makes the analysis of the results considerably easier. However, if an arbitrary distribution of refractive index is to be considered, an approximate method from Schardin (2) and Weinberg (3) may be used which allows the required refractive index distribution to be obtained from the derivatives dn/dy and d^2n/dy^2 by iteration.

According to Eq. (10a) in Section II, the differential equation describing the path of the light ray is

$$\frac{d^2y/dz^2}{[1 + (dy/dz)^2]^{3/2}} = \frac{1}{R} = \left(\frac{1}{n}\right)\left(\frac{dn}{dy}\right)$$

The z coordinate, as stipulated, lies in the direction of the light ray. The slope of the light ray $(dy/dz)_0$ at entrance into the refractive index field (subscript zero) is zero for parallel rays which enter perpendicularly; the slope $\epsilon_l = (dy/dz)_l$ as it emerges (subscript l) should be small so that $(dy/dz)^2$ is negligible compared to unity. The differential equation then reduces to

$$d^2y/dz^2 = (1/n)(dn/dy) \tag{46}$$

and

$$(dy/dz)_l = \int_{z=0}^{z=l} (1/n)(dn/dy)\, dz \tag{46a}$$

therefore,

$$\tan \epsilon_l = (dy/dz)_l = \int_{z=0}^{z=l} (1/n)(dn/dy)\, dz \qquad (46\text{b})$$

As in Section III, the integral appearing should be treated as a line integral with integration along the path of the light ray. If the deflections are small, e.g., as encountered for moderate refractive index gradients and short model lengths in the z direction or for very small refractive index gradients and larger model lengths, the integration along the path of the light ray from 0 to l may be replaced by integration in the z direction alone:

$$\epsilon_l = \int_{z=0}^{z=l} (1/n) \cdot (dn/dy)_0\, dz \qquad (46\text{c})$$

With the assumption, very often valid in the region of the ray path, $dn/dy = \text{const}$,

$$\epsilon_l = (1/n) \cdot (dn/dy)_0 \cdot l \qquad (46\text{d})$$

In the following the refractive index n is set equal to a constant since its variation is small compared to that of dn/dy. As seen in the following example for a concentration boundary layer (Section IV,D), the differences in refractive index in practice are of the order $\Delta n < 10^{-3}$ for the total region of the boundary layer. For the region traversed by the (curved) light ray, the calculation may be carried out using an average value of the refractive index.

For further illustration of the approximation procedure, the quantities used are shown in Fig. 20. The ray enters the region of interest at y_0.

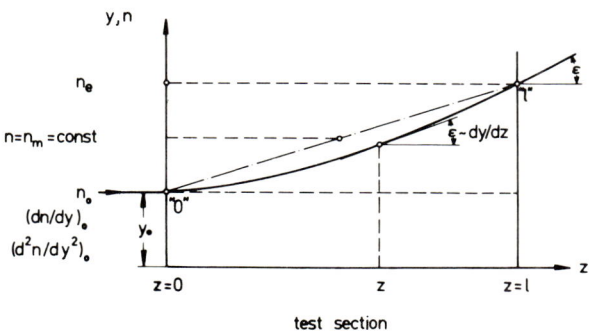

FIG. 20. Light ray path in two-dimensional strata.

1. First Approximation

$$(dn/dy)_{z=0} = (dn/dy)_0 = \text{const}$$

According to Eq. (46a),

$$dy/dz = \int_0^z (1/n)(dn/dy)_0 \, dz = (1/n)(dn/dy)_0 z \quad (47)$$

$$y = (1/n)(dn/dy)_0 z^2/2 + y_0 \quad (48)$$

According to the first Approximation, for $z = l$, the angle is

$$\epsilon_l = (dy/dz)_l = (1/n)(dn/dy)_0 \cdot l \quad (49)$$

2. Second Approximation

The parabolic path resulting from the first approximation is now used in place of the linear path whereby, similar to the first approximation, $(d^2n/dy^2)_0$ is set equal to a constant:

$$(1/n)(dn/dy) = (1/n)[(dn/dy)_0 + (d^2n/dy^2)_0(y - y_0)] \quad (50)$$

Substituting Eq. (48) into Eq. (50),

$$(1/n)(dn/dy) = (1/n)[(dn/dy)_0 + (d^2n/dy^2)_0(1/n)(dn/dy)_0 z^2/2]$$

$$(1/n)(dn/dy) = (1/n)(dn/dy)_0[1 + (1/n)(d^2n/dy^2)_0 z^2/2] \quad (51)$$

Integrating, as in the first approximation, yields

$$dy/dz = (1/n)(dn/dy)_0[z + (1/n)(d^2n/dy^2)_0 z^3/6]$$
$$y = (1/n)(dn/dy)_0[z^2/2 + (1/n)(d^2n/dy^2)_0 z^4/24] + y_0 \quad (52)$$

According to the second approximation, the angle as the ray emerges (for $z = l$) is

$$\epsilon_l = (dy/dz)_l = (1/n)(dn/dy)_0[l + (1/n)(d^2n/dz^2)_0 l^3/6]$$

The dependence of the angle ϵ on the coordinate y, i.e., $\epsilon \sim (dn/dy)_0 = f(y)$, is obtained from experiment. Therefore, successively using the first approximation and then the second, the distribution of refractive index, $n = f(y)$, can be obtained.

In practical applications the calculation is begun at those positions of the refractive index distribution whose locations can best be measured; in the concentration boundary-layer example given below, this is the turning point of the refractive index distribution, characterized in the experiment (deflection photographs, Figs. 25 and 26, p. 181, 183) by the maximum deflection.

In general, the refractive index gradient is large only in one coordinate direction, e.g., dn/dy. In the present case, only this gradient must be considered. If, however, the gradient in the other direction dn/dx cannot be neglected, the deflection angle in both directions, ϵ_x and ϵ_y, must be determined and vectorially added:

$$\epsilon_y \approx (1/n)(dn/dy)l \tag{53a}$$

$$\epsilon_x \approx (1/n)(dn/dx)l \tag{53b}$$

$\epsilon = (\epsilon_x^2 + \epsilon_y^2)^{1/2}$ is the angle between the emerging ray and the z axis.

C. Diffraction by a Parallel Slit and at the Model Boundary, $\lambda/2$ Phase Plate

As can be seen, e.g., from the schlieren methods already described and from Figs. 14–17 in Section III, diffraction phenomena restrict the sensitivity and, in some cases, the applicability of optical methods.

A considerable part of theoretical optics is devoted to a consideration of diffraction problems [e.g., Born and Wolf (1)]. These will be discussed here only basically and to the extent required for experimental application.

As already mentioned in Section II, the methods of geometrical optics (the special case of vanishingly small wavelengths) break down if the wave field experiences sudden changes or very large gradients. In these cases the assumption that the wavelength may be neglected is no longer valid, and reference must be made to the differential equation of wave optics (1). These so called classical diffraction problems are considered using the concept of a scalar spherical wave. This is Huygens' principle, introduced in Section II,D, which, as shown by Kirchhoff, follows exactly from the optical differential equations. The so-called rigorous diffraction solutions (Sommerfeld) have Maxwell's electrodynamic differential equations as a starting point; in this case the basic nonscalar electrodynamic nature of the light wave is considered.

1. *Discussion of the Diffraction Image from a Slit*

For this special case, i.e., the diffraction of a cylindrical wave front entering a very long slit of width $2b$ (Fig. 21), the Fresnel diffraction

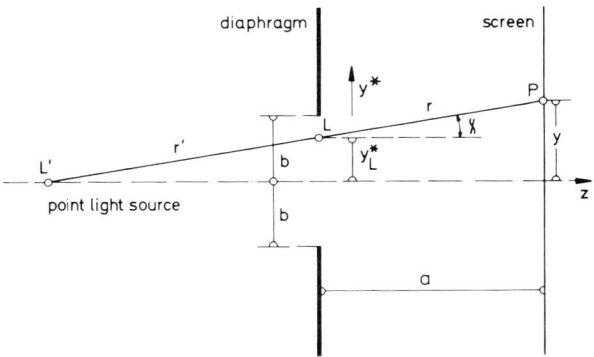

FIG. 21. Diffraction of a cylindrical wave front at a parallel slit.

patterns will be illustrated. The distances r and r' along a given ray, i.e., the distances from the light source L' to the point of intersection with the plane of the slit opening L and from L to the point of observation P on the screen, respectively, are assumed large compared to the slit width $2b$. The local dimensions in this two-dimensional problem are defined as y^* in the plane of the slit and y in the plane of the screen. The coordinate of the penetrating point is y_L^*. All coordinates on the screen are expressed as those of the corresponding penetrating points: $y_L^* = (y \cdot r')/(r + r')$.

The direction cosine of the ray $\gamma = \cos \chi$ approaches unity for large values of r', i.e., for a distant light source L'. This describes the special case of the diffraction of a plane wave front at a parallel slit. A "focal length" f, defined as $1/f = 1/r + 1/r'$, allows the imaging relations of geometrical projection to be expressed. For the case of a very distant light source L' ($r' \to \infty$), the "focal length" f approaches the now perpendicular distance a between the planes of the slit and screen.

Only the results of classical diffraction theory will be given here, the details of calculation being omitted (9). The term L, in Fig. 21, represents an elementary point light source as used in Huygens' principle; in computing the light amplitude at P, the integration extends to both sides of L in the y^* direction. The well-known result for a parallel slit is a linear combination of so-called Fresnel integrals:

$$u = [(1 - i)/2]\, u_0 (F(w_2) - F(w_1)) \tag{54}$$

where u is the amplitude of the light wave, u_0 is the amplitude of the primary undiffracted light (incoming wave) at the point P, without the parallel slit

$$F(w) = \int_0^w e^{i\pi t^2/2}\, dt$$

$$F(w) = C(w) + i\, S(w) \tag{54a}$$

$$C(w) = \int_0^w \cos\left(\frac{\pi}{2} t^2\right) dt; \quad S(w) = \int_0^w \sin\left(\frac{\pi}{2} t^2\right) dt$$

and

$$\begin{aligned} w_1 &= \gamma[-(b - y_L^*)/(\lambda \cdot f/2)^{1/2}] \\ w_2 &= \gamma[(b - y_L^*)/(\lambda \cdot f/2)^{1/2}] \end{aligned} \tag{54b}$$

are parameters in terms of quantities defined in Fig. 21. The maxima and minima of the light intensity falling on the screen, now sought, are given by the condition

$$d|u^2|/dy_L^* = 0 \tag{55}$$

since the square of the amplitude is proportional to the intensity. Squaring Eq. (54) and differentiating one obtains

$$\frac{d}{dy_L^*}(F(w_2) - F(w_1)) = \left(-\frac{\gamma}{(\lambda \cdot f/2)^{1/2}}\right)(F'(w_2) - F'(w_1)) = 0 \tag{56}$$

where

$$dw_1/dy_L^* = dw_2/dy_L^* = -\gamma/(\lambda \cdot f/2)^{1/2} = \text{const}$$

To satisfy Eq. (56), $F'(w_2) = F'(w_1)$, from which follows the relation

$$\exp(i\pi w_2^2/2) = \exp(i\pi w_1^2/2)$$

or

$$\frac{\pi}{2}(w_2^2 - w_1^2) = -2\pi m \quad (m = 0, \pm 1, \pm 2, \ldots) \tag{56a}$$

$$(w_2 - w_1) \cdot (w_2 + w_1) = -4m$$

According to Eq. (54b), the linear combinations of arguments are

$$w_2 - w_1 = 2\gamma b/(\lambda \cdot f/2)^{1/2}; \quad w_1 + w_2 = -2\gamma y_L^*/(\lambda \cdot f/2)^{1/2} \tag{56b}$$

Combining Eqs. (56a) and (56b) yields the coordinates y_L^* at which a maximum or minimum in the intensity appears:

$$y_L^* = \frac{\lambda \cdot f}{2b\gamma^2} \cdot m, \qquad \begin{matrix} m = 0, \pm 2, \pm 4, ... & \text{for maxima} \\ m = \pm 1, \pm 3, ... & \text{for minima} \end{matrix} \qquad (57)$$

The distance between two extreme values (for $\Delta m = 1$) is $\Delta y_L^* = (\lambda \cdot f)/2b\gamma^2$. The distance Δy_L^* decreases with increasing b and increases with increasing λ and f.

2. Graphical Representation (Cornu's Spiral)

If $F = C + iS$ is interpreted as a point in the complex F plane having coordinates C and S, a conformal representation of the complex w plane, with $F = F(w)$, is given. The real w axis is represented by a curve (Cornu's spiral) in the F plane. The representation is accurate in both length and angle; it is a question, therefore, of a simple unwinding. From the relation $dF/dw = e^{i\pi w^2/2}$, $|dF/dw| = 1$ and, therefore, $|dF| = |dw|$.

In constructing Cornu's Spiral, the points at $w = \pm\infty$ and $w = 0$ may be located immediately. The coordinates are

$$F(+\infty) = \frac{1+i}{2}; \qquad F(-\infty) = -\frac{(1+i)}{2}; \qquad F(0) = 0$$

This is shown from the Laplace integral, $\int_0^\infty e^{-pt^2} dt = \frac{1}{2}(\pi/p)^{1/2}$. Considering, for example, $F(+\infty)$,

$$F(+\infty) = \int_0^\infty e^{i\pi t^2/2} dt = 1/2[2/(-i)]^{1/2} = (1+i)/2$$

where p in the Laplace integral has been replaced by $-i\pi/2$.

The curve lies antisymmetric about the zero point in the F plane, the remaining part described by $F(w) = C(w) + i\, S(w)$, according to Eq. (54a). Not only the various diffraction patterns from slits but also those for the limiting case of a slit, i.e., the infinite half plane (model boundary), and for the case of a $\lambda/2$ phase plate (Wolter) may be constructed from this curve.

3. Construction of the Slit Diffraction Pattern

From Eq. (54), the ratio of the amplitudes in the diffraction pattern to the amplitude u_0 of the primary undiffracted wave is, for this case, given by

$$|(u)/(u_0)| = (1/\sqrt{2}) \cdot |F(w_2) - F(w_1)| \qquad (58)$$

The magnitude of the expression $\sqrt{2}\,|(u)/(u_0)|$ is equal to the length of the chord joining the points $F(w_2)$ and $F(w_1)$.

In Fig. 22, such a chord, corresponding to $y_L{}^* = 0$ in the diffraction

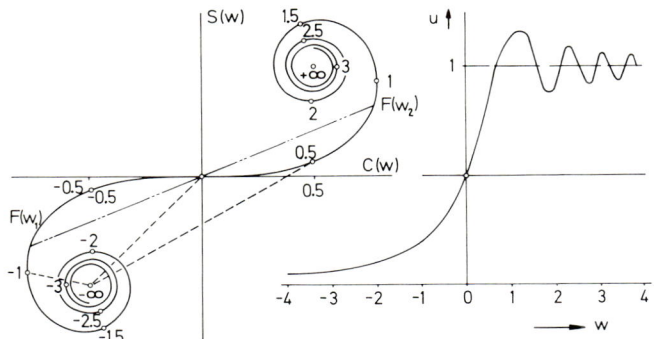

FIG. 22. Left: Cornu's spiral in the complex F plane. Parameter in the positive quadrant, $w_2 = \gamma[(b - y_L{}^*)/(\lambda \cdot f/2)^{1/2}]$; parameter in the negative quadrant, $w_1 = \gamma[(-b - y_L{}^*)/(\lambda \cdot f/2)^{1/2}]$. Right: Diffraction image from a screen edge (model boundary) constructed using Cornu's spiral. (Length of the broken lines corresponds to ordinates on right.)

pattern, has been indicated (dot-dash line). The length of the curve segment corresponding to this line is constant since, according to Eq. (56b),

$$w_2 - w_1 = 2b\gamma/(\lambda \cdot f/2)^{1/2} = \text{const}$$

it is independent of $y_L{}^*$ and, therefore, of the coordinate y of the observation point P. This is valid for all points $P(y)$ for which the intensity ratio is to be determined. The curve segment is therefore shoved along the spiral curve to the position corresponding to a given $y_L{}^*$. The calculated w_1 and w_2 form the end points of a new chord. The plotted length of this chord against w, for example, gives the desired diffraction pattern.

4. Construction of the Diffraction Pattern of a Half Plane (Model Boundary)

The diffraction pattern from a half plane, equally important for shadow and interference techniques and especially important in boundary-layer investigations, is shown in Fig. 22 for the special case of a plane wave $f = a$, $\gamma = 1$. In this case one of the edges of the slit is moved to $-\infty$ while the other remains fixed.

According to Eq. (58),

$$|(u)/(u_0)| = (1/\sqrt{2}) \cdot |F(w) - F(-\infty)|$$

$w = y_L^*/(\lambda \cdot a/2)^{1/2}$ where now y_L^* is measured outward from the edge of the unmoved slit. The construction of the diffraction pattern is analogous to that in the previous example where, however, the point $w = -\infty$ is now fixed. For example, three of the lines, corresponding to $w = -1$, $w = 0$, and $w = +0.5$, have been drawn as dashed lines in the left diagram of Fig. 22, and shown in the neighboring diagram as ordinate. For the important point $F(w) = 0$ (boundary of the geometrical shadow),

$$|u/u_0| = \tfrac{1}{2}$$

since

$$|u/u_0| = (1/\sqrt{2}) \cdot |F(-\infty)| = (1/\sqrt{2})|-(1+i)/2| = \tfrac{1}{2}$$

The geometrical shadow (position of the wall) in an interference pattern has, therefore, one quarter the intensity of the undiffracted light. The positions of the first maxima ($m = 0, \pm 2, \pm 4,...$) and minima ($m = \pm 1, \pm 3,...$) are given by Francon (15, p. 372):

1. Maximum	1. Minimum	2. Maximum	2. Minimum
$w = 1.2172$	1.8725	2.3445	2.7390

For the specific data $f = a = 0.5$ m, $\lambda = 0.5 \cdot 10^{-6}$ m (medium wavelength, plane wave), the position of the first maximum is $y_L^* = 0.45 \cdot 10^{-3}$ m.

It now becomes evident that only boundary layers having thicknesses many times that of that diffraction region, resulting from the boundary or wall of the model, are suitable for investigation using schlieren or interference methods. On the other hand, a very narrow boundary layer in liquids acts as a slit which appears on the schlieren pattern intermixed with diffraction interference lines.

5. Construction of the Interference Pattern at a Phase Edge ($\lambda/2$ Layer), after Wolter

Two different diffraction patterns, this time represented in a log scale, corresponding to how the eye perceives or to the sensitivity of a photographic film, are shown in Fig. 23. Figure 23a again shows the diffraction

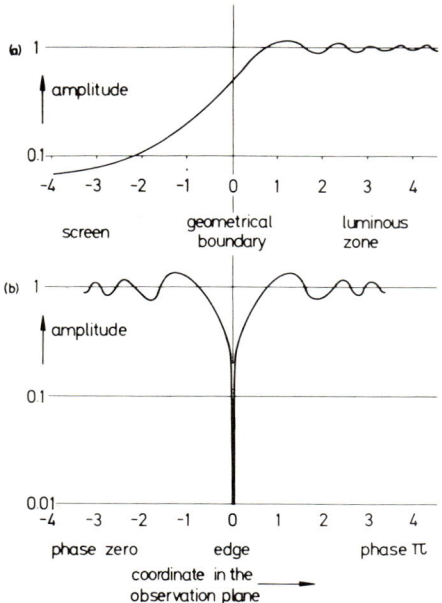

Fig. 23. Interference patterns to a logarithmic scale: (a) Model boundary (half plane, semi-infinite screen); (b) phase plate ($\lambda/2$ layer) (Wolter).

pattern from a half plane where the intensity ratios correspond to the diffraction patterns [see Wolter (16)]. In Fig. 23b a very sharp minimum may be seen in a symmetric diffraction pattern. This was obtained using Wolter's (16) experimental arrangement where a portion of the wave front, assumed here to be a plane, suffered a phase change of $\lambda/2$, compared to the other part of the front. This was produced by using a thin film perpendicular to the ray direction (phase plate).

This diffraction pattern may also be constructed using Cornu's spiral if, for the case of a half plane, the zone giving the shadow is replaced by the half of the wave front having the shifted phase, i.e., by the negative of the undisturbed half of the wave front. According to Eq. (54),

$$(u/u_0) = [(1-i)/2][F(w_2) - F(-\infty)]$$
$$(u'/u_0) = [(1-i)/2][F(w_2') - F(-\infty)]$$ (58a)

Furthermore,

$$w_2 = y_L{}^*/(\lambda \cdot a/2)^{1/2}$$
$$w_2' = (-y_L{}^*)/(\lambda \cdot a/2)^{1/2} = -w_2$$

Since one partial wave is shifted in phase by π relative to the other, it follows, for the total light disturbance at a specific point, that

$$(u/u_0)_{\text{tot}} = (u/u_0) + (-1)(u'/u_0) = [(1 - i)/2][F(w_2) - F(-w_2)]$$
$$| u/u_0 |_{\text{tot}} = (1/\sqrt{2}) | F(w_2) - F(-w_2)|$$
(58b)

The relative amplitude may clearly be obtained from the length of the line connecting the two symmetric points $w_1 = -w_2$. This line must always pass through the point of antisymmetry $w = 0$, as it is indicated in Fig. 22 by the dash-dotted line.

D. SCHLIEREN METHODS WITHOUT IMAGE FORMATION (WIENER, 1893)

1. *Measurement Ray Principle*

An example which fundamentally illustrates this methods has already been given in Section IV,A, i.e., Wiener's diffusion investigation.

The schlieren region is probed in this case using a narrow bundle of light (measurement ray). The deflection $\epsilon = (\epsilon_x^2 + \epsilon_y^2)^{1/2}$ is measured at every point (x_0, y_0) in the plane at the entrance of the light bundle into the schlieren region and, from this, a field consisting of lines of constant deflection is obtained. From this field, using one of the above methods, which are based on the differential equation describing the light ray path, the refractive index field may be computed. If the refractive index field is stationary, the region may be investigated point by point, e.g., by moving the diaphragm in known intervals in the y and or x direction and in each case recording the deflection photographically. An example of this type of probing is given by Sperling (17) who measured the temperature distribution in a vertically burning carbon arc. In this work a vertical slit diaphragm was moved horizontally by a screw mechanically coupled to the advance mechanism of the camera; each time the film was advanced, the diaphragm was moved by a fixed interval. In this way the deflection curves were recorded in the resulting photographs.

For a nonstationary process in which the above method cannot be used, the next step would be to place parallel slits of a screen in front of the schlieren object. In this way one obtains in a single photograph the distortion of this grid. However, because of the diffraction at each slit, the grid dimensions must be kept large to avoid the superposition of maxima from neighboring slits. Therefore, especially for thin boundary layers, little information can be obtained.

For one-dimensional fields, the method using the angularly placed slit described at the beginning or, what is equivalent, a relatively large triangular-shaped bundle of parallel rays whose outline is formed by a mask can be used. The straight line or side of the triangle ($x/y = $ const) establishes, for each deflected point of light in the photograph, the original coordinates (x_0, y_0), where the light ray entered the schlieren region. In this case also, the diffraction patterns set bounds on the sensitivity and the ability to analyze. The slit and edge of the triangle appearing in the photograph are flanked by diffraction maxima which limit the information obtainable, especially when the deflection is great and on the slit boundary formed by the wall of the model.

These secondary disturbing influences from diffraction may be suppressed by breaking the slit into a line of single points. The superposition of the diffraction on the point then appears less disturbing. A comparison of deflection patterns of these kinds is shown in Fig. 26, p. 183.

It is recommended for taking schlieren photographs that the light used be not strictly monochromatic but rather consist of a mixture of wavelengths in a band $\Delta\lambda$. This may be obtained, for example, by passing the continuous spectrum light from a carbon arc through a color filter with a relatively large band width. The diffraction patterns from the light of different wavelengths λ in the band are superimposed and therefore appear weaker.

2. Ray Identification Using Intensity Maxima

The effectiveness of using a bundle of rays as a "measurement ray" can be determined qualitatively also for other ray boundaries, e.g., circular diaphragms, from the known diffraction by a slit or half plane (triangular mask).

The position of maximum y_L^* in a diffraction pattern such as from a slit is given by Eq. (57). With the notation of Fig. 21,

$$y_L^* = (\lambda f/2b\gamma^2)m, \quad m = 0, \pm 2, \pm 4, \ldots$$

For thin boundary layers, it would be often desirable to keep the slit breadth $2b$ very small; on the other hand, however, the diffraction pattern then expands according to the above equation, the position of the first strong intensity maximum being especially affected. As a result, for a grid consisting of slits there is a limiting minimum grid constant.

For a single slit, there is evidently an optimum slit breadth for which the width of the first-order maximum is smallest. For smaller slit breadths, according to the above equation, the lateral displacement of the primary maximum becomes greater.

In addition, the property of the photographic materials (steepness of the density curve) used in recording the diffraction pattern is decisive.

Especially, the intensity of the slit illumination is set so that, with the given exposure duration, the undesired neighboring maxima are to fall below the threshold value of exposure. These factors have been more closely investigated by Wolter (*18*).

3. *Ray Identification Using Intensity Minima*

The light energy in a bundle of rays is always scattered into the regions to the side of the light path. The bundle is, therefore, never sharply defined.

The method of ray identification using intensity maxima is inherently limited in accuracy. As shown by Wolter (*18*), the sharpness of the indication can be much improved by using the intensity minimum of a $\lambda/2$ phase plate. For example, if a $\lambda/2$ phase plate is substituted for an optimum slit in a rotating-mirror instrument (galvanometer) the sharpness of the light pointer can be improved by a factor of 25. However, this degree of sharpness can only be attained if the light beam passes through a homogenic medium after leaving the phase plate and if the phase plate is imaged on the scale. If, e.g., in the case of the schlieren methods, the plane of the phase plate is not imaged on a screen in order to register the deflection ϵ, the distance between the plate and the screen should not be too great because the slope of the flanks is then flattened. With greater spacings between the plate and the screen, the larger "focal lengths" f lead to smaller parameters w for the same screen coordinates. The diffraction pattern is then widened as can be seen in Eq. (58b). Furthermore, it should be considered that the light beams traverse optically inhomogenic regions in the schlieren apparatus. Nevertheless, using a $\lambda/2$ phase plate in Wiener's diffusion experiment (Section IV,A,1), for example, leads to improved accuracy.

The suitability of the phase plate method is better for decreasing spacings between the phase plate and the screen. This method is therefore better suited for ray indication in liquid boundary layers, with their small dimensions and strong deflection, than in gas boundary layers; i.e., if the distance from the phase plate to the observation plane can be kept smaller than the several meter length of the light pointer usually used in schlieren instruments for use with a gas.

It is possible to produce a phase plate by coating one half of a plane, parallel, schlieren-free glass plate with a thin zapon lacquer film, or a suitable transparent evaporated quartz film. The thickness of this film is such that the plate splits a plane wave front passing through it into two separate plane wave fronts 180° out of phase. The thickness of the layer is then $\lambda/2(n - 1)$, n is the refractive index of the lacquer film. A suitably thinned lacquer solution on a vertically placed glass plate produces, for a given amount of wetting, films which are reproducible. These can be tested for their effectiveness in shifting the phase by approximately $\lambda/2$ in an interference system; however, a purely visual examination is often sufficient. In order to obtain an approximately step-shaped separating edge between the coated and the uncoated portions of the plate, the lacquer film must be cut.

The interference pattern of the phase plate, Fig. 23b, shows that theoretically an interference minimum of any degree of sharpness can be obtained by using an exposure of suitable duration. The darkening of the photographic film is approximately proportional to the logarithm of the exposure intensity, where the exposure intensity is the product of the luminous intensity and the duration of the exposure. The effect of a step by step overexposure can be seen in Fig. 23b. A certain exposure time would produce exactly the threshold value of the exposure at unit amplitude. A photograph then reveals symmetrical maxima (amplitudes >1) and minima with a broad interference minimum at the position of the separating edge. A tenfold exposure duration would produce the threshold value for an amplitude 0.1; in this case, the light ray marker is a very narrow unexposed strip set off against the exposed background. A further decrease in the width of this minimum is limited by the film grain size and especially by light scattering which reaches the threshold point of the photographic film at a certain exposure time, exposing it at the position of the minimum. Wolter gives an optimum exposure duration of half the time required for the scattered light to reach the threshold point. However, for other reasons, this rather long exposure time is not desired in schlieren experiments.

Over a certain range of the light path, sharp interference minima can also be produced using a Fresnel biprism which, therefore, can also serve as a ray marker. Such a prism replaces the double mirror in Fresnel's well-known classical mirror experiment on interference production. Two very flat 90° prisms set together with two short sides of the prisms adjacent replace the function of the mirror. The center angle thereby produced is close to 180°. The development of interference lines with straight-line light propagation is easy to comprehend [cf. Born (*1*), p. 262]. However, for an exact analysis of schlieren photographs, the curvature of the light ray must be considered.

For the tracing of light rays, one could consider using the characteristic diffraction pattern of a rectangular opening (with a wide slit), even though the minima are not as distinct as those of a phase plate or a biprism. The investigation here (Fig. 24) was carried out only to demonstrate such an influence of a refractive index field on the formation of diffraction patterns

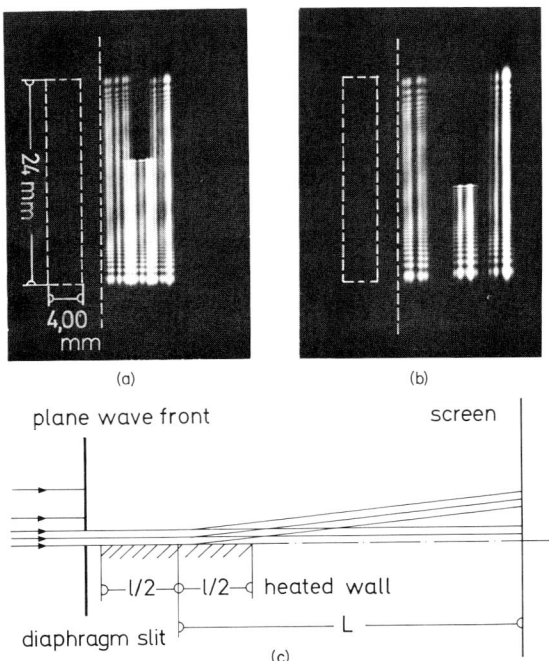

FIG. 24. Diffraction pattern of a slit diaphragm with deflections resulting from respective passage through the refractive index fields of two thermal boundary layers in air (natural convection). The vertical wall of the model is indicated by a broken line. (a) $\Delta\vartheta = 0, 25, 50°C$; $L = 2.2$ m; $l = 0.35$ m. (b) $\Delta\vartheta = 0, 38, 65°C$; $L = 3.2$ m; $l = 0.35$ m.

or the interference lines of a Fresnel biprism. The arrangement, consisting of a rectangular slit in front of a heated vertical wall, may be seen in Fig. 24. The diffraction patterns have, in each case, been recorded on a single photographic plate so that the deflection of the whole diffraction pattern as well as the influence of the increasing ray curvature can be seen. The broken line shows the position of the heated walls. The formation of a focal line, described in Section III,B,3, especially for the case of the largest deflection, can be recognized.

4. *Schlieren Patterns from Various Boundary Layers*

a. Thermal Boundary Layer. As an example showing the application of Wolter's phase plate, the thermal boundary layer on a heated vertical wall in a fluid (water) at rest is considered.

The optical system corresponds to that in Fig. 13 where, however, the triangular mask is replaced by the dividing edge of a diagonally placed $\lambda/2$ phase plate. The model is a plate (heated with a thermostat fluid) whose dimension in the direction of the light ray is 0.1 m. The test section has the dimensions 0.8×0.8 m. A carbon arc light was used to produce a parallel ray illumination; as usual, the light was made parallel by passing it through a condenser, a green filter, and a circular diaphragm on to a concave mirror ($f = 3$ m). The distance from the middle of the plate to the plane of observation is $L = 0.7$ m. In contrast to the analogous photographs using air, Figs. 14–17, this distance was held constant. According to Eq. (35), as a consequence of the dimensionless temperature profile, similar patterns on the screen are produced by either a variation in the temperature difference $\Delta\vartheta = \vartheta_w - \vartheta_\infty$, or distance to the screen L.

The deflection by the refractive index distribution in the boundary layer is demonstrated for comparison purposes using a triangular mask and a phase plate at six temperature differences, $\Delta\vartheta = \vartheta_w - \vartheta_\infty$. The broken lines indicate the position of the wall. The exposure time is 3 sec. The system corresponds approximately to that in Fig. 13. In comparing the patterns, the growth of the boundary layer in the x direction must be considered.

The quality of the deflection pattern using the phase plate can be compared in Fig. 25 to that obtained using a triangular mask. The phase plate is covered except for a slit-shaped area around the separating edge

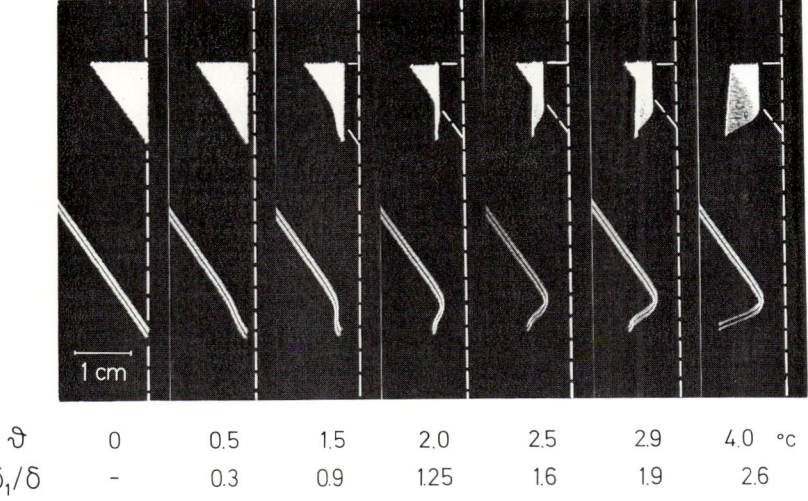

ϑ	0	0.5	1.5	2.0	2.5	2.9	4.0	°C
δ_1/δ	–	0.3	0.9	1.25	1.6	1.9	2.6	

FIG. 25. Thermal, free convection boundary layer on a vertical wall in water.

so that the deflected portions from the separating edge do not become illuminated from the remaining bright field. The reversal in direction of that end of the trace corresponding to the rays closest to the wall back into the original direction of the diagonally placed phase edge (a result of the boundary layer "wandbindung"), can be seen more clearly when the phase plate is used ($\delta_1/\delta = 0.3$, $\delta_1/\delta = 0.9$). In addition, the formation of focal lines becomes visible in the region of the refracted curve running parallel to the wall ($\delta_1/\delta = 0.9$).

b. Concentration Boundary Layer. A substantially different type of boundary layer develops through the isothermal injection of a fluid (water) through a porous, vertical wall into an enveloping fluid (0.1% glycerin solution) at rest. The same test chamber (model length in ray direction $l = 0.1$ m) and also the same optical setup as in the preceding case are used.

The fluids chosen are such that, as with a thermal boundary layer on a heated wall, the light is deflected away from the wall. Also, the expected refractive index gradients are of approximately the same order of magnitude. The velocity of water leaving normal to the wall is 2 mm/min; this velocity is such that, for a stationary process, a region of pure water exists close to the wall, around which is a narrow mixing region and finally the unmixed glycerin solution. This experiment has a certain similarity to Wiener's diffusion experiment (Section IV,A,1), however, the mixing zones developing from diffusion are considerably larger than those for the boundary-layer conditions considered here.

The refractive index distribution of the boundary layer, described below, and obtained using the grid consisting of holes (Fig. 27), is drawn in Fig. 26. The refractive index field corresponds to the concentration field if the gradient $dn/dC = (n_2 - n_1)/(C_2 - C_1)$, $dn/(n_2 - n_1) = dC/(C_2 - C_1)$, is assumed constant. The term n_1, n_2 and C_1, C_2 are defined, respectively, as the refractive indices and concentrations of the two components.

In Fig. 26, a and c are, respectively, the zones of the pure component 1 (water) and the pure component 2 (0.1% glycerin solution). In between lies a mixing zone with the steepest gradient and the largest deflection at b. The zones a and c contain no gradients; the portion of the triangle near the wall and its apex, as well as the corresponding portions of the interference minimum from the phase plate, are not deflected. The very narrow mixing zone appears black in the triangle since the correspondingly small portion of light becomes so strongly deflected by the steep gradients that considerable difficulties develop in taking the photograph; the intensity of the deflected light is too weak compared to the

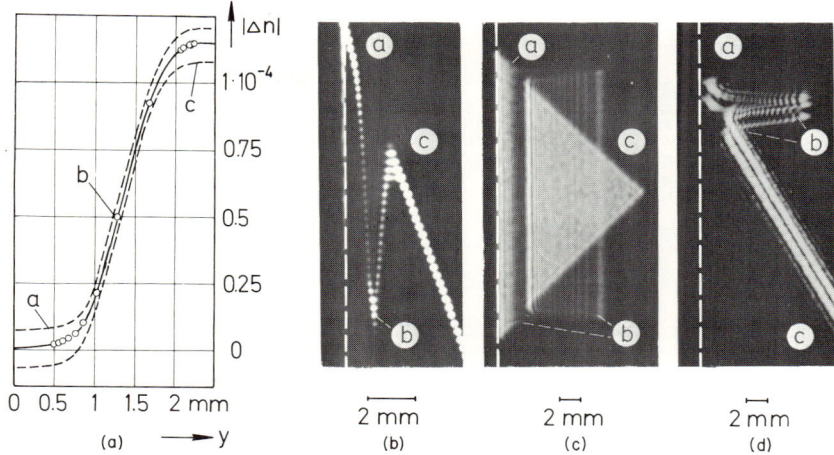

FIG. 26. A Comparison of various ray traces for a concentration boundary layer. Region a is a zone containing Component 1 (water) which enters through a porous wall ($v_{H_2O} = 2$ mm/min); next to this is the mixing zone with the steepest gradient at b; c is the region containing Component 2 (0.1% glycerin solution). The white broken line shows the position of the wall. (a) Variation of the refractive index; (b) grid (evaluated); (c) triangular aperture; (d) interference minimum (Wolter).

background illumination to visibly stand out well or to form the interference minimum from the phase plate.

A setup, using higher-intensity light, where the shift on the photograph is directed parallel rather than perpendicular to the wall, is shown in Fig. 27. A "slit," consisting of a line of small holes, provides the trace; a phase plate could have been used as well.

The grid consisting of holes is set diagonal to the porous wall (x axis) and illuminated with diffuse directed light (i.e., from each diaphragm point of the grid, a cone of light rays emerges of such a size that the boundary layer zone of the test section becomes illuminated) from a lamp and condenser. The light is reflected by a concave mirror, serving as a long focal length camera lens, so that it emerges from the test section, passes through a circular diaphragm located at the focal point of the mirror, and falls on the film. The diaphragm therefore allows only those narrow ray bundles emerging perpendicularily from the test section to pass.

In this case, the ray pattern is practically the reverse of that in the previous experiment. From the rays drawn in Fig. 27, one sees that various regions of the refractive index field could have a given hole in

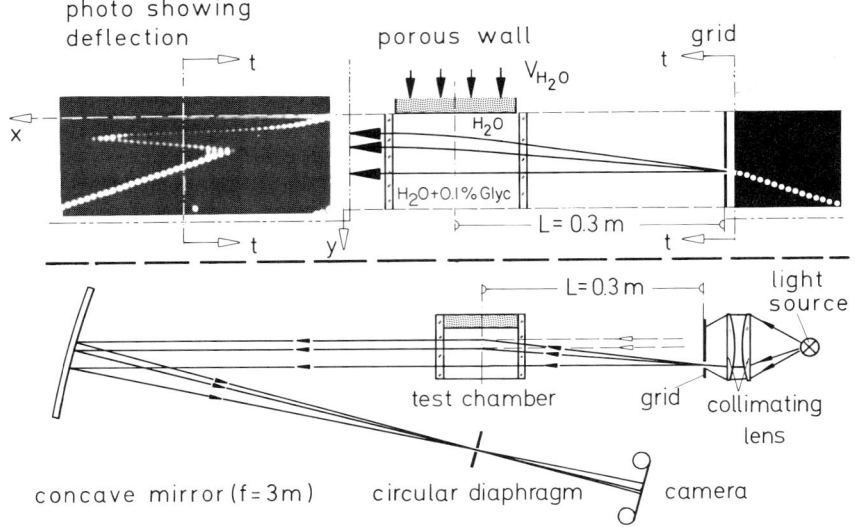

FIG. 27. Schlieren setup with diffuse lighting of the grid.

the grid as a light source. Using the broad light cone from this light source leads to a more intensive illumination of every region.

It is true that the objective plane, the conjugate of the film plane, does not lie at infinity, although it is very far behind the plane of the grid (the telecentric path of the rays from the concave mirror). It follows that grid points projected through the homogeneous medium in the test section are not sharply focused, i.e., they are enlarged by diffraction displacement (Fraunhofer–Fresnel diffraction). For the region in which the boundary layer acts as a "schlieren lens," with variable positive "focal length" dependent on the refractive index gradient, the conjugate objective plane becomes distorted. This plane lies closer to the plane of the grid.

If in this experimental setup one imagines that the light source (which can be thought of as a point source), condenser, and the grid (e.g., a crossline grid) are exchanged with the camera so that the former film plane becomes the plane of the grid, then the image of the grid by means of the finite aperture of the hole diaphragm (see also the following section on Abbe's image theory) forms behind the test section. The narrow bundle of light entering the test section perpendicularily has a sufficient aperture to bring about the formation of the image. This investigation, more suited for demonstrating the "schlieren lens," permits the distortion and the distorted plane of focus to be clearly seen on a screen.

The upper half of Fig. 27 shows the relation of a point on the grid in the $x = $ const plane, seen from above, to two deflected points, each corresponding to one half of the symmetric refractive index profile. The boundary-layer photo and the upper portion of the grid appear in the figure rotated about the t–t axis into the plane of the drawing ($x = $ const).

If, as in this case, the boundary layer over the height of the plate is uniform enough, a second grid, whose distance L_1 can be varied, could be used. The slope of the ray could then be additionally obtained from the difference in the deflection pattern,

$$y' = \Delta e / \Delta L = (e - e_1)/(L - L_1) = \tan \epsilon$$

and, in the analysis of the results, the positions of ray entrance onto and emergence from the test section would be known to a first approximation. The refractive index profile (Fig. 26a) obtained by this diffusion boundary-layer experiment has an uncertainty of $\pm 12\%$ (indicated by broken lines). It is known empirically that the accuracy of boundary-layer measurements using schlieren methods is comparable to that of normal calorimetric experiments.

E. Schlieren Methods with Image Formation (Toepler, 1864)

Toepler's schlieren method is known in ballistics and glass technology simply as "the" schlieren method. It can be always advantageously employed if, e.g., in the cases mentioned, the local position of shock waves or inhomogeneous zones in glass plates are to be determined, but the magnitudes of the refractive index distribution need not be known in great detail. This method surpasses even the detection sensitivity of other methods, including interference methods, and is sometimes the only usable optical method, e.g., in the case of very small refractive index gradients in processes occurring in rarefied gases.

In contrast to methods in which images are not formed and from which the field of lines of constant deflection $\epsilon = $ const is obtainable only through calculations, the image-forming methods permit the recording of this field in the pattern developed by the inhomogeneous region.

1. *Toepler's Schlieren Apparatus*

The Toepler method, with some variations, will now be illustrated for the case of a test section having one vertical wall heated and one cooled.

The schlieren apparatus in Fig. 28 contains a vertical slit diaphragm which is usually illuminated using light from an arc. In the horizontal

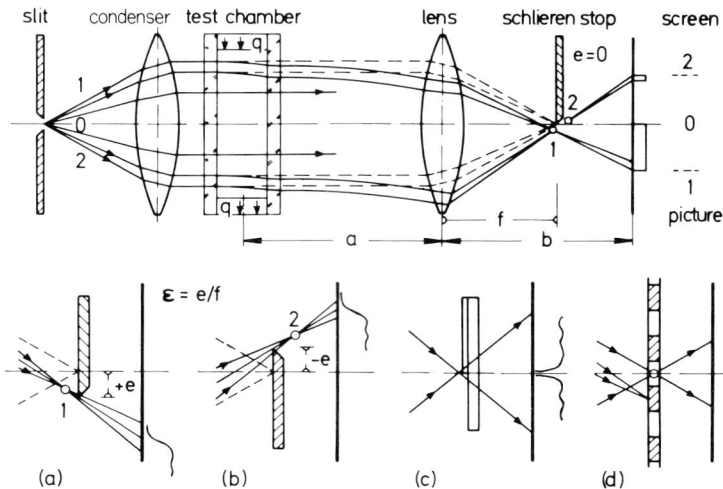

FIG. 28. Schematic of Toepler's schlieren apparatus. The test section has one heated wall and one cooled wall: (a,b) Knife edge, displaced a measured distance e from optical axis; (c) knife edge replaced by Wolter's phase plate; (d) knife edge replaced by a grid of transparent and opaque zones (or colored transparent zones). [(a–d) The intensity distribution on the screen, representing a line $\epsilon =$ const, is shown schematically.]

plane shown, the condenser lens produces parallel rays in whose path the object under investigation (in this case the test section) is located. The lens images the central plane of the test section on the screen.

Next, suppose that the knife edge is removed and the test section is neither heated nor cooled. In this case, the broken-line light path is obtained. This forms the image of the light-source slit in the focal plane of the lens, and of the test section outline on the plane of the screen. The rays diffracted at the model boundaries (diffraction by a half plane) are recombined by the "ideal" lens at the plane of the screen giving the projected outline of the model (stigmatic image). In the focal plane the diffracted rays run alongside the image of the light-source slit; these are not, however, shown in Fig. 28.

What has just been said serves only to illustrate how an image is formed in a parallel light path according to Abbe's theory; a pure shadow projection is no way implied. The image of the test section on the screen is indicated by 1 and 2.

A knife edge is now introduced at the focal plane of the lens so that one half the image of the light-source slit is covered. This position of the knife edge on the optical axis is given the designation $e = 0$.

The test section remains, however, in its original condition, i.e., with no thermal boundary layer. A portion of the test section image appears dimmed on the screen (rays 0–2); the other (rays 0–1) remains illuminated since one half of the light-source image in the focal plane is not covered by the knife edge.

Let one of the vertical walls of the test section be weakly heated or the other cooled, as is shown in Fig. 28 by the direction of the heat flux q. Through natural convection, thick boundary layers which act as "schlieren lenses" develop. The bundles of parallel rays at 1 and 2 are, respectively, deflected toward the denser medium, the boundary layer on the heated wall acting as a weak collecting lens. The corresponding ray paths are illustrated by the solid lines in Fig. 28. The image of the light-source slit in the focal plane of the lens becomes slightly distorted into a three-dimensional surface. The portion of the rays from the heated wall (1) are combined at position 1 on one side of the optical axis and in front of the knife edge located in the focal plane of the lens. Analogously, the portion coming from the vicinity of the cooled wall (2) is combined behind the schlieren edge at position 2. The powers of refraction of the "schlieren lenses" and that of the lens itself are additive. The thermal boundary layer on the heated wall, 1, has a positive power of refraction, therefore shortening the focal length of the lens; that on the cooled wall, 2, with a negative power of refraction, increases the focal length. The local deflection in the thermal boundary layer, and therefore the power of refraction, varies locally, the latter varying from zero to its maximum value at the wall. The distorted light-source image in the focal plane of the lens spreads over a spatially curved surface from 1 to 2. The unrefracted rays passing through the central portion of the test section are reunited at the knife edge.

The following may be observed on the screen: The portion of rays from the heated wall (1) falls in the illuminated region of the pattern and produces only a displacement of the model contour. The portion from the cooled wall (2) produces a brightening of the region in shadow; depending on the magnitude of the deflection, more and more light bundles get by the knife edge. The deflection produces a gradual lightening and is transformed into a photographically recordable intensity distribution to be densimetrically analyzed.

The dependence of the intensity distribution on the deflection is calculable (2, 4). However, in practical cases, the influence of scattered light, inexact adjustment of the knife edge, and, not least, the fact that the response of the film is not linear limit the accuracy of the method.

As in the densimetric method, it is best to use a neutral gray wedge which, by its gradations of gray, gives an index for the deflection. For this purpose, e.g., a thin planoconvex lens may be used together with the test section. The focal length of this lens and therefore its refraction properties over the radius are known; these appear in the image as gradations of gray. For an investigation using a gas, a soap bubble filled with the test gas of known density may be used instead in the test section.

For more intense heating and stronger deflections, corresponding to the sketch in Fig. 28, very specific areas of the focal surface can be scanned through measurable displacements of the knife edge, as shown in Fig. 28a and b. In Fig. 28a, where the knife edge is displaced a distance e, only the bundles with deflections $\epsilon > +e/f$ are detected, the remainder being cut off. A bright–dark zone develops on the screen indicating all portions having deflection $\epsilon > +e/f$. In other words, the bright–dark path represents a line of constant deflection. Successive movement of the knife edge in a positive or negative direction (Fig. 28b) allows the field lines, with $\epsilon = $ const, to be obtained. However, this analysis is not exact since the shadow of the knife edge on the screen depicts the diffraction pattern of a half plane instead of a sharp light–dark zone. Theoretically, the boundary of the shadow is given by the location where the brightness is one quarter of the intensity outside the diffraction zone (see Section IV,C, diffraction).

Replacement of the knife edge used here by Wolter's phase plate in this case also leads to an improvement in the results. Using the phase plate, no dark zones are exibited on the screen; only a line $\epsilon = $ const traced out by Wolter's interference minimum, Fig. 28c, appears.

2. *The Grid-Diaphragm Method*

In the case of strong deflection, and especially for reducing the light from self-illuminated objects, a slit, whose position is successively altered, may be used instead of the knife edge (*17*). The line $\epsilon = $ const is traced out in this method by a bright line in a dark field. In the analysis of the data, the diffraction pattern of the slit must be taken into consideration.

The next step would be to replace the movable slit diaphragm by a grid, consisting of parallel opaque and transparent strips, placed in the focal plane of the lens (Fig. 28d). Each transparent strip would allow all or part of the deflected portion of the light-source image at specified distances e to pass through. In this way, bright zones of constant deflection appear on the screen having boundaries of graduated brightness. Analogous to the preceding example, the data must be analyzed densimetrically. The grid constant is limited by the condition that the light-source image with the first diffraction maxima must be allowed to pass through the slit unhindered. A more detailed description of this method, together with many modifications of the schlieren setup, has been given by Schardin (2). This grid-diaphragm method was applied by Schardin and Gaebler to the case of natural convection on a horizontal heated cylinder. For this rotationally symmetric application, the deflections must be obtained using two photographs, giving ϵ_x and ϵ_y components, obtained from a horizontally and vertically placed slit.

3. *The Colored-Schlieren Method*

Instead of the transparent and opaque slits used in the grid-diaphragm method, a color-zone filter can be utilized. This consists of parallel, colored slits for a slit-shaped light source, and of concentric colored rings for a point light source. A one-dimensional (ϵ_x, ϵ_y deflection) or two-dimensional [$\epsilon = (\epsilon_x^2 + \epsilon_y^2)^{1/2}$] pattern showing regions of constant deflection are obtained, the total deflections being indicated, respectively, by the colors of the filter strips. If complementary colors are successively used, this rather qualitative method yields also intermediate values between the major gradations in color concentration [see Schardin (2)]. A continuous sequence of colors is obtained using a dispersion prism.

F. SHADOW METHOD (DVOŘAK, 1880)

The shadow of an inhomogeneity or a refractive index field in a parallel or weakly divergent light path (as produced by a distant point light source) is not object preserving and is comparable to the nonimage-forming methods described above. In these, the origin of the light ray bundles was traced, this being the basis of a method for quantitatively analyzing the results. This is not the case in this method if one disregards the focal line formation by the boundary layer acting as a "schlieren lens."

The outer border of the heart-shaped curve in Fig. 29a is the area of uniformly deflected light rays from the vicinity of the wall of the heated cylinder, where the refractive index gradient is constant. From the considerations in Section III, the local heat transfer can be determined to a certain degree of accuracy from this pattern. The accuracy is comparable to that using nonimage-forming methods, i.e., of the order of 10%.

Schmidt (20), who demonstrated this possibility of the shadow method, carried out investigations on heat transfer by natural convection from various horizontally placed cylindrical and prismatic bodies in air. For model dimensions and temperature gradients corresponding to those in Fig. 29, Schmidt has given an optimum distance from the screen to the center of the model for which this focal line has its optimum sharpness.

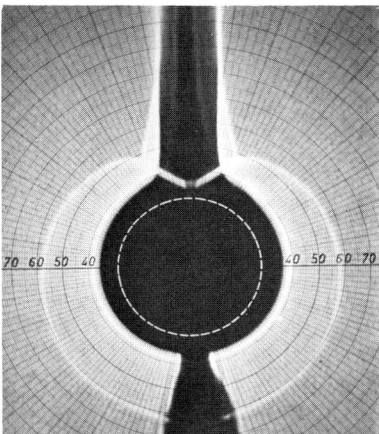

Fig. 29a. Focal-line formation of a thermal boundary layer in a parallel light ray. The dotted line indicates the contour of a heated horizontal cylinder in free convection. Cylinder length, $l = 0.3$ m; temperature difference, $\Delta\vartheta = 80°C$; distance to screen, $L = 9$ m. [Photograph by Killermann (19).]

This distances, empirically based, is $L = 8$–15 m. The compensating effects of the dispersion in the region of the "schlieren lens" close to the wall and the diffraction of the wall, together with the small coherence of white light from an arc, favor the formation of more sharply defined focal lines.

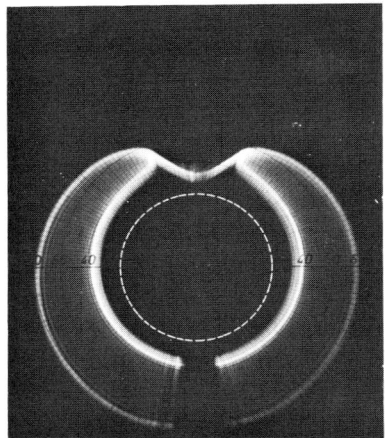

Fig. 29b. Horizontal heated cylinder with ring-shaped diaphragm allowing passage only to the incoming light in the region close to the wall. Data as in Fig. 29a. [Photograph by Killermann (19).]

In Fig. 29b, the entire bright field, with the exception of a narrow ring-shaped zone around the cylinder, is covered by a diaphragm at the position where the rays enter the boundary layer. For this reason, the outer focal line appears more distinctly on the dark background. The inner focal line and the shadow of the core are approximately the bounds of the thermal boundary layer (compare with Fig. 12).

V. Theory of Interference Methods

Compared with shadowgraph and schlieren methods the interference methods offer more detailed information about the model which is to be investigated (which in the following is termed the phase object). These allow a greater accuracy. However, they are more often applied to quantitative measurement in spite of the added complexity and cost and the usually more restricted range of measurement compared with the shadow and schlieren methods.

The two-beam interferometers, which are used for measurements on transparent objects, greatly differ in constructional details and sophistication. Accordingly the versatility for practical quantitative measurements varies. The most common, but also the most expensive, is the Mach–Zehnder interferometer. In the following this instrument will be considered exclusively and the behavior of the phase objects, e.g., thermal boundary layers, which is principally the same for all two-beam interferometers, will be demonstrated with it.

A. Methods of Two-Beam Interference

1. *Principle of Two-Beam Interference*

As has been set forth in Section II,B, all of the information about a schliere (phase object) is contained in the deformation of an originally plane wave front (Fig. 1). According to the premises of geometrical optics this wave front (surface of constant phase) is an equipotential surface (eikonal). The change of ray direction (normal to the wave front) is essential for the shadowgraph and schlieren methods.

The structure of such a wave front can be made visible by interference methods. The phase differences of these deformed wave fronts (compared with a reference wave front) effect, by interference, changes in the intensity and are thereby made visible. Two of the most important possibilities for interference will be put forward in the following.

a. Normal Two-Beam Interference (Mach–Zehnder). Figure 1 shows the interference pattern of the variation of temperature (index of refraction) of the cross section of a tube of 40 mm diam. The interference pattern—denoting maximum and minimum intensity—are lines of constant phase difference with reference to the original plane wave front and are—in this special case—isotherms.

Interference lines in Fig. 1 are contour lines of the spatial wave front in the z direction, which in this case also indicate the spatial temperature distribution. The contour lines are the intersections of parallel equidistant plane surfaces ($g = S \cdot \lambda_0$) with the distorted wave front. The profile of the center of the wave front depicted in Fig. 2 is starkly exaggerated in the z direction. The same profile of the spatial wave front is utilized in Fig. 30a (the scale of the z axis has been multiplied by a factor of 5250),

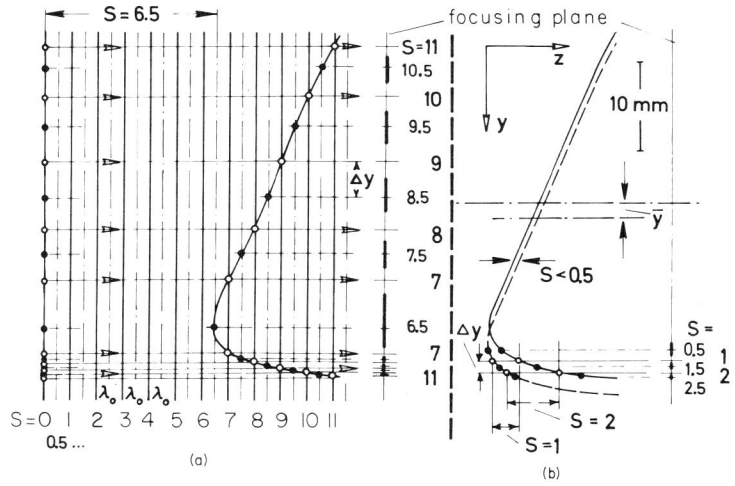

FIG. 30. Interfering wave fields of the measuring beam and the reference beam in two-beam interference: (a) Normal two-beam interference (Mach–Zehnder); (b) differential interference. The wave front of the measuring beam corresponds to that of Fig. 2.

neglecting the very small light ray deflections ϵ_z (cf. Fig. 2). The path of the light rays is assumed to be parallel to the z axis, which is a condition for "ideal two-beam interference," which will be discussed in detail below. Figure 30a depicts the superposition of two wave fields at a certain time and a certain position in space. The parallel lines serve to illustrate portions of the plane wave fronts of the reference beam. Their mutual distance in this case is $\lambda_0 = 0.546 \cdot 10^{-6}$ m.

The profile belongs to the "measuring beam" whose deformed wave fronts were plane before entering the temperature field of the test section. For the sake of clarity only one wave front of the measuring beam is depicted.

The propagation direction is denoted by arrows. The phase relationship is constant so that interference results. The optical path differences $g = S \cdot \lambda$ are expressed as multiples of the wavelength λ. For integer values of S, maxima of intensity result, which are denoted by small circles. For odd multiples of $\lambda_0/2$, minima of intensity (black dots) result. The interference pattern of Figs. 1 and 2 is the intersection of the interference structure with the stationary plane (focusing plane). In this case only the changes in the refractive index in the temperature field are responsible for the path differences. The greatest path difference $S = 11.3$ at the wall of the tube indicates the smallest refractive index and the greatest difference of temperature in the test section. The smallest path difference $S = 6.5$ appears in the inner part of the cross section of the tube (maximum refractive index, smallest temperature difference). For comparison, a profile of an interference pattern obtained in a similar case is shown in Fig. 77. In this example the greatest path difference $S = 11.5$ (greatest temperature difference) at the wall is approximately equal to that of the example in Fig. 2. The zone of the smallest path difference $S = 0$ corresponds to a temperature difference of zero. The temperature here is equal to the original temperature.

b. *Differential Interference.* In Fig. 30b differential interference is demonstrated on the same wave front profile as Fig. 2 (solid curve). In this case the reference wave front (dashed curve) is not a plane wave front, but the same deformed wave front with a lateral displacement $\bar{y}$. In the differential interferometer both the measurement beam and the reference beam pass through the measuring zone with a lateral displacement $\bar{y}$ and then interfere. The lateral displacement can also be introduced in the x direction, according to the construction of the interferometer. Its magnitude and direction are known.

It can be seen in Fig. 30b how interference patterns arise in the region of greater temperature gradient for a given value $\bar{y}$. They should be interpreted—as for normal two-beam interference (preceding paragraph) —as the intersection of the interfering wave fields of the measuring and reference beams, respectively, with the focusing plane. In the region of the weaker gradient of the wave front, the path difference S between the measuring and the reference beam is smaller than 0.5.

The first interference minimum does not yet appear for $S = 0.5$;

however, a value of intensity corresponding to the path difference can be registered photometrically and evaluated. If large values of $\bar{y}$ are chosen, interference lines can also be seen in the region of weaker gradient. It can be seen that the value of $\bar{y}$ corresponding to an optimum number of interference lines is highly dependent on the form of the wave fronts to be analyzed [Smeets (21)]. For a complete analysis it is even necessary under certain circumstances to evaluate a second interference pattern; in this case, however, for a displacement in the x direction. In this respect differential interference methods are similar to schlieren methods (Section IV,E); i.e., they are dependent on the direction of the displacement. In addition, the interference lines are a measure of the gradient of the deformed wave front in both cases. To illustrate the last point imagine that the displacement $\bar{y}$ is reversed so that the two profiles coincide.

Note that maxima and minima of the dashed profile do not have the same positions after the displacement as their counterparts on the solid profile. It is now possible to construct a right-angled triangle formed by the chord connecting two corresponding maxima (or minima) as the hypotenuse and the displacement $\bar{y}$. The length of the third side is known to be $g = S \cdot \lambda_0$. The first interference minimum $S = 0.5$ therefore characterizes that position of the wave front where the quotient of the differential length Δz and Δy is equal to $g/\Delta y = S \cdot \lambda_0 / \bar{y} = 0.5 \cdot \lambda_0 / \bar{y}$.

In this example, the path difference S is—as in normal two-beam interference described in Section V,A,1,a—proportional to the temperature difference (difference of refractive indices) so that it is possible to measure the difference quotient $\Delta \vartheta / \Delta y$ of the temperature directly. Especially in heat transfer experiments, knowledge of the temperature gradient $d\vartheta/dy$ (and also $d\vartheta/dx$) is important; hence, differential interference appears to be the most suitable method of measurement.

Generally, however, the measuring accuracy is smaller than that obtainable with normal two-beam interference (Mach–Zehnder) because of the smaller number of interference lines. As in the shadowgraph and schlieren methods, the result of the differential interference and the distribution of the difference quotients $g/\Delta y = S \cdot \lambda_0 / \bar{y} = f(x, y)$ must be integrated to obtain the wave front. Normal two-beam interference yields the wave front directly. In this case, gradients must be obtained by differentiation. However, accuracy is not impaired by this process because the interference lines can be considered as "elevation contours," which is practically equivalent to differentiation. Even if only the gradients of the wave front are required, normal two-beam interference is superior to differential interference methods.

2. Types of Two-Beam Interferometers

The wave fields of the measuring and reference beam must be able to interfere. They must therefore be coherent, i.e., their phase relationship must be constant. This is achieved with interferometers by splitting the beam of a light source into a measuring and a reference beam and later recombining. Separate physical light sources with the same frequency and phase are difficult to attain, even with lasers.

Two-beam interferometers can be classified according to the method of beam splitting. The complexity increases with the diameter of the beam and the spatial displacement between measuring and reference beam. Here we shall discuss only a few types of interferometers. Comprehensive presentations can be found, e.g., in Born and Wolf, Chapter VII (*1*); Françon, p. 452 (*15*); Weinberg (*3*); Tolansky (*6*); Krug *et al.* (*7*); Ladenburg (*22*); Holder (*23*); Françon (*24*).

a. Division of Amplitude. Classic mirror interferometers utilize semitransparent layers (mirrors) as beam-splitting elements. The parallel beam, which is most commonly used, is split into a measuring beam and a reference beam by means of such a semitransparent mirror and eventually recombined. Usual measuring beam diameters (0.1–0.3 m) require large and expensive interferometer mirrors with highly refined surfaces ($\lambda/20$). Furthermore, the semitransparent plates must be very accurately parallel. The beam-splitting effect of semitransparent metalized mirrors is independent of the wavelength; however, the great absorption is detrimental. Dielectric layers are more suitable for splitting the beam into two beams of equal amplitude, but the effect is dependent upon the wavelength.

The mirror interferometers are universally applicable instruments with great light intensity because large light sources are used, which are not point sources in the strict sense of the word. The Mach–Zehnder interferometer (*25, 26*) is suitable for transparent objects. This is comprised of two reflecting and two beam-splitting mirrors positioned as a rectangle or a parallelogram (Fig. 31). The displacement between the measuring and the reference beam can be practically as large as required. It can also be constructed for large model lengths (4 m) (*27*).

The Mach–Zehnder interferometer was preceded by the Jamin interferometer (1856) (*28*), which consists of two inclined mirrors (thick plane-parallel glass plates; Fig. 31). Each glass plate has a semitransparent layer, respectively splitting or recombining the beam. There is also a completely reflecting layer on the rear surface. These four layers have the same effect as the four separate mirrors in the Mach–Zehnder interferometer. The displacement between the measuring and

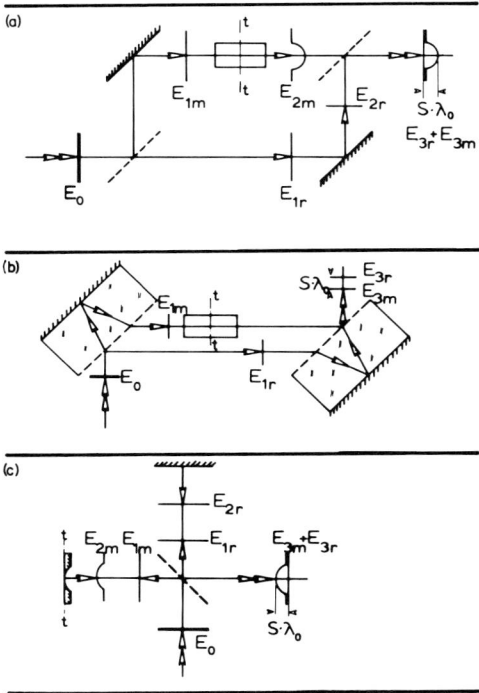

FIG. 31. Beam division schemes in mirror interferometers (two-beam interference): E_0, undivided plane wave front which is then split into the wave fronts E_{1m} and E_{1r} by a semitransparent mirror; E_{1m}, wave front of the measuring beam (index m) before traversing the test zone t–t (schlieren region); E_{1r}, wave front of the reference beam (index r) (same reference time as E_{1m}); E_{2m}, deformed wave front (in the case of the Jamin interferometer: displaced plane wave front) of the measuring beam; E_{2r}, original plane wave front of the reference beam; and $E_{3m} + E_{3r}$, interfering wave front recombined by the semitransparent mirror. The path difference of the wave fronts E_{3m} and E_{3r}, made visible by interference, is $S \cdot \lambda_0$. (a) Mach–Zehnder; (b) Jamin; (c) Michelson.

the reference beam is small and given by the thickness of the glass plate. The Jamin interferometer can be used as a differential interferometer; however, it has been more commonly used for the measurement of refractive indices in gases and the measurement of diffusion coefficients in liquids.

The most well-known mirror interferometer is the Michelson interferometer (1882) (*29, 30*) (Fig. 31) which is mainly employed in measuring lengths and in surface investigations. It is not very suitable for measurements on transparent objects, excepting measurements of refractive coefficients in gases and liquids. The measuring beam traverses the test object twice along different paths, a consequence of beam deflection by the refractive coefficient gradient in the test object. This complicates the quantitative evaluation of such interferograms. Further important applications are: investigation of the fine structure of atomic spectra and the classic experiments to detect motion in the ether (*31*). Modified mirror interferometers are mainly used for the appraisal of optical elements (lenses, mirrors), e.g., the Twyman–Green interferometer (*32*) (similar to the Michelson interferometer) and the wave front shear interferometer [Bates (*33*)] (similar to the Mach–Zehnder interferometer).

Figure 31 shows schematically the path of a plane wave front E for different time intervals (index 0, 1, 2, 3) for the mirror interferometers discussed above. The index of the measuring beam is m, that of the reference beam is r. The path difference $S \cdot \lambda_0$ originating in the test zone t–t (schlieren region) is made visible by interference of the recombined partial wave fronts E_{3m} and E_{3r}.

In the Mach–Zehnder interferometer (Fig. 31), the path difference $S \cdot \lambda_0$ stems from the greater propagation velocity in the test zone (smaller optical density, i.e., smaller coefficient of refractive index).

In the case of the Jamin interferometer (Fig. 31) it is assumed that the test zone is of greater optical density: The plane wave front is conserved but the propagation velocity in the test section is smaller.

In the Michelson interferometer (Fig. 31) the delay is produced by surface reflection on the test object.

b. Division of the Wave Front. Working on the basis of Young's interference experiments, Lord Rayleigh (1896) (*34*) developed a similar design for the measurement of refractive indices. With the addition of a device compensating the path difference (Haber–Löwe interferometer) it has found wide-spread use for measuring the concentrations of two- and three-component mixtures of gases or liquids. The refractive indices of the components must be known in this case (*35*); compared with the Jamin interferometer Rayleigh's scheme exhibits fundamental shortcomings (*36*).

c. Birefringent Splitting of Polarized Light. Interferometers in which (small) beam displacements are produced by birefringent prisms have attained significance mainly in interference microscopy (*37, 38*). In this context, relatively simple optical arrangements can be realized for macroscopic models. These schemes function as differential interferometers and are similar to Toepler's schlieren apparatus (Fig. 28). Their mode of operation can be demonstrated on the interferometer described by Smith (1947) (*38*) in Fig. 32: The polarizer P serves to polarize the

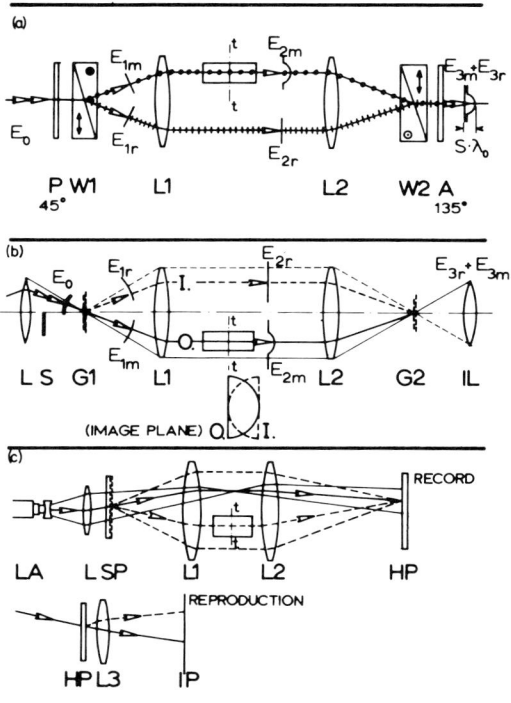

FIG. 32. Two-beam interference arrangements similar to schlieren methods with image formation (Toepler, Fig. 28): Notation of the wave fronts E, the test zone t–t, and the path difference $S \cdot \lambda_0$ as in Fig. 31. (a) Smith interferometer (division by birefringence): P, (45°) polarizer; $W1$, Wollaston prism for angular beam division (cross-lined beam: oscillation direction parallel to the drawing plane, dotted beam: oscillation direction perpendicular to the drawing plane); $L1$, $L2$, objectives; $W2$, Wollaston prism; A, analyzer. (b) Kraushaar interferometer (division by diffraction grating): L, illumination lens; S, stop for half beam; $G1$, diffraction grating for beam division; $L1$, $L2$, objectives; 0., zero-order diffraction maximum (measuring beam); I., first-order diffraction maximum (reference beam); $G2$, diffraction grating for beam recombination; IL, image-forming lens. (c) Interference holography: *Exposure*: LA, laser with concave lens producing divergent beam; L, illumination lens; SP, scattering plate for diffuse illumination; HP, photographic plate for hologram reproduction. *Reproduction*: HP, hologram illuminated by reproduction beam; $L3$, lens, which combines the light diffracted by the hologram HP in the image plane forming the interferogram. Zero-order diffraction maxima, solid lines; diffracted light, dashed lines.

wave front at a 45° angle to the reference plane and impinges on the Wollaston prism $W1$. This consists of two birefringent uniaxial crystals, usually Iceland spar or quartz, glued together. The crystal axes are perpendicular with respect to one another, as is implied in illustration. The first half of the prism splits the impinging wave front into two polarized partial wave fronts, one of which oscillates in the plane of the illustration. The undivided wave fronts are split angularly in the second half of the prism. In the objective $L1$ the split beams are then deflected so that they are parallel. The angular splitting of the beam (inversely proportional to the wavelength), and therefore the lateral displacement $\bar{y}$ ($\bar{y} < 13$ mm) of the measuring and reference beams, is in most cases very small (*21*). The object t–t in the diagram is assumed to be small so that only normal two-beam interference and no differential interference occurs. The wave fronts E_{3m} and E_{3r} are recombined by the objective $L2$ and the Wollaston prism $W2$, which are usually identical in construction to $L1$ and $W1$. At this stage they do not yet interfere; they are linearly polarized and their planes of oscillation are perpendicular. The difference in polarization is canceled in the analyzer A and the wave fronts E_{3m} and E_{3r} can then interfere. Their plane of oscillation is now inclined 135° (−45°) to the image plane.

Other interference schemes have been conceived by Jamin (1868), Jamin–Lebedeff (1930), Françon (1951), Fleischmann (1951), Lindberg (1951), Nomarski (1952), Philpot (1951), and are described, e.g., in (*15*).

d. Division by Diffraction Gratings. To obtain a measuring and a reference beam Kraushaar (*39*) utilizes the zero-order and the first-order maximum of a diffraction grating. The schematic diagram (Fig. 32) of the optical configuration is, as that in the preceding section, similar to Toepler's schlieren apparatus (Fig. 28). The image of the light source on the diffraction grating $G1$ is formed by a condenser lens L. Half of the light cone is screened off by a stop S. The remaining half impinges on the diffraction grating (100 lines per millimeter). The zero-order maximum passes through the bottom half (measuring beam), the first-order maximum through the top half (reference beam) of the objective $L1$. In this case, the angular splitting is proportional to the wavelength. An objective $L2$ which is identical to $L1$ deflects the beams so that they are recombined when impinging on the diffraction grating $G2$ (identical to $G1$, copy). The two beams are diffracted by the grating $G2$ and are refocused by the image-forming lens IL and interfere on the image plane. Instead of the objectives $L1$ and $L2$, Kraushaar uses large-diameter concave mirrors since only a portion (about one-third) of the aperture of the objective serves as the observation field (overlaps). The diffraction grating inteferometer can be considered as a further development of the Ronchi schlieren system. Ronchi has given a comprehensive summary in (*40*).

e. Interference Holography. The utilization of lasers as light sources in optical systems with a large coherence length has opened a wide range of applications for holography [Gabor *(41)*]. The further development of this method by Leith and Upatnieks *(42–44)* for practical applications makes it possible to store the information of object wave fronts in a hologram. For reproduction, these object wave fronts are reconstructed with the aid of the hologram. The image thereby formed is equivalent to that which would have been obtained by the original object wave fronts.

The object wave fronts emanating from an object illuminated by diffuse laser light would produce a normal image in the case of a camera lens projecting the image on a photographic film or plate (cf. Abbe's theory of image reproduction). If a photographic plate is placed in the line of sight, the information on the object given by the phase relationships and the intensity of the object wave fronts can be stored in the form of microscopic interference structures (hologram). These interference structures on the photograph are produced by interference of the object wave fronts with the reference wave fronts impinging at an angle. The reference beam is derived from the original laser beam (which serves to illuminate the object), e.g., by a semitransparent mirror. In reproducing the developed holographic plate, it is illuminated at an angle under the same circumstances as the exposure. The light diffracted by the interference structures corresponds to the original object wave fronts: It is possible to "see" the original object behind the hologram. In interference holography two holograms of a surface object or transparent object are superposed. The two holograms are recorded on the same photographic plate by double exposure (two-step method). In the reproduction of an interference hologram of this kind, the phase differences of the two exposures can be seen as interference lines superposed on the image of the object. The two superposed holograms (interference hologram) serve as reference and measuring wave fronts as in two-beam interference. For measurements on surface objects it is possible to make visible small deformations of the surface by mechanical strains, e.g., as in the case of the Michelson interferometer *(45)*. When making measurments on transparent objects, two-beam interference, similar to the Mach–Zehnder type, can be produced with an interference hologram, if the hologram of the optical configuration is first exposed without the object and subsequently the hologram of the configuration with the object is taken. It is, however, possible to combine two holograms of a nonstationary process taken at different times to form an interference hologram. It is not possible to obtain such "time differential interference" with conventional interferometers.

In the two-step method all unwanted movements of the object or the photographic plate relative to each other cause additional interference fringes, which can impose severe limitations on this method.

The configuration suggested by Burch (46) for interference holography on transparent objects makes it possible to obtain an interference hologram of a phase object by only one exposure, although the quality of the interferograms is inferior to the two-step method. Furthermore it functions as a "scattering-plate interferometer" and exhibits characteristics similar to the diffraction-grating interferometer described by Kraushaar: The small-diameter parallel light beam emanating from the laser source diverges after traversing a concave lens (or microscope objective). The virtual focal plane of this diverging beam is projected on the plane of the test zone t–t by the lens L and the objective $L1$. The beam is partially scattered by the scattering plate SP which is positioned in the focal plane of $L1$ and has the effect of a beam splitter. The primary beam (solid lines) bypasses the phase object and serves as a reference beam. The scattered light (dashed lines) passes through the phase object and is shifted in phase. The photographic plate HP for the exposure of the hologram is positioned in the focal plane of the objective $L2$. The plane of the scattering plate is projected on the plane of the photographic plate by $L1$ and $L2$. The combination of the components of the primary beam and the phase-shifted diffracted light results in the interference hologram. To obtain the interference patterns the developed interference hologram is placed in its original position in the optical system (omitting the phase object). A camera lens, e.g., $L3$, reproduces the interferogram on the image plane IP (e.g., photographic plate), the conjugate of the object plane t–t, (Fig. 32).

If a hologram of the configuration without the phase object is inserted —analogous to the diffraction grating $G2$ in the Kraushaar interferometer—the arrangement can be considered to be a scatter-plate interferometer where direct observation of the interference pattern is possible. The hologram is simultaneously an accurate reproduction of the scattering plate SP and all lens defects; both together have an effect similar to the identical diffraction gratings $G1$ and $G2$ in the Kraushaar interferometer. The holographic method has the advantage—compared with the scatter-plate interferometer—that lens defects have no effect on the interference pattern since only phase shifts occurring between the first and second exposures result in interference patterns.

f. Alteration of the Diffraction Pattern in the Focal Plane. The phase contrast method devised by Zernike (4, 47) is an important expedient in microscopy, where extremely small phase differences of an object are to be detected. In the focal plane of the projecting objective, the diffraction pattern is altered, distorting the image in the image plane and changing contrast distribution. The undiffracted light (zero-order maximum) is either stopped by an absorbing circular disk and only the diffracted light produces the image (dark-field observation method) or a thin plate is inserted, shifting the phase of the zero-order maximum; together with the diffracted light the image is then formed (phase contrast method).

Other interference arrangements related to this principle have been described utilizing the extremely straightforward configuration given by Toepler (Fig. 28). The light intensity is comparatively weak in these arrangements since the light source must be smaller than the zero-order maximum of its image. Here and in the following the phase distortion will be introduced in the plane where Toepler's knife edge is placed (Fig. 28) in schlieren methods.

In the interferometer, according to Erdmann (*48*), the light diffracted by the phase object is partially absorbed in a suitable manner and then interferes with the undisturbed light of the zero-order maximum.

Gayhart and Prescott (*49*) produce "schlieren interference" by placing a wire of suitable diameter in the zero-order maximum. The light deflected by the phase object then interferes with light diffracted by the wire. A theory for this method has been given by Temple (*50*).

Rottenkolber (*51*) has modified this method with a biprism interferometer. The sharp edge of a Fresnel biprism is adjusted to the zero-order maximum whose light is split into two parts and leaves the prism refracted towards the optical axis. The light deflected by the phase object, which is also split into two parts, enters the biprism beside the zero-order maximum and is refracted towards the optical axis. Both components produce interference lines of higher contrast than the method described above. However, the image is split into two halves, symmetric to the edge of the biprism.

A drawback of these methods is that the gradient of the refractive index is only reproduced correctly in one given direction, similar to the image-forming schlieren methods.

B. The Mach–Zehnder Interferometer

1. *Ideal Interferometry*

The properties of a "real" instrument shall now be explained by means of an idealized model of a Mach–Zehnder interferometer—referred to in the following as MZI—and an idealized phase object. The properties of a real instrument and a real phase object can then be obtained by the addition of correction terms.

In ideal interferometry of the MZI, the following assumptions are made:

a. Planar Interferometer. Their mirror planes M_1, M_2, M_1', and M_2' are exactly parallel. Their midpoints Q_1, Q_2, Q_1', and Q_2' form an exact parallelogram (or rectangle) and are in one plane (hence, planar interferometer). This position of the mirrors is termed the "geometric basic position" (Fig. 33).

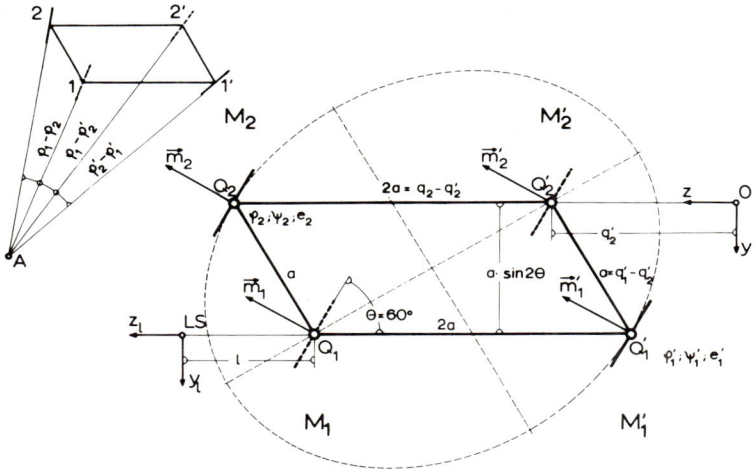

Fig. 33. Mach–Zehnder interferometer (in 60° parallelogram configuration): M_1, M_2', beam splitters; M_2, M_1', reflecting mirrors; $\mathbf{m}_i$, normal to the mirror in the illustrating plane; Q_i, mirror midpoints; φ_i, angle at which the mirror is rotated around the normal to the illustrating plane Q_i; ψ_i, angle at which the mirror is tilted around the axis through Q_i formed by the mirror plane and the plane of the parallelogram; e_i, displacement of the mirror midpoint in the direction of the normal to the mirror $\mathbf{m}_i$; $\theta = 60°$, angle between the mirror plane and the light beam; $2a$, a, sides of the parallelogram; O, origin on the central ray (x axis normal to the illustrating plane in O is not shown); L, extended light source (coordinates x_1, y_1, z_1); A, common rotational axis of the mirror planes normal to the illustrating plane for a practical basic position ($\psi_i = 0$). (Letters with arrows correspond to boldface in text.)

b. Mathematical Mirror Planes. The mirrors are considered to be mathematical planes. The semitransparent layers of the beam splitters M_1 and M_2' have no thick glass plate as a support for the reflecting layer.

c. Strictly Monochromatic Light Source. The light source is a point source of undamped spherical wave fronts which are transformed by means of an ideal lens into plane wave fronts of a parallel beam. The central ray of the parallel beam passes through the midpoints Q_1, Q_2, Q_1', and Q_2' of the mirrors.

d. Absence of Aberration. The beam path of the interferometer contains no optical elements whatsoever, including particularly the phase object, which could lead to a deflection of the beam aberration. The rays remain parallel while traversing the phase object (length l). The image-forming process does not lead to any distortion.

e. Interferometer Equation in Ideal Interferometry. In the following the parallel beam M_1–M_2–M_2' and the parallel beam M_1–M_1'–M_2' will be referred to as the measuring beam (subscript m) and the reference beam (subscript r), respectively, in accordance with Fig. 33. The Cartesian coordinate system is—as in the discussion of the schlieren method—so oriented that the z axis is parallel to the light beam. In the interference pattern, the difference $S_i \cdot \lambda$ of the optical paths is registered for each point $P(x_i, y_i)$ of the beam cross section. The path difference is the difference between the optical path length $n_\mathrm{m}(x_i, y_i) \cdot l$ in the test section (length l) and the optical length $n_\mathrm{r}(x_i, y_i) \cdot l$ of the corresponding section in the path of the reference beam (two-dimensional problem):

$$S_i(x_i, y_i) \cdot \lambda = n_\mathrm{r}(x_i, y_i) \cdot l - n_\mathrm{m}(x_i, y_i) \cdot l$$
$$= \Delta n(x_i, y_i) \cdot l \qquad (59)$$

Assuming $n_\mathrm{r}(x_i, y_i) = \mathrm{const}$, $S \cdot \lambda = \Delta E$, and propagation of the light in a straight line, Eq. (59) can be derived from Eq. (13a).

The loci of constant phase difference S_i in the two-dimensional phase object are, e.g., points of constant temperature difference $\Delta \vartheta_i = T(x_i, y_i) - T_\infty$, if the gradient of the refractive index is considered as a negative constant $(dn/dT = \mathrm{const})$. With Eq. (59), it then follows that

$$\left(\frac{dS}{dT}\right) = \left(\frac{l}{\lambda} \cdot \frac{dn}{dT}\right); \quad S_i(x_i, y_i) = \frac{l}{\lambda} \cdot \frac{dn}{dT} \int_{T_\infty}^{T_i} dT$$
$$S_i(x_i, y_i) = \frac{l}{\lambda} \cdot \frac{dn}{dT} \cdot \Delta T = \frac{l}{\lambda} \cdot \frac{dn}{dT} \cdot \Delta \vartheta_i \qquad (59\mathrm{a})$$

Along these lines an interference pattern can be considered as an isothermal field, whose interpretation requires knowledge of the temperature at at least one point of the cross section. For the wavelength $\lambda = 0.5461 \cdot 10^{-6}$ m and a model length $l = 0.5$ m in air (20°C, 760 Torr), the temperature difference $\Delta \vartheta$ between two interference lines ($S = 1$) is 1.2 deg; in water (20°C) with $l = 0.05$ m, $\Delta \vartheta = 0.12$ deg.

2. Basic Position of the Mach–Zehnder Interferometer

The condition that the path lengths of the measuring and reference beams are equal in an idealized MZI is met by the parallelogram configuration of the planar interferometer (Fig. 33: $Q_1Q_2Q_2' = Q_1Q_1'Q_2'$). Furthermore, in the reflection by a plane mirror, the impinging beam, the reflected beam, and the normal to the mirror surface in the reflection point Q_i are in one plane, which is, in this case, the plane of the parallelogram of the planar interferometer. The planar interferometer described in Section V,B,1,a is one special case meeting these requirements. One general locus for the reflecting points Q_i is a rotational symmetric ellipsoid, as indicated in Fig. 33. The reflecting points Q_1 and Q_2' of the beam splitters are the focal points of the ellipsoid. The points Q_1' and Q_2 are the points of contact of the tangential planes to the ellipsoid, which are the mirror planes in this case. These requirements are met by the properties of the ellipsoid. The sum of the focal ray lengths is constant and therefore the points Q_1' and Q_2 on the surface of the ellipsoid can be chosen arbitrarily. The parallelogram in the case of the planar interferometer is then a generalized quadrangle whose triangular halves $Q_1Q_2Q_2'$ and $Q_1Q_1'Q_2'$ can be rotated around the main axis Q_1Q_2'. The second requirement that the impinging ray, the reflected ray, and the normal be coplanar is met by the planes of the triangular halves formed by the focal rays. In the special case of the Michelson interferometer the locus of the reflecting points is a sphere. The beam splitters M_1 and M_2' overlap and the mirror points Q_1 and Q_2' are the center of the sphere. A theory for this general case has been given by Kahl and Bennett (52) in a vectorial representation which was introduced by Silberstein (53) as a general method of ray tracing in optical systems. If, however, the MZI is adjusted with white light and if the glass bases of the beam-splitting reflectors M_1 and M_2' are of equal thickness, the symmetric configuration of the planar interferometer results, as shown by Kahl and Bennett (Fig. 34).

a. Planar Interferometer. To obtain conditions for the general basic positions of a real MZI, the planar interferometer is considered to be in the geometric basic position as in Section V,B,1,a; however, small deviations are permissible (Fig. 33). This representation by Fromme and Hannes (54) also utilizes vector calculus and has been outlined by Lamla (55) in a form which is similar in principle but less comprehensive. In the parallelogram configuration, the area of the interferometer mirror is more fully exploited (larger beam cross section) than in the rectangular configuration. In the following, the special case of a planar interferometer is considered in which the mirrors are inclined at an angle $\theta = 60°$ to

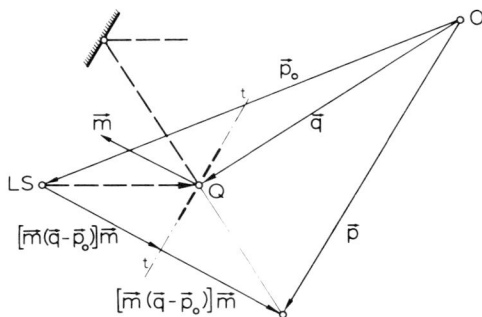

FIG. 34. Vector representation of single reflection: $\mathbf{p}_0$, vector of a single point of the light source; $\mathbf{p}$, vector of the light source image; $\mathbf{m}$, normal to mirror plane; $\mathbf{q}$, vector of the point of incidence on the mirror surface t–t. (Arrows over letters correspond to boldface in text.)

the incident ray. Furthermore, the relation of the sides to one another is 2 : 1, which offers certain advantages [single-mirror adjustment by Kinder (56)].

In Fig. 33 the mirror midpoints Q_i—through which the central ray of the parallel light bundle passes—are the corners of a plane parallelogram. The mirror midpoints in the geometric basic position can be represented by vectors $\mathbf{q}$, originating at 0. As can be seen in the illustration, their components are as follows:

Position vectors $\mathbf{q}_i$, in the geometric *basic position* ($\theta = 60°$):

	x	y	z	
q_1	0	$a \cdot \sin 2\theta$	$q_2 + a \cdot \cos 2\theta$	
q_1'	0	$a \cdot \sin 2\theta$	$q_2' + a \cdot \cos 2\theta$	(60a)
q_2	0	0	q_2	
q_2'	0	0	q_2'	

In the geometric basic position, the unit vectors $\mathbf{m}_i$ normal to the mirror plane lie in the plane of the parallelogram.

Normal to the mirror plane $\mathbf{m}_i$, $|\mathbf{m}_i| = 1$ (Fig. 35):

$$\mathbf{m}_i = (0, -\cos\theta, \sin\theta) \tag{60b}$$

General positions other than the basic position can be described by additional small correction terms φ_i, ψ_i, and e_i. Only first-order correction terms are considered. The mirror midpoints Q_i are to be displaced

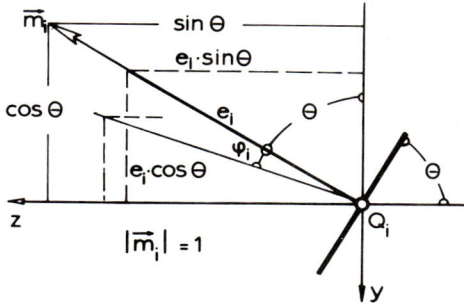

FIG. 35. Vector normal to mirror plane $\mathbf{m}_i$ in a 60° planar interferometer (representation in the plane of the parallelogram). (Arrows over letters correspond to boldface in text.)

by the amounts e_i in the direction of the normal to the mirror plane $\mathbf{m}_i$ so that the following vectors result (Fig. 35):

Position vectors $\mathbf{q}_i$ in the *general basic position*:

	x	y	z	
q_1	0	$a \cdot \sin 2\theta - e_1 \cos \theta$	$q_2 + a \cdot \cos 2\theta + e_1 \sin \theta$	
q_1'	0	$a \cdot \sin 2\theta - e_1' \cos \theta$	$q_2' + a \cdot \cos 2\theta + e_1' \sin \theta$	(60c)
q_2	0	$- e_2 \cos \theta$	$q_2 + e_2 \sin \theta$	
q_2'	0	$- e_2' \cos \theta$	$q_2' + e_2' \sin \theta$	

Furthermore, the mirror planes in the general basic position are tilted by the angles φ_i and ψ_i (cf. caption of Fig. 33). The following normal vectors (to the mirror plane) then result (Fig. 35):

Normal to the mirror plane $\mathbf{m}_i$ in the general position $|\mathbf{m}_i| = 1$:

$$\mathbf{m}_i = (-\psi_i, -\cos \theta + \varphi_i \sin \theta, \sin \theta + \varphi_i \cos \theta) \tag{60d}$$

In Fig. 33 a general light source L is assumed the spatial dimensions of which are characterized by the coordinates x_l, y_l, z_l of the single point light sources.

Coordinates of a point light source:

$$\mathbf{p}_0 = (x_l, a \cdot \sin 2\theta + y_l, q_2 + a \cdot \cos 2\theta + l + z_l)$$

The following equation describes the spacing of the mirrors (cf. Fig. 33):

$$q_1 - q_2 = q_1' - q_2' = 2a \tag{60e}$$

All relevant quantities have thus been considered. For the following calculation, the vector operation for reflection is required.

b. Single Reflection in Three-Dimensional Vector Representation. The reflection of a source point $\mathbf{p}_0$ in the mirror plane t–t and in the point of incidence $\mathbf{q}$ of a light ray, the direction of which is given by $(\mathbf{q} - \mathbf{p}_0)$, is shown in Fig. 34. The virtual image point of $\mathbf{p}_0$ behind the mirror plane is given by the vector $\mathbf{p}$. The points $\mathbf{p}_0$ and $\mathbf{p}$ are both equally distant from the mirror plane. The distance between the source point $\mathbf{p}_0$ and the mirror plane is given by the scalar product of the vector $(\mathbf{q} - \mathbf{p}_0)$ of the light beam and the normal vector $\mathbf{m}$ to the mirror plane. The virtual image point $\mathbf{p}$ is arrived at by adding $\mathbf{p}_0$ and the vector $2\mathbf{m}(\mathbf{m}(\mathbf{q} - \mathbf{p}_0))$:

$$\mathbf{p} = \mathbf{p}_0 - 2\mathbf{m}(\mathbf{m} \cdot \mathbf{p}_0) + 2\mathbf{m}(\mathbf{m} \cdot \mathbf{q}) \tag{61}$$

This vector operation can be written in matrix form, where $\bar{E}$ is the unit matrix and $\bar{M}\mathbf{v} = \mathbf{m}(\mathbf{m}\mathbf{v})$ (cf. calculation, Section VII,A):

$$\mathbf{p} = (\bar{E} - 2\bar{M})\mathbf{p}_0 + 2\bar{M}\mathbf{q} \tag{61a}$$

Since $(\bar{E} - 2\bar{M})$ is an orthogonal matrix, it only effects a rotation of the vector $\mathbf{p}_0$. The second summand $2\bar{M}\mathbf{q}$ effects a lateral translation of the image point $\mathbf{p}$.

c. Mirror Images of the Light Source in a Mach–Zehnder Interferometer. The following calculations are based upon the assumption that the source point $\mathbf{p}_0$ produces two images—denoted by $\mathbf{p}_1$ and $\mathbf{p}_1'$—by means of the mirrors M_1 and M_1'. These have, themselves, two mirror images $\mathbf{p}_2$ and $\mathbf{p}_2'$, respectively, produced by the mirrors M_2 and M_2'.

In the basic position of the MZI it is required that the images of both ray paths coincide:

$$\mathbf{p}_2 = \mathbf{p}_2' \tag{62}$$

Single reflection (M_1 or M_1', respectively) [Eq. (61a)]:

$$\mathbf{p}_1 = (\bar{E} - 2\bar{M}_1) \cdot \mathbf{p}_0 + 2\bar{M}_1 \mathbf{q}_1$$

Double reflection (M_2 or M_2'):

$$\begin{aligned}\mathbf{p}_2 &= (\bar{E} - 2\bar{M}_2)(\bar{E} - 2\bar{M}_1) \cdot \mathbf{p}_0 + 2(\bar{E} - \bar{M}_2)\bar{M}_1\mathbf{q}_1 + 2\bar{M}_2\mathbf{q}_2 \\ \mathbf{p}_2' &= (\bar{E} - 2\bar{M}_2')(\bar{E} - 2\bar{M}_1') \cdot \mathbf{p}_0 + 2(\bar{E} - \bar{M}_2')\bar{M}_1'\mathbf{q}_1' + 2\bar{M}_2'\mathbf{q}_2'\end{aligned} \tag{63}$$

Considering the basic position [Eq. (62)], it now follows that in Eq. (63) the terms denoting the rotational matrix [Eq. (64a)], which contains only components of the unit normal vectors $\mathbf{m}_i$, and the translational terms [Eq. (64b)], which appear in the case of double reflection, must be equal:

$$(\bar{E} - 2\bar{M}_2)(\bar{E} - 2\bar{M}_1) = (\bar{E} - 2\bar{M}_2')(\bar{E} - 2\bar{M}_1') \tag{64a}$$

$$2(\bar{E} - \bar{M}_2)\bar{M}_1\mathbf{q}_1 + 2\bar{M}_2\mathbf{q}_2 = 2(\bar{E} - \bar{M}_2')\bar{M}_1'\mathbf{q}_1' + 2\bar{M}_2'\mathbf{q}_2' \tag{64b}$$

The calculation of these expressions containing the terms explained in Eq. (60) is performed in Section VII,A.

d. Conditions for the Basic Position. For the planar interferometer assuming mathematical mirror planes the following general conditions for the basic position of a MZI result from Eqs. (64a) and (64b):

$$\varphi_1 - \varphi_2 - \varphi_1' + \varphi_2' = 0 \tag{65a}$$

$$\psi_1 - \psi_2 - \psi_1' + \psi_2' = 0 \tag{65b}$$

$$e_1 - e_2 - e_1' + e_2' = 0 \tag{65c}$$

$$2\varphi_1 - \varphi_2 - \varphi_2' = 0 \tag{65d}$$

$$2\psi_1 - \psi_2 - \psi_2' = 0 \tag{65e}$$

The case of the geometric basic position corresponds to the trivial solution $\varphi_i = \psi_i = 0$, $e_i = 0$. Furthermore, an infinite number of general basic positions exist, which satisfy the five equations with the twelve variables given above.

Equation (65c) is fulfilled by a reflector (M_2') adjustable to various parallel positions, because only an adjustment of the path length is necessary and no specific values of the sides $2a$ and a of the parallelogram are required (Fig. 33).

If Eqs. (65a) and (65b) are rewritten in the form

$$(\varphi_1 - \varphi_2) = (\varphi_1' - \varphi_2')$$

$$(\psi_1 - \psi_2) = (\psi_1' - \psi_2')$$

it can readily be seen that the mirror pairs M_1, M_2 and M_1', M_2' can be rotated by the same amount without violating the basic conditions. The terms $(\varphi_1 - \varphi_2)$, $(\varphi_1' - \varphi_2')$, $(\psi_1 - \psi_2)$, and $(\psi_1' - \psi_2')$ are the angles which the respective mirror pairs form. Eqs. (65d) and (65e) mean, furthermore, that the mirror planes must all intersect in one line. This case is shown in Fig. 33 for $\psi_i = 0$; the common intersection line A is perpendicular to the plane of the illustration and moves towards infinity if φ_i approaches zero (geometric basic position). In order to fulfill the conditions given by Eqs. (65a–e), it is necessary, apart from the already-mentioned mirror M_2' for path length alignment, to make two other mirrors rotatable $[M_2(\varphi_2, \psi_2), M_1'(\varphi_1', \psi_1')]$, so that five variables can now be adjusted to fulfill the five equations (65a–e).

In a real MZI, the two beam splitters M_1 and M_2' are realized in the form of two equally thick, plane-parallel glass plates, which serve as a basis for the reflecting layer. The parallel displacement of the beam by these plane-parallel plates is dependent upon the wavelength λ of the light used (dispersion of glass). One further requirement for adjustment with polychromatic or white light is apparent: The respective path lengths in the glass of the beam splitters M_1 and M_2' must be equal in order to obtain the symmetric conditions required for the planar interferometer (Fig. 36). One light ray entering M_1 and M_2', whose direction is charac-

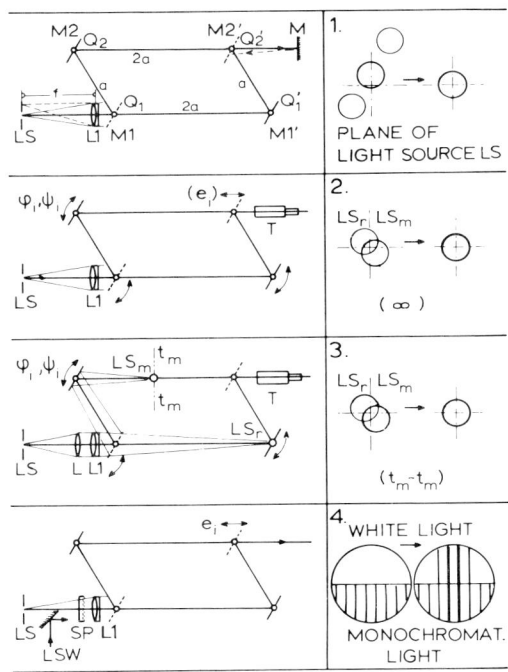

FIG. 36. Basic adjustment of an MZI: LS, light source: filtered ($\lambda = 0.546\,\mu$) or unfiltered light of a Hg low-pressure lamp, projected on a medium-large perforated diaphragm by means of a condensor; $L1$, achromatic lens or concave reflector; M_1, M_2', beam splitters; M_1', M_2, mirrors; M, auxiliary mirror; T, observation telescope with cross lines, possibly with level; SP, scattering plate: one half illuminated by diffuse white light, other half by unfiltered Hg light (Hg low-pressure lamp).

terized by a vector **p**, will now be considered. After traversing the beam splitters, the directions of the beams must be the same. This requirement can be generally formulated as

$$\mathbf{p} \cdot \mathbf{m}_1 = [(\bar{E} - 2\bar{M}_2)(\bar{E} - 2\bar{M}_1) \cdot \mathbf{p}]\mathbf{m}_2'$$

since the transformation matrix $(\bar{E} - 2\bar{M}_2)(\bar{E} - 2\bar{M}_1)$ gives the direction of the ray after being reflected by M_1 and M_2. The calculation (Section VII,A) results in two additional equations:

$$2\varphi_2 - \varphi_1 - \varphi_2' = 0 \tag{65f}$$

$$2\psi_2 - \psi_1 - \psi_2' = 0 \tag{65g}$$

The conditions [Eqs. (65a–g)] for a basic position for white light are summarized in the following equations:

$$e_1 - e_2 = e_1' - e_2' \tag{66a}$$

$$\varphi_1 - \varphi_2 = \varphi_1' - \varphi_2'; \quad \varphi_1 = \varphi_2; \quad \varphi_1' = \varphi_2' \tag{66b}$$

$$\psi_1 - \psi_2 = \psi_1' - \psi_2'; \quad \psi_1 = \psi_2; \quad \psi_1' = \psi_2' \tag{66c}$$

This set of equations (66) requires three rotatable and one sliding mirror in order to obtain the general basic position in white light. The glass plate of the beam splitters must be plane parallel, of equal thickness, and must exhibit identical refractive properties in order to obtain, with white light, interference patterns of high contrast.

Mechanical designs of MZI's have been detailed by Eckert et al. (*57*), Gebhart and Knowles (*58*), Johnstone and Smith (*59*), and Hansen (*60*).

e. Basic Adjustment Procedures for a Mach–Zehnder Interferometer. Previous adjustment methods tried to realize the geometric basic position within the measuring accuracy of elaborate instruments such as levels, pentaprisms, etc. [Kinder (*56*), Lamla (*55*)]. However, it is less difficult to attain a general basic position by observing the interference patterns only [Clark et al. (*61*), Price (*62*)]. The series of adjustment steps leading to a general basic position can be understood by considering the points given above [Hannes (*54*)]. The schematic diagram in Fig. 33 is the basis for the following considerations. The optical axes indicated in the diagram represent the central rays of parallel bundles $(Q_1Q_2Q_2'$ and $Q_1Q_1'Q_2')$. The mirrors M_1, M_2, M_1' can be rotated (rotation angle φ_i) and tilted (tilt angle ψ_i). In practice it seems to be important that both axes coincide with the mirror surface. The mirror M_2' serves for path length alignment; its position can only be changed laterally.

Preparation. The four mirrors are adjusted in the form of the parallelogram chosen (angle θ) and are illuminated by a parallel beam. The semitransparent, dielectric layers of the beam splitters must be especially calculated for the angle chosen (in this case, $\theta = 60°$; Fig. 33) and for the wavelength λ of the light to be primarily used, in order to obtain a ratio of the intensities of the reference to the measuring beam of 1 : 1. The distances between the mirrors, $2a$ and a, must be measured with an accuracy of at least ± 1 mm (cf. the following, Section V,B,3, coherence lengths of physical light sources). Figure 36 is a schematic representation of the adjustment steps.

(1) At the outset the parallel beams of the measuring and the reference beams are completely malaligned, i.e., they meet at an arbitrary angle after M_2', and corresponding wave fronts, emanating from the beam splitter M_1, are displaced with respect to one another. First, the severe angular displacement is corrected by means of an auxiliary mirror M mounted behind M_2' [Price (62)]. This mirror reflects the two parallel beams back in their original direction. A second splitting of the two beams at M_2' now produces four parallel beams. The objective $L1$ then projects four focused images of the light source LS onto the plane of the light source. However, in the plane of the light source only three images can be distinguished, since two of the four images coincide even when the mirrors are completely malaligned. By rotating and tilting (φ_i, ψ_i) the mirrors—mainly M_2' and M_1—these images can be made to coincide near the light source. Reducing the diaphragm aperture (from 3 to 0.2 mm) facilitates this adjustment. The auxiliary mirror M also permits the alignment of the parallel beam by adjusting the distance between the objective L_1 and the light source LS to produce sharp images of the light source in the plane of the light source. The distance thereby found is equal to the focal length of the objective. However, this procedure is not precise enough; in addition, it is necessary to check the diameter of the beam at several points, which should be as far apart as possible.

(2) A telescope T in the optical axis is now substituted for the auxiliary mirror M, permitting the observation of the apparently infinitely distant light source with a suitable magnification. By means of a level it is possible to adjust the optical axis so that it is perpendicular to the direction of the gravitational force. This is important, for example, in convection experiments on a horizontal plate, which on the one hand should be perpendicular to gravitational force and on the other hand parallel to the beam direction. The light source images LS_m and LS_r corresponding, respectively, to the measuring beam (m) and the reference beam (r) are made to coincide by adjusting the mirrors M_2 and M_1'. In general, the light-source image is interlaced by fine interference patterns, the contrast

and fringe width of which one tries to enhance. By appraising the enhancement of the color contrast in unfiltered Hg light it can be estimated how well the general basic position has been approximated. If no interference fringes are visible, the mirror spacings $2a$ and a should be rechecked. If Na light is employed, the beat formed by the two components of the D-line doublet can obliterate the interference pattern. In both cases this can usually be alleviated by a small displacement (e_i) of the mirror M_2'.

(3) A lens L is placed between the light source LS and the objective $L1$, so that real image of the light source LS_m is formed in the plane t_m-t_m, and an additional image LS_r is formed near M_1'. The telescope is focused on these images, which are then made to coincide by suitable adjustment (M_2, M_1'). Observation (and attendant adjustment) of LS_m and LS_r at infinity, as detailed in (2), and then in the plane t_m-t_m, as detailed in (3), is repeated over and over, until LS_m and LS_r coincide in both cases and are interlaced by wide interference fringes of high contrast. The angular adjustment is now nearly completed.

(4) The length adjustment is observed without the telescope T. A scattering plate is placed before the objective $L1$. One half of the plate is illuminated by unfiltered Hg light (LS), the other half by diffuse white light LSW of approximately equal intensity (incandescent lamp). Interference fringes are now visible in the Hg light and can be adjusted by M_2 and M_1' into a vertical position and medium spacing, so that they can be conveniently observed. Displacing the mirror M_2' causes these to move through the field of vision. In doing this, the vertical position and the spacing of the fringes must be rechecked. If the mirror M_2' is displaced in the correct direction (found empirically), the contrast of the fringes of the Hg light is enhanced, and a so-called achromatic stripe appears in the white light, flanked by several colored fringes. The basic adjustment procedure is completed. The objective $L2$ (Fig. 33) forms the image of the plane t_m-t_m and the achromatic stripe, whose contrast is to be enhanced as much as possible by fine adjustment $(M_2, M_1'$, and, if necessary, $M_1)$. If the spacing between fringes is increased so much that the whole field of vision (beam cross section) is taken in by the achromatic stripe (white light), the general basic position is attained.

3. *Positions for Practical Use; Coherence*

The adjustment described above leads to a general basic position, which approximates the geometric basic position of ideal interferometry as well as possible, in order to obtain high contrast (or depth of color) using quasi-monochromatic light or white light, respectively. The real symmetric interferometer with beam-splitting plates $M1$ and $M2'$ of finite thickness is similar to the planar interferometer in ideal interferometry (Section V,B,1). For the following description of the adjustment for practical use, a point light source (as in ideal interferometry) will be assumed. The source produces the plane wave fronts of the measuring (m) and the reference beam (r) by means of the lens $L1$ (Fig. 37).

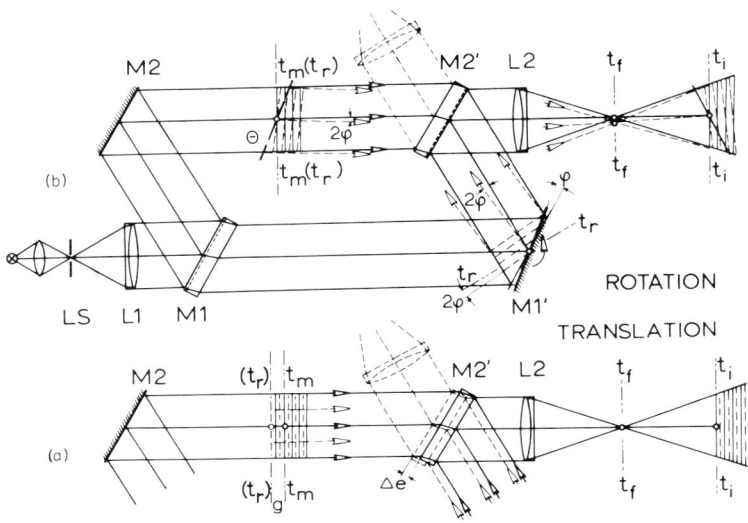

FIG. 37. Positions for practical use of the MZI: (a) Displacement of the beamsplitter $M2'$ (magnitude e) in the direction of the mirror normal (increasing or reducing the order of interference); (b) rotation of the mirror $M1'$ by the angle φ (single-mirror adjustment) produces virtual wedge. Observation field interlaced by parallel fringes. LS, point light source; $L1, L2$, objectives; $M1, M2'$, beam splitters; $M1', M2$, mirrors; t_m–t_m, focusing plane of $L2$ in measuring beam, conjugate of image plane t_i–t_i; t_r–t_r, focusing plane of $L2$ in reference beam, conjugate of image plane t_i–t_i; t_i–t_i, image plane (film); (t_r)–(t_r), focusing plane of reference beam, which can also be assumed to be positioned in the path of the measuring beam; t_f–t_f, focal plane of objective $L2$.

a. General Basic Position, Fringe Field, Field of Infinite Fringe. It is typical for the basic position that the plane wave fronts of the measuring and the reference beams recombine identically. This is equivalent to Eq. (62) (basic adjustment conditions: image vectors $\mathbf{p}_2$ and $\mathbf{p}_2'$ of equal magnitude and direction). The whole beam cross section exhibits uniform (maximum) intensity. When using white light, the field of vision is of uniform white color because the path length is exactly aligned for all wave lengths ($\sum e_i = 0$). In the following, this observation field will be referred to as the "field of infinite fringe."

In Fig. 37, the objective $L2$ forms both the image of the plane t_m–t_m (in the measuring beam path, normally occupied by the test object) and of the plane t_r–t_r (in the reference beam path) in the image plane (film) t_i–t_i. The interference patterns develop in the plane t_i–t_i. They represent the wave fronts in their momentary form in the planes t_r–t_r (plane wave front) and t_m–t_m (eikonal of the phase object). The wave fields in the vicinity of these two object planes can be considered to interfere virtually at the location of the model t_m–t_m, as if the plane t_r–t_r were transferred to the position (t_r)–(t_r) symmetric to the plane of the beam splitter $M2'$. This representation has already been used in Section V,A and will be retained in the following discussion.

The general basic position, meeting the requirements of Eq. (66), is a special position compared with the twelve degrees of freedom of the four mirrors. Deviating mirror positions lead to positions with additional path differences, superimposed on those caused by the phase object.

b. Phase-Shifted Field of Infinite Fringe. Translating the beam splitter $M2'$ by the amount Δe normal to its plane shortens the measuring beam path and lengthens the reference beam path, both by the same amount (Fig. 37a). The planes t_m–t_m and (t_r)–(t_r), with two individual conjugate image planes, are displaced parallel to one another, causing recurring dark and bright phase-shifted fields of infinite fringe in the observation plane t_i–t_i. If the measuring and the reference beams are of equal intensity, complete destructive interference occurs for a path difference of $\lambda/2$. The light is reflected by the semitransparent reflecting layer of the beam splitter $M2'$. By means of a second objective (represented by dashed lines in Fig. 37) maximum intensity can now be observed.

This can be used as an *alignment aid* to obtain the exact adjustment of the field of infinite fringe. In practical cases, for a field of maximum intensity (general basic position) the adjustment for maximum and, above all, uniform, intensity (image plane t_i–t_i) can be better appraised visually by simultaneously observing the field by means of the auxiliary lens (dashed lens in Fig. 37a). The field viewed here is uniformly darkened. Deviations from this position ($>\lambda/8$) lead to a brightening of the dark field where the contrasts can be more easily detected visually than in the infinite fringe field of maximum intensity. If a white light source is used, displacing the beam splitter M_2' results in an infinite fringe field of uniform color (corresponding to the spectrum of white light) rather than alternating bright and dark fields. This is the case because for each displacement Δe in the region of the geometric basic positions (which is characterized by uniform distribution of white light in the observation field), one specific wavelength is phased out. The

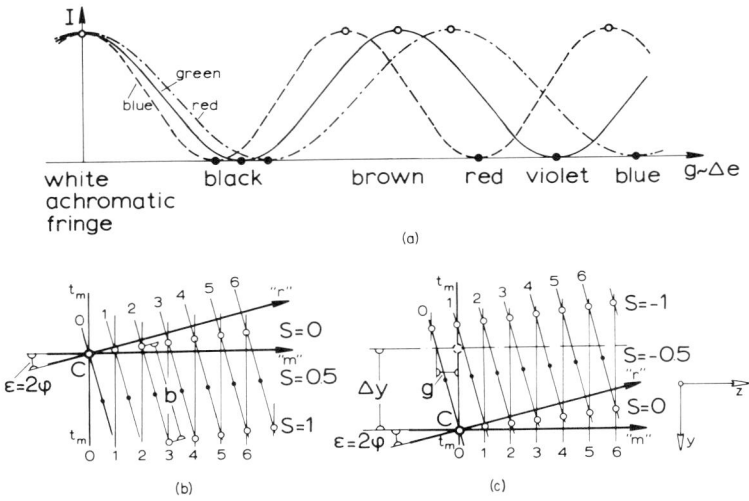

FIG. 38. (a) Formation of phase-shifted infinite fringe field of uniform color using a white light source; Δe displacement of the beamsplitter $M2'$, I intensity. (b) Fringe field in MZI (rotation): rotation of the wave fronts belonging to the reference beam (r) by the angle $\epsilon = 2\varphi$ (φ = angle of rotation of the mirror M_1' in Fig. 37). The axis C of the virtual wedge is in the center of the field of observation (the z axis corresponds to the central ray of the parallel beam). (c) Fringe field in MZI (rotation and translation): Rotation ($\epsilon = 2\varphi$) and additional displacement (Δe) of the wave fronts belonging to the reference beam. The axis C of the virtual wedge is displaced from the central position [as in (b)] by Δy. It can be positioned far outside of the field of vision. Circles, maxima of intensity; dots, minima of intensity. For white illumination the axis C corresponds to the achromatic fringe.

remaining spectral components result in the observed color (Fig. 38a). For greater displacements $\Delta e(>2\mu m)$ a specific color can no longer be observed ("white light of higher order"). If the "white" light of an unfiltered Hg lamp, consisting of discrete spectral lines with continuous background radiation, is used, the observed sequence of the spectral components recurs periodically, but with diminishing contrast.

c. Fringe Field, Virtual Wedge. In Fig. 37 the mirror $M1'$ is rotated by an angle φ from its orientation in the general basic position. The rotational axis, denoted by a circle, is also the axis of the twin wedge formed by the rotated mirror plane and the original one. The wave fronts (r) of the reference beam are deflected by an angle 2φ with respect to the original orientation t_r–t_r (dashed lines). These wave fronts (r) are virtually superimposed on the wave fronts (m) of the measuring beam at the position of the phase object (plane t_r–t_r transferred to the plane t_m–t_m).

Interference fringes parallel to the rotational axis are now formed and interlace the whole field of observation (Fig. 38b). In the field of observation, the rotational axis corresponds to the interference fringe with the path difference $S = 0$. In white light this fringe is also white (achromatic fringe), since the conditions of the general basic position ($e = 0$) are met in the rotational axis, independent of the rotational angle φ. In monochromatic light, this fringe is flanked by dark and bright fringes corresponding to the increasing path difference (Fig. 38b). In white light, colored neighboring fringes appear according to the sequence of Newton's color scale. This is analogous to the representation of phase-shifted infinite fringe fields in Fig. 38a, since the path difference increases with the distance of the rotational axis.

The fringe field can also be considered as the interference pattern of a phase object at the position t_m-t_m corresponding to the virtual wedge (formed by the original and the rotated mirror plane). For example, a suitable phase object would be a real, extremely thin twin wedge of glass, which is inclined—as the interferometer mirrors—by an angle θ with respect to the optical axis (indicated in dashed lines at the position t_m-t_m in Fig. 37a). Since this discussion has been based on an ideal point light source, emitting undamped, continuous waves, it is possible to observe the virtual interferences at any position in the path of the measuring beam. The focusing plane t_m-t_m can therefore be placed in positions other than in Fig. 37b. When considering real light sources, which do not exhibit the above quality, the virtual locus of interference is limited to the vicinity of the virtual wedge (or, in the other case, to the vicinity of the phase object), symmetric to the rotational axis C. This will be discussed in detail in the following.

If the mirror $M1'$ is now tilted by an angle ψ about an axis perpendicular to the axis C (Fig. 37b), a new virtual wedge results, skewed with respect to the optical axis. The interference fringes are now perpendicular to the previous ones. Rotating (φ) and tilting (ψ) the mirror simultaneously produces a virtual wedge whose axis is arbitrarily oriented. However, the axis still intersects the optical axis in t_m-t_m or t_r-t_r (Fig. 37b).

Figure 38c shows a fringe pattern resulting from a rotation φ of the mirror $M1'$ and supplementary displacement $\Delta e \sim g$ of the mirror $M2'$. The axis C of the virtual wedge now no longer intersects the optical axis (z axis), but is lateraly displaced by Δy and can be positioned far outside the observation field. In white light, the position of the axis C is represented by the achromatic fringe and the symmetrically positioned adjacent colored fringes. The position of the achromatic fringe in the observation field can be altered by actuating the shifting mechanism of $M2'(\Delta e)$.

d. Single-Mirror Adjustment ("Einspiegeleinstellung" by Kinder (56)).
The localization of the interference (using real light sources) in the mirror plane $M1'$ or at the position of the phase object t_m–t_m is given by a relation of the sides of the parallelogram of 2 : 1. The position of the angle and its wedge (spacing and orientation of the parallel interference fringes) can be set by adjusting the mirror $M1'$. Together with the phase object at the position of the plane t_m–t_m, the wedge is imaged in the image plane (photographic film) by the objective $L2$ (Fig. 37b).

A fringe field can also be obtained by rotating mirrors other than $M1'$. However, the position of the resulting virtual wedge would then be arbitrary. The virtual wedge would have to be brought to the zone t_m–t_m by special adjustment of the mirrors. Each change of the fringe field, e.g., the fringe spacing (wedge angle φ), is accompanied by a simultaneous change in the position of the wedge, which would have to be corrected. This drawback is circumvented by the "single-mirror adjustment" with $M1'$.

The path difference of the phase-shifted field of infinite fringe and of the fringe field can serve to compensate the path differences at specific points of the phase object (cf. Section VI,B,3).

The interference pattern of the phase object in the field of infinite fringe (Fig. 1) should be interpreted as the vertical intersections of the reference wave fronts and the eikonal of the measuring beam. The pattern appears in the fringe field as oblique intersections of the plane wave fronts, which are, however, rotated by $\epsilon = 2\varphi$, and the eikonal of the measuring beam(cf. Section V,B,1).

4. *Coherence*

a. Coherence in Time of Real Light Sources. In the preceding discussion it was assumed that the light sources were monochromatic point sources. In the following, however, achromatic point sources are considered. The spectrum of a Na vapor lamp, for example, consists of two components of slightly different frequency (D-line doublet). If this source is utilized for illumination of the MZI, adjusted to produce a fringe field, each component causes a separate fringe field. The superposition of the two separate interference patterns results in the observed pattern. Since the fringe width is proportional to the wavelength, as can be seen in Fig. 38b, the fringe width differs slightly in the two primary patterns. The axis of the wedge (zero-order fringe), however, is the same for both fields. As shall be shown in the following, the intensity distribution of both fields is sinusoidal and the superposition of both results in beats, due to the small difference in frequency. The contrast is greatest for zero-order fringe (wedge axis): The first node with practically zero contrast appears for $S = \pm 981$. The contrast K is defined as

$$K = (I_{\max} - I_{\min})/(I_{\max} + I_{\min})$$

where $I_{\max}$ and $I_{\min}$ are the intensities of adjacent minima and maxima. Proceeding further, the contrast increases, reaching another maximum when the maxima of both patterns coincide. The observed fringe width is—similar to the mean frequency formed in the beat process—the arithmetic mean of the respective fringe widths of the two primary fields. This is, however, only valid with the limitation that the intensities of the two primary fields are approximately equal and constant, as they would be if they were produced by two strictly monochromatic waves of slightly different frequency. However, a sodium vapor lamp, as considered here, does not emit two continuous monochromatic waves, but a multitude of damped wave trains of finite length. The magnitude of the contrast maxima then diminishes with increasing path difference $g = S \cdot \lambda$ of the virtual wedge (or of the phase-shifted zero-order field) and the contrast disappears fully after a few extrema of contrast. This corresponds to the "higher-order white" which appears in the fringe field with illumination by white light.

As a quasi-monochromatic light source for the MZI, one filtered spectral component of a thermal light source is usually employed (e.g., the green line of the Hg spectrum, corresponding to $\lambda = 0.546 \cdot 10^{-6}$ m).

For the duration of the observation, a statistical mean wavelength or corresponding mean frequency can be assigned to the large number of different finite wave trains. The wavelengths are distributed statistically, e.g., according to a Gauss distribution function so that—at least for low-pressure vapor lamps—the intensity can be described by $I \sim \exp[-a^2(\lambda - \lambda_m)^2]$. This distribution is characterized by the half-width $\Delta\lambda$ or $\Delta\nu$ (bandwidth) which is an indication of the mean coherence length Δl and the mean coherence time Δt of a wave train. The coherence time and the bandwidth are correlated by the reciprocal relation

$$\Delta t \cdot \Delta \nu \geqslant 1/4\pi$$

[Born (1), p. 318]; i.e., the effective bandwidth $\Delta\nu$ of a spectral line is of the magnitude of the reciprocal of the duration of a single wave train. The following relation then results:

$$\Delta l = c \cdot \Delta t \simeq c/\Delta\nu = \lambda_m^2/\Delta\lambda \tag{67}$$

The bandwidth $\Delta\lambda$ is dependent on the physical nature of the emitting gas and increases with increasing pressure and temperature. The coherence length is then correspondingly smaller (pressure broadening of spectral lines). Wave trains of finite length cannot be monochromatic; they always have a finite bandwidth since the finite wave train can be described by a sum of Fourier components centered in ν_m. Even an imagined monochromatic and continuous wave train exhibits a certain bandwidth since it is not infinite but originated at a certain time.

The interference pattern of a virtual wedge results—analogous to illumination by a Na vapor lamp—as a superposition of the interference patterns of all spectral components of the emitted line. If the intensity distribution is Gaussian, the interference contrast of the fringe field diminishes exponentially with increasing order of the fringe. Other spectral line shapes result in different interference contrast distributions [Born (1), Bennett (63)]. The usefulness of a light source can be appraised by comparing the coherence length Δl with the path difference $g = \lambda_m \cdot S$. If $g \ll \Delta l$, the emitted light can be considered to be quasi-monochromatic; in more general terms, it is required that the bandwidth be small compared to the center frequency:

$$\Delta \nu / \nu_m \ll 1$$

or

$$\Delta \lambda / \lambda_m \ll 1$$

(narrow spectral line). The following is valid for all interfering patterns at one point of the observed pattern.

$$S = g/\lambda; \quad dS = -g\, d\lambda/\lambda^2 = -S\, d\lambda/\lambda \tag{68}$$

If the bandwidth $\Delta \lambda$ is inserted instead of the differential wavelength $d\lambda$, the resulting interference contrast is approximately $\tfrac{1}{2}$ if two interference fringes corresponding to the wavelength λ_m and $\lambda_m + \Delta \lambda$ are shifted by one-fourth order with respect to another. According to Eq. (68), this is the case for the order S^*:

$$\begin{aligned}\tfrac{1}{4} &= |\Delta S| = S^* \Delta\lambda/\lambda_m \\ S^* &= \tfrac{1}{4} \lambda_m/\Delta\lambda\end{aligned} \tag{68a}$$

Or, if both sides of Eq. (67) are approximately equal, an average coherence length Δl is required:

$$\Delta l = \lambda_m^2/\Delta\lambda = |\lambda_m \cdot S/\Delta S| = |4\lambda_m \cdot S| \tag{68b}$$

The 0.546-μ line of a medium pressure mercury vapor lamp exhibits $\lambda_m/\Delta\lambda = 2500$. The resulting maximum permissible path difference is $g = \lambda \cdot S^*$, the corresponding value of S being $S^* = \frac{1}{4}\lambda_m/\Delta\lambda = 625$; the mean coherence length is $\Delta l = 1.35$ mm. A low-pressure mercury vapor lamp filled with the isotope Hg^{198} (purity 99.9%) exhibits a mean coherence length $l = 0.6$ m (corresponding to about $10^6\lambda$) for the same spectral line. The line widths of high-pressure mercury lamps (≈ 130 atm) are much greater. The spectrum also exhibits a continuum. The line width is then determined by the band width of the filter. Typical values for absorption filters are $\Delta\lambda = 0.012 \cdot 10^{-6}$ m for 50% light absorption; $\Delta\lambda = 0.008 \cdot 10^{-6}$ for 85% absorption. The light transmission compared to the bandwidth is greater for combinations of interference filters; the transmission frequency is, however, dependent upon the exact orientation of the filter in the parallel beam.

Gas lasers exhibit great coherence lengths, so that compensation cuvettes are not required in the reference beam. A drawback of laser illumination of the MZI is that distinct diffraction phenomena arise from dust particles, etc., which impair the evaluation of the desired patterns. Because of the great coherence lengths it is not necessary to align the interference length accurately, so that simpler configurations with concave mirrors are also possible [Goldstein (*64*), Grigull, and Rottenkolber (*65*)]. However, the requirements imposed upon the optical elements and the angular adjustment mechanisms are as severe as in the case of the MZI. In concave mirror configurations, astigmatic errors, caused by slanting illumination of the mirrors as in schlieren arrangements, must be compensated for.

b. Spatial Coherence of Real Light Sources. The coherence lengths Δl of the light sources discussed thus far are generally much greater than the path differences occurring in practice ($S \leqslant 100$). Nevertheless, areas of zero interference contrast can be observed if the fringe density is great (large number of interference fringes per unit length). This effect is caused by the spatial coherence of the real light source, which is a consequence of the finite dimensions of the source.

In order to illustrate this effect it shall be assumed that the MZI is illuminated by a dual light source (two separate monochromatic point sources of equal intensity and the same frequency) in order to produce a fringe field (in contrast to the two spectral components of the sodium lamp discussed in the previous section); cf. Fig. 40. As in the previous arrangement, one light source LS_1 is positioned on the optical axis; the other source LS_2 is displaced by the amount r perpendicularly to the axis of the virtual wedge (not indicated in the illustration). The respective central beams of both light sources therefore enclose the small angle $\omega = r/f$. Although these two light sources are monochromatic to a sufficient degree (quasi-monochromatic), no constant phase relation exists between them. A light source of this kind can be realized by means of a screen with two pinholes placed before a filtered Hg source (66).

The light source LS_1 on the optical axis produces the usual spatial interference fringe system (cf. Fig. 38b). Circles denote maxima; dots denote minima and the intersection with the focusing plane t_m–t_m, which shall also contain the axis C of the virtual wedge, results in the fringe field. Since the adjustment conditions of the MZI as detailed in Section V,B are also valid for light sources not on the optical axis, a second identical interference pattern results from the source LS_2 (Fig. 40) which is parallel to the first. Both patterns have the axis C in common. Both patterns are, however, rotated by the angle ω with respect to one another around the axis C since the two parallel bundles emanating from LS_1 and LS_2, which traverse the MZI independently, enclose the angle ω.

Both fringe systems superpose to form the observed spatial interference pattern. Maximum interference contrast results in the immediate proximity of the axis C, since the maxima and minima of both wave trains coincide here ("principle of coinciding orders"). The displacement of the maxima and minima increases with the distance from the axis C.

Regions symmetric to the axis C then appear where one maximum of one system coincides with a minimum of the second system. As in the preceding section, (coherence in time) extrema in the interference contrast distribution can be observed whose position is dependent upon the wedge angle $\epsilon/2$ and the wavelength λ and the angular displacement ω between the light sources.

A circular extended light source is now substituted for the dual light source. It shall be assumed that this source consists of an infinite number of uncorrelated point sources with an isotropic emission pattern. This can be realized in practice by an illuminated pinhole. The observed interference contrast distribution results from the integration of all of the interference patterns produced by the separate point sources.

Central interference field. The point source in the center of the light-emitting circle, which lies on the optical axis, produces the interference fringe field in ideal interferometry as discussed previously. This central interference fringe field serves to normalize the real pattern as is detailed in the following.

In Fig. 39, the geometric correlation of the central interference fringe field and the coherent wave trains of the reference and measuring beams is shown for the case of a virtual wedge (Section V,B,3,*a*). The interfering wave fronts are plane and rotated with respect to one another by an angle ϵ. The angle $\epsilon/2 = \varphi$ can be produced either by rotation of the

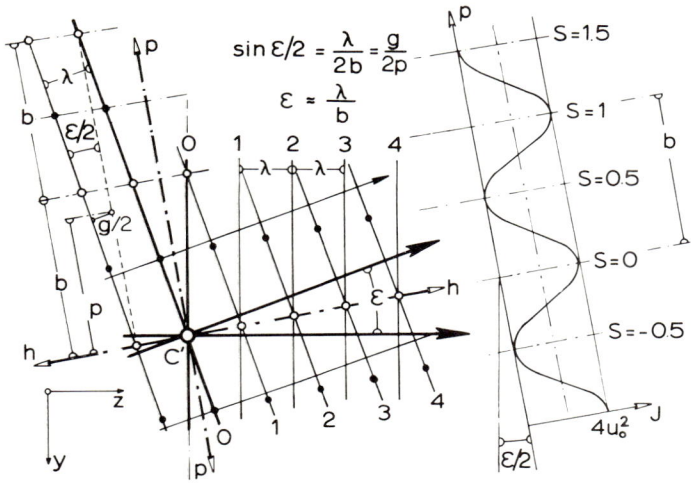

FIG. 39. Spatial distribution of the interference intensity of two monochromatic wave trains (virtual wedge): C, axis of the virtual wedge, also rotation axis of the measuring and reference beam (plane wave fronts); ϵ, angle between the measuring and reference wave fronts ($\epsilon = 2\varphi$, where φ is the wedge angle); z, optical axis; h, axis of the interference field rotated with reference to the z axis by $\epsilon/2$; y, axis perpendicular to the interference fringes; p, axis of the interference field perpendicular to the interference fringes; I, intensity; S, fringe order; circles, intensity maxima ($S = \pm 0; 1,...$); dots, intensity minima ($S = \pm 0.5, 1.5,...$); right: intensity distribution in the focusing plane.

mirror M_1' (Fig. 37) or by a constant gradient of refractive index in a phase object. Here C is the axis of the virtual wedge corresponding to the interference order $S = 0$ and is also the axis of symmetry of the fringe field which is rotated with respect to the optical axis (z axis) by $\epsilon/2$. The interference lines develop as the intersections of the spatial interference field and the focusing plane, which in the optimum case contain the p axis and is orientated perpendicularly to the illustrating plane. The fringe width b can be derived by means of the following expression (cf. Fig. 39):

$$\sin \epsilon/2 = \lambda/2b \qquad (69)$$

or

$$\epsilon \approx \lambda/b \qquad (69a)$$

In principle, the position of the focusing plane can be chosen arbitrarily. In practice, the focusing plane t_m–t_m is positioned parallel to the y axis and perpendicular to the illustrating plane. In the case of phase objects, the angle ϵ is small so that the fringe width as observed on the focusing plane is practically equal to the fringe width b (cf. Fig. 44a and b). In the case of a fringe field produced by rotating the mirror M_1', the virtual wedge is inclined by $\theta = 60°$ relative to the optical axis, as can be seen in Fig. 37. Now the p axis of the interference field and the h axis also enclose the angle $\theta = 60°$. The focusing plane t_m–t_m intersects the interference system, which is now oblique [Kinder (56), Stamm (67)].

The intensity distribution of two interfering monochromatic wave trains with a path difference $g = S \cdot \lambda$ is equal to the square of the light disturbance:

$$I = u^2 = u_{0m}^2 + u_{0r}^2 + 2u_{0m}u_{0r} \cos\left(\frac{2\pi}{\lambda} \cdot g\right) \tag{70}$$

where the separate wave trains of the reference and measuring beams are described by

$$u_r = u_{0r} \cos\left[\frac{2\pi}{\lambda}(c \cdot t)\right]; \quad u_m = u_{0m} \cos\left[\frac{2\pi}{\lambda}(c \cdot t + g)\right] \tag{70a}$$

The vectorial addition of u_{0m} and u_{0r} in the complex plane results in Eq. (70). Insertion of $g/2 = p \cdot \sin \epsilon/2 = p \cdot \lambda/2b$ (as can be seen in Fig. 39) yields

$$I = 2u_0^2\left[1 + \cos\left(\frac{2\pi}{\lambda} \cdot g\right)\right] = 2u_0^2\left[1 + \cos\left(\frac{4\pi \sin \epsilon/2}{\lambda} p\right)\right] \tag{71}$$

assuming that the amplitudes u_{0r} and u_{0m} are equal, as is usually the case for an MZI.

The intensity of the interference pattern varies periodically (with period b) between zero and $4u_0^2$. Maxima result for

$$[(4\pi \sin \epsilon/2)/\lambda] \cdot p = 2\pi S$$

where $S = \pm 1, 2,...$ and $p = \pm b, 2b,...$.

Fringe field of an extended light source. A circular extended light source (radius r) is depicted in Fig. 40. The light source LS_1 of the dual light source in the introduction is now the central point light source; LS_2 is a point light source at the perimeter of the circular diaphragm. In addition, an infinite number of incoherent point light sources exist whose central beams all traverse the center of the lens L_1 and form the aperture cone A enclosing the angle $2\omega = 2r/f$. Instead of two, an infinite number of parallel ray bundles must be considered which intersect under the maximum angle 2ω and whose respective interference fields superpose to form the spatial interference pattern in the vicinity of the wedge axis C.

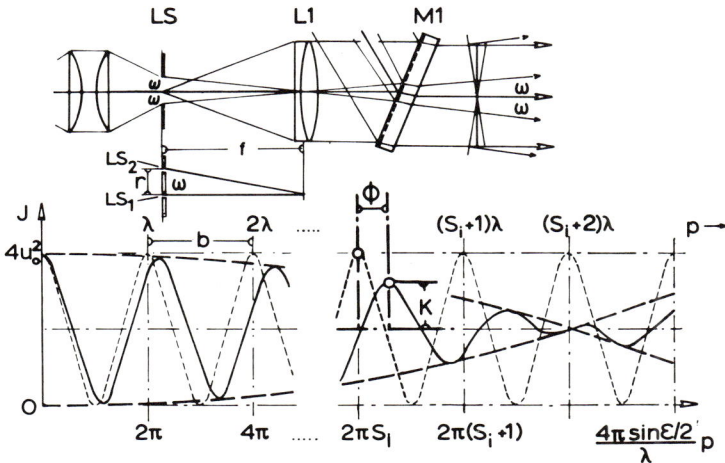

FIG. 40. Illumination of the MZI by an extended light source (diaphragm) and a condenser: LS, extended circular light source: diameter $2r$, aperture cone enclosing the angle 2ω, $\omega = r/f$; L_1, objective; M_1, beam splitter; LS_1, LS_2, point light sources with spacing r (in the illustrating plane). Intensity distribution of the interference pattern formed by a virtual wedge and an extended light source: I, intensity; p, coordinate axis as in Fig. 39; S, order of interference; K, modulation functions.

The calculation [Schulz (68, 69), Minkwitz and Schulz (70)] is similar to that given for the central point light source in the preceding section; however, the path difference g for a point in the vicinity of the wedge axis C is, in addition, a function of the light-source coordinates. The solid angle of the aperture cone is Ω; the light intensity corresponding to an element $d\Omega$ is dI. The intensities of the corresponding wave trains in the reference and the measuring beams are given by

$$dI_r = \mu_{0r}^2 \, d\Omega \quad \text{and} \quad dI_m = \mu_{0m}^2 \, d\Omega$$

Analogous to Eq. (70), the intensity distribution is given by

$$dI = \mu_{0r}^2 \, d\Omega + \mu_{0m}^2 \, d\Omega + 2\mu_{0r}\mu_{0m} \cos\left(\frac{2\pi}{\lambda} g\right) d\Omega$$

$$I = \iint_\Omega \mu_{0r}^2 \, d\Omega + \iint_\Omega \mu_{0m}^2 \, d\Omega + 2 \iint_\Omega \mu_{0r}\mu_{0m} \cos\left(\frac{2\pi}{\lambda} g\right) d\Omega$$

For equal intensity in the two ray paths in the MZI $\mu_{0r} = \mu_{0m} = \mu_0$, it follows that

$$I = 2\mu_0^2 \Omega \left[1 + \frac{1}{\Omega} \iint_\Omega \cos\left(\frac{2\pi}{\lambda} g\right) d\Omega\right]$$

The solution of this equation is of a similar form as the expression (71) for the point light source:

$$I = 2\mu_0^2 \Omega \left[1 + K(\tilde{\omega}, \tilde{z}) \cos\left(\frac{4\pi \sin \epsilon/2}{\lambda} p - \phi(\tilde{\omega}, \tilde{z}) \right) \right] \tag{72}$$

Here again the cosine function represents the spatial interference field where the locus dependent modulation parameters $K(\tilde{\omega}, \tilde{z})$ (interference contrast) and $\phi(\tilde{\omega}, \tilde{z})$ (phase modulation) are superposed on the spatial distribution corresponding to a point light source.

The variables $\tilde{\omega}$, $\tilde{z}$ are proportional to the coordinates p and h (Fig. 39) and contain, besides the fringe width $1/b = (2/\lambda) \cdot \sin \epsilon/2$ (Eq. 69), the aperture angle ω.

$$\tilde{\omega} = \text{const} \cdot p = [(4\pi \sin \epsilon/2)/\lambda] \, \omega^2 \cdot p \tag{73a}$$

$$\tilde{z} = \text{const} \cdot h = [(4\pi \sin \epsilon/2)/\lambda] \, \omega \cdot h \tag{73b}$$

Figure 40 shows the qualitative intensity distribution (solid line) produced by the intersection of the spatial interference field with the

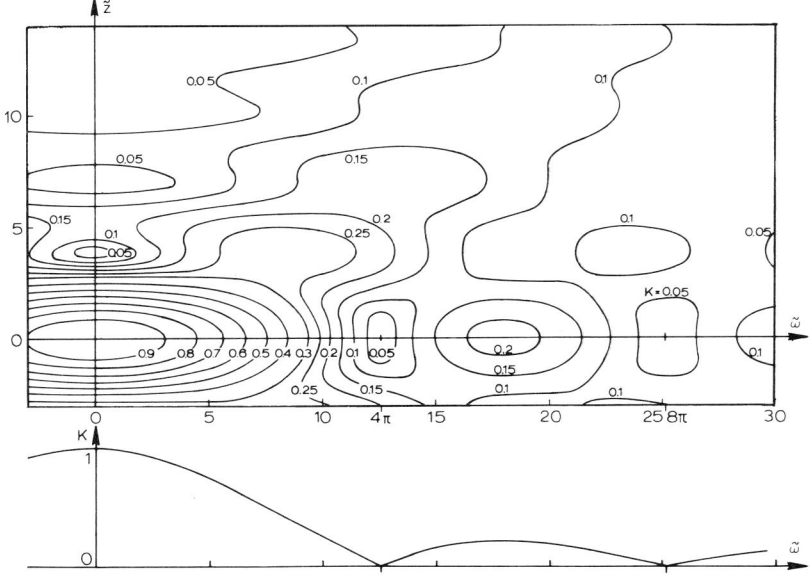

FIG. 41. Spatial distribution of the interference contrast $K(\tilde{\omega}, \tilde{z})$ in the dimensionless coordinates $\tilde{\omega} = (4\pi/\lambda) \sin(\epsilon/2) \, \omega^2 p$ and $\tilde{z} = (4\pi/\lambda) \sin(\epsilon/2) \, \omega h$. The coordinate origin is the intersection of the wedge axis C and the $\tilde{\omega}$ and $\tilde{z}$ axes, which coincide with the p and h axes, respectively. The distribution of the interference contrast along the $\tilde{\omega}$ axis (p axis) for $\tilde{z} = 0$ ($h = 0$) is shown separately: $K(\tilde{\omega}, 0) = |\sin(\tilde{\omega}/4)/(\tilde{\omega}/4)|$. (From Minkwitz and Schulz (70)).

focusing plane. The focusing plane usually contains both the wedge axis C and the p axis (Fig. 39). The interference contrast is greatest along the wedge axis C (interference order $S = 0$) and is here equal to the uniform interference contrast of a point light source. The intensity distribution of a point light source is indicated by dashed lines in Fig. 40.

The interference contrast and the modulation function K change with increasing magnitude of $2\pi S = [(4\pi \sin \epsilon/2)/\lambda] p$, where $p = S \cdot b$. The argument of the cosine function $2\pi S_i$ [Eq. (72)] for $p = S_i b$ is reduced by ϕ so that the resulting fringe width is larger than in the case of a point light source (ideal interferometry).

The modulation functions $K(\tilde{\omega}, \tilde{z})$ and $\phi(\tilde{\omega}, \tilde{z})$ are shown in Figs. 41 and 42. They are symmetric to the planes defined by $\tilde{\omega} = 0$ and $\tilde{z} = 0$; only one quadrant is depicted. In the dimensionless representation as in Eq. (73), the $\tilde{\omega}$ and $\tilde{z}$ axes correspond to the p and h axes, respectively. The modulation functions $K(\tilde{\omega}, \tilde{z})$ and $\phi(\tilde{\omega}, \tilde{z})$ can be expressed by means of Lommel's functions $U_1(\tilde{\omega}, \tilde{z})$ and $U_2(\tilde{\omega}, \tilde{z})$, which were found

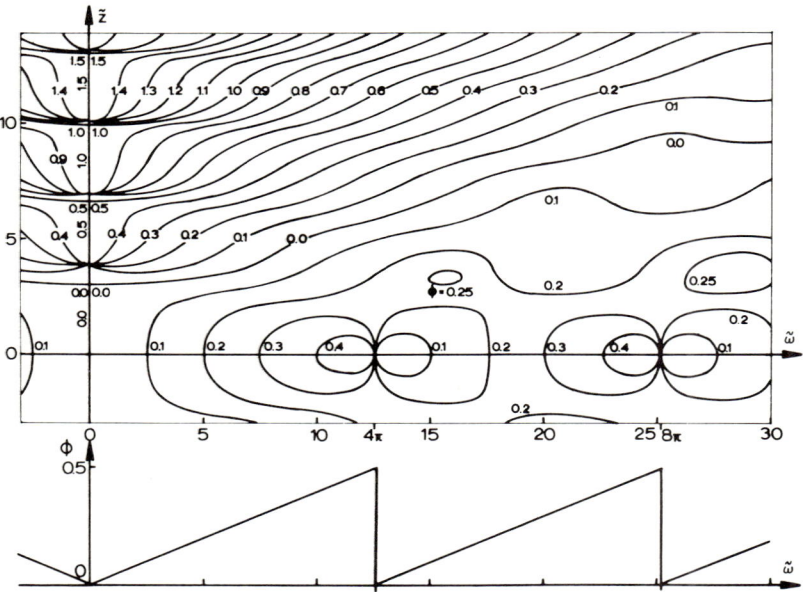

FIG. 42. Spatial distribution of the phase modulation $\phi(\tilde{\omega}, \tilde{z})$ in the dimensionless coordinates $\tilde{\omega} = (4\pi/\lambda) \sin(\epsilon/2) \omega^2 p$ and $\tilde{z} = (4\pi/\lambda) \sin(\epsilon/2) \omega h$. The coordinate origin is the intersection of the wedge axis C and the $\tilde{\omega}$ and $\tilde{z}$ axes, which coincide with the p and h axes, respectively. The phase shift $\phi(\tilde{\omega}, \tilde{z})$ with respect to $2\pi S$ in the central interference field is given in fractions S of the interference order. The phase distribution ϕ on the $\tilde{\omega}$ axis (p axis) is shown separately. (From Minkwitz and Schulz (70)).

by Lommel while calculating the intensity distribution of an image point in imaging by circular bundles: (71)

$$K = \left| \left(\left[\frac{2}{\tilde{\omega}} U_1(\tilde{\omega}, \tilde{z}) \right]^2 + \left[\frac{2}{\tilde{\omega}} U_2(\tilde{\omega}, \tilde{z}) \right]^2 \right)^{1/2} \right| \tag{74}$$

$$\phi = \tilde{\omega}/2 - \arccos \frac{2\, U_1(\tilde{\omega}, \tilde{z})}{\tilde{\omega}\, K(\tilde{\omega}, \tilde{z})} = \tilde{\omega}/2 - \arcsin \frac{2\, U_2(\tilde{\omega}, \tilde{z})}{\tilde{\omega}\, K(\tilde{\omega}, \tilde{z})} \tag{75}$$

$$U_1 = U_1(\tilde{\omega}, \tilde{z}) = \sum_{k=0}^{\infty} (-1)^k (\tilde{\omega}/\tilde{z})^{2k+1} J_{2k+1}(\tilde{z}) \tag{76a}$$

$$U_2 = U_2(\tilde{\omega}, \tilde{z}) = \sum_{k=0}^{\infty} (-1)^k (\tilde{\omega}/\tilde{z})^{2k+2} J_{2k+2}(\tilde{z}) \tag{76b}$$

J_ν are Bessel functions of the νth order. The spatial interference field of an extended light source results from the interference field of the central point light source in Fig. 39, where the planes of constant fringe order S (circles, dots) are parallel to the h axis. The modulation functions $K(\tilde{\omega}, \tilde{z})$ and $\phi(\tilde{\omega}, \tilde{z})$ shown in Figs. 41 and 42 are imposed on the interference field. The values of K_i and ϕ_i for a point $P_i(p_i, h_i)$ of the interference order S_i can be determined with the dimensionless coordinates $\tilde{\omega}_i$ and $\tilde{z}_i$ of Eq. (73) from Eqs. (74) and (75) or Figs. 41 and 42:

$$\tilde{\omega} = \frac{4\pi \sin \epsilon/2}{\lambda} \omega^2 p; \qquad \tilde{z} = \frac{4\pi \sin \epsilon/2}{\lambda} \omega h \tag{73}$$

For smaller values of the aperture angle $\omega \to 0$, i.e., smaller diaphragm diameters, a point light source is approximated; $\tilde{\omega}$ and $\tilde{z}$ are then limited to the immediate vicinity of the origin (axis of the virtual wedge). The spatial extent of the interference field, expressed in terms of the coordinates p and h, or the fringe density $1/b = (2 \sin \epsilon/2)/\lambda$ [Eq. (69)] can then be correspondingly large. For $\tilde{\omega} \approx 0$ and $\tilde{z} \approx 0$ the corresponding values $K \approx 1$ and $\phi \approx 0$ can be derived from Figs. 41 and 42. If these values are inserted in the expression for the intensity distribution of the real interference field [Eq. (72)] the simplified equation (70) for the intensity distribution of the central point light source results.

For decreasing wedge angles ($\epsilon/2 \to 0$), $\tilde{\omega}$ and $\tilde{z}$ are also approximately zero, which means that optimum interference contrast ($K \approx 1$) and vanishing phase modulation ($\phi \approx 0$) can be obtained for great fringe widths $1/b = (2 \sin \epsilon/2)/\lambda$. According to Eq. (73), the light source ($\omega = r/f$) may in this case be large. For $\epsilon/2 = 0$, i.e., infinite fringe width, the range of $\tilde{\omega}$ and $\tilde{z}$ values is confined to the wedge axis C. The interference contrast is then only defined by the large coherence in time assumed here (quasi-monochromatic light).

The magnitude of the light source aperture ω is dictated in practice by the chosen fringe density $1/b = (2 \sin \epsilon/2)/\lambda$, since the extent of the spatial interference field, expressed in terms of the coordinates p and h, must be sufficiently large. However, a lower limit for the interference contrast exists, i.e., the dimensionless coordinates should be chosen in the vicinity of the wedge axis so that no points with $K = 0$ (Fig. 41) appear:

$$0 < |\tilde{\omega}| < 4\pi; \qquad 0 \leqslant |\tilde{z}| < 3.83$$

(the first order Bessel function vanishes for $\alpha_1 = 3.83$).

The structure of the interference field, i.e., the locations of the planes $S = $ const, differs from that of the central point source. Because of the phase modulation ϕ, the parallel planes $S = $ const are curved around the wedge axis. These are normal to the $\tilde{\omega}$, $\tilde{z}$ plane (p–h plane) and can be developed. The function is shown in Fig. 42 in the $\tilde{\omega}$, $\tilde{z}$ plane with fractions of the interference order as a parameter. The interference pattern of the virtual wedge results as the intersection of the interference field with an arbitrarily orientated focusing plane. However, the fringe width is generally no longer constant (cf. also Fig. 40).

For phase objects the discussion pertaining to the virtual wedge (Section V,C) can readily be extended in a qualitative manner. The focusing plane is then practically parallel to the p axis ($\tilde{\omega}$ axis), and one will seek to position the focusing plane so that it contains the wedge axis C (origin) in order to obtain optimum conditions. The interference contrast $K(\tilde{\omega}, \tilde{z} = 0)$ and the phase modulation $\phi(\tilde{\omega}, \tilde{z} = 0)$ along the $\tilde{\omega}$ axis is shown in a separate diagram in Figs. 41 and 42. According to Eqs. (76a) and (76b) it follows for $\tilde{z} = 0$ that

$$U_1 = \sum_{k=0}^{\infty} (-1)^k (\tilde{\omega}/\tilde{z})^{2k+1} \frac{(\tilde{z}/2)^{2k+1}}{(2k+1)!} = \sin(\tilde{\omega}/2)$$

and

$$U_2 = \sum_{k=0}^{\infty} (-1)^k (\tilde{\omega}/\tilde{z})^{2k+2} \frac{(\tilde{z}/2)^{2k+2}}{(2k+2)!} = 1 - \cos(\tilde{\omega}/2)$$

so that, with Eq. (74), the interference contrast K results (Fig. 41):

$$K = [\sin(\tilde{\omega}/4)]/(\tilde{\omega}/4) \tag{77}$$

The phase modulation $\phi(\tilde{\omega}, \tilde{z} = 0)$ is a periodically linear function (saw tooth) of $\tilde{\omega}$ (Fig. 42). For $0 \leqslant \tilde{\omega} < 4\pi$ it can be written as

$$\phi = \tilde{\omega}/4 = \frac{4\pi \sin \epsilon/2}{\lambda} \frac{\omega^2}{4} p = 2\pi S \frac{\omega^2}{4} \tag{78}$$

If a new interference order S^* of the real interference field is defined, the argument of the cosine function in Eq. (72) can be written as

$$2\pi S^* = \frac{4\pi \sin \epsilon/2}{\lambda} p - \phi = 2\pi S - \phi \tag{78a}$$

Together with Eq. (78), a correction equation for the interference order of the central interference field is obtained. This expression is valid along the $\tilde{\omega}$ axis (p axis) and in the adjoining region where ϕ is still approximately linear; for $\tilde{\omega} \leqslant 4\pi$,

$$2\pi S^* = 2\pi S(1 - \omega^2/4)$$

or

$$S^* = S - \Delta S = S(1 - \omega^2/4); \qquad \Delta S/S = \omega^2/4 \tag{79}$$

Correction for the phase difference is

$$\Delta S_0 = (\omega^2/4)S \tag{79a}$$

which is only valid for $p < \lambda/(\omega^2 \sin \epsilon/2)$ and $h \approx 0$. The p axis corresponds to the axis perpendicular to the fringes of the interference figure. Since the angle $\epsilon/2$ is usually very small the p and h axes are practically equivalent to the y and z axes.

C. Two-Dimensional Phase Object in a Mach–Zehnder Interferometer

1. *Imaging of a Phase Object*

a. Focusing Effects. In Fig. 43 a Mach–Zehnder interferometer is shown with a specific phase object in the measuring beam (m). The parallel beam of the measuring path is deflected by an angle ϵ with reference to the optical axis while traversing the test section TS. The wave fronts are then plane again and the interference pattern corresponds to one caused by a virtual wedge.

The position of this virtual wedge differs from that in Fig. 37 where the wedge was produced by rotating the mirror M_1'. In the latter case, the wedge forms an angle of about 60° with respect to the optical axis; in this case, however, the virtual wedge is practically perpendicular to the optical axis (z axis). The angle of inclination of the interference field with respect to the focusing plane (Fig. 39) is so small that it can be neglected. The position of the focusing plane t_m–t_m is so chosen that it coincides with the wedge.

A phase object of this kind was realized in the example of Section VI,B,3. Between a heated $(\vartheta_\infty + \Delta\vartheta/2)$ and a cooled $(\vartheta_\infty - \Delta\vartheta/2)$ surface there is a linear temperature drop (Fig. 43) in the test medium where the

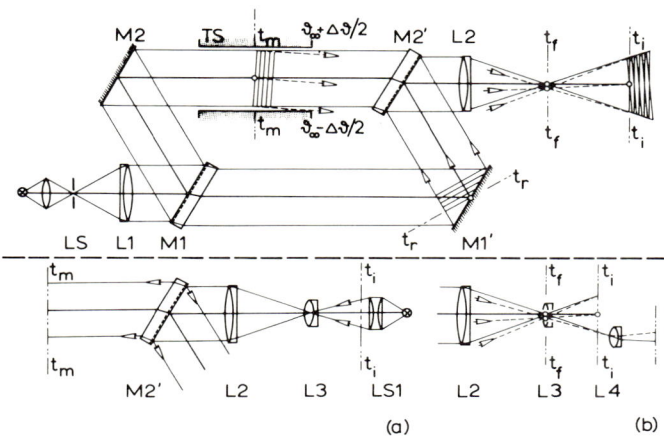

FIG. 43. Phase object corresponding to a virtual wedge, caused by a constant temperature gradient (leading to constant refractive index gradient) in the test section: TS, test section; $t_m - t_m$, focusing plane in the measuring path; $t_r - t_r$, focusing plane in the reference path; $t_f - t_f$, focal plane of the objective L_2 containing the images of the light source from the measuring and the reference beams which are displaced by $e \approx \epsilon f$ with respect to one another; $t_i - t_i$, image plane (photographic film) conjugated to the focusing planes $t_m - t_m$ and $t_r - t_r$. Focusing aids; (a) Focusing plane $t_m - t_m$ considered as the projection of a transparent raster slide in the image plane $t_i - t_i$ (reversal of the ray path): L_2, objective; L_3, camera lens; LS_1, illumination of the raster by means of an auxiliary light source and a dual condenser. (b) Adjustment of the image plane $t_i - t_i$ so that it is conjugated to the focusing plane $t_m - t_m$ (which contains the axis of the virtual wedge): L_2, objective; L_3, camera lens in the focal plane $t_f - t_f$ of L_2; L_4, auxiliary objective with screen for subsequent magnification of the interference pattern in the image plane $t_i - t_i$, which consists of a transparent raster slide. Both the interference pattern and the raster image should be sharply imaged on the auxiliary screen.

gradient of refractive index dn/dT is assumed to be constant. The linear temperature profile in the test section TS is, therefore, proportional to the profile of the refractive index. The constant refractive index gradient causes in the whole a deflection of the wave fronts (m) which can be interpreted as the effect of a virtual wedge, which in the following will serve as a substitute phase object. The (superposed) wave fronts of the focusing planes $t_m - t_m$ and $t_r - t_r$ are imaged in the imaging plane $t_i - t_i$ by the objective L_2. In the focal plane $t_f - t_f$ of L_2 the deflection ϵ of the plane wave fronts belonging to the measuring beam appears as a displacement $e = \epsilon \cdot f$ (f = focal length of L_2) between the light source images of the reference beam (r) and the measuring beam (m), respectively (cf. also Fig. 37).

The correlation between the deflection ϵ of the light beam in the phase object exhibiting a linear gradient of refractive index is shown in Fig. 44 along with the formation of the interference pattern in the image plane $t_i - t_i$. The length of the heated wall in the test section is l and its coordi-

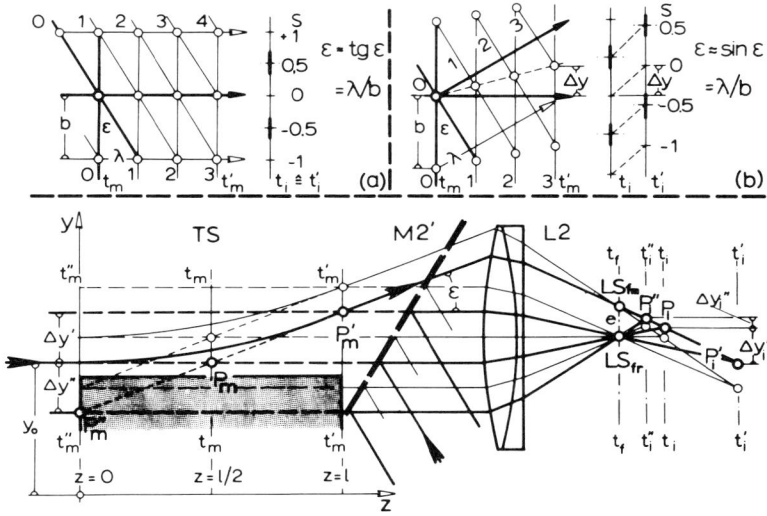

FIG. 44. Effect of focusing on the formation of an interference pattern. A linear refractive index profile is assumed above the heated wall. The MZI is in a general basic position: TS, test section; M_2', beam splitter; L_2, objective; z, coordinate in the test section ($0 \leqslant z \leqslant l$); l, model length; y_0, the y coordinate of the light beam entering the test section (solid line); y_0, $y_0 + \Delta y'$, $y_0 - \Delta y''$, the y coordinates of separate rays belonging to the reference beam (dashed lines), which interfere with the deflected measuring beam in the image plane $t_i - t_i$, $t_i' - t_i'$, and $t_i'' - t_i''$, according to the position of the focusing planes ($t_m - t_m$, $t_m' - t_m'$, $t_m'' - t_m''$); P_m, P_m', P_m'', virtual object points corresponding to the image points; P_i, P_i', P_i'', image points; $t_f - t_f$, focal plane; LS_{fr}, image of the light source in the focal point formed by the reference beam; LS_{fm}, image of the light source produced by the measuring beam, displaced by $e = \epsilon \cdot f$ from the optical axis; $t_m - t_m$, $t_m' - t_m'$, $t_m'' - t_m''$, focusing planes in the path of the measuring beam conjugated to the image planes. (a) Interference field of the virtual wedge in ideal interferometry: The intersection with two different focusing planes t_m and t_m' results in the same interference pattern $t_i = t_i'$ (the rotation ϵ has been neglected). (b) Interference field of the virtual wedge: The intersections of different focusing planes with the interference field result in interference patterns t_i and t_i' where the fringes are displaced by Δy. (The slightly inclined position of the interference field (angle $\epsilon/2$) with respect to the focusing plane in Fig. 39 has been neglected.)

nate in the direction of the light beam is z ($0 \leqslant z \leqslant l$). A light beam entering the test section perpendicularly at the height y_0 (thick solid line) assumes a parabolic path curvature under the influence of the constant gradient of refractive index (dn/dy), as has already been detailed in Section III for the immediate vicinity of the wall in a thermal boundary layer.

Outside of the test section TS the light beam propagates in a straight line and traverses the beam splitter M_2' (for simplification, depicted only as a semitransparent layer without the glass base) and the objective L_2. A second beam indicated by a thin line, entering the test section somewhat above the first beam, remains parallel to it. After passing through the objective L_2, both rays intersect in the focal plane t_f–t_f at a distance $e = \epsilon \cdot f$ from the optical axis. The sum of all rays belonging to the measuring beam form the image of the light source LS_{fm}. The rays of the reference beam are reflected at M_2' and form the image of the light source LS_{fr} on the optical axis. Here it has been assumed that the MZI is in a general basic position.

In order to explain the process of image formation better, it is assumed that the reference beam (r) is transferred to the measuring beam (m) symmetrically to the mirror plane M_2'. The dashed rays (r) then coincide with the corresponding rays of the measuring beam before entering the test section; i.e., ray pairs are formed as existed in the original beam before it was split by the first beam splitter M_1.

If the position t_i–t_i is selected for the image plane (from the positions t_i–t_i, t_i'–t_i', t_i''–t_i''), it then corresponds to the conjugate focusing plane t_m–t_m, and the image point P_i along with the interference pattern at this position then corresponds to the object point P_m.

No image distortion arises since the deflected image-forming beam (measuring beam) and the virtual image-forming beam (reference beam) both seem to emanate from the virtual object point P_m (the coordinate of P_m being equal to the y coordinate y_0 on entering the test section). The deflected rays follow a parabolic curve in the test section. The point P_m is situated in the middle of the test section. This is similar to the case in Section III,A, where the situation in the vicinity of the wall of a thermal boundary layer was described. The interference pattern at P_i is caused by the path difference between the beams (m) and (r).

If, for example, the image plane t_i'–t_i' and the conjugate focusing plane t_m'–t_m' at the exit of the test section are chosen, the image point P_i' corresponds to the object point P_m'. The measuring beam (m), whose y coordinate is y_0, and the reference beam (r), whose y coordinate is $y_0 + \Delta y'$, now interfere in the image point P_i'. The same is valid if the image plane is chosen at t_i''–t_i'' with the focusing plane t_m''–t_m'' at the entrance of the test section. The object point P_m'', whose y coordinate is $y_0 - \Delta y''$, corresponds to the image point P_i''.

According to the position of the focusing plane t_m-t_m "image distortion" results, i.e., displacement ($\Delta y'$, $\Delta y''$) of the object point $P_m(P_m', P_m'')$ corresponding to the respective measuring ray. The image points P_i' and P_i'' are displaced, respectively, by the amounts $\Delta y_i'$ and $\Delta y_i''$, not considering the corresponding imaging scale factor.

If only the fringe width is of interest (the fringes in Fig. 44 can be assumed to be represented by the two parallel measuring rays), no error results from the ray curvature. The two object points have the same spacing in all focusing planes; this is also valid for the image points which are all displaced by the same amount if the respective imaging scale is considered. In this case, the prerequisite for true imaging of the fringe width is that the measuring rays remain parallel in the test section, which is the case in a thermal boundary layer in the immediate vicinity of the wall ($dn/dy = $ const).

In practice, it is more advantageous to position the focusing plane t_m-t_m in the center of the test section ($z = l/2$), where the displacement Δy is equal to zero, if an approximately parabolic path curvature can be assumed (where the measuring rays need not necessarily be parallel). The rays of the measuring and reference beams are then pairs which were produced from one beam by means of the beam splitter (same y coordinate y_0 at the entrance). This improves the interference contrast (Section V,C,1,b). The whole refractive index field is imaged in this manner without displacement errors as in the case of ideal interferometry. Small displacement errors due to deviations from parabolic ray curvature shall be discussed later for several boundary-layer models.

The following serves to elucidate the formation of the interference field. For this, imagine the wave fronts with a spacing perpendicular to the rays. The wave fronts of the measuring rays (m) are curved (cf. Fig. 7). After leaving the test section they are planar but are inclined by an angle ϵ with respect to the reference beam (r). The intersection of the focusing plane t_m-t_m and the interference field results in an interference pattern identical to one due to a virtual wedge assumed as a substitute for the phase object. One can also imagine that the deflected measuring beam emanating from the test section comes from the point P_m with corresponding plane wave fronts on the rear extension (indicated by a dashed line). By this means, the interference field corresponding to a virtual wedge develops, consisting of the reference wave fronts (r) and the inclined wave fronts of the measuring beam (m).

The displacement Δy due to focusing is shown in Figs. 44a and 44b for the case of an interference field of a virtual wedge. In the case of ideal interferometry (Fig. 44a), the small deflection angle is neglected. The intersections of the focusing planes t_m–t_m and t_m'–t_m' with the interference field result in identical interference patterns at t_i–t_i and t_i'–t_i' (presuming identical imaging scales). In the situation shown in Fig. 44b, the interference fringes ($S =$ const) shift when the focusing is adjusted (plane t_m–t_m, t_m'–t_m' and plane t_i–t_i, t_i'–t_i') by the amount Δy with respect to an imaginary reference point located in the focusing plane and outside of the test section where it is always imaged correctly, i.e., without any displacement. This comparatively small displacement of the interference fringes should not be confused with the lateral movement of the interference fringes while shifting the phase of a fringe field (Section V,B,3, Fig. 38c).

Experimental verification. Interference patterns (field of infinite fringe, the interference lines correspond approximately to isotherms) of a vertical cuvette filled with water are shown in Fig. 45 for different positions of the focusing plane t_m–t_m. As in Fig. 44, the z axis points in the direction of propagation, its origin coinciding with the beginning of the test section ($l = 50$ mm). The y axis is perpendicular to the (heated) right-hand wall pointing to the left and the x axis is parallel to the heated wall in flow direction. The interference lines cannot be distinguished in the vicinity of the cooled wall on the left-hand side. The interference pattern cannot be evaluated here since the light rays are deflected towards the wall and are reflected there so that the pattern in the immediate vicinity of the wall is washed out. The gradient of the temperature field in the circulating current is greatest in the bottom right corner (stagnation point), diminishing along the wall in the direction of the x axis.

The central pattern in Fig. 45 ($z = 0.55l$) is correctly focused; i.e., according to the above discussion the displacement Δy is zero and independent of the deflection angle ϵ, which is characterized by the fringe density $1/b$ ($\sin \epsilon/2 = \lambda/2b$; $\epsilon = \lambda/b$). Furthermore, it can be assumed—within the limits set by the measuring accuracy—that the curvature of the ray path is parabolic. The profile of the heated wall is identical to the profile in an exposure of the unheated wall, which in practice can serve as a criterion for correct focusing. Usually the center of the cuvette $z = l/2$ is sharply focused. This is facilitated by a marker affixed at this position. Better yet is a transparent calibration scale which furnishes the imaging scale in the photograph of the unheated wall. The focusing plane is not positioned exactly at $z = 0.5l$ because of the beam displacement in the window where the beam emerges from the cuvette and the displacement in the beam splitter M_2'. In the case of the left-hand picture ($z = 0$: plane t_m–t_m at the entrance of the test section), the displacement ($-\Delta y$) decreased in the direction of the x axis with the fringe density $1/b = \epsilon/\lambda$.

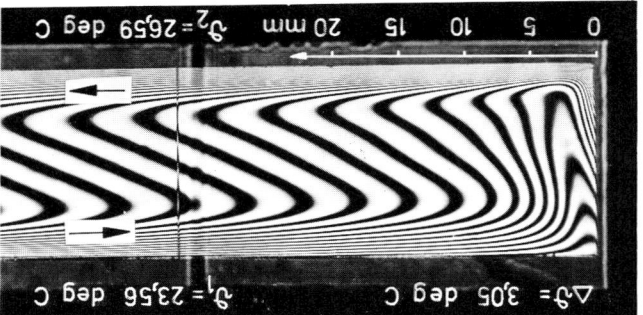
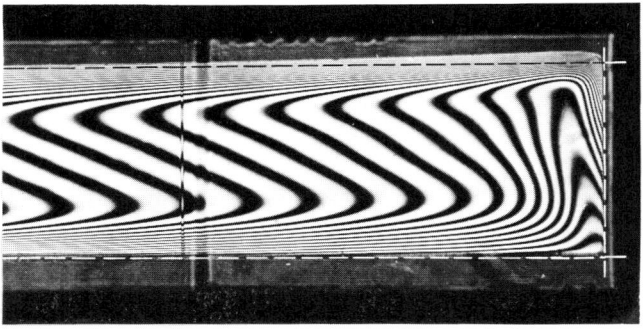

FIG. 45. Effect of focusing in the case of a thermal boundary layer with a variable gradient $d\vartheta/dy = f(x)$, viz. $dn/dy = f(x)$. The position of the heated wall $y = 0$ (x axis) is indicated by a dashed line. The left-hand (cooled) wall is not considered. The interference lines approximate isotherms with a mean graduation $\Delta\vartheta \approx 0.1$ deg (for water, model length $l = 50$ mm). All photographs are taken in a infinite fringe field. Left: focusing plane t_m-t_m at $z = 0$ (entrance of the test section), displacement $-\Delta y$. Center: focusing plane t_m-t_m at $z = 0.55l$ (approximately center of the test section; influence of the cuvette window), displacement $\Delta y = 0$. Right: focusing plane t_m-t_m at $z = 2l$ (outside of test section), displacement $+\Delta y$ whereby a focal line appears at the point of the maximum gradient (stagnation point of the circulating current).

The position of the not yet heated wall is indicated by a black dashed line.

In the right-hand picture ($z = 2l$: plane t_m–t_m outside of the test section) a focal line appears at the point of maximum gradient (stagnation point of the circulating current). The displacement ($+\Delta y$) again decreases with decreasing fringe density (in the x direction), i.e., decreasing gradient $dn/dy = f(x)$ at the wall. The unheated wall ($y = 0$) is indicated by a white dashed line.

The position of the focusing plane, e.g., can be found by inverting the direction of the rays as carried out in Fig. 45. A transparent raster positioned at t_i–t_i (camera screen) is illuminated by an auxiliary light source LS_1. The real image of the raster is produced on a screen (plane t_m–t_m) in the path of the measuring beam. For a given position of the plane t_m–t_m (screen) with the coordinate z(e.g., $z = l/2$) the corresponding conjugate plane t_i–t_i can be found by focusing (Fig. 43a).

However, it is better if the interference pattern, which appears undisplaced (undistorted) in the plane t_i–t_i (which is to be found), is projected enlarged on an additional screen by means of an auxiliary objective L_4. Here it is expedient to select the region with the greatest fringe density, i.e., with a maximum displacement for an incorrect position of t_i–t_i. When the correct plane t_i–t_i ($\Delta y = 0$) is found, the raster is sharply focused on the auxiliary screen so that the plane of the photographic plate (raster slide) coincides with the plane t_i–t_i. An additional criterion for correct focusing ($\Delta y = 0$) is maximum interference contrast, as is shown in the following (Fig. 43b).

b. Interference Contrast. Figure 46 is an enlarged portion of the interference pattern of Fig. 45, showing the vicinity of the stagnation point. Here the refractive index gradient dn/dy is greatest at the wall, which also applies to its x dependence (the x axis is parallel to the wall). The effect of focusing errors can be seen in the left ($z = 0$) and the right ($z = 2l$) hand pictures: The interference fringes are displaced by $-\Delta y$ and $+\Delta y$, respectively. Furthermore the interference contrast is smaller than in the case of correct focusing (center).

The interference pattern in the vicinity of the wall is equivalent to that of a virtual wedge, as was shown in Fig. 44. In the left-hand picture ($z = 0$) the focusing plane intersects the interference field in front of the wedge axis (as viewed in the direction of the beam); in the right-hand picture ($z = 2l$) the intersection is behind the wedge axis. In the center picture (correct focusing) the focusing plane, the wedge axis, and the p axis of the interference field practically coincide and therefore the contrast is optimum.

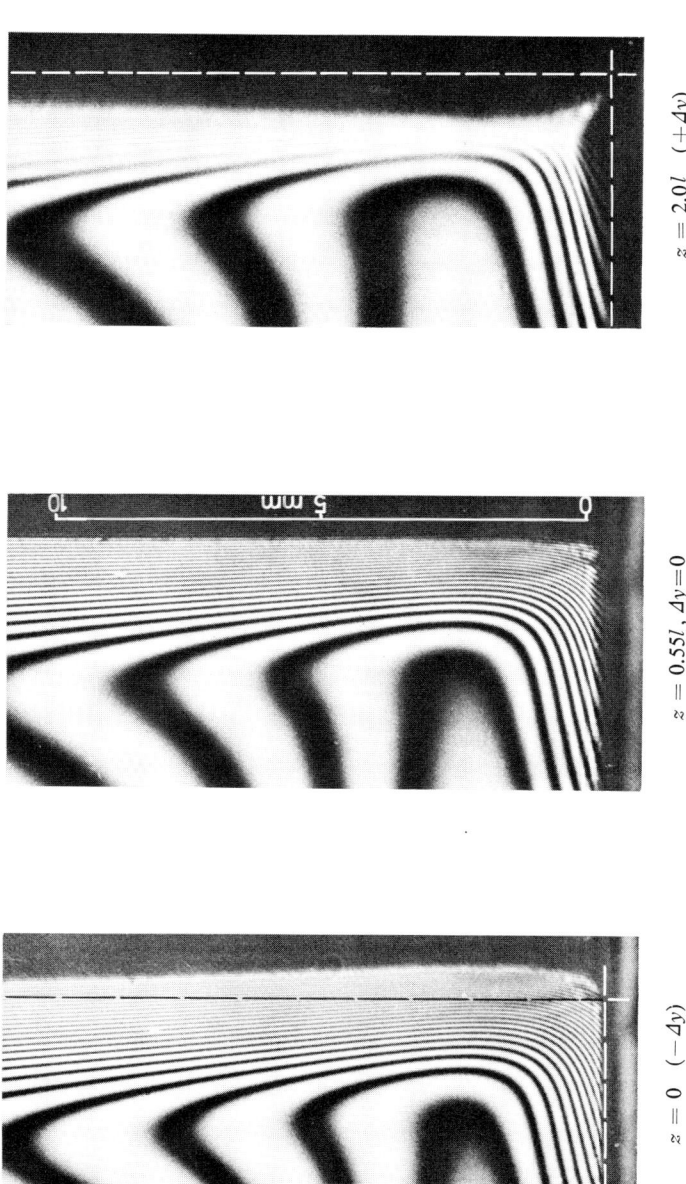

FIG. 46. Effect of focusing on the interference contrast in the case of a thermal boundary layer (enlarged portion of Fig. 45: stagnation point). In the vicinity of the wall the phase object is comparable to a virtual wedge. Left: the focusing plane at the entrace to the test section intersects the interference field of the virtual wedge in front of the wedge axis, as viewed in the direction of the beam ($z = 0$ in Fig. 44): diminished interference contrast, displacement $-\Delta y$. Center: the wedge axis is in the focusing plane: optimum contrast, $z = 0.55l$, no displacement ($\Delta y = 0$). Right: the focusing plane is outside of the stet section and intersects the interference field of the virtual wedge far behind the wedge axis: poor interference contrast, appearance of a focal line, $z = 2.0l$, displacement $+\Delta y$.

This can be utilized as a focusing aid if the spatial coherence is temporarily limited by selecting a large diaphragm at the light source (large angle ω of aperture). The spatial interference field is restricted to the immediate vicinity of the wedge axis. The correct position of the focusing plane can be found more easily by this means. If the gradient is small (small wedge angle $\epsilon/2$), it is possible to reduce the coherence further yet by using, e.g., unfiltered Hg light in order to adjust the focusing plane in the vicinity of the wedge axis [Eq. (73)].

In Fig. 47, the interference field of a virtual wedge is compared with the field corresponding to a thermal boundary layer with a constant gradient in the vicinity of the wall (referred to the case in Fig. 48).

The interference field of the virtual wedge of Fig. 39 is shown again in Fig. 47; however, for the sake of clarity only every sixth maximum is indicated. The wedge axis C and the p, h coordinate system describing the interference contrast at the object point P_m (p_m, h_m) are the same as in the discussion of spatial coherence in Section V,B,4,b. The "correct" focusing plane t_m–t_m contains the wedge axis and an object point P_m in the vicinity of the wall, similar to Fig. 44. The coordinate h_m is small compared to p_m since the magnitude of the angle ϵ is exaggerated.

The interference field of the thermal boundary layer in Fig. 47 with deformed wave fronts in the measuring path is identical to that of a virtual wedge in the vicinity of the wall, as detailed on page 234, Fig. 44. Here—constant refractive index gradient assumed—the distribution of the interference contrast corresponds to the case of a virtual wedge if the wedge axis C and the corresponding p–h coordinate system are assumed as in Fig. 47. This distribution (for $\tilde{z} = 0$ or $h = 0$) is described by Eq. (77), e.g., for the point $P_m(p_m, h_m \approx 0)$:

$$K = \sin(\tilde{\omega}/4)/(\tilde{\omega}/4)$$

where

$$\tilde{\omega} = [(4\pi \sin \epsilon/2)/\lambda] \, \omega^2 \cdot p_m$$

It is possible to obtain the interference contrast distribution of the complete interference system of the thermal boundary layer—and also the phase modulation ϕ—by considering the deformed wave front as a composition of plane wave fronts belonging to interference systems of individual virtual wedges with different wedge angles ($\epsilon/2$) and individual p, h coordinates. The p axis appears stretched in the boundary region according to the varying wedge angle. This analogy to the virtual wedge can be extended to describe qualitatively general phase objects with spatially deformed wave fronts, just as a lens can be considered to be composed of a large number of small prisms.

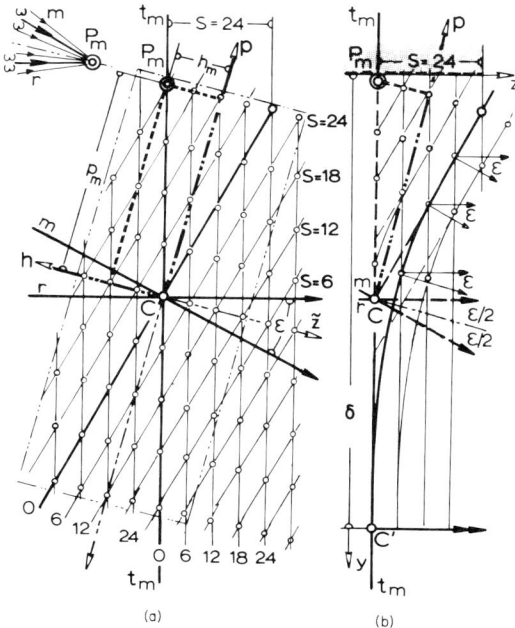

FIG. 47. Interference field of a virtual wedge compared to that of a thermal boundary layer with constant gradient near the wall. Spatial coherence is assumed. (a) *Virtual wedge*. The schematic diagram and the designations correspond to Fig. 39. Only every sixth wave front of the reference and measuring beams is indicated. White circles: maxima of the interference field. t_m–t_m, focusing plane containing the wedge axis C; P_m, object point corresponding to an interference order $S = 24$ of the central interference field. Coordinates p_m and h_m; h_m is small compared to p_m since the magnitude of the angle ϵ is exaggerated. The rays emanating from the constituent point light sources of an extended light source intersect in P_m and comprise a cone with the axis angle 2ω. (b) *Boundary layer*. The interference field in the vicinity of the heated wall (shaded) is identical to that of a virtual wedge if the wave fronts are plane again after traversing the test section (only two wave fronts are indicated): t_m–t_m, focusing plane containing the axis C of the virtual wedge in the vicinity of the wall; P_m, object point; δ, thickness of the boundary layer. The deformed wave front can be represented as a combination of small plane wave fronts belonging to interference systems of different virtual wedges with decreasing angles ϵ and individual wedge axis, e.g., wedge axis C' at the end of the boundary layer where $\epsilon \to 0$.

The interference contrast distribution was recorded photometrically for the case of a thermal boundary layer at a heated horizontal cylinder (Fig. 48). The distorted contrast distribution K with the first minimum ($S \approx 20$) can be recognized clearly. Greatest contrast appears at the end of the boundary layer ($S = 0$), corresponding to the origin of the p, h coordinate system.

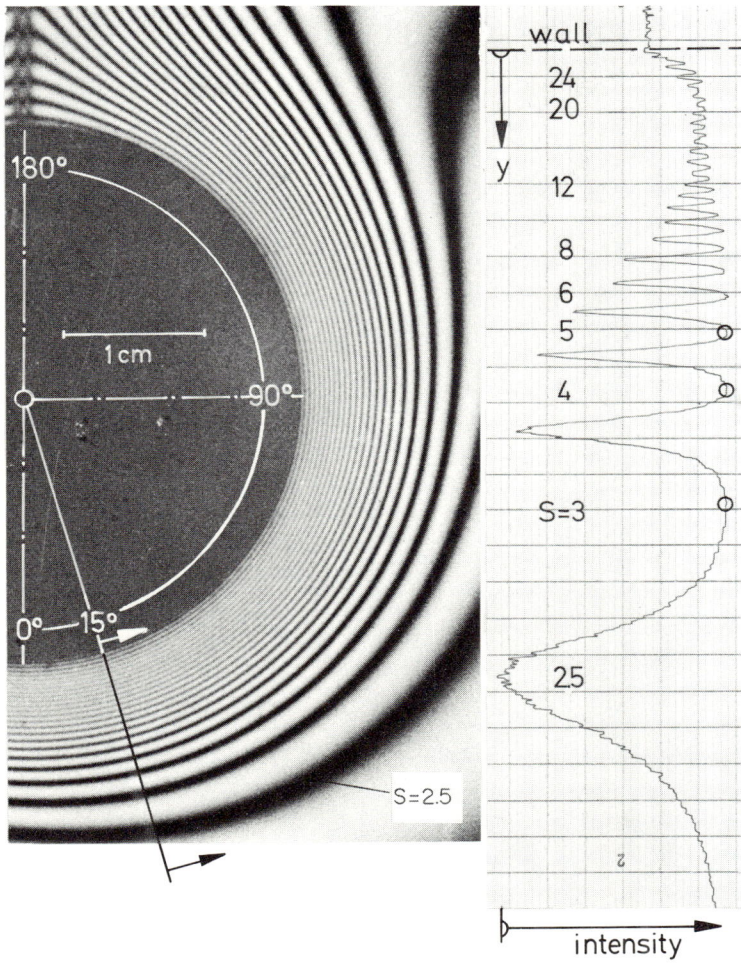

FIG. 48. Interference contrast at a thermal boundary-layer wedge angle at the wall $(\epsilon/2)_w = 0.9 \cdot 10^{-3}$, $\Delta\vartheta = \vartheta_w - \vartheta_\infty = 33.4$ deg, corresponding fringe density $1/b = 3.2/\text{mm}$. The photogram (position 15°) shows a distorted distribution K of the fringe density. The light source was extremely large.

This pattern was recorded under unfavorable conditions (extremely large light source without a diaphragm) in order to obtain the first minimum of K with a relatively small number of fringes. For thermal boundary layers at plane vertical walls (free convection), this contrast minimum usually appears at a fringe density $1/b \approx 20$ mm^{-1}. Assuming a focal length $f = 0.5$ m of the objective L_1 and radius $r = 1$ mm of the diaphragm, the coordinate p_m of the interference minimum is approximately [considering Eqs. (69) and (73)]

$$\tilde{\omega}_{\min} = 4\pi = [(4\pi \sin \epsilon/2)/\lambda] \, \omega^2 \cdot p_{\min}$$
$$p_{\min} = 2b/\omega^2 = (2 \cdot 0.05 \cdot 10^{-3})/(0.04 \cdot 10^{-4}) = 25 \quad \text{mm} \tag{80}$$

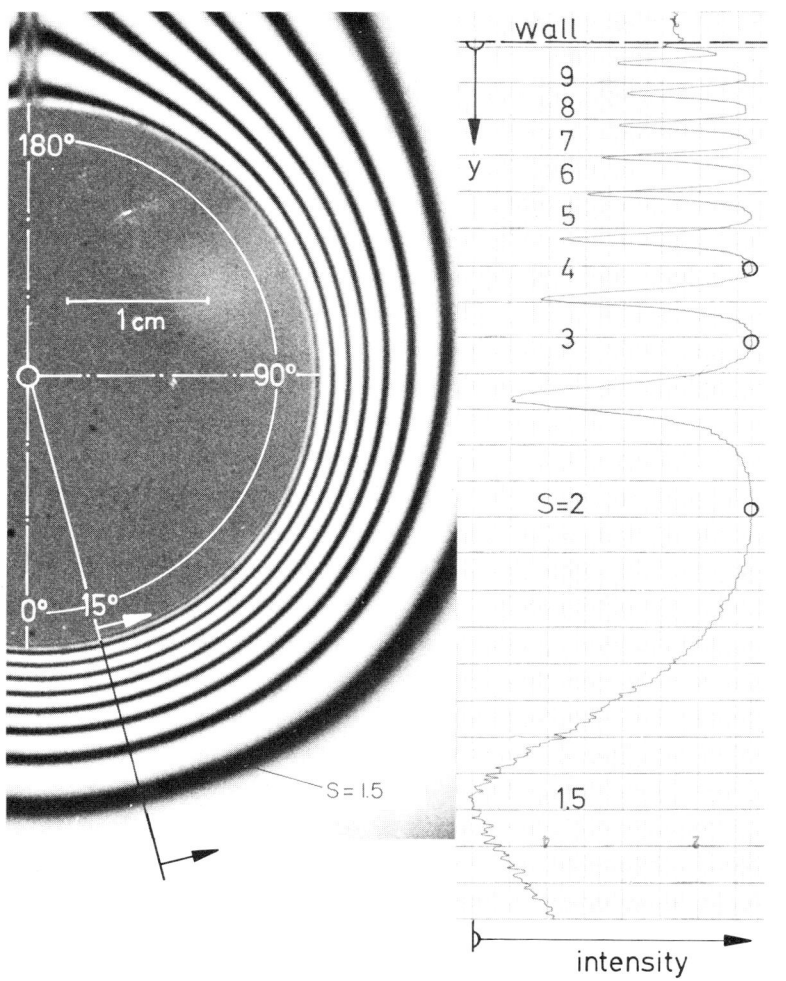

FIG. 49. Interference contrast at a thermal boundary layer. The circumstances are the same as in Fig. 48; however, the wedge angle is smaller $(\sin \epsilon/2)_w = 0.3 \cdot 10^{-3}$. The interference contrast is sufficiently good, $\Delta\vartheta = \vartheta_w - \vartheta_\infty = 12.8$ deg.

This is about equal to the thickness of the boundary layer between the wedge axis ($p = 0$) at the end of the boundary layer and the interference contrast minimum in the vicinity of the wall. If the image on the photographic plate is approximately $\frac{1}{3}$ of its natural size, the fringe density $1/b = 20$ mm^{-1} approaches the resolution of standard film emulsions (60 lines/mm, 200 lines/mm in the case of repro film).

The circumstances of the exposure shown in Fig. 49 are the same as in Fig. 48. However, the gradients of the refractive index are smaller, corresponding to smaller temperature gradients. The interference contrast $K(\tilde{\omega}, 0)$ does not yet assume its first minimum since $\tilde{\omega} = (4\pi/\lambda) \omega^2 \cdot p \cdot \sin \epsilon/2 < 4\pi$, because of the smaller wedge angle $\epsilon/2 = \lambda/2b$.

The schlieren lens of a phase object can also be considered as an element of an optical imaging system composed of the schlieren lens, the objective $L2$, and the beam splitter $M2'$ (Fig. 50). The objective $L2$ is

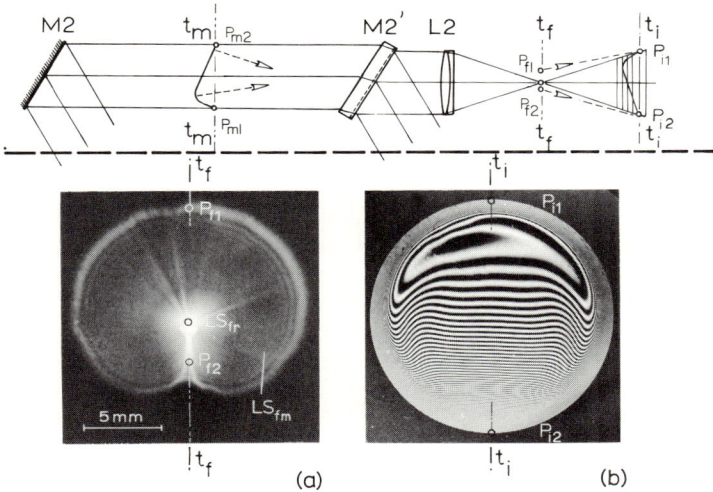

Fig. 50. Schlieren lens in the measuring path of the MZI considered as an element of an imaging system. Cylindrical cuvette filled with alcohol in transient convection. The temperature difference between the interference lines is $\Delta \vartheta = 0.03$ deg. Diameter $d = 30$ mm. Interference taken in the infinite fringe field. $M2, M2'$, mirror, beam splitter; $L2$, objective; t_m–t_m, focusing plane; t_f–t_f, focal plane of $L2$; t_i–t_i, image plane; $P_{m1}P_{m2}$, correspond to the points $P_{f1}P_{f2}$ and $P_{i1}P_{i2}$ in the corresponding planes and the central profile of the interferogram. (a) Exposure in the focal plane t_f–t_f: The image LS_{fr} of the light source (reference beam) is indicated in natural size (black circle). The analogous image LS_{fm} of the light source (measuring beam) is starkly deformed and appears as the large luminous region. (b) Interferogram of the phase object: fringe density in P_{i1}, $1/b = 30$ lines/mm; fringe density in P_{i2}, $1/b = 10$ lines/mm (cf. example in Section VI,B,1).

focused on the schlieren lens (t_m–t_m) and images it in the interferogram (t_i–t_i) with minimum distortion. The beam splitter produces an astigmatic error in the measuring path. The effect of the schlieren lens (cf. Toepler's schlieren method) can be recognized in Fig. 50 (left) in the large luminous region which is the deformed image of the light source LS_{fm} of the undisturbed measuring beam in the focal plane t_f–t_f. The analogous source image LS_{fr} of the reference beam is indicated in natural size (black circle). Segments of the central profile of the schlieren lens (convection in a cylindrical cuvette) P_{m1} and P_{m2} correspond to points P_{f1} and P_{f2} in the focal plane t_f–t_f and to points P_{i1} and P_{i2} in the interferogram. The deflection in the focal plane $e = \epsilon \cdot f$ is proportional to the fringe density ($\sin \epsilon/2 = \lambda/2b$): The region near the wall in the test section, which is characterized in the interferogram by a large gradient (fringe density), therefore corresponds to the outer zone of the deformed image LS_{fm} of the light source in the focal plane. The bright diamond-shaped spot of light is produced by the rays coming from the inner part of the test section.

If a camera objective is positioned in the focal plane t_f–t_f of $L2$ (Fig. 50; cf. Fig. 43b), in order to obtain a desired imaging scale one must ascertain that the diameter of the objective is large enough to take up the complete deformed image of the light source LS_{fm}. For the same reason, the camera objective may not be stopped down. The light intensity must be adjusted with the diaphragm of the light source. If the light rays from the perimeter are blocked by the lens mounting, the corresponding interference fringes do not appear in the interferogram. This must be especially observed in the case of cameras of small focal length where the diameter of the front lens is usually not very large.

2. *Two-Dimensional Phase Object (Thermal Boundary Layer)*

a. Displacement of the Interference Fringes. In the case of the parabolic ray paths in Fig. 44 the focusing plane must be positioned in the center of the test section in order to produce distortion-free images. For an arbitrary continuous refractive index distribution in the thermal boundary layer (nonparabolic ray path), the displacement Δy of the interference fringes is calculated relative to the position they would assume in ideal interferometry with negligible ray deflection (72–76).

The refractive index gradient in the boundary layer is greatest in the vicinity of the wall; it is taken to be zero at the end of the boundary layer. The ray next to the wall ($\eta = 0$) in Fig. 51 suffers the greatest deflection and its deflection angle is ϵ_{lw} when leaving the test section. The object point P_{mw} (cf. Fig. 44) is imaged by the objective $L2$ and produces the

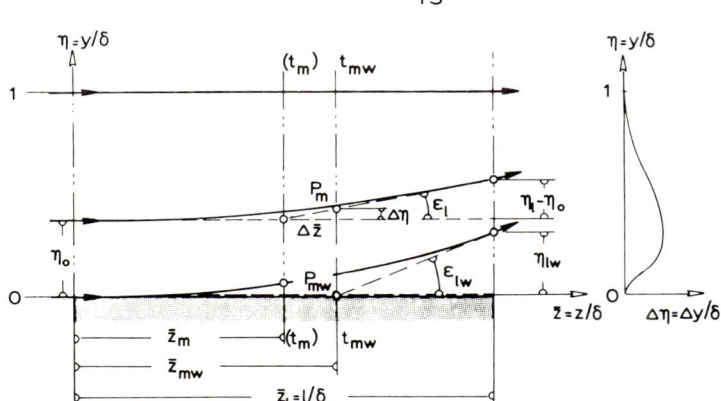

Fig. 51. Calculation of the displacement error $\Delta\eta$ in a thermal boundary layer: δ, thickness of the boundary layer; $\eta = y/\delta$, boundary layer coordinate; $\bar{z} = z/\delta$, dimensionless coordinate measured in the direction of the light ray; $\bar{z}_l = l/\delta$, model length expressed in multiples of the boundary layer thickness; $\bar{z}_{mw}$, position of the focusing plane t_{mw}–t_{mw} in the case of correct focusing; $\bar{z}_m$, deviating position of the focusing plane t_m–t_m; P_{mw}, P_m, object points representing specific interference fringes; η_0, η_l, coordinates of the positions where the light ray enters (0) and leaves (l) the test section; $\Delta\eta$, displacement error. The rays of the measuring beam are indicated by solid lines, those corresponding to the reference beam (transferred to the test section) by dashed lines.

image point of the ray nearest to the wall. In the case of "correct" focusing, no displacement $\Delta\eta$ is introduced if the focusing plane is positioned at $\bar{z}_{mw}$.

A ray entering the boundary layer at the position η_0 is deflected by a smaller amount. It seems to emanate from the object point P_m whose coordinate in the focusing plane is displaced by $\Delta\eta$ relative to the entrance coordinate η_0 of the ray. This displacement error would not appear if the plane t_m–t_m were focused correctly; however, then the ray nearest to the wall would be imaged incorrectly, i.e., displaced. At the end of the boundary layer ($\eta = 0$), the ray propagates in a straight line and corresponding object points are always imaged without displacement errors.

If the "correct" focusing plane t_{mw}–t_{mw} is selected, the vicinity of the wall and the outer limit of the boundary layer are imaged without displacement errors. The inner region is imaged with a displacement error $\Delta\eta = f(\eta)$, shown schematically in Fig. 51.

If the focusing plane is positioned at the end of the test section, the displacement error $\Delta\eta$ is equal to the ray deflection $\eta_l - \eta_0$. Therefore, the configuration must be focused very carefully.

For the calculation of $\Delta\eta$, the coordinates y and z are expressed as multiples of the thickness δ of the boundary layer. Of course, then the length l of the test section is also expressed as a multiple of δ ($l = \bar{z}_l \cdot \delta$).
The following relation can be derived from Fig. 51:

$$\tan \epsilon_l = (d\eta/d\bar{z})_l = \frac{\eta_l - \eta_0}{\bar{z}_l - \bar{z}_m}$$

or

$$\bar{z}_m = \bar{z}_l - \frac{1}{\tan \epsilon_l}(\eta_l - \eta_0) \tag{81}$$

and

$$\bar{z}_{mw} = \bar{z}_l - \eta_{lw}/\tan \epsilon_{lw} \tag{82}$$

$\Delta\eta$ can then be expressed as

$$\Delta\eta = \frac{\bar{z}_{mw} - \bar{z}_m}{\bar{z}_l - \bar{z}_m}(\eta_l - \eta_0) \tag{83}$$

or

$$\Delta\eta = (\eta_l - \eta_0) - (\tan \epsilon_l/\tan \epsilon_{lw}) \cdot \eta_{lw} \tag{84}$$

In general, the displacement error $\Delta\eta$ is small ($\Delta\eta < 1\%$) relative to δ, if correct focusing is assumed (cf. Section V,C,4,b). If refractive index fields are to be determined exactly, either a hypothetical distribution of the refractive index in the test section must be assumed or the corrections are calculated by iteration from the experimentally determined distribution.

Correction term:

$$\Delta\eta_1 = (\eta_l - \eta_0) - \frac{(d\eta/d\bar{z})_l}{(d\eta/d\bar{z})_{lw}} \cdot \eta_{lw} \tag{84a}$$

b. Interferometer Equation. The interference order S of an interference fringe represented by an image point P_i (Fig. 44) differs from that in ideal interferometry [Eq. (59)] because of the deflection of the measuring beam. The image point P_{iw} e.g., corresponds to the object point P_{mw} in Fig. 52. The following discussion, however, is valid for arbitrary object points P_m in the focusing plane t_m-t_m.

The path difference $S \cdot \lambda$ in the image point P_{iw} is to be determined. This is the intersection of the respective measuring and reference rays in the image plane t_i-t_i. These rays seem to emanate from the virtual object point P_{mw} and are denoted by thick lines in Fig. 52.

If the optical paths of the measuring and reference beams are traced

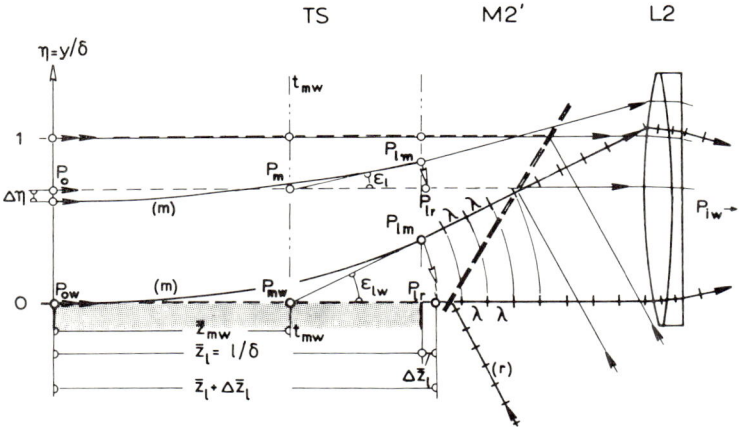

Fig. 52. Phase difference in the case of a curved ray path (thermal boundary layer): TS, test section; M2', beam splitter; L_2, objective; P_0, P_{0w}, points of ray entrance at the test section (index w = wall); P_m, P_{mw}, object points in the test section; P_{lm}, P_{lr}, points where the rays leave the test section (m, measuring beam; r, virtual reference beam); P_{iw}, image point (not indicated), conjugated to P_{mw}; $\eta = y/\delta$, and $\bar{z} = z/\delta$, dimensionless boundary-layer coordinates (δ: thickness of the boundary layer); $\bar{z}_{mw}$, coordinate of the "correct" focusing plane t_{mw}–t_{mw}; $\bar{z}_l = l/\delta$, length of the test section in multiples of the thickness of the boundary layer; $\bar{z}_l + \Delta\bar{z}_l$, reference length of the reference beam; $\eta_0 = y_0/\delta$, coordinate of the entrance point of the ray; $\Delta\eta$, ray displacement.

backwards beginning at the image point P_{iw}, it can be seen that these are equal outside of the test section, i.e., up to the points P_{lm} and P_{lr}. This is a consequence of the lens property that arbitrary rays emanating from object points P_{mw} all intersect in the image point P_{iw} and then trace optical paths of equal length because of the ideal properties of L_2. This is indicated in Fig. 6 for the case of an infinitely distant object point (plane wave fronts) and the corresponding image point (spherical wave fronts); it also applies in principle to Fig. 52.

The phase difference S at P_{iw} is, therefore, equal to the phase difference at the points P_{lm} and P_{lr} which are located on a circle whose center is P_{mw}. Since the measuring beam (m) and the virtual reference beam (r), indicated by dashed lines, do not exhibit a relative phase shift before entering the test section, the following expression results (Fig. 52):

Reference beam: $P_0 P_{lr}$ (refractive index, $n = n_\infty$)
Measuring beam: $P_0 P_{lm}$ [refractive index, $n = n(\eta)$]

$$S \cdot \lambda = n_\infty(\bar{z}_l + \Delta\bar{z}_l) - \int_{P_0}^{P_{lm}} n \, d\bar{s} \tag{85}$$

$\bar{z} = z/\delta$, $\bar{\lambda} = \lambda/\delta$, $\bar{s} = s/\delta$ are again dimensionless coordinates (δ, thickness of the boundary layer).

The following geometrical relation can be derived from Fig. 52:

$$\bar{z}_l + \Delta\bar{z}_l = \bar{z}_{mw} + (\bar{z}_l - \bar{z}_{mw})/\cos\epsilon_l \tag{86a}$$

$$= \bar{z}_{mw} + (\bar{z}_l - \bar{z}_{mw})[1 + (d\eta/d\bar{z})_l^2]^{1/2} \tag{86b}$$

With $d\bar{s} = d\bar{z}\,[1 + (d\eta/d\bar{z})^2]^{1/2}$ the general interferometer equation results:

$$S \cdot \bar{\lambda} = n_\infty[\bar{z}_{mw} + (\bar{z}_l - \bar{z}_{mw})(1 + (d\eta/d\bar{z})_l^2)^{1/2}]$$

$$- \int_0^{\bar{z}_l} n(\bar{z})[1 + (d\eta/d\bar{z})^2]^{1/2} \cdot d\bar{z} \tag{87}$$

The phase difference S is dependent on the position of the focusing plane (coordinate $\bar{z}_{mw}$) and changes, e.g., by the amount ΔS if the focusing plane is placed at the entrance ($\bar{z}_m = 0$) or at the exit ($\bar{z}_m = \bar{z}_{ml}$) of the test section instead of its "correct" position:

$$S \cdot \bar{\lambda} = n_\infty \cdot \bar{z}_l[1 + (d\eta/d\bar{z})_l^2]^{1/2} - \int_0^{\bar{z}_l} n(\bar{z})\,[1 + (d\eta/d\bar{z})^2]^{1/2}\,d\bar{z}$$
(entrance)

$$S \cdot \bar{\lambda} = n_\infty \cdot \bar{z}_l - \int_0^{\bar{z}_l} n(\bar{z})\,[1 + (d\eta/d\bar{z})^2]^{1/2}\,d\bar{z}$$
(exit)

$$\Delta S \cdot \bar{\lambda} = n_\infty \cdot \bar{z}_l[(1 + (d\eta/d\bar{z})^2)^{1/2} - 1] \tag{87a}$$

The effect of focusing on the phase difference is small since usually $(d\eta/d\bar{z})_l^2 \ll 1$. In retrospect it therefore seems justified to define the "correct" focusing position by $\bar{z}_m = \bar{z}_{mw}$ so that the displacement error $\Delta\eta_1$ is small. In order to express the phase difference $S \cdot \bar{\lambda}$ in terms of ideal interferometry (straight-line light propagation), the light path of the measuring beam in the region $\eta_l - \eta_0$ is assumed to be a parabolic curve (correction parabola). For this, a linear distribution of the refractive index is assumed in the region $\eta_l - \eta_0$ (as was already done in Section III): The profile of the refractive index is, therefore, replaced by a series of straight lines which is equivalent to a substitution of the schlieren lens by a multitude of virtual wedges.

A profile of the refractive index [Eq. (18)] expressed in dimensionless coordinates,

$$n(\eta) = n_0 + (dn/d\eta)_0(\eta - \eta_0) \tag{88a}$$

Optical Methods in Heat Transfer

where $(dn/d\eta)_0 = $ const is assumed at the entrance of the test section. For a ray entering the test section at an arbitrary angle, the light path is described by the following equation, which is calculated in Section VII,C:

$$\eta - \eta_0 = (d\eta/d\bar{z})_0 \cdot \bar{z} + (dn/d\eta)_0(1 + (d\eta/d\bar{z})_0^2) \cdot \bar{z}^2/2n_0 \quad (89a)$$

$$(d\eta/d\bar{z}) = (d\eta/d\bar{z})_0 + (dn/d\eta)_0(1 + (d\eta/d\bar{z})_0^2) \cdot \bar{z}/n_0 \quad (89b)$$

Considering Eq. (88a), the distribution of the refractive index along the measuring beam can be expressed as

$$n(\bar{z}) = n_0 + (dn/d\eta)_0(d\eta/d\bar{z})_0 \cdot \bar{z} + (dn/d\eta)_0^2 \cdot [1 + (d\eta/d\bar{z})_0^2]\, \bar{z}^2/2n_0 \quad (88b)$$

The general equation, Eq. (87), can be integrated with Eq. (88b) where the focusing plane is positioned at the center of the test section, corresponding to the parabolic path of the measuring beam ($\bar{z}_{mw} = \bar{z}_l/2$):

$$S \cdot \bar{\lambda} = n_\infty \cdot \bar{z}_l/2 \cdot [1 + (1 + (d\eta/d\bar{z})_l^2)^{1/2}]$$

$$- \int_0^{\bar{z}_l} [n(\bar{z})(1 + (d\eta/d\bar{z})^2)^{1/2}]\, d\bar{z} \quad (87b)$$

The calculation is performed in Section VII,C,4. With Eq. (22), the path difference $S \cdot \bar{\lambda}$ for the correction parabola is

$$S \cdot \bar{\lambda} = (n_\infty - n_0)\bar{z}_l + \frac{(dn/d\eta)_0^2}{4n_\infty}\bar{z}_l^3 - \frac{(dn/d\eta)_0^2}{3n_0}\bar{z}_l^3$$

$$S \cdot \bar{\lambda} = (n_\infty - n_0)\bar{z}_l - \frac{(dn/d\eta)_0^2}{n_0}\bar{z}_l^3\left(\frac{1}{3} - \frac{n_0}{4n}\right)$$

Assuming $n_0/n \approx 1$,

$$S \cdot \bar{\lambda} = (n_\infty - n_0)\bar{z}_l - \frac{(dn/d\eta)_0^2}{12n_0}\bar{z}_l^3 \quad (90a)$$

With Eq. (89b) for $(d\eta/d\bar{z})_0 \approx 0$, the following expression for the direction of the ray at the end of the test section results:

$$(d\eta/d\bar{z})_l = [(dn/d\eta)_0/n_0]\,\bar{z}_l$$

and, therefore, from Eq. (90a)

$$S \cdot \bar{\lambda} = (n_\infty - n_0)\,\bar{z}_l - [(d\eta/d\bar{z})_l^2/12]\, n_0 \cdot \bar{z}_l \quad (90b)$$

(valid for rays entering the test section at about 90°).

The first term of Eqs. (90a) and (90b) is the phase difference in ideal interferometry [Eq. (59)]; the second term describes the deviation for a parabolic ray path. The latter serves to calculate the phase difference S for ideal interferometry from the interferogram:

$$\Delta S_1 \cdot \bar{\lambda} = (d\eta/d\bar{z})_l^2 \cdot (n_0/12) \cdot \bar{z}_l \tag{91}$$

$(d\eta/d\bar{z})_l = \tan \epsilon_l$ can be taken directly from the interferogram. It can be derived from the fringe width according to Eq. (69) and Fig. 39:

$$\sin(\epsilon_l/2) = \bar{\lambda}/2\bar{b} \quad \text{or} \quad (d\eta/d\bar{z})_l = \tan \epsilon_l = \bar{\lambda}/\bar{b}$$

In practice the following form of the correction term may be used:

$$\Delta S_1 \cdot \bar{\lambda} = (n_0/12) \cdot (\bar{\lambda}/\bar{b})^2 \cdot \bar{z}_l \tag{92}$$

where $\bar{\lambda} = \lambda/\delta$, $\bar{b} = b/\delta$, $\bar{z}_l = l/\delta$, and $n_w \leqslant n_0 \leqslant n_\infty$. In general, $n_0 \approx n_w$ or $n_0 \approx n_\infty$, since $\Delta n = n_\infty - n_w$ is of the order 10^{-4}–10^{-5}.

3. Model Boundary Layers

For demonstration purposes, the ray path $\eta = f(\bar{z})$ in the test section, the displacement error $\Delta \eta$, and the path difference $S \cdot \bar{\lambda}$ will be calculated for a few selected boundary-layer profiles. The following expressions are all given in dimensionless coordinates (δ = thickness of the boundary layer):

$$\eta = y/\delta; \qquad \bar{z} = z/\delta; \qquad \bar{z}_l = l/\delta; \qquad \bar{\lambda} = \lambda/\delta; \qquad \bar{b} = b/\delta$$

The necessary calculations are performed in Section VII,B.

a. Thermal Boundary Layer; Quadratic and Exponential Refractive Index Distribution:

Thermal boundary layer. According to Section III (Boundary-Layer Optics) the distribution of the refractive index in a model boundary layer is [Eq. (32a)]

$$n = n_w + \Delta n \, (2\eta - 2\eta^3 + \eta^4) \tag{93a}$$

$$dn/d\eta = 2 \, \Delta n \, (1 - 3\eta + 2\eta^3) \tag{93b}$$

The gradient at the wall ($\eta_0 = \eta_{0w} = 0$) is of the magnitude

$$(dn/d\eta)_w = 2 \, \Delta n = 2(n_\infty - n_w) \tag{93c}$$

Furthermore, two additional profiles will be considered for comparison which also contain the boundary condition $(dn/d\eta)_w = 2 \, \Delta n$.

Quadratic refractive index profile:

$$n = n_w + \Delta n \,(2\eta - \eta^2) \tag{94a}$$

$$dn/d\eta = 2\Delta n \cdot (1 - \eta) \tag{94b}$$

Exponential refractive index profile:

$$n = n_w + \Delta n \,(1 - \exp(-2\eta)) \tag{95a}$$

$$dn/d\eta = 2\Delta n \cdot \exp(-2\eta) \tag{95b}$$

The exponential profile is not a boundary-layer profile in the strict sense of the word since it attains the value n_∞ asymptotically ($\eta \to \infty$). However, for $\eta = 1$ the profile deviates by only 13.5% ($n = n_w + 0.87\,\Delta n$) from its limit $n = n_\infty$ [Fig. (53)]. The refractive index

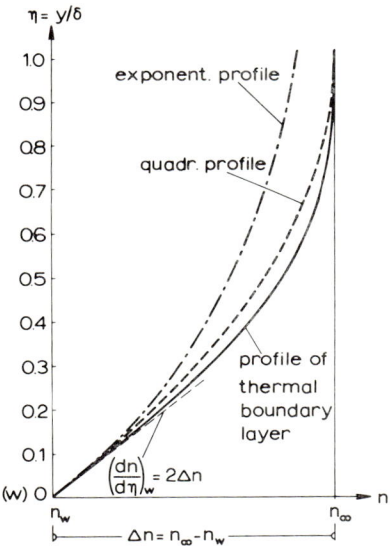

FIG. 53. Refractive index distribution of selected model boundary layers, $n = n(\eta)$: thermal boundary-layer profile, quadratic refractive index profile, exponential refractive index profile, linear profile of the refractive index gradient at the wall $(dn/d\eta)_w = 2\Delta n$.

gradient is finite at the end of the boundary layer ($\eta = 1$). In contrast, it is zero by definition for the profile of a thermal boundary layer and the quadratic profile.

These refractive index profiles corresponding to thermal boundary layers and diffusion boundary layers are shown in Fig. (53).

b. Ray Path, Displacement Errors in the Model Boundary Layers. In the test section the refractive index profiles of the model boundary layers correspond to ray paths which will be calculated in Section VII,B,2 and are shown in Fig. 54. These functions are of the form

$$\eta - \eta_0 = f[(2\,\Delta n/n)^{1/2}\bar{z}] \tag{96}$$

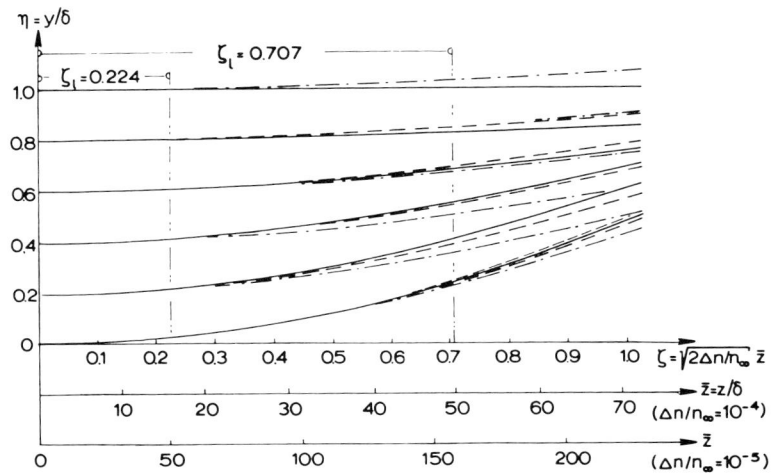

FIG. 54. Ray paths $\eta - \eta_0 = f(\zeta) = f[(2\,\Delta n/n)^{1/2}\,\bar{z}_l]$ according to the selected profiles of refractive index in Fig. 53: (———) measuring beam of the thermal boundary layer; (- - -) measuring beam of the quadratic profile; (- · -) measuring beam of the exponential profile; (– – –) parabolic ray path (correction parabola) according to the refractive index gradient $(dn/d\eta)_0$ at the entrance of the test section (only depicted for the coordinate $\eta_0 = 0$ and $(dn/d\eta)_w = 2\,\Delta n$). The region $0 \leqslant \zeta_l \leqslant 0.707$ corresponds to Example 1; the region $0 \leqslant \zeta_l \leqslant 0.224$ corresponds to Examples 2 and 3.

where η_0 is the coordinate at the point where the ray enters the boundary layer and $\Delta n = n_\infty - n_w$ is the total difference of the refractive index of the boundary-layer profile. The light paths are shown in Fig. 54 for the corresponding refractive index profiles in Fig. 53 and the coordinates $\eta_0 = 0, 0.2, 0.4,..., 1$.

The curves are similar for different values of Δn, which is expressed in terms of the coordinate $\zeta = (2\,\Delta n/n_\infty)^{1/2}\,\bar{z}$. For comparison, the values of $\bar{z} = z/\delta$ are given for typical values $\Delta n/n_\infty = 5 \cdot 10^{-5}$ and 10^{-4}. The region of the test section of length l (for a total difference of the refractive index Δn) is $0 \leqslant z \leqslant z_l$ or $0 \leqslant \zeta \leqslant \zeta_l$, where $\bar{z}_l = l/\delta$ and $\zeta_l = (2\,\Delta n/n_\infty)^{1/2}\,\bar{z}_l$. Outside of the test section ($\zeta > \zeta_l$ or $\bar{z} > \bar{z}_l$) the deflected rays propagate in a straight line. Therefore, Fig. 54 depicts in part the path of the measuring rays in the test section for the refractive index profiles given above with given model length l and various values of Δn.

The path of the measuring rays for the three profiles assumed above are described by the following equations which are valid under the assumption that the rays enter the test section parallel to the wall $[(d\eta/d\bar{z})_0 = 0]$ (see Section VII,B,2).

Thermal boundary layer (ray equation):

$$\int_0^{\eta_l} d\eta/[(\eta^4 - \eta_0^4) - 2(\eta^3 - \eta_0^3) + 2(\eta - \eta_0)]^{1/2} = (2\,\Delta n/n_\infty)^{1/2}\bar{z} \quad (97a)$$

$$d\eta/d\bar{z} = (2\,\Delta n/n_\infty)^{1/2}[(\eta^4 - \eta_0^4) - 2(\eta^3 - \eta_0^3) + 2(\eta - \eta_0)]^{1/2} \quad (97b)$$

Results of numerical calculation of the function $\eta - \eta_0 = f[(2\,\Delta n/n_\infty)^{1/2}\,\bar{z}]$ are in Fig. 54 (elliptic integral).

Quadratic refractive index profile (ray equation):

$$\eta - \eta_0 = (1 - \eta_0)\{1 - \cos[(2\,\Delta n/n_\infty)^{1/2}\bar{z}]\} \quad (98a)$$

$$d\eta/d\bar{z} = (2\,\Delta n/n_\infty)^{1/2}(1 - \eta_0)\sin[(2\,\Delta n/n_\infty)^{1/2}\bar{z}] \quad (98b)$$

Exponential refractive index profile (ray equation):

$$\eta - \eta_0 = \ln\cosh[\exp(-2\eta_0)(2\,\Delta n/n_\infty)^{1/2}\bar{z}] \quad (99a)$$

$$d\eta/d\bar{z} = (2\,\Delta n/n_\infty)^{1/2}\exp(-2\eta_0)\tanh[\exp(-2\eta_0)(2\,\Delta n/n_\infty)^{1/2}\bar{z}] \quad (99b)$$

For discussion of the ray paths in Fig. 54 given by Eqs. (97)–(99), the corresponding refractive index profiles in Fig. 53 [Eqs. (93)–(95)] are considered.

Rays nearest to the wall, $\eta_0 = 0$. The deflection of these rays is the greatest and is approximately the same for the above profiles since these rays pass through regions where the refractive index gradients do not differ greatly from the gradient at the wall $(dn/d\eta)_w = 2\,\Delta n$, which is the same in all three cases. Only after further deflection do the rays pass through regions where the gradient differs appreciably for the three profiles.

The thermal boundary layer profile differs only slightly from the linear profile of the wall gradient (indicated by thin dashed line). Therefore, the deflection is greater for increasing values of ζ than in the other two cases. As an exception, the parabolic ray path alone (correction parabola; Section VII,B,4,e), which is indicated only for $\eta_0 = 0$ by means of thin dashed lines, assumes larger values $\eta - \eta_0$. It corresponds to the linear profile with a gradient equal to the gradient at the wall and the assumptions of Eqs. (88a) and (88b).

The ray deflection is smallest in the case of the exponential profile since the gradient decreases more rapidly with increasing distances from the wall than in the case of the other profiles.

Ray path $\eta_0 = 0.2$–0.4. The differences between the respective gradients are greatest, hence the deflections also differ the most.

Ray path $\eta_0 = 0.4$–0.8. In the proximity of the outer limit of the boundary layer ($\eta = 1$), the quadratic profile and the profile of the thermal boundary layer decrease more rapidly than the gradient of the exponential profile. In contrast to the region near the wall, the ray deflection $\eta - \eta_0$ for the exponential profile assumes greater values than the quadratic profile and the profile of the thermal boundary layer.

Ray path $\eta_0 = 1$. For the quadratic profile and the profile of the thermal boundary layer the gradient of the refractive index vanishes, $(dn/d\eta)_{\eta_0=1} = 0$. The rays propagate undisturbed in straight lines. Since the gradient of the exponential profile still assumes a finite value, the ray is deflected by a small amount.

In order to estimate the magnitude of the ray deflection $\eta - \eta_0$ and the displacement error $\Delta\eta$ the following examples will be considered:

$$\text{Model boundary layer in gases:} \quad n_\infty \approx 1$$

$$\text{Model length:} \quad l = 0.5 \text{ m}$$

$$\text{Light wave length:} \quad \lambda = 0.5 \cdot 10^{-6} \text{ m}$$

$$\text{Fringe order at the wall}$$
$$\text{[maximum fringe order Eq. (59)]:} \quad S_\text{w} \approx \Delta n \cdot l/\lambda.$$

With Eq. (59) and Eq. (93c) the following fringe density $1/b$ (infinite fringe field) results:

$$1/b = (dS/dy) = (dn/dy) \cdot l/\lambda = (dn/d\eta)(d\eta/dy) \cdot l/\lambda$$
$$1/b = (dn/d\eta) \cdot l/\lambda \cdot \delta$$

Fringe density at the wall: $1/b_\text{w} = 2 \Delta n \cdot l/\lambda \cdot \delta$.

Example 1: $\tilde{z}_l = 50$; $\Delta n = 10^{-4}$; $\delta = 0.01$ m
$$[\zeta_l = (2 \Delta n/n_\infty)^{1/2} \tilde{z}_l = 0.707]$$

Temperature difference (Table XII) at 760 Torr, 18°C: $\Delta\vartheta = \vartheta_\text{w} - \vartheta_\infty = 160°\text{C}$
Maximum fringe order: $S_\text{w} = 100$
Fringe density at the wall: $1/b_\text{w} = 20 \text{ mm}^{-1}$

(The region of the test section $0 \leqslant \zeta \leqslant \zeta_l$ is indicated in Fig. 54.) This example represents an upper limit of the measuring range since the interference contrast assumes its first minimum at this fringe density. This is due to the limits of spatial coherence of average light sources (see example in Section V,B,4,*b*).

Example 2: $\bar{z}_l = 50$; $\Delta n = 10^{-5}$; $\delta = 0.01$ m; $\zeta_l = 0.224$

Temperature difference (Table XII) at 760 Torr, 18°C: $\Delta \vartheta = \vartheta_w - \vartheta_\infty = 12°C$
Maximum fringe order: $S_w = 10$
Fringe density at the wall: $1/b_w = 2$ mm^{-1}

These values are typical of an average interferogram.

Example 3: $\bar{z}_l = 16$ (15.8); $\Delta n = 10^{-4}$; $\delta = 0.032$ m; $\zeta_l = 0.224$

Temperature difference (Table XII) at 760 Torr, 18°C: $\Delta \vartheta = \vartheta_w - \vartheta_\infty = 160°C$
Maximum fringe order: $S_w = 100$
Fringe density at the wall: $1/b_w = 6$ mm^{-1}

In this example, the difference of the refractive index, i.e., the temperature difference, is the same as in Example 1; however, the thickness of the boundary layer is greater. The ray path (in ζ coordinates) in Fig. 54 is the same as in the Example 2 since the values for ζ_l are the same ($\zeta_l = 0.224$).

The region of the test section is indicated in Fig. 54 for the examples given above. For the ray nearest to the wall, the ray deflection at the end of the test section is $\eta_l - \eta_0 = f(\zeta)_l = f[(2 \Delta n/n_\infty)^{1/2} \bar{z}_l]$, Eqs. (97a)–(99a):

Example 1   (extreme values)

Region of the test section: $0 \leqslant \zeta \leqslant 0.707$
or for $\Delta n = 10^{-4}$: $0 \leqslant \bar{z} \leqslant 50$

Ray deflection ($\eta_0 = 0$) at the end of the test section $\bar{z}_l$:
Correction parabola: $\eta_{lw} = 0.25$
Thermal boundary layer: $\eta_{lw} = 0.243$
Quadratic profile: $\eta_{lw} = 0.239$
Exponential profile: $\eta_{lw} = 0.232$

Examples 2 and 3 (normal values)

	Region of the test section:	$0 \leqslant \zeta \leqslant 0.224$
	or for $\Delta n = 10^{-5}$:	$0 \leqslant \tilde{z} \leqslant 50$
	or for $\Delta n = 10^{-4}$:	$0 \leqslant \tilde{z} \leqslant 16$
	Ray deflection ($\eta_0 = 0$) at the end of the test section $\tilde{z}_l$:	
	Correction parabola:	$\eta_{lw} = 0.0250$
	Thermal boundary layer:	$\eta_{lw} = 0.0249$
	Quadratic profile:	$\eta_{lw} = 0.0248$
	Exponential profile:	$\eta_{lw} = 0.0247$

The displacement error assumes the values η_l if the focusing plane is positioned at the end of the test section ($\tilde{z}_m = \tilde{z}_l$), and approximately $-\eta_l$ if the focusing plane is positioned at the entrance of the test section ($\tilde{z}_m = 0$; see Fig. 44). In Example 1, this distortion is about 25% with reference to the ray at the end of the boundary layer ($\eta_0 = 1$) (extreme value). For Examples 2 and 3 (normal values) displacement is still about 2.5% of the boundary-layer thickness δ. These displacement errors (Fig. 55) can be reduced by a factor of 100 by "correct" focusing, i.e., by judicious selection of $\tilde{z}_{mw}$ [Eq. (82)].

Position of the focusing plane (coordinate $\tilde{z}_{mw}$). The coordinates $\tilde{z}_{mw}$

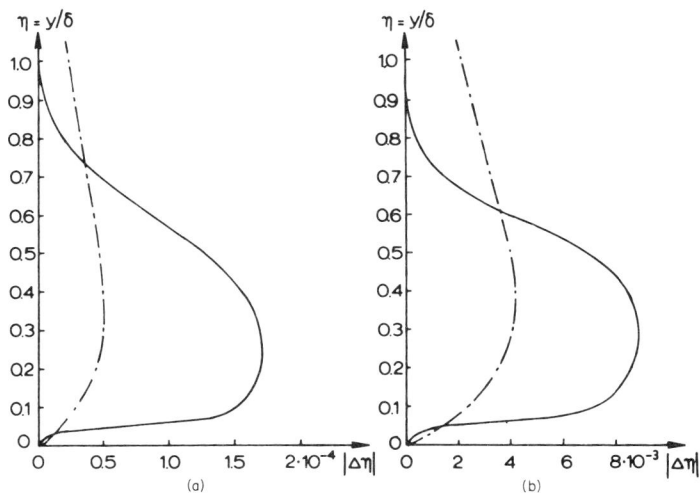

FIG. 55. Displacement error in the model boundary layers: (———) thermal boundary layer; (– · –) exponential refractive index profile. (a) Examples 2 and 3, $\zeta_l = 0.224$; (b) Example 1, $\zeta_l = 0.707$ (extreme values).

which deviate somewhat from one another will be determined in Section VII,B,3 for the refractive index profiles given above.

Example 1: $\Delta n/n_\infty = 10^{-4}$; $\bar{z}_l = 50$; $\zeta_l = 0.707$

	$\bar{z}_{mw}$	ζ_{mw}
Correction parabola:	25	0.353
Thermal boundary layer:	24.63	0.349
Quadratic profile:	23.96	0.340
Exponential profile:	23.11	0.327

Example 2: $\Delta n/n_\infty = 10^{-5}$; $\bar{z}_l = 50$; $\zeta_l = 0.224$

	$\bar{z}_{mw}$	ζ_{mw}
Correction parabola:	25	0.112
Thermal boundary layer:	24.8	0.111
Quadratic profile:	24.6	0.110
Exponential profile:	24.4	0.109

Example 3: $\Delta n/n_\infty = 10^{-4}$; $\bar{z}_l = 16$; $\zeta_l = 0.224$

	$\bar{z}_{mw}$	ζ_{mw}
Correction parabola:	8.0	0.112
Thermal boundary layer:	7.93	0.111
Quadratic profile:	7.87	0.110
Exponential profile:	7.83	0.109

The coordinate $\bar{z}_{mw}$ for "correct" focusing has been selected so that the displacement error of the ray nearest to the wall ($\eta_0 = 0$) is zero ($\Delta \eta = 0$).

Displacement error $\Delta \eta = f(\eta)$ *in the focusing plane* $t_m - t_m$ *(coordinate* $\bar{z}_{mw}$*).* The calculation $\Delta \eta = f(\eta)$ is given in Section VII,B,3 using Eq. (84).

Correction parabola: If the focusing plane is positioned in the center of the test section ($\bar{z}_{mw} = \bar{z}_l/2$), the displacement error $\Delta \eta$ is zero for parabolic ray path. For illustration, the correction parabola for the ray nearest the wall ($\eta_0 = 0$) is indicated in Fig. 54. It corresponds to the linear refractive index profile of the wall gradient in Fig. 53. If all of the ray paths shown in Fig. 54 were to be represented by parabolas, it would be necessary to replace the refractive index profile of the boundary layers by an approximation consisting of a series of straight lines. This is equivalent to representing the schlieren lens as a composition of a multitude of virtual wedges, as was already noted in connection with the discussion of interference contrast (Section V,C,1,b).

Quadratic profile: For a continuous distribution of the refractive index in the boundary layers, a displacement error $\Delta\eta = f(\eta)$ is to be expected as is indicated schematically. However, in the case of a quadratic profile the displacement error $\Delta\eta = (\eta)$ is zero for correct focusing. This is due to the cosine dependence, as is shown in Eq. (98a) (Section VII,B,3,d). The focusing plane, however, is no longer positioned in the center of the test section ($\tilde{z}_{\mathrm{mw}} < \tilde{z}_l/2$), as can be seen in the examples given above.

Therefore, the focusing plane is imaged without distortion as in ideal interferometry if the distribution of the refractive index—in the region of interest of the interferogram—can be described by a function of second order.

Profile of the thermal boundary layer and the exponential profile: The functions $\Delta\eta = f(\eta)$ are calculated in Section VII,B,3,c and e and are shown in Fig. 55 for the specific cases given above. In Fig. 55, the displacement error $\Delta\eta$ is shown in an exaggerated scale as a function of η. The coordinate of an interference fringe corresponding to a measuring ray entering the test section at η_0 is $\eta_0 + \Delta\eta_0$ in the case of a thermal boundary layer and $\eta_0 - \Delta\eta_0$ for the exponential profile.

For the three examples considered, the resulting maximum displacement errors are given below (see Fig. 55a, b):

Example 1: $\Delta n/n_\infty = 10^{-4}$; $\tilde{z}_l = 50$; $\zeta = 0.707$

	Thermal boundary layer:	$\Delta\eta_{\max} = 8.8 \cdot 10^{-3}$
	Exponential profile:	$\Delta\eta_{\max} = -4.2 \cdot 10^{-3}$

Example 2: $\Delta n/n_\infty = 10^{-5}$; $\tilde{z}_l = 50$; $\zeta_l = 0.224$

Example 3: $\Delta n/n_\infty = 10^{-4}$; $\tilde{z}_l = 16$; $\zeta_l = 0.224$

	Thermal boundary layer:	$\Delta\eta_{\max} = 1.7 \cdot 10^{-4}$
	Exponential profile:	$\Delta\eta_{\max} = -0.5 \cdot 10^{-4}$

Result: If the configuration has been focused carefully this error can be neglected. Even in Example 1, it amounts to only about 1% of the thickness of the boundary layer. As a consequence of wandbindung, the displacement $\Delta\eta$ in the narrow zone near the wall $0 \leqslant \eta \leqslant 0.04$ is small in the case of a thermal boundary layer. The interference fringes are imaged practically without distortion and the measurement of the fringe width $\bar{b}_w$ yields the correct gradient. In the adjoining region $0.04 < \eta_0 < 0.1$, the displacement error increases rapidly so that the fringe width appears too large (Fig. 55a, b). The real fringe width

$$\bar{b} = \eta_{0k} - \eta_{0i} \qquad (\bar{b} = b/\delta)$$

and the incorrectly imaged fringe width

$$\bar{b}_f = (\eta_{0k} + \Delta\eta_{0k}) - (\eta_{0i} + \Delta\eta_{0i})$$

are compared for the specific examples given above:

$$\bar{b}_f/\bar{b} = 1 + \frac{\Delta\eta_{0k} - \Delta\eta_{0i}}{\eta_{0k} - \eta_{0i}} = 1 + \frac{d(\Delta\eta/\eta)}{d\eta} \qquad (100)$$

$$(\bar{b}_f - \bar{b})/\bar{b} = d(\Delta\eta/\eta)/d\eta$$

The gradient of the refractive index, which is inversely proportional to the fringe width as derived from the interferogram, is incorrect. The following maximum deviations result for the zone $0.04 \leqslant \eta_0 \leqslant 0.1$, where $d(\Delta\eta/\eta)/d\eta \approx$ const in Fig. 55a, b.

Example 1: $\Delta n/n_\infty = 10^{-4}$; $\tilde{z}_l = 50$; $\zeta_l = 0.707$
$(\bar{b}_f - \bar{b})/\bar{b} = 7.5 \cdot 10^{-2}$ (7.5%)

Example 2: $\Delta n/n_\infty = 10^{-5}$; $\tilde{z}_l = 50$; $\zeta_l = 0.224$

Example 3: $\Delta n/n_\infty = 10^{-4}$; $\tilde{z}_l = 16$; $\zeta_l = 0.224$
$(\bar{b}_f - \bar{b})/\bar{b} = 0.34 \cdot 10^{-2}$ (0.3%)

The region $0 \leqslant \eta_0 \leqslant 0.04$ of the thermal boundary layer corresponds to a certain amount of fringes near the wall:

Example 1:   8 fringes

Example 2:   1 fringe

Example 3:   7 fringes

In this region the gradient is correctly obtained.

The effect of the displacement error $\varDelta\eta$ on the determination of the wall gradient increases rapidly with increasing values of the model length $\zeta_l = (2\,\varDelta n/n_\infty)^{1/2}\,\bar{z}_l$. The numerical values are typical of experimental results; the values of the first example are therefore approximately an upper limit of the measuring range (error 7.5%).

The exponential profile already exhibits an increase $[d(\varDelta\eta/\eta)/d\eta]$ in the displacement error in the region nearest to the wall $0 \leqslant \eta_0 \leqslant 0.04$ leading to the following deviations $(\bar{b}_f - \bar{b})/\bar{b}$ of the fringe width:

Example 1: $\qquad (\bar{b}_f - \bar{b})/\bar{b} = -0.15 \cdot 10^{-2}$

Examples 2, 3: $\qquad (\bar{b}_f - \bar{b})/\bar{b} = -0.03 \cdot 10^{-2}$

In this case, the fringe width obtained from the correctly focused image is too large by these amounts.

c. *Path Differences $S \cdot \bar{\lambda}$ Model of the Boundary Layers.* The path difference $S \cdot \bar{\lambda}$ for a curved ray path is calculated in Section VII,B,4 for the individual refractive index profiles. The general equation of interferometry [Eq. (87)] is used in the form $(d\eta/d\bar{z} \ll 1)$

$$S \cdot \bar{\lambda} = n_\infty \cdot \bar{z}_l - \int_0^{\bar{z}_l} n(\bar{z})\,d\bar{z} \tag{87c}$$

in order to simplify the calculation.

Correction parabola:

$$S \cdot \bar{\lambda} = (n_\infty - n_0)\bar{z}_l - \frac{(dn/d\eta_0)^2}{6n_0}\,\bar{z}_l^{\,3} \tag{101}$$

Quadratic profile:

$$S \cdot \bar{\lambda} = \varDelta n\,(1 - \eta_0)^2 \left[\frac{\bar{z}_l}{2} + \frac{1}{4(2\,\varDelta n/n_\infty)^{1/2}} \sin\left(2\left(2\,\frac{\varDelta n}{n_\infty}\right)^{1/2}\bar{z}_l\right)\right] \tag{102}$$

or

$$\left(2\,\frac{\varDelta n}{n_\infty}\right)^{1/2} \cdot S \cdot \bar{\lambda} = \varDelta n\,\frac{(1-\eta_0)^2}{2}\,[\zeta_l + \tfrac{1}{2}\sin(2\zeta_l)]$$

Profile of the thermal boundary layer: The numerical calculation is performed with Eq. (93a) and the ray path $\eta = f(\bar{z})$.

Exponential profile:

$$S \cdot \bar{\lambda} = \Delta n \frac{\exp(-2\eta_0)}{(2\Delta n/n_\infty)^{1/2}} \tan[\exp(-\eta_0)(2\Delta n/n_\infty)^{1/2} \bar{z}_l] \tag{103}$$

or

$$(2\Delta n/n_\infty)^{1/2} \cdot S \cdot \bar{\lambda} = \Delta n \exp(-2\eta_0) \tan[\exp(-\eta_0) \cdot \zeta_l]$$

In Fig. 56, the path difference $S \cdot \bar{\lambda}$ of the profiles (indicated in the same manner as in Fig. 53) is compared with a linearly increasing path difference as in ideal interferometry $[S \cdot \bar{\lambda} = (n_\infty - n_0) \bar{z}_l]$ and the path

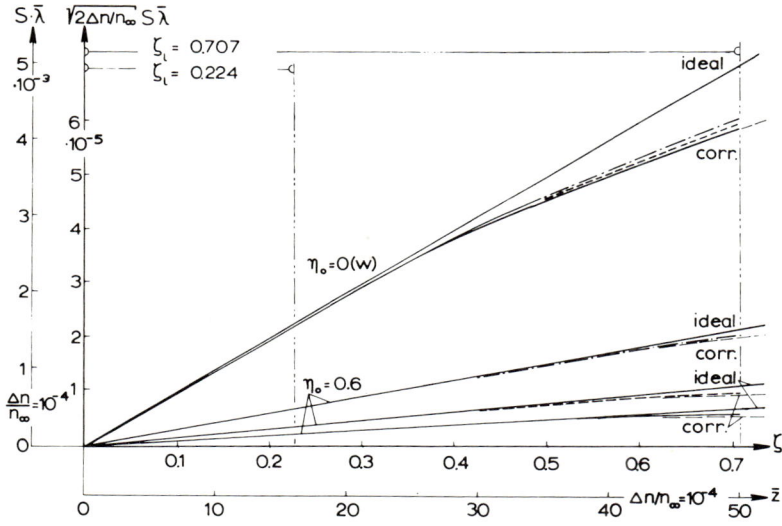

FIG. 56. Phase difference $S\bar{\lambda}$ of the model boundary layers, calculated for the points $\eta_0 = 0$ (wall) and $\eta_0 = 0.6$: (———) ideal interferometry (straight lines); (– – –) correction parabola; (- - -) quadratic refractive index profile; (– · –) exponential refractive index profile; (———) thermal boundary layer.

difference of the correction parabola (index correction). The measuring ray nearest to the wall ($\eta_0 = 0$) and the ray $\eta_0 = 0.6$ are considered in this manner. In examples 2 and 3 ($\zeta_l = 0.224$), the deviation from ideal interferometry can only be detected for the ray $\eta_0 = 0$. In Example 1 ($\zeta_l = 0.707$), the differences relative to ideal interferometry are large. However, the path difference of the real ray path can be approximated satisfactorily by the correction parabola.

Special values of the path difference for the three examples considered result (Fig. 56):

Example 1: $\Delta n/n_\infty = 10^{-4}$; $\bar{z}_l = 50$; $\zeta_l = 0.707$

	$S \cdot \bar{\lambda}$
$\eta_0 = 0$	
Ideal interferometry	$5 \quad \cdot 10^{-3}$
Correction parabola	4.18
Thermal boundary layer	4.19
Quadratic profile	4.24
Exponential profile	4.30
$\eta_0 = 0.6$	
Thermal boundary layer	0.415
Ideal interferometry	0.5
Correction parabola	0.39
$\eta_0 = 0.6$	
Quadratic profile	0.697
Ideal interferometry	0.8
Correction parabola	0.670
$\eta_0 = 0.6$	
Exponential profile	1.79
Ideal interferometry	1.87
Correction parabola	1.75

Example 2: $\Delta n/n_\infty = 10^{-5}$; $\bar{z}_l = 50$; $\zeta_l = 0.224$

Example 3: $\Delta n/n_\infty = 10^{-4}$; $\bar{z}_l = 16$; $\zeta_l = 0.224$

	$S \cdot \bar{\lambda}$
$\eta_0 = 0$	
Ideal interferometry	$1.6 \quad \cdot 10^{-3}$
Correction parabola, profiles	1.56
$\eta_0 = 0.6$	
Thermal boundary layer	0.156
Ideal interferometry	0.158
Correction parabola	0.156
$\eta_0 = 0.6$	
Quadratic profile	0.237
Ideal interferometry	0.25
Correction parabola	0.237
$\eta_0 = 0.6$	
Exponential profile	0.475
Ideal interferometry	0.480
Correction parabola	0.475

4. Corrections to the Interferograms

a. Influence of the Window of the Cuvette. An additional displacement Δy_2 and an additional path difference $\Delta S_2 \cdot \lambda$ is to be expected because of the different path lengths of the reference and the deflected measuring beams in the window of the cuvette and in the beam splitter $M2'$ of the interferometer. In the general basic position of the MZI, corresponding rays of the reference and the measuring beams suffer the same displacement in the beam splitters $M1$ and $M2'$ so that they coincide in phase and direction at the ray exit of the MZI (Fig. 37).

In Fig. 57, the path influenced by the beam splitter $M2'$ of the ray

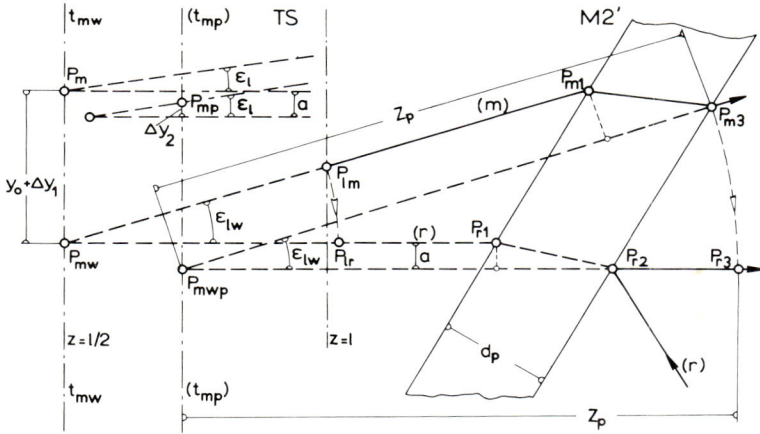

Fig. 57. Thick-plate effect of the beam splitter $M2'$ (indication as in Fig. 52): Δy_1, displacement error, caused by the boundary layer; Δy_2, displacement error, caused by the beam splitter $M2'$; d_p, thickness of the beam-splitter plate.

nearest to the wall is shown, similar to Figs. 51 and 52. Points P_{mw} and P_m are the object points in the focusing plane t_m–t_m which correspond to the image points P_{iw} and P_i, as in Fig. 51, for the case of the theoretically ideal mirror $M2'$, i.e., neglecting the influence of the glass carrier plates. As can be seen in Fig. 57, the deflected measuring beam and the reference beam leave the beam splitter $M2'$ at the points P_{m3} and P_{r2}. They do this by forming the same deflection angle ϵ_{lw} but their displacement in the y direction is different. A new apparent position of the focusing plane t_{mp}–t_{mp} results and the point P_{mw} now corresponds to the point P_{mwp}. At the end of the boundary layer ($y = \delta$), an undeflected measuring ray is displaced by the same amount a as the rays of the reference beam. The distance between the object point nearest to the wall P_{mwp} and the object point at the end of the boundary layer (not indicated) is therefore the same in the apparent focusing plane t_{mp}–t_{mp} as in the original focusing plane t_{mw}–t_{mw} (viz. δ). No distortion appears at these points except a negligible change in the imaging scale. However, this is not valid for object points $P_{\text{m}}(y_0 + \varDelta y_1)$ within the boundary layer which are imaged by a measuring ray deflected by $0 < \epsilon_l < \epsilon_{lw}$ and a corresponding reference ray with the entrance coordinate $y + \varDelta y_1$. In Fig. 57, P_{m} corresponds to the point P_{mp} in the focusing plane, where the latter point is displaced by the amount $a - \varDelta \eta_2$ or has the displacement error $\varDelta \eta_2$ relative to P_{mwp}.

It is assumed in the calculation that the path difference $S \cdot \lambda$ is known at the point P_{mw} or at $P_{l\text{m}}$ and $P_{l\text{r}}$ (Fig. 52), so that an additional deviation $\varDelta S_2 \cdot \lambda$ results when calculating the difference of the optical path of the measuring ray

$$n_\infty \overline{(P_{\text{mw}}P_{\text{m1}})} + n_{\text{gl}} \overline{(P_{\text{m1}}P_{\text{m3}})}$$

and the reference ray

$$n_\infty \overline{(P_{\text{mw}}P_{\text{r1}})} + n_{\text{gl}} \overline{(P_{\text{r1}}P_{\text{r2}})} + n_\infty \overline{(P_{\text{r2}}P_{\text{r3}})}$$

The points P_{m3} and P_{r3} are located in analogy to Fig. 52, on the perimeter of a circle around P_{mwp} with radius Z_{p}. Because of the assumed ideal properties of the objective $L2$, these have the same optical path to the image point P_{iw} (not indicated); see Fig. 44. Calculation, which shall not be performed here, results in the error of the thickness of the plate d_{p}, the refractive index of the plate n_{gl}, the angle θ, and the deflection angle ϵ_l for the following specific values which correspond approximately to those in the first example (extreme values):

$$\tan \epsilon_l = 10^{-2}; \quad n_{\text{gl}} = 1.5$$

$$\theta = 60°: \quad \varDelta S_2 \cdot \lambda = 0.544 \cdot 10^{-7} d_{\text{p}}$$

$$\theta = 45°: \quad \varDelta S_2 \cdot \lambda = 0.574 \cdot 10^{-7} d_{\text{p}}$$

For $\theta = 90°$, this description is also valid for the cuvette window:

$$\theta = 90°: \quad \varDelta S_2 \cdot \lambda = 0.433 \cdot 10^{-9} d_{\text{p}}$$

Compared with the deviations which can arise because of incorrect focusing these effects are smaller by several orders of magnitude and can therefore be neglected. Similar considerations apply to the displacement error $\varDelta y_2$ which assumes the value $\varDelta y_2 = 0.6 \cdot 10^{-5}$ mm at the same position, which is very small compared to the value $\varDelta y_{\text{max}} = 0.09$ mm in Example 1.

The influence of the splitter plates can be avoided by exchanging the measuring and the reference paths (Fig. 37) (79, 80). Then the deflected ray of the measuring path only impinges on the reflecting layers of the mirrors $M1'$ and $M2'$ but no longer traverses the splitter plate $M2'$. This configuration increases the interference contrast, especially in white light or filtered light, e.g., produced by spark gaps (small degree of coherence).

A drawback of this method is that the single-mirror adjustment procedure (Kinder) is no longer applicable.

b. *End Effects.* In Fig. 58, methods are indicated by which deviations at the outer limit of the test zone can be allowed for. In this case, the end effect leads to an apparent additional length Δl. This effect is more important for small model lengths l.

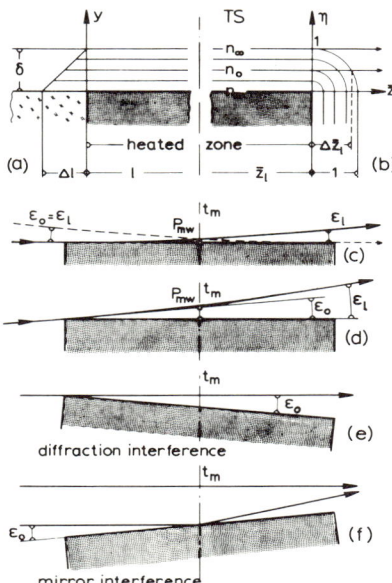

FIG. 58. Corrections of the interferogram: (a, b) end effects of a boundary-layer model; (c–f) adjustment of the model in respect to the parallel beam; ϵ_0, finite angle at the entrance (normally $\epsilon_0 \approx 0$); $\Delta l, \Delta \bar{z}_l$, additional lengths of the model caused by end effects.

First assumption (Fig. 58a). The extension of the heated wall consists of insulating material. The limit of the boundary layer in the region Δl is indicated in the illustration. The additional path difference $\lambda \cdot \Delta S_l$

at y_0 caused by the end effects of the two ends of the plate can then be calculated according to Eq. (59):

$$\Delta S_l \cdot \lambda = 2[n_\infty - n(y)] \cdot \Delta l \,(1 - y/\delta) \qquad (104)$$

Since a first approximation of $n_0(y)$ is known from the interferogram

$$n(y_0) = n_\infty - S_0 \cdot \lambda/l$$

the path correction term can be expressed as

$$\Delta S_l = 2 \cdot S(y_0) \cdot \frac{\Delta l}{l} (1 - y/\delta)$$

Δl must be estimated (e.g., $\Delta l \approx \delta$) or be determined by means of additional temperature measurements by thermocouples in the end region.

Second assumption (Fig. 58b). The heated test section is surrounded by a uniform thermal boundary layer with a cylindrical region at the ends. According to Eq. (106) for cylindrical phase objects the additional path difference $\Delta S_l \cdot \lambda$ (calculated in boundary-layer coordinates $\eta = y/\delta$) results:

$$\Delta S_l \cdot \lambda = 2 \int_0^{\Delta \bar{z}_l} [n_\infty - n(\eta)] \, d\bar{z}$$

or

$$\Delta S_l \cdot \lambda = 2 \int_{\eta_0}^1 [n_\infty - n(\eta)] \cdot \eta \cdot d\eta/(\eta_0^2 - \eta^2)^{1/2} \qquad (105)$$

c. *Angular Deviations.* Large measuring errors can result if the wall of the model is not exactly aligned with reference to the parallel beam. In order to estimate the accuracy with which this alignment must be performed, a measuring ray is indicated in Fig. 58c which leaves the test section with deflection ϵ_l. Since the propagation direction can be reversed without further changes, it can be seen that a measuring ray (dashed line) with the same path difference entering the test section at an angle $\epsilon_0 = |\epsilon_l|$ (angular deviation of the model wall with reference to the optical axis) leaves the test section parallel to the wall ($\epsilon_l = 0$). Since $\tan \epsilon_l$ is of the order $\tan \epsilon_l \leqslant 10^{-3}$, the alignment of the test section must be performed with comparable accuracy. This also applies to the degree of eveness, which must be preserved during the heating process, of the model wall.

Alignment of the model. The angular alignment is facilitated by an autocollimator, similar to the method shown in Fig. 36(1). Here the model wall, which must be a sufficiently good reflector (if not, a mirror can be affixed to it), has the same effect as the auxiliary mirror M in Fig. 36(1). A pentaprism, which deflects the rays of the measuring beam by ninety degrees independent of its position relative to the wall of the model, is placed before the plate. The deflected ray bundle impinges on the model wall and is reflected, enclosing an angle $2\epsilon_0$ with its original direction. The ray returns to the measuring path via the pentaprism and appears as an image of the light source in the focal plane of the objective $L1$. The orientation of the plate is adjusted until the source image in the focal plane of $L1$ coincides with the light source.

It is simpler—and in some cases more exact—to perform the alignment with magnified observation of the interference patterns at the wall which are caused by diffraction interference or by interference between rays reflected by the plate (Lloyd's mirror experiment). The reference path is blocked during this procedure and the magnified image of the plane t_m–t_m is observed in an image plane t_i–t_i (Fig. 58e, f).

Beginning with the position of the wall as in Fig. 58e, where normal Fresnel diffraction interference is observed, the orientation of the unheated plate is changed until (Fig. 58f) an illuminated border and interference lines as in Lloyd's mirror experiment appear. By carefully turning the plate back toward its original position, the position can be found where the illuminated border merges with the contour of the wall and the characteristic diffraction pattern of a semi-infinite plane appears with the first principle maximum in the immediate vicinity of the wall. In the case of long models in gaseous media, very small dust particles can be detected in the model zone, which shine brightly when the incident rays are nearly parallel to the surface of the model (Fig. 58e).

If the reference path is again opened, a phase-shifted image of the diffraction interference pattern appears where the position of the first principle maximum relative to the wall appears more prominently.

With the aid of the methods described above careful adjustment yields an adjustment precision of $\epsilon_0 \leqslant 0.5 \cdot 10^{-3}$.

D. Cylindrical and Irregular Phase Objects

1. *Cylindrical Phase Object*

Cylindrical phase objects with radial refractive index distribution are of considerably lesser importance than two-dimensional phase objects but are also relatively common. Cylindrical objects occur especially in applications of optical methods in gas dynamics and in flame investigations. In order to simplify the following calculations, it will be assumed that the parallel measuring rays enter the model perpendicular to the cylinder axis and that they traverse the phase object undeflected. This is valid for small objects or objects without large local gradients of the refractive index. It is appropriate to focus on the plane of symmetry containing the axis (P_m in Fig. 59).

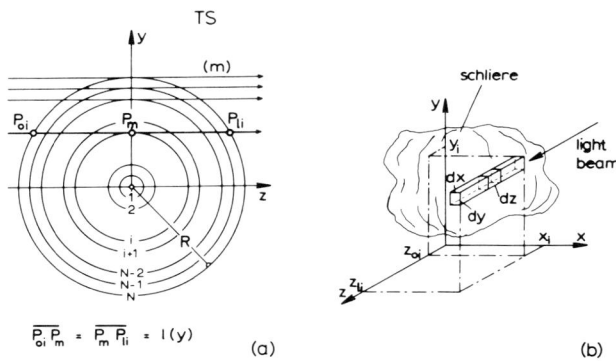

FIG. 59. Schematic representation of a cylindrical (a) and an irregular (b) phase object; straight-line ray propagation is assumed in the measuring path (ideal interferometry).

The calculation of the path difference in a circular cross section of the cylindrical object can be extended to the evaluation of spherical objects. In this case, the circular cross section of the cylinder corresponds to a cross section of the sphere containing its center (great circle).

a. Step Function. The step-by-step calculation of a radial refractive index distribution in a cylindrical object has already been performed by Mach (25) and Schardin (2) and will be introduced here in a form given by Van Voorhis (78).

The phase object is divided into equidistant concentric zones where the refractive index n is assumed to be constant. The radial distribution $n(r)$ can then be expressed as a step function with N steps (Fig. 59), where the corresponding radius increases by the constant amount $\Delta r = r_k - r_i$:

$$0 = r_0 < r_1 < r_2 \cdots < r_i < r_{i+1} \cdots < r_{N-1} < r_N = R$$

A measuring ray entering the cylindrical phase object at the point P_{0i} and leaving it again at P_{li} integrates the optical path segments of the individual zones step by step along its (variable) path $P_{0i}P_{0m} + P_{0m}P_{li} = 2l(y)$. The equation of ideal interferometry (Eq. 59) can then be expressed, utilizing the circular symmetry, in radial coordinates $z = (r^2 - y^2)^{1/2}$, $dz = r\, dr/(r^2 - y^2)^{1/2}$, and $n(y) = n(r)$, as

$$S(y) \cdot \lambda = 2 \int_0^{l(y)} [n_\infty - n(y)]\, dz$$

or

$$S(y) \cdot \lambda = 2 \int_y^R [n_\infty - n(r)](r \cdot dr)/(r^2 - y^2)^{1/2} \tag{106}$$

In the zone $r_i \leqslant r \leqslant r_{i+1}$, the constant difference of the refractive index is $n_\infty - n(r_i) = \Delta n_i$, vanishing for $r > R$. Equation (106) then transforms into a sum of linear equations:

$$S(y) \cdot \lambda = S(r_i) \cdot \lambda = S_i \cdot \lambda = \sum_{k=i}^{N-i} 2\, \Delta n_k \int_{r_k}^{r_{k+1}} r\, dr/(r^2 - r_i^2)^{1/2} \tag{106a}$$

$$S_i \cdot \lambda = 2 \sum_{k=i}^{N-1} \Delta n_k \, [(r_{k+1}^2 - r_i^2)^{1/2} - (r_k^2 - r_i^2)^{1/2}] \tag{106b}$$

Since the radius of the zone is $r_i = \Delta r \cdot i$, this expression can be simplified further,

$$S_i \cdot \lambda = 2\Delta r \sum_{k=i}^{N-1} \Delta n_k \, \{[(k+1)^2 - i^2]^{1/2} - (k^2 - i^2)^{1/2}\}$$

and the final evaluation formula results,

$$S_i \cdot \lambda = 2\, \Delta r \sum_{k=i}^{N-1} \Delta n_k \cdot A(k, i) \tag{107}$$

The function $A(k, i)$ is tabulated in Table II for a maximum of 24 steps. At the start of the evaluation of the interferogram, the radius is divided into the selected number of steps; Δr and the values of $S_i(r_i)$ are determined.

Beginning at the perimeter ($r = R$) the individual steps in the calculation result from Eq. (107):

$$S_N \cdot \lambda = 0 \qquad (\Delta n_N = 0)$$

$$S_{N-1} \cdot \lambda = 2\, \Delta r \sum_{k=N-1}^{N-1} \Delta n_k \cdot A(k, N-1)$$

$$= 2\, \Delta r \cdot \Delta n_{N-1} \cdot A(N-1, N-1)$$

$$S_{N-2} \cdot \lambda = 2\, \Delta r \sum_{N-2}^{N-1} \Delta n_k \cdot A(k, N-2)$$

$$= 2\, \Delta r\, (\Delta n_{N-2} \cdot A(N-2, N-2) + \Delta n_{N-1} \cdot A(N-1, N-2))$$

$$S_{N-3} \cdot \lambda = \cdots$$

TABLE II

Cylindrical Models; Step Function: $A(i, k) = [(k + 1)^2 - i^2]^{1/2} - (k^2 - i^2)^{1/2}$

$i =$	0	1	2	3	4	5	6
$k = 0$	1						
1	1	1.7321					
2	1	1.0964	2.2361				
3	1	1.0446	1.228	2.6458			
4	1	1.026	1.1185	1.3542	3		
5	1	1.0171	1.0743	1.1962	1.4721	3.3166	
6	1	1.0121	1.0513	1.1284	1.2724	1,5824	3.6056
7	1	1.0091	1.0378	1.0916	1.1836	1.346	1.686
8	1	1.007	1.029	1.0691	1.1341	1.2383	1.4167
9	1	1.0056	1.023	1.0541	1.1029	1.1769	1.2918
10	1	1.0046	1.0187	1.0436	1.0818	1.1377	1.2195
11	1	1.0038	1.0155	1.0359	1.0668	1.1108	1.1728
12	1	1.0032	1.0131	1.0302	1.0556	1.0913	1.1403
13	1	1.0028	1.0112	1.0257	1.0471	1.0767	1.1165
14	1	1.0024	1.0097	1.0221	1.0404	1.0654	1.0986
15	1	1.0021	1.0084	1.0193	1.0351	1.0565	1.0847

TABLE II (*continued*)

$i =$	0	1	2	3	4	5	6
$k = 16$	1	1.0018	1.0074	1.017	1.0308	1.0494	1.0736
17	1	1.0016	1.0066	1.015	1.0272	1.0435	1.0646
18	1	1.0015	1.0059	1.0134	1.0242	1.0387	1.0572
19	1	1.0013	1.0053	1.0121	1.0217	1.0346	1.051
20	1	1.0012	1.0048	1.0109	1.0196	1.0312	1.0458
21	1	1.0011	1.0044	1.0099	1.0178	1.0282	1.0414
22	1	1.001	1,004	1.009	1.0162	1.0257	1.0376
23	1	1.0009	1.0036	1.0083	1.0148	1.0234	1.0343
24	1	1.0008	1.0034	1.0076	1.0136	1.0215	1.0314

$i =$	7	8	9	10	11	12
$k = 7$	3.873					
8	1.7839	4.1231				
9	1.4846	1.8769	4.3589			
10	1.3439	1.5498	1.9657	4,526		
11	1.2615	1.3944	1.6127	2.0507	4.7958	

TABLE II (*continued*)

$i =$	7	8	9	10	11	12
$k = 12$	1.2077	1.3027	1.4436	1.6734	2.1324	5
13	1.1699	1.2422	1.343	1.4913	1.7321	2.2111
14	1.1421	1.1995	1.2762	1.3824	1.5378	1.7889
15	1.121	1.1678	1.2288	1.3097	1.4209	1.583
16	1.1044	1.1436	1.1934	1.2577	1.3425	1.4586
17	1.0912	1.1245	1.1663	1.2189	1.2863	1.3748
18	1.0804	1.1092	1.1447	1.1889	1.2441	1.3145
19	1.0715	1.0966	1.1274	1.165	1.2114	1.2691
20	1.064	1.0862	1.1131	1.1457	1.1853	1.2337
21	1.0577	1.0774	1.1012	1.1297	1.164	1.2054
22	1.0522	1.07	1.0912	1.1164	1.1465	1.1823
23	1.0476	1.0636	1.0826	1.1051	1.1317	1.1632
24	1.0435	1.058	1.0752	1.0955	1.1192	1.1471

TABLE II (*continued*)

$i =$	13	14	15	16	17	18
$k = 13$	5.1962					
14	2.2872	5.3852				
15	1.8441	2.3608	5.5678			
16	1.6271	1.8977	2.4322	5.7446		
17	1.4954	1.6701	1.9499	2.5016	5.9161	
18	1.4065	1.5315	1.712	2.0007	2.5692	6.0828
19	1.3423	1.4376	1.5669	1.753	2.0504	2.635
20	1.2937	1.3696	1.4682	1.6015	1.7932	2.0989
21	1.2558	1.3181	1.3965	1.4982	1.6354	1.8325
22	1.2254	1.2777	1.3421	1.423	1.5277	1.6687
23	1.2006	1.2453	1.2994	1.3658	1.4491	1.5567
24	1.1799	1.2187	1.265	1.3208	1.3892	1.4748

$i =$	19	20	21	22	23	24
$k = 19$	6.245					
20	2.6993	6.4031				
21	2.1463	2.762	6.5574			
22	1.8709	2.1927	2.8234	6.7082		
23	1.7014	1.9087	2.2381	2.8835	6.8557	
24	1.5852	1.7335	1.9457	2.2827	2.9423	7

With each step an increment Δn_i is obtained and step-by-step the function $n(r)$ results.

This method converges very well and is insensitive to inaccuracies, e.g., in S_i and rounded figures. If an uncertainty $\Delta S_i = 0.03$ in the determination of the phase difference is assumed for small gradients and $\Delta S_i = 0.07$ is assumed for large gradients, the mean error is about 5% and the maximum error is smaller than 10%.

b. Linear Approximation. A greater accuracy is offered by a method also given by Van Voorhis (78), where a step-by-step evaluation is performed as above. However, a linear distribution of the refractive index is assumed within the individual zones. But this method is very time consuming and is more suitable for computer evaluation.

The difference of the refractive index in a zone $r_i < r < r_{i+1}$ is now

$$n_\infty - n(r_i) = \Delta n_i + \frac{r - r_i}{r_{i+1} - r_i}(\Delta n_{i+1} - \Delta n_i)$$

$$= \Delta n_i + \frac{r - r_i}{\Delta r}(\Delta n_{i+1} - \Delta n_i)$$

It is zero for $r = R$. Equation (106) now assumes the form

$$S_i \cdot \lambda = 2 \int_{r=r_k}^{r=r_{k+1}} \frac{\Delta n_i \, \Delta r + (r - r_i)(\Delta n_{i+1} - \Delta n_i)}{\Delta r} \frac{r \, dr}{(r^2 - r_i^2)^{1/2}} \tag{108}$$

Integration yields a sum of the following relations:

$$S_i \cdot \lambda = \Delta r \cdot \Delta n_i \, A^*(i, i) + \Delta r \sum_{k=i+1}^{N-1} \Delta n_k \, [A^*(k, i) - A^*(k-1, i)]$$

The indices $A^*(k, i)$ can be determined by

$$A^*(k, i) = (k + 1)[(k + 1)^2 - i^2]^{1/2} - k(k^2 - i^2)^{1/2}$$
$$- i^2 \ln \frac{k + 1 + [(k + 1)^2 - i^2]^{1/2}}{k + (k^2 - i^2)^{1/2}}$$

2. *Irregular Phase Object (Schliere)*

Density and temperature fields can be determined quantitatively from interference patterns of two-dimensional or cylindrical phase objects. In the irregular bounded region of a schliere, the distribution of the refractive index usually can not be derived theoretically. Accordingly, it is not possible to determine temperature or density distributions from the interference pattern of a schliere. However, general information pertaining to the whole region can be gained, e.g., the enthalpy of the schliere [Hannes (*81*)].

a. Eikonal of a Schliere. If straight-light propagation (i.e., small refractive index gradients in the schlieren region which is illuminated by the parallel ray bundle of the measuring beam) in the direction of the z axis is assumed (Fig. 59b), the following equation of ideal interferometry is valid:

$$S(x_i, y_i) \cdot \lambda = n_\infty(x_i, y_i) \cdot (z_{i1} - z_{i0}) - \int_{z_{i0}}^{z_{i1}} n(x_i, y_i)\, dz \qquad (109)$$

A light beam entering the schliere at (x_i, y_i, z_{i0}) and leaving it at (x_i, y_i, z_{i1}) integrates over all elements $n\, dz$ of the optical path length. The distribution of the density and therefore that of the refractive index is arbitrary in the ray path as was presumed. The phase difference $S(x_i, y_i)$ in the schlieren zone can be determined from an interferogram. All the path differences $S(x_i, y_i) \cdot \lambda$ together can be considered as the eikonal of the schliere, assuming negligible ray curvature. With the knowledge of the function $S(x, y)$ and the integral $I = \iint S(x, y)\, dx\, dy$, the integral $\iiint \psi(x, y, z)\, dx\, dy\, dz$ of a physical quantity ψ can be determined which is directly proportional to the refractive index of the medium within the schliere. The integration along the z axis has already been accomplished by the light ray. Integration over the x and y coordinates within the limits of the schliere must be performed in the interferogram.

b. Enthalpy of a Schliere. The enthalpy H of a schliere in air relative to the surrounding air shall be determined in detail: The enthalpy H is directly proportional to the change of refractive index (density) in the schlieren region (Section V,E):

$$H(x, y, z) = k_n \cdot \Delta n(x, y, z) \qquad (110)$$

where H is the enthalpy, k_n a factor of proportionality, and Δn the change of the refractive index with respect to the undisturbed medium:

$$H_{\text{tot}} = \iiint H(x, y, z)\, dx\, dy\, dz = k_n \iiint \Delta n(x, y, z)\, dx\, dy\, dz \qquad (110a)$$

The integration has already been performed in part by the light ray. This part appears in the form of the phase difference:

$$S(x, y) = \frac{1}{\lambda} \int_{\text{lightpath}} \Delta n\, dz$$

Equation (110) can then be written as

$$H_{\text{tot}} = k_n \iint \left(\int \Delta n \, dz \right) dx \, dy$$

$$= k_n \cdot \lambda \iint S(x, y) \, dx \, dy \tag{111}$$

The factor k_n can be derived from the physical properties of the medium, as detailed in Section V,E:

$$H = \rho \cdot c_p (T - T_\infty) \tag{112}$$

$$\Delta n = (n - n_\infty) = (dn/dT)(T - T_\infty) \tag{113}$$

c_p is the specific heat (averaged over the schlieren region), ρ the density (averaged over the schlieren region); n, T the refractive index and temperature in a volume element of the schliere; n_∞, T_∞ the refractive index of the undisturbed surroundings; and dn/dT the temperature gradient of the refractive index, a physical constant of the medium. Taking Eqs. (110), (112), and (113) into account,

$$k_n = \frac{H}{\Delta n} = \frac{\rho \cdot c_p (T - T_\infty)}{(dn/dT)(T - T_\infty)} = \frac{\rho \cdot c_p}{dn/dT} \tag{114}$$

This is valid for an arbitrary medium where k_n, i.e., ρ, c_p, dn/dT, can be considered to be constant. Generally this is correct—within the limits of the measuring accuracy—for small temperature changes. In this specific case (cf. example in Section VI,A,3), the medium is air at normal pressure, considered as an ideal gas. For gases ($n \approx 1$), the refractive index as a function of the density is given by the Gladstone–Dale equation, Eq. (117):

$$\bar{r} \cdot \rho = \tfrac{2}{3}(n - 1)$$

where $\bar{r}$ is the specific refractivity; for air $\bar{r} = 0.1505$ cm³/gm. Considering the ideal gas equation, ρ can be expressed as

$$\rho = p/R_0 \cdot T$$

so that

$$(n - 1) = \tfrac{3}{2}(\bar{r} p / R_0 T)$$

where R is the gas constant. It then follows that

$$\partial n / \partial T = -\tfrac{3}{2}(\bar{r} p / R_0)(1/T^2)$$

This relation can be expanded in powers of $\Delta T = T - T_\infty$:

$$\partial n/\partial T = -\frac{3}{2}\frac{\bar{r}p}{R_0}\frac{1}{T_\infty^2} + 3\frac{\bar{r}p}{R_0}\frac{\Delta T}{T_\infty^3} - \frac{9}{2}\frac{\bar{r}p}{R_0}\frac{\Delta T^2}{T_\infty^4}$$

It is sufficient to consider only the first term since the others hardly contribute to the result; e.g., for $\Delta T = 5$ deg and $T_\infty = 293°K$ their contribution is less than 1%. For $T_\infty = 293°K$ and $p_\infty = 760$ Torr, k_n results to

$$k_n = \text{const} = -\frac{2}{3}\frac{\rho \cdot c_p \cdot R_0 \cdot T_\infty^2}{p_\infty \cdot \bar{r}} = -\frac{2}{3}\frac{\rho \cdot c_p}{p_\infty \cdot \bar{r}} T_\infty \tag{114a}$$

For a mean density ρ in the schliere, which can be estimated from the volume and the heat content of the schliere, we can write

$$H_{\text{tot}} = -\frac{2}{3}\frac{\rho \cdot c_p}{p_\infty \cdot \bar{r}} T_\infty \cdot \lambda \iint S(x, y)\, dx\, dy \tag{111a}$$

For small temperature increases, ρ/ρ_∞ is approximately unity.

This method can also be applied if the proportion of one component of a binary mixture in a schliere is to be determined and if the variation of the refractive index is due to changes in the concentration. The second component (e.g., solvent) is then the surrounding fluid. The total system must be kept at a constant temperature in order to detect only the effect of the concentration. Factor k_n of proportionality:

Enthalpy:

$$k_n = \frac{\rho \cdot c_p}{dn/dT}$$

Concentration:

$$k = \frac{2\rho_2 n_1(n_2^2 + 2)}{(n_2^2 - n_1^2)(n_1^2 + 2)}$$

The latter relation applies to fluid mixtures whose volume is independent of the concentration (e.g., a diluted mixture of glycerine and water), where n_1, n_2 denote the refractive indices of the components.

This method is fairly sensitive. If it is assumed that phase differences in the interference pattern can be determined to an accuracy of $S = 0, 1$, the minimum measurable heat content of various substances for an area $A = 1$ cm² (natural size) is

Air	0.06 J
Water	2.3 J
Organic fluids (mean value)	0.3 J

E. THE INTERFEROGRAM; TEST MEDIA

1. Evaluation Techniques of Interferograms

Kennard (77) and Schardin (2) have pointed out the possibility of applying the Mach–Zehnder interferometer to the investigation of heat transfer problems and have also described the evaluation of the interference patterns.

In the following, the interpretation of an interferogram—both of the fringe field and the field of infinite fringe—will be demonstrated for the case of a thermal boundary layer, produced by a heated vertical wall.

a. Infinite Fringe Field. In the left-hand portion of Fig. 60, part of the

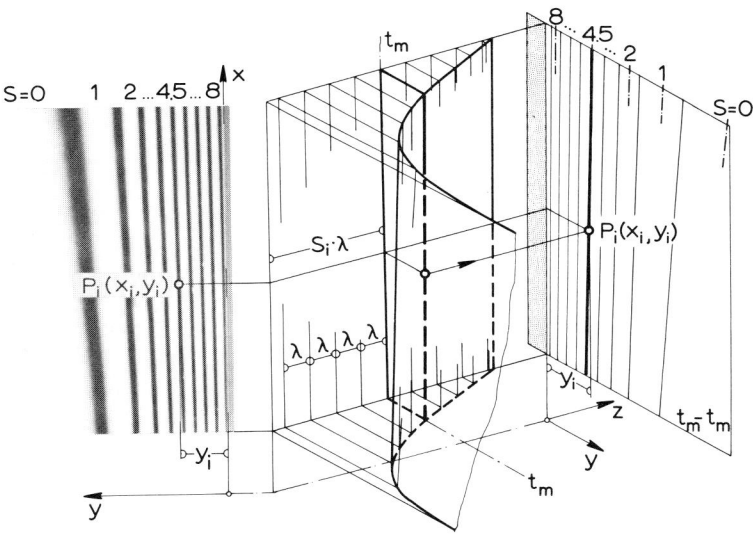

FIG. 60. Interferogram of a thermal boundary layer in an infinite fringe field (perpendicular intersection). Schematic representation of the virtual, spatial interference field consisting of the reference and the measuring wave fronts. This illustration corresponds to the interferogram in the left-hand portion of the figure; $t_m - t_m$, focusing plane.

interferogram of a boundary layer is shown. The fact that the interference fringes are not exactly parallel to the wall ($y = 0$)—isotherms—indicates the increasing thickness of the boundary layer in the direction of the x axis. In the right-hand portion of the illustration, it is shown how the interferogram originates as the intersection of the displaced focusing plane and the spatial virtual interference field in the test section. This has already been described in Section V,A,1, Fig. 30. The interference field is produced as the intersection of the plane reference wave fronts (r) and the deformed wave fronts (eikonals) of the measuring beam (m), between which a constant phase relationship exists. Only one wavefront of the measuring beam is shown in Figs. 60 and 61.

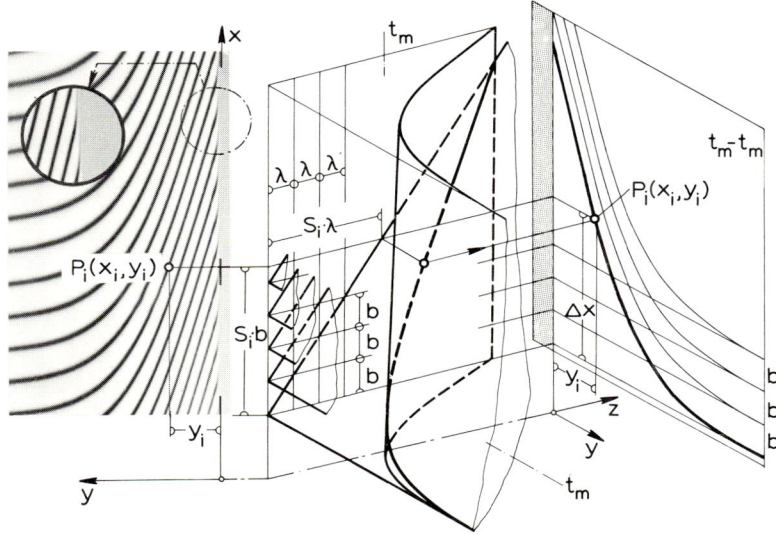

FIG. 61. Interferogram of a thermal boundary layer with a superimposed fringe field (oblique intersection). Schematic representation of the virtual spatial interference field consisting of the tilted reference wave fronts and the measuring wave fronts. It corresponds to the interferogram in the left-hand portion of the figure. The effect of diffraction interference is shown in the circular enlarged cutout; t_m–t_m , focusing plane.

The magnitude of the path difference $S_i \cdot \lambda$ (in this case equal to 4.5) at the point $P_i(x_i, y_i)$ is the result of the fact that the plane, undistorted wave front of the measuring and the reference beams in the infinite fringe field coincide, as is the case in the undisturbed region outside of the boundary layer ($S = 0$). The deformation of the wave fronts in the measuring beam relative to the corresponding wave fronts in the reference beam, which define the reference plane, is due to the phase object.

In general, no path difference exists between the plane undisturbed wave fronts (r, m), and an interferogram is produced as shown ($S = 0$ to $S_w = 8.5$ at the wall). In the case of a phase-shifted infinite fringe field (Section V,B,3), the wave fronts (m, r) are displaced (e.g., by eight interference orders), remaining parallel to one another. An identical interference pattern of the interference orders ($S = -8$, $S = 0$,

$S_w = 0.5$) is produced with increased interference contrast at the wall. (The axis of the virtual wedge corresponds to the interference order $S = 0$ (cf. Section V,C,1,b.)

Advantages of the infinite fringe field. The interferogram can easily be interpreted as a field of lines corresponding to equal density, i.e., temperature. The temperature fields in the vicinity of the wall (which usually resemble boundary layers) produce parallel interference fringes whose relative positions can be measured with the aid of a photometer with a long, and correspondingly narrow, slit.

Drawbacks of the infinite fringe field. The information $S(y; x)$ is gained in a discontinuous manner; i.e., the maxima ($S = 0, 1, 2,...$) and minima ($S = 0.5, 1.5, 2.5,...$) are evaluated point by point. For this reason, it can be expected that a reliable evaluation is only possible for about five or more fringes ($S \geqslant 5$).

The adjustment error of the infinite fringe field (similar to the case of a phase-shifted fringe field is superimposed additively on the desired interferogram. If the error $\Delta S \approx 0.1$ ($\lambda/8$), the resulting interference orders are $S = 0.1, 1.1,..., S_w \approx 8.6$ (cf. Section V,B,3a). This error, which is equivalent to an increase of the individual temperature levels, is less important if the number of fringes is large.

In experiments conducted over a long period of time, the sensitive adjustment for the infinite fringe field can be disturbed by vibrations, changes in room temperature, etc.

b. Fringe Field. Figure 61 shows the interferogram at the same phase object with the superposed fringe field. Outside of the boundary layer, the undisturbed fringe field with fringe density $1/b$ can be seen. This fringe field is produced by the plane undisturbed wave fronts of the reference and the measuring beams which are rotated by an angle relative to one another. The representation of the spatial interference system shows the preserved wave front (eikonal) of the measuring beam which is intersected obliquely by the plane wave fronts of the reference beam. Interference lines are formed which interlace the whole region of the boundary layer and appear as deflected interference fringes of the undisturbed field. As can be seen in the illustration, the interference order S_i at the point $P_i(x_i, y_i)$ results from the deflection Δx as

$$S_i = \Delta x / b \qquad (115)$$

The wave fronts can also be rotated in the opposite sense: The deflection Δx is then in the direction of the negative x axis. In both cases, the rotation axis of the fringe field (wedge axis) is parallel to the y axis. If it is positioned parallel to the x axis, interference fringes result which are parallel to the wall. The fringe density and the number of fringes can be increased or reduced by corresponding selection of the rotation angle. For example, it is possible to compensate the effect of a phase object by these means. The effect of this fringe field is then somewhat like a superimposed temperature gradient, and deviations in the phase object from this linear gradient can be detected sensitively (cf. example in Section VI,B,3).

Advantages of a fringe field. For small values of the interference order ($S < 5$) and irregular objects (schlieren) the interferogram in a fringe field can be evaluated more easily since an optimum value of the deflection in the region of the phase object can be chosen by judicious selection of b. Changes in the fringe field caused by vibrations, etc., are of no importance since the (altered) fringe width b at an undisturbed part of the interferogram can be used as a scale.

Drawbacks of the fringe field. In order to evaluate the interference pattern, the value of b must be determined in the vicinity of the region of interest with fairly high accuracy since the fringe width is not exactly constant due to imperfections in the reflecting plates and also because of the limitations of spatial coherence. In practice, a fringe field perpendicular to the plane wall of the model is advantageous (as in Fig. 62) for the evaluations of boundary layers. In the case of cylindrical objects, for example, the fringe field is both parallel and perpendicular to the wall. Some regions at the perimeter of the object then offer optimum conditions for evaluation, as in the case of plane models. At other regions, however, the aforementioned compensation effect appears.

In the left-hand part of Fig. 61, a magnified circular segment of the vicinity of the wall is shown where the effect of diffraction interference can be seen. The wall at the ray entrance of the model has the effect of a diffracting edge; the focusing plane in the center of the model plays the role of a screen. The interference order S_w at the wall cannot be determined directly from an interferogram and usually must be extrapolated from adjacent parts of the field or must be determined by additional measurements of the wall temperature.

c. Evaluation Techniques. The interferogram (photonegative) can be evaluated most accurately by means of a photometer equipped with a micrometer gage. Another, less precise, method is to measure a magnified print of the interferogram with a measuring microscope. The interference lines appear broader because of saturation effects in the photographic emulsion; the exact positions of maximum darkening are difficult to determine.

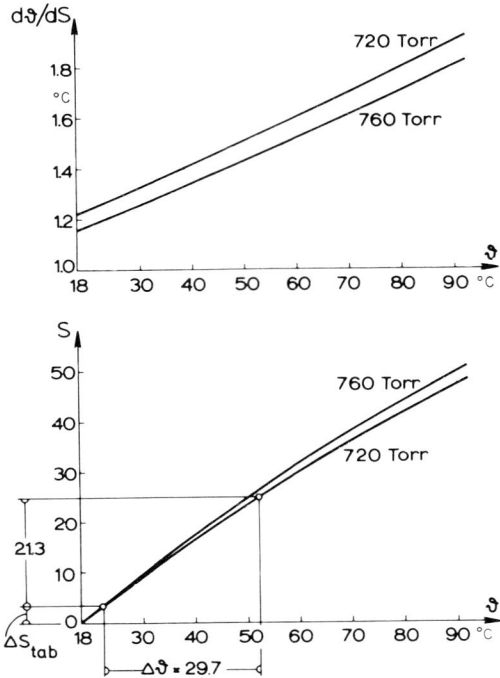

FIG. 62. Fringe temperature in air. The numerical values correspond to the example in Section VI,A,1 (circular annulus): $l = 0.5$ m, $\lambda = 0.5461 \cdot 10^{-6}$ m, $\vartheta_\infty = 18°C$.

This can be facilitated by application of equidensitometric processes. These correspond to a differentiation of the darkening distributions on the photographic plate so that the flanks are enhanced (Figs. 71 and 80, pp.322 and 337). The interference fringe is then represented by a pair of thin lines where the center of the two, i.e., the point of maximum darkening, can be determined much more easily. This pair of lines corresponds to a specific degree of darkening on the two flanks of the distribution. The formation of these two lines can be visualized in the following manner: The negative of an interferogram is placed in exact alignment on top of a transparent positive. If the degrees of darkening in the negative and in the positive are equal the superposition of the two appears uniformly grey. However, the degree of darkening can be altered in the reproduction by using "hard" or "soft" transparent prints. If a negative and a corresponding positive with steeper slopes of the darkening distribution are superimposed in exact alignment, the result no longer appears uniformly grey, and rather distinct minima corresponding to a specific degree of darkening can be seen. These can be registered photographically.

In practice, optical alignment of the two prints is too imprecise; however, special films are available with which this process can be conducted in one step. Equidensitometric negatives can be produced with "hard" reproduction emulsions in the following manner: The reprofilm is exposed so as to produce high contrast. The darkening in the negative should penetrate the whole emulsion. A certain amount of overexposure is not harmful in this case. Before fixing, the carefully rinsed photoplate is exposed for a few minutes in diffuse light. During the second development, the equidensitometric lines are formed by solarization and diffusion effects. A high degree of darkening is advantageous; the transparence of the negative can then be improved by uniformly reducing the darkness of the whole print.

2. Test Media

With the correction terms discussed heretofore, the interferogram is interpreted as a distribution of the phase difference $S(x, y)$ in ideal interferometry—or, analogously, as a distribution of the refractive index $n(x, y)$ according to the (ideal) equation of interferometry, Eq. (59):

$$S(x, y) = (l/\lambda)[n_r - n_m(x, y)]$$

When the medium in the reference path (reference cuvette) is the same as the undisturbed medium in the model zone, the above expression can be written as

$$S(x, y) = (l/\lambda)[n_\infty - n_0(x, y)] \tag{59b}$$

The desired temperature distribution can then be determined by means of the refractive index gradient (dn/dT) of the test medium, according to Eq. (59a):

$$\Delta\vartheta(x, y) = (\lambda/l) \cdot (dT/dn) \cdot S(x, y) \tag{59c}$$

The most important physical property is the refractive index gradient (dn/dT). The refractive index n itself is necessary in order to determine the ray deflection and the resulting correction terms. Both values are listed here for several substances which are suitable for this application. The relative accuracy of all of these values is of the same magnitude. In practical work it is advantageous to list temperature difference per interference fringe instead of the refractive index gradient:

$$d\vartheta/dS = (\lambda/l) \cdot (dT/dn) = f(T, p) \tag{59d}$$

If dT/dn can be considered constant within the measuring range, the temperature field is directly proportional to the interference field:

$$\vartheta_w - \vartheta_\infty = (\lambda/l)(dT/dn) \cdot S \tag{59e}$$

$S = 0$ corresponds to the temperature ϑ_∞. If the gradient of refractive index is a function of the temperature, it is better to consider the temperatures ϑ corresponding to the interference orders S, where an arbitrary temperature can be selected as a zero level ($S = 0 \stackrel{\wedge}{=} \vartheta = \vartheta_\infty$), and to calculate these by means of the inverse function $T(n)$ according to Eq. (59d) considering that $d\vartheta = dT$:

$$S = l/\lambda \cdot \int_{T_\infty}^{T_0} (dn/dT)\, dT = (l/\lambda)[n(T_\infty) - n(T_0)]$$

or, with the temperature values:

$$n(\vartheta_0) = n(\vartheta_\infty) - (\lambda/l) \cdot S \tag{59f}$$

expressed as a function $\vartheta_0 = f(S)$.

a. Refractive Indices and Physical Properties; The Lorentz–Lorenz Equation. The values of refractive indices and their gradients are expressed here as functions of the standardized model constants l/λ in Eq. (59a–f), which can easily be converted to other model lengths.

Wavelengths:

$\lambda = 0.5461 \cdot 10^{-6}$ m,   mercury vapor lamp

$\lambda = 0.6328 \cdot 10^{-6}$ m,   He–Ne Laser

Model constant l/λ:

	Hg light	He–Ne laser	l
Gases	$0.91558 \cdot 10^6$	$0.79014 \cdot 10^6$	0.5 m
Fluids	$0.091558 \cdot 10^6$	$0.079014 \cdot 10^6$	0.05 m

In Tables III and IV, physical properties (*83–85*) of gases and fluids at 20°C are compiled in order to determine their suitability for specific application. The measuring range $\Delta\vartheta = \vartheta_w - \vartheta_\infty$ for a maximum number of fringes $S = 60$ for several test media is listed in Table V.

TABLE IIIA

REFRACTIVE INDICES OF GASES AT 20°C AND 760 TORR[a,b]

Gas	n [—]	dn/dT [deg^{-1}]	$\bar{r}$ [m^3/kg]
Air	1.000,2724	$0.929 \cdot 10^{-6}$	$0.1508 \cdot 10^{-3}$
Nitrogen	1.000,2793	$0.953 \cdot 10^{-6}$	$0.1599 \cdot 10^{-3}$
Oxygen	1.000,2531	$0.864 \cdot 10^{-6}$	$0.1269 \cdot 10^{-3}$
Carbon dioxide	1.000,4197	$1.432 \cdot 10^{-6}$	$0.1529 \cdot 10^{-3}$
Argon	1.000,2630	$0.897 \cdot 10^{-6}$	$0.1056 \cdot 10^{-3}$
Water vapor	1.000,2354	$0.803 \cdot 10^{-6}$	$0.2095 \cdot 10^{-3}$

[a] Values calculated from data in the Critical Tables (84). (The values for water vapor were calculated for 760 Torr.)
[b] $\lambda_{Hg} = 0.5461 \cdot 10^{-6}$ m (mercury vapor lamp).

TABLE IIIB

REFRACTIVE INDICES OF GASES AT 20°C AND 760 TORR[a,b]

Gas	n [—]	dn/dT [deg^{-1}]	$\bar{r}$ [m^3/kg]
Air	1.000,2716	$0.927 \cdot 10^{-6}$	$0.1504 \cdot 10^{-3}$
Nitrogen	1.000,2781	$0.949 \cdot 10^{-6}$	$0.1592 \cdot 10^{-3}$
Oxygen	1.000,2516	$0.858 \cdot 10^{-6}$	$0.1261 \cdot 10^{-3}$
Carbon dioxide	1.000,4174	$1.424 \cdot 10^{-6}$	$0.1521 \cdot 10^{-3}$
Argon	1.000,2618	$0.894 \cdot 10^{-6}$	$0.1052 \cdot 10^{-3}$
Water vapor	1.000,2337	$0.798 \cdot 10^{-6}$	$0.2081 \cdot 10^{-3}$

[a] Values calculated from data in the Critical Tables (84). (The values for water vapor were calculated for 760 Torr.)
[b] $\lambda_{laser} = 0.6328 \cdot 10^{-6}$ m (He–Ne Gaslaser).

TABLE IIIC

PROPERTIES OF GASES AT 20 °C AND 760 TORR[a]

Gas	ν [m^2/sec]	k_c [W/m · deg]	Pr [—]
Air	$15.05 \cdot 10^{-6}$	0.02571	0.710
Nitrogen	$15.05 \cdot 10^{-6}$	0.02555	0.714
Oxygen	$15.22 \cdot 10^{-6}$	0.02625	0.710
Carbon dioxide	$7.95 \cdot 10^{-6}$	0.01605	0.773
Argon	$13.54 \cdot 10^{-6}$	0.01733	0.677
Water vapor	$510.15 \cdot 10^{-6}$	0.0194	0.847

[a] Values of water vapor are taken at saturation state.

TABLE IVA

Refractive Indices of Fluids at 25°C [a]

Fluid	$\lambda_{Hg} = 0.5461 \cdot 10^{-6}$ m		$\lambda_{laser} = 0.6328 \cdot 10^{-6}$ m	
	n [—]	dn/dT [deg^{-1}]	n [—]	dn/dT [deg^{-1}]
Water	1.3341	$1.00 \cdot 10^{-4}$	1.3314	$0.985 \cdot 10^{-4}$
Methyl alcohol	1.3280	$4.05 \cdot 10^{-4}$	1.3253	$4.0 \cdot 10^{-4}$
Ethyl alcohol	1.3612	$4.05 \cdot 10^{-4}$	1.3583	$4.0 \cdot 10^{-4}$
Isopropyl alcohol	1.3757	$4.15 \cdot 10^{-4}$	1.3726	$4.15 \cdot 10^{-4}$
Benzene	1.5030	$6.42 \cdot 10^{-4}$	1.4950	$6.40 \cdot 10^{-4}$
Toluene	1.4986	$5.55 \cdot 10^{-4}$	1.4901	$5.55 \cdot 10^{-4}$
Nitrobenzene	1.5579	$4.68 \cdot 10^{-4}$	1.5458	$4.68 \cdot 10^{-4}$
n-Hexane	1.3742	$5.43 \cdot 10^{-4}$	1.3711	$5.4 \cdot 10^{-4}$
c-Hexane	1.4260	$5.46 \cdot 10^{-4}$	1.4224	$5.43 \cdot 10^{-4}$
Acetone	1.3576	$5.31 \cdot 10^{-4}$	1.3542	$5.31 \cdot 10^{-4}$
Chloroform	1.4477	$5.98 \cdot 10^{-4}$	1.4435	$5.98 \cdot 10^{-4}$
Carbon tetrachloride	1.4613	$5.99 \cdot 10^{-4}$	1.4547	$5.98 \cdot 10^{-4}$
Carbon disulfide	1.6347	$7.96 \cdot 10^{-4}$	1.6185	$7.96 \cdot 10^{-4}$

[a] Refractive indices calculated from data in the Critical Tables (84). Refractive index gradients from G. Schödel (85) (measurement accuracy: ±1%).

TABLE IVB

Transport Properties of Fluids at 25°C

	ν [m^2/sec]	k_c [W/m · deg]	β [deg^{-1}]	Pr [—]
Water	$0.8967 \cdot 10^{-6}$	0.6065	0.000256	6.15
Methyl alcohol	$0.6948 \cdot 10^{-6}$	0.2070	0.00119	6.73
Ethyl alcohol	$1.384 \cdot 10^{-6}$	0.191	0.00110	14.53
Isopropyl alcohol	$2.618 \cdot 10^{-6}$	0.1533	0.00106	33.73
Benzene	$0.6948 \cdot 10^{-6}$	0.1568	0.00123	6.74
Toluene	$0.6373 \cdot 10^{-6}$	0.1542	0.00106	6.26
Nitrobenzene	$1.5247 \cdot 10^{-6}$	0.160	0.00083	17.4
n-Hexane	$0.4771 \cdot 10^{-6}$	0.143	0.00135	4.9
c-Hexane	$1.157 \cdot 10^{-6}$	—	—	—
Acetone	$0.403 \cdot 10^{-6}$	0.179	0.00143	3.80
Chloroform	$0.3649 \cdot 10^{-6}$	0.10	0.00128	5.3
Carbon tetrachloride	$0.5677 \cdot 10^{-6}$	0.1058	0.00122	7.28
Carbon disulfide	$0.2826 \cdot 10^{-6}$	0.1604	0.00119	2.20

TABLE V
Test Media Suitable for Similarity Experiments

	dn/dT (deg^{-1})	$d\vartheta/dS$ (deg)	$\Delta\vartheta^a$ (deg)	Pr
	($l = 0.5$ m; $\lambda = 0.5461 \cdot 10^{-6}$ m)			
Air (20°C, 760 Torr)	$0.927 \cdot 10^{-6}$	1.17	107	0.710
	($l = 0.05$ m; $\lambda = 0.5461 \cdot 10^{-6}$ m)			
Water (20°C)	$1.00 \cdot 10^{-4}$	0.109	10.9	6.15
Methanol (20–30°C)	$4.05 \cdot 10^{-4}$	0.027	2.69	6.73
Ethanol (20–30°C)	$4.05 \cdot 10^{-4}$	0.027	2.69	14.53
Propanol (20–30°C)	$4.15 \cdot 10^{-4}$	0.026	2.63	33.73

^{a} Temperature difference corresponding to a (maximum) fringe number $S = 60$.

For water and air, (dn/dT) is a function of temperature. In the indicated temperature range, dn/dT is constant for the other substances.

The advantage of optical methods in experiments to measure heat transfer by conduction and convection independent of radiation is realized for the listed test media. The experimental results can then be expressed as dimensionless values (Nu, Gr, Pr). Gases are practically pervious to heat radiation. The absorption coefficient of the above-listed fluids, however, is so large that even a practically undetectable thin film is sufficient to absorb all of the heat radiated from the wall. All of the other fluids listed in Table IV have medium absorption coefficients, so that the effect of radiation must be considered. They are, therefore, not suitable for similarity experiments (cf. example, Section VI,B,3). The general physical relation between the refractive index n and the density ρ is given by the Lorentz–Lorenz equation:

$$[(n^2 - 1)/\rho(n^2 + 2)] = \bar{r}(\lambda) = \text{const} \tag{116}$$

$\bar{r}$ is a function of the substance and is termed specific refractivity. For gases with refractive index near unity, Eq. (116) can be written in a simplified form (Gladstone–Dale equation):

$$[2(n - 1)/3\rho] = \bar{r} = \text{const} \quad (n \approx 1) \tag{117}$$

If isothermal concentration fields are to be determined, Eq. (116) can also be applied. However, this is only possible if the components do not undergo any change in volume. Often this is only the case for dilute solutions. Experimentally determined values of the refractive index must be used for greater concentrations. In the case of mixtures consisting of

TABLE VI

Fringe Temperatures in Air
$\vartheta_\infty = 18°C; \quad l = 0.5 \text{ m}; \quad \lambda = 0.5461 \cdot 10^{-6} \text{ m}$

	p [Torr]	710	720	730	740	750	760
$S =$	0	18	18	18	18	18	18
	0.5	18.623	18.614	18.606	18.598	18.59	18.582
	1	19.249	19.232	19.215	19.198	19.182	19.166
	1.5	19.877	19.851	19.826	19.801	19.777	19.753
	2	20.509	20.474	20.439	20.406	20.374	20.342
	2.5	21.143	21.099	21.056	21.014	20.973	20.934
	3	21.779	21.726	21.674	21.624	21.575	21.526
	3.5	22.419	22.356	22.296	22.237	22.18	22.124
	4	23.061	22.989	22.92	22.852	22,787	22.723
	4.5	23.706	23.625	23.547	23.47	23.396	23.324
	5	24.354	24.264	24.176	24.091	24.008	23.927
	5.5	25.005	24.905	24.808	24.714	24.622	24.533
	6	25.658	25.549	25.443	25.34	25.239	25.142
	6.5	26.314	26.196	26.08	25.968	25.859	25.753
	7	26.974	26.845	26.72	26.599	26.481	26.366
	7.5	27.636	27.498	27.363	27.233	27.106	26.982
	8	28.301	28.153	28.009	27.869	27.733	27.601
	8.5	28.969	28.811	28.658	28.508	28.363	28.222
	9	29.64	29.472	29.309	29.15	28.996	28.846
	9.5	30.314	30.136	29.963	29.795	29.631	29.472
	10	30.991	30.803	30.62	30.442	30.269	30.101
	10.5	31.671	31.473	31.28	31.092	30.91	30.732
	11	32.354	32.145	31.942	31.745	31.553	31.367
	11.5	33.041	32.821	32.608	32.401	32.199	32.003
	12	33.73	33.5	33.276	33.059	32.848	32.643
	12.5	34.422	34.181	33.948	33.72	33.5	33.285
	13	35.118	34.866	34.622	34.385	34.154	33.93
	13.5	35.816	35.554	35.299	35.052	34.811	34.577
	14	36.518	36.245	35.98	35.722	35.471	35.228
	14.5	37.223	36.939	36.663	36.395	36.134	35.881
	15	37.931	37.636	37.349	37.07	36.8	36.537
	15.5	38.643	38.336	38.038	37.749	37.468	37.195

TABLE VI (continued)

p [Torr]	710	720	730	740	750	760
S = 16	39.358	39.04	38.731	38.431	38.14	37.857
16.5	40.076	39.746	39.426	39.116	38.814	38.521
17	40.797	40.456	40.125	39.804	39.491	39.188
17.5	41.522	41.169	40.827	40.494	40.172	39.858
18	42.25	41.885	41.532	41.188	40.855	40.531[
18.5	42.981	42.605	42.24	41.885	41.541	41.207
19	43.716	43.328	42.951	42.585	42.23	41.885
19.5	44.454	44.054	43.666	43.289	42.923	42.567
20	45.196	44.783	44.383	43.995	43.618	43.251
20.5	45.941	45.516	45.104	44.704	44.316	43.939
21	46.69	46.253	45.829	45.417	45.018	44.63
21.5	47.442	46.992	46.556	46.133	45.722	45.323
22	48.198	47.735	47.287	46.852	46.43	46.02
22.5	48.957	48.482	47.021	47.574	47.141	46.719
23	49.72	49.232	48.759	48.3	47.854	47.422
23.5	50.486	49.985	49.5	49.029	48.572	48.128
24	51.256	50.742	50.244	49.761	49.292	48.837
24.5	52.03	51.503	50.992	50.498	50.016	49.549
25	52.808	52.267	51.744	51.235	50.742	50.264
25.5	53.589	52.035	52.498	51.978	51.473	50.982
26	54.374	53.807	53.257	52.723	52.206	51.704
26.5	55.163	54.582	54.019	53.473	52.943	52.429
27	55.955	55.36	54.784	54.225	53.683	53.157
27.5	56.752	56.143	55.553	54.981	54.426	53.888
28	57.552	56.929	56.326	55.741	55.173	54.623
28.5	58.356	57.719	57.102	56.504	55.924	55.36
29	59.164	58.513	57.882	57.27	56.677	56.102
29.5	59.976	59.311	58.666	58.041	57.434	56.846
30	60.792	60.112	59.453	58.814	58.195	57.594
30.5	61.612	60.917	60.244	59.592	58.959	58.346
31	62.437	61.727	61,039	60.373	59.727	59.1
31.5	63.265	62.54	61.838	61.158	60.498	59.859

TABLE VI (*continued*)

p [Torr]	710	720	730	740	750	760
$S =$ 32	64.097	63.357	62.64	61.946	61.273	60.62
32.5	64.933	64.178	63.447	62.738	62.051	61.385
33	65.774	65.003	64.257	63.534	62.834	62.154
33.5	66.619	65.832	65.071	64.334	63.619	62.926
34	67.467	66.666	65.889	65.137	64.409	63.702
34.5	68.321	67.503	66.711	65.945	65.202	64.482
35	69.178	68.344	67.537	66.756	65.999	65.265
35.5	70.04	69.19	68.368	67.571	66.799	66.051
36	70.906	70.04	69.202	68.39	67.604	66.842
36.5	71.777	70.894	70.04	69.213	68.412	67.636
37	72.652	71.752	70.882	70.04	69.224	68.433
37.5	73.531	72.615	71.729	70.871	70.04	69.235
38	74.415	73.482	72.579	71.706	70.86	70.04
38.5	75.303	74.353	73.434	72.545	71.684	70.849
39	76.196	75.229	74.293	73.388	72.511	71.662
39.5	77.094	76.109	75.157	74.235	73.343	72.478
40	77.996	76.995	76.025	75.087	74.179	73.299
40.5	78.903	77.883	76.897	75.942	75.018	74.124
41	79.814	78.776	77.773	76.802	75.862	74.952
41.5	80.73	79.675	78.654	77.666	76.71	75.784
42	81.651	80.577	79.539	78.534	77.562	76.621
42.5	82.577	81.485	80.429	79.407	78.418	77.461
43	83.508	82.397	81.323	80.284	79.279	78.306
43.5	84.444	83.314	82.222	81.165	80.143	79.154
44	85.384	84.235	83.125	82.051	81.012	80.007
44.5	86.329	85.161	84.033	82.941	81.886	80.863
45	87.28	86.093	84.945	83.836	82.763	81.724
45.5	88.235	87.029	85.863	84.735	83.645	82.59
46	89.196	87.969	86.784	85.639	84.531	83.459
46.5	90.162	88.915	87.711	86.547	85.422	84.332
47	91.132	89.866	88.643	87.46	86.317	85.21
47.5	92.108	90.822	89.579	88.378	87.216	86.093

TABLE VI (*continued*)

p [Torr]	710	720	730	740	750	760
$S =$ 48	93.09	91.782	90.52	89.3	88.12	86.979
48.5	94.076	92.748	91.466	90.227	89.029	87.87
49.	95.068	93.719	92.417	91.159	89.942	88.766
49.5	96.065	94.696	93.373	92.095	90.86	89.665
50	97.068	95.677	94.334	93.037	91.782	90.57
50.5	98.076	96.664	95.3	93.983	92.71	91.479
51	99.09	97.656	96.271	94.934	93.642	92.392
51.5	100.109	98.653	97.247	95.89	94.578	93.31
52	101.134	99.655	98.229	96.851	95.52	94.233
52.5	102.164	100.664	99.215	97.817	96.466	95.16
53	103.200	101.677	100.207	98.788	97.417	96.092
53.5	104.242	102.696	101.204	99.764	98.373	97.028
54.	105.290	103.721	102.207	100.745	99.334	97.97
54.5	106.343	104.751	103.215	101.732	100.300	98.916
55	107.403	105.787	104.228	102.724	101.271	99.867
55.5	108.468	106.828	105.247	103.721	102.247	100.823
56	109.539	107.876	106.271	104.728	103.228	101.784
56.5	110.617	108.929	107.301	105.731	104.214	102.750
57	111.700	109.988	108.336	106.744	105.206	103.721
57.5	112.790	111.052	109.378	107.762	106.203	104.697
58	113.885	112.123	110.424	108.786	107.205	105.677

several components it is conceivable to utilize the dependence of the refractive index on the wavelength of light in order to distinguish the individual components by means of separate interferograms taken at different wavelengths.

However, this dependence $n(\lambda)$ is very small and only assumes larger proportions in the vicinity of an absorption line ν_{abs} [see Born (1), p.94]:

$$(n^2 - 1) \sim \mathrm{const}/(\nu_{abs}^2 - \nu^2)$$

For the majority of substances suitable as test media, the absorption levels are not in the visible range (ionic excitation in the infrared and electronic excitation in the ultraviolet range).

b. Fringe Temperatures in Air (Gladstone–Dale). The temperatures ϑ corresponding to specific fringes according to Eq. (59a) are compiled in Table VI in air for different pressures, $\vartheta_\infty = 18°C$ has been selected for $S = 0$.

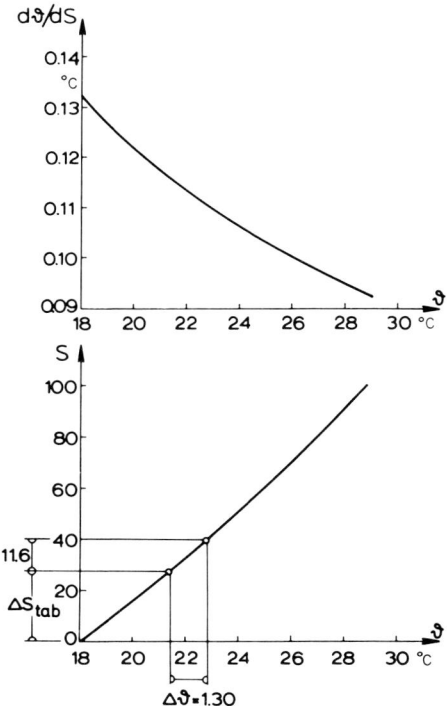

FIG. 63. Fringe temperature in water. The numerical values correspond to the example in Section VI,B,1 ($l = 0.05$ m, $\lambda = 0.5461 \cdot 10^{-6}$ m, $\vartheta_\infty = 18°C$).

Considering the equation of state for air, $1/\rho = R_0 \cdot T/p$, the following expression results from Eq. (117):

$$(n_\infty - 1)/(n_0 - 1) = \rho_\infty/\rho_0 = T_0/T_\infty$$

With the equation of ideal interferometry [Eq. (59)], $n - n_0 = S \cdot \lambda/l$, it follows that

$$T_0/T_\infty = 1/(1 - aS), \quad a = \text{const} = 2(\lambda/l)\, p/(2\bar{r} \cdot R_0 \cdot T)$$

$$\vartheta_0 = T_\infty[aS/(1 - aS)]$$

$\lambda = 0.5461 \cdot 10^{-6}$ m, $l = 0.5$ m, $p = 720$–760 Torr, $T_\infty = 291.2°$K ($\vartheta_\infty = 18°$C), $\bar{r} = 0.1506 \cdot 10^{-3}$ m³/kg. The tabulated range of $\vartheta_0(S)$ for various pressures p is shown in Fig. (62). The values of the example of Section VI,A,1 are denoted.

c. Fringe Temperatures in Water (Tilton–Taylor). Similar to the above, fringe temperatures in water will be calculated with the aid of the empirical relation for $n(\lambda, \vartheta)$ given by Tilton and Taylor (*86*). These are listed in Table VII, the following parameters apply: $\lambda = 0.5461 \cdot 10^{-6}$ m, $l = 0.05$ m, $\vartheta_\infty = 18°$C. The function $\vartheta(S)$ is shown in Fig. 63.

Equation of Tilton and Taylor:

$$n = n(\vartheta, \lambda) = n(20°C; \lambda) - n[(\vartheta - 20°C); \lambda]$$

$$n = \left(a_{20}^2 - k_{20}\lambda^2 + \frac{m_{20}}{\lambda^2 - l_{20}^2}\right)^{1/2}$$

$$- 10^{-7} \frac{[B^* - b^*(\Delta\lambda^*)^3/(\lambda - l)]\,\Delta\vartheta_{20}^3 + [A^* - a^*(\Delta\lambda^*)(1 + a^{**}/(\lambda - l))]\,\Delta\vartheta_{20}^2}{\vartheta + D^*}$$

$$- 10^{-7} \frac{[C^* - c^*(\Delta\lambda^*)(1 + c^{**}/(\lambda - l))]\,\Delta\vartheta_{20}}{\vartheta + D^*}$$

For ϑ [°C], $\Delta\vartheta_{20} = \vartheta - 20$, $\Delta\lambda^* = (\lambda - \lambda_{\text{Na}})$, especially;

$$\Delta\lambda^* = \lambda_{\text{Hg}} - \lambda_{\text{Na}} = 0.043188 \cdot 10^{-6} \text{ m}.$$

1st term		2nd and 3rd term	
$a_{20}^2 = 1.7616316$	$A^* = 2352.12$	$a^* = 143.63,$	$a^{**} = 0.4436$
$k_{20} = 0.0119882$	$B^* = 6.3449$	$b^* = 10.562$	
$l_{20}^2 = 0.0149119$	$C^* = 76087.9$	$c^* = 12504,$	$c^{**} = 0.08430$
$m_{20} = 0.00644277$	$D^* = 65.7081$	$\lambda - l = \lambda_{\text{Hg}} - l_{20} = 0.4239595$	

TABLE VII

Fringe Temperatures in Water

$\vartheta_\infty = 18°C$; $l = 0.05$ m; $\lambda = 0.5461 \cdot 10^{-6}$ m

S (−)	ϑ (°C)	$d\vartheta/dS$ (deg)	n (−)	$dn/d\vartheta$ (deg^{-1})
0.0	18.00	0.1323	1.3346 381	$0.8254 \cdot 10^{-4}$
0.2	18.028	0.1322	359	0.8263
0.4	18.055	0.1320	337	0.8272
0.6	18.081	0.1319	315	0.8281
0.8	18.108	0.1317	293	0.8290
1.0	18.134	0.1316	1.3346 271	0.8300
1.2	18.160	0.1314	250	0.8309
1.4	18.187	0.1313	228	0.8318
1.6	18.213	0.1311	206	0.8328
1.8	18.239	0.1310	184	0.8337
2.0	18.265	0.1308	1.3346 162	0.8346
2.2	18.291	0.1307	140	0.8355
2.4	18.317	0.1306	118	0.8364
2.6	18.344	0.1304	097	0.8374
2.8	18.369	0.1303	075	0.8383
3.0	18.395	0.1301	1.3346 053	0.8392
3.2	18.422	0.1300	031	0.8401
3.4	18.448	0.1298	009	0.8410
3.6	18.474	0.1297	1.3345 987	0.8420
3.8	18.500	0.1295	966	0.8429
4.0	18.525	0.1294	1.3345 944	0.8638
4.2	18.551	0.1293	922	0.8447
4.4	18.577	0.1291	900	0.8456
4.6	18.603	0.1290	878	0.8465
4.8	18.628	0.1288	856	0.8474
5.0	18.655	0.1287	1.3345 835	0.8483
5.2	18.680	0.1286	813	0.8492
5.4	18.706	0.1284	791	0.8502
5.6	18.732	0.1283	769	0.8511

TABLE VII (continued)

S	ϑ	$d\vartheta/dS$	n	$dn/d\vartheta$
(–)	(°C)	(deg)	(–)	(deg^{-1})
5.8	18.757	0.1282	1.3345 747	$0.8520 \cdot 10^{-4}$
6.0	18.783	0.1280	725	0.8529
6.2	18.808	0.1279	703	0.8538
6.4	18.834	0.1278	682	0.8547
6.6	18.860	0.1276	660	0.8556
6.8	18.885	0.1275	638	0.8565
7.0	18.911	0.1274	1.3345 616	0.8574
7.2	18.936	0.1272	594	0.8583
7.4	18.961	0.1271	572	0.8592
7.6	18.987	0.1270	551	0.8601
7.8	19.012	0.1268	529	0.8610
8.0	19.038	0.1267	1.3345 507	0.8619
8.2	19.063	0.1266	485	0.8628
8.4	19.088	0.1264	463	0.8637
8.6	19.114	0.1263	441	0.8646
8.8	19.139	0.1262	420	0.8655
9.0	19.164	0.1261	1.3345 398	0.8664
9.2	19.189	0.1259	376	0.8672
9.4	19.214	0.1258	354	0.8681
9.6	19.240	0.1257	332	0.8690
9.8	19.265	0.1256	310	0.8699
10.0	19.290	0.1254	1.3345 288	0.8708
10.2	19.315	0.1253	267	0.8717
10.4	19.340	0.1252	245	0.8726
10.6	19.365	0.1251	223	0.8734
10.8	19.390	0.1249	201	0.8743
11.0	19.415	0.1248	1.3345 179	0.8752
11.2	19.440	0.1247	157	0.8760

TABLE VII (*continued*)

S (−)	ϑ (°C)	dϑ/dS (deg)	n (−)	dn/dϑ (deg^{-1})
11.4	19.465	0.1246	1.3345 136	0.8769·10^{-4}
11.6	19.490	0.1244	114	0.8778
11.8	19.514	0.1243	092	0.8786
12.0	19.539	0.1242	1.3345 070	0.8795
12.2	19.564	0.1241	048	0.8804
12.4	19.589	0.1239	026	0.8812
12.6	19.614	0.1238	004	0.8821
12.8	19.639	0.1237	1.3344 983	0.8830
13.0	19.663	0.1236	1.3344 961	0.8838
13.2	19.688	0.1235	939	0.8846
13.4	19.713	0.1233	917	0.8854
13.6	19.737	0.1232	895	0.8962
13.8	19.762	0.1231	873	0.8871
14.0	19.787	0.1230	1.3344 852	0.8880
14.2	19.811	0.1229	830	0.8888
14.4	19.836	0.1228	808	0.8896
14.6	19.860	0.1226	786	0.8905
14.8	19.885	0.1225	764	0.8914
15.0	19.909	0.1224	1.3344 742	0.8922
15.2	19.934	0.1223	721	0.8930
15.4	19.958	0.1222	699	0.8938
15.6	19.983	0.1221	677	0.8947
15.8	20.007	0.1220	655	0.8956
16.0	20.031	0.1218	1.3344 633	0.8964
16.2	20.056	0.1217	611	0.8972
16.4	20.080	0.1216	589	0.8980
16.6	20.104	0.1215	568	0.8989
16.8	20.129	0.1214	1.3344 546	0.8998

TABLE VII (continued)

S (–)	ϑ (°C)	$d\vartheta/dS$ (deg)	n (–)	$dn/d\vartheta$ (deg^{-1})
17.0	20.153	0.1213	1.3344 524	0.9006·10^{-4}
17.2	20.177	0.1212	502	0.9014
17.4	20.201	0.1210	480	0.9022
17.6	20.226	0.1209	458	0.9030
17.8	20.250	0.1208	437	0.9039
18.0	20.274	0.1207	1.3344 415	0.9048
18.2	20.298	0.1206	393	0.9056
18.4	20.322	0.1205	371	0.9064
18.6	20.346	0.1204	349	0.9072
18.8	20.370	0.1203	327	0.9081
19.0	20.394	0.1202	1.3344 306	0.9088
19.2	20.418	0.1201	284	0.9096
19.4	20.442	0.1200	262	0.9104
19.6	20.466	0.1199	240	0.9112
19.8	20.490	0.1198	218	0.9120
20.0	20.514	0.1197	1.3344 196	0.9128
20.2	20.538	0.1195	174	0.9136
20.4	20.562	0.1194	153	0.9144
20.6	20.585	0.1193	131	0.9152
20.8	20.610	0.1192	109	0.9160
21.0	20.634	0.1191	1.3344 087	0.9168
21.2	20.657	0.1190	065	0.9176
21.4	20.681	0.1189	043	0.9184
21.6	20.705	0.1188	022	0.9192
21.8	20.729	0.1187	000	0.9200
22.0	20.752	0.1186	1.3343 978	0.9208
22.2	20.776	0.1185	956	0.9216
22.4	20.800	0.1184	934	0.9224

TABLE VII (*continued*)

S (−)	ϑ (°C)	dϑ/dS (deg)	n (−)	dn/dϑ (deg⁻¹)
22.6	20.824	0.1183	1.3343 912	0.9232·10⁻⁴
22.8	20.847	0.1182	890	0.9240
23.0	20.871	0.1181	1.3343 869	0.9248
23.2	20.894	0.1180	847	0.9256
23.4	20.918	0.1179	825	0.9264
23.6	20.942	0.1178	803	0.9272
23.8	20.965	0.1177	781	0.9280
24.0	20.989	0.1176	1.3343 759	0.9288
24.2	21.012	0.1175	738	0.9296
24.4	21.036	0.1174	716	0.9304
24.6	21.059	0.1173	694	0.9312
24.8	21.083	0.1172	672	0.9320
25.0	21.106	0.1171	1.3343 650	0.9328
25.2	21.129	0.1170	628	0.9336
25.4	21.153	0.1169	607	0.9344
25.6	21.176	0.1168	585	0.9352
25.8	21.200	0.1167	563	0.9360
26.0	21.223	0.1166	1.3343 541	0.9367
26.2	21.246	0.1165	519	0.9374
26.4	21.269	0.1164	497	0.9382
26.6	21.293	0.1163	475	0.9390
26.8	21.316	0.1162	454	0.9398
27.0	21.339	0.1161	1.3343 432	0.9406
27.2	21.362	0.1160	410	0.9414
27.4	21.386	0.1159	388	0.9421
27.6	21.409	0.1158	366	0.9428
27.8	21.432	0.1157	344	0.9436
28.0	21.455	0.1156	1.3343 323	0.9444

TABLE VII (*continued*)

S (–)	ϑ (°C)	$d\vartheta/dS$ (deg)	n (–)	$dn/d\vartheta$ (deg^{-1})
28.2	21.478	0.1155	1.3343 301	$0.9452 \cdot 10^{-4}$
28.4	21.501	0.1155	279	0.9460
28.6	21.524	0.1154	257	0.9468
28.8	21.547	0.1153	235	0.9476
29.0	21.570	0.1152	1.3343 213	0.9483
29.2	21.594	0.1151	192	0.9490
29.4	21.617	0.1150	170	0.9498
29.6	21.639	0.1149	148	0.9506
29.8	21.662	0.1148	126	0.9514
30.0	21.685	0.1147	1.3343 104	0.9521
30.2	21.708	0.1146	082	0.9528
30.4	21.731	0.1145	060	0.9536
30.6	21.754	0.1144	039	0.9544
30.8	21.777	0.1143	017	0.9551
31.0	21.800	0.1143	1.3342 995	0.9558
31.2	21.823	0.1142	973	0.9566
31.4	21.846	0.1141	951	0.9574
31.6	21.868	0.1140	929	0.9581
31.8	21.891	0.1139	908	0.9588
32.0	21.914	0.1138	1.3342 886	0.9596
32.2	21.937	0.1137	864	0.9604
32.4	21.959	0.1136	842	0.9611
32.6	21.982	0.1136	820	0.9618
32.8	22.005	0.1135	798	0.9626
33.0	22.027	0.1134	1.3342 777	0.9634
33.2	22.050	0.1133	755	0.9641
33.4	22.073	0.1132	733	0.9648
33.6	22.100	0.1131	711	0.9656
33.8	22.118	0.1130	689	0.9664

TABLE √II (continued)

S (−)	ϑ (°C)	$d\vartheta/dS$ (deg)	n (−)	$dn/d\vartheta$ (deg^{-1})
34.0	22.141	0.1129	1.3342 667	$0.9672 \cdot 10^{-4}$
34.2	22.163	0.1129	645	0.9679
34.4	22.186	0.1128	624	0.9686
34.6	22.208	0.1127	602	0.9694
34.8	22.231	0.1126	580	0.9701
35.0	22.253	0.1125	1.3342 558	0.9708
35.2	22.276	0.1124	536	0.9716
35.4	22.298	0.1123	514	0.9723
35.6	22.321	0.1123	493	0.9730
35.8	22.343	0.1122	471	0.9737
36.0	22.366	0.1121	1.3342 449	0.9744
36.2	22.388	0.1120	427	0.9752
36.4	22.411	0.1119	405	0.9759
36.6	22.433	0.1118	383	0.9766
36.8	22.455	0.1118	361	0.9773
37.0	22.477	0.1117	1.3342 340	0.9780
37.2	22.499	0.1116	318	0.9788
37.4	22.522	0.1115	296	0.9794
37.6	22.545	0.1114	274	0.9802
37.8	22.567	0.1113	252	0.9810
38.0	22.589	0.1112	1.3342 230	0.9817
38.2	22.611	0.1112	209	0.9824
38.4	22.634	0.1111	187	0.9830
38.6	22.656	0.1110	165	0.9838
38.8	22.678	0.1109	143	0.9845
39.0	22.700	0.1108	1.3342 121	0.9852
39.2	22.722	0.1107	099	0.9860
39.4	22.744	0.1107	078	0.9867

TABLE VII (*continued*)

S (−)	ϑ (°C)	$d\vartheta/dS$ (deg)	n (−)	$dn/d\vartheta$ (deg⁻¹)
39.6	22.767	0.1106	1.3342 056	0.9874·10⁻⁴
39.8	22.789	0.1105	034	0.9882
40.0	22.811	0.1104	1.3342 012	0.9889
40.2	22.833	0.1103	1.3341 990	0.9896
40.4	22.855	0.1103	968	0.9903
40.6	22.877	0.1102	946	0.9910
40.8	22.899	0.1101	925	0.9917
41.0	22.921	0.1100	1.3341 903	0.9924
41.2	22.943	0.1100	881	0.9931
41.4	22.965	0.1099	859	0.9938
41.6	22.987	0.1098	837	0.9945
41.8	23.009	0.1097	815	0.9952
42.0	23.031	0.1097	1.3341 794	0.9959
42.2	23.053	0.1096	772	0.9966
42.4	23.075	0.1095	750	0.9973
42.6	23.097	0.1094	728	0.9980
42.8	23.118	0.1094	706	0.9987
43.0	23.140	0.1093	1.3341 684	0.9994
43.2	23.162	0.1092	663	1.0001
43.4	23.184	0.1091	641	1.0008
43.6	23.206	0.1091	619	1.0015
43.8	23.228	0.1090	597	1.0022
44.0	23.249	0.1089	1.3341 575	1.0029
44.2	23.271	0.1088	553	1.0036
44.4	23.293	0.1088	531	1.0043
44.6	23.315	0.1087	510	1.0050
44.8	23.336	0.1086	488	1.0057
45.0	23.358	0.1085	1.3341 466	1.0064
45.2	23.380	0.1084	444	1.0071

TABLE VII (continued)

S (–)	ϑ (°C)	$d\vartheta/dS$ (deg)	n (–)	$dn/d\vartheta$ (deg^{-1})
45.4	23.401	0.1084	1.3341 422	$1.0078 \cdot 10^{-4}$
45.6	23.423	0.1083	400	1.0085
45.8	23.445	0.1082	379	1.0092
46.0	23.466	0.1081	1.3341 357	1.0099
46.2	23.488	0.1081	335	1.0106
46.4	23.509	0.1080	313	1.0113
46.6	23.531	0.1079	291	1.0120
46.8	23.553	0.1078	269	1.0127
47.0	23.574	0.1078	2.3341 247	1.0134
47.2	23.600	0.1077	226	1.0141
47.4	23.617	0.1076	204	1.0148
47.6	23.639	0.1076	182	1.0155
47.8	23.661	0.1075	160	1.0162
48.0	23.682	0.1074	1.3341 138	1.0168
48.2	23.703	0.1073	116	1.0175
48.4	23.725	0.1073	095	1.0182
48.6	23.746	0.1072	073	1.0188
48.8	23.768	0.1071	051	1.0195
49.0	23.789	0.1070	1.3341 029	1.0202
49.2	23.811	0.1070	007	1.0208
49.4	23.832	0.1069	1.3340 985	1.0215
49.6	23.853	0.1068	964	1.0222
49.8	23.875	0.1068	942	1.0228
50.0	23.896	0.1067	1.3340 920	1.0236
50.2	23.917	0.1066	898	1.0242
50.4	23.939	0.1066	876	1.0249
50.6	23.960	0.1065	854	1.0256
50.8	23.981	0.1064	832	1.0262
51.0	24.003	0.1064	1.3340 811	1.0269

TABLE VII *(continued)*

S (–)	ϑ (°C)	$d\vartheta/dS$ (deg)	n (–)	$dn/d\vartheta$ (deg^{-1})
51.2	24.024	0.1063	1.3340 789	$1.0276 \cdot 10^{-4}$
51.4	24.045	0.1062	767	1.0282
51.6	24.066	0.1062	745	1.0289
51.8	24.087	0.1061	723	1.0296
52.0	24.109	0.1060	1.3340 701	1.0302
52.2	24.130	0.1060	680	1.0309
52.4	24.151	0.1059	658	1.0316
52.6	24.172	0.1058	636	1.0322
52.8	24.194	0.1058	614	1.0329
53.0	24.215	0.1057	1.3340 592	1.0336
53.2	24.236	0.1056	570	1.0342
53.4	24.257	0.1056	549	1.0349
53.6	24.278	0.1055	527	1.0356
53.8	24.299	0.1054	505	1.0362
54.0	24.320	0.1054	1.3340 483	1.0369
54.2	24.341	0.1053	461	1.0376
54.4	24.362	0.1052	439	1.0382
54.6	24.382	0.1052	417	1.0389
54.8	24.404	0.1051	396	1.0396
55.0	24.425	0.1050	1.3340 374	1.0402
55.2	24.446	0.1050	352	1.0409
55.4	24.467	0.1049	330	1.0416
55.6	24.488	0.1048	308	1.0422
55.8	24.509	0.1048	286	1.0428
56.0	24.530	0.1047	1.3340 265	1.0435
56.2	24.551	0.1046	243	1.0442
56.4	24.572	0.1046	221	1.0448
56.6	24.593	0.1045	199	1.0454
56.8	24.614	0.1044	177	1.0461

TABLE VII (*continued*)

S (−)	ϑ (°C)	$d\vartheta/dS$ (deg)	n (−)	$dn/d\vartheta$ (deg^{-1})
57.0	24.635	0.1044	1.3340 155	$1.0468\cdot10^{-4}$
57.2	24.656	0.1043	134	1.0474
57.4	24.676	0.1042	112	1.0480
57.6	24.697	0.1042	090	1.0487
57.8	24.718	0.1041	068	1.0494
58.0	24.739	0.1040	1.3340 046	1.0500
58.2	24.760	0.1040	024	1.0506
58.4	24.780	0.1039	002	1.0513
58.6	24.801	0.1038	1.3339 981	1.0520
58.8	24.822	0.1038	959	1.0526
59.0	24.843	0.1037	1.3339 937	1.0532
59.2	24.863	0.1036	915	1.0538
59.4	24.884	0.1036	893	1.0544
59.6	24.905	0.1035	871	1.0550
59.8	24.926	0.1035	850	1.0557
60.0	24.946	0.1034	1.3339 828	1.0564
60.2	24.967	0.1033	806	1.0570
60.4	24.988	0.1033	784	1.0576
60.6	25.008	0.1032	762	1.0582
60.8	25.029	0.1031	740	1.0588
61.0	25.050	0.1031	1.3339 718	1.0595
61.2	25.070	0.1030	697	1.0602
61.4	25.091	0.1030	675	1.0608
61.6	25.111	0.1029	653	1.0614
61.8	25.132	0.1028	631	1.0620
62.0	25.152	0.1028	1.3339 609	1.0626
62.2	25.173	0.1027	587	1.0633
62.4	25.194	0.1027	566	1.0640
62.6	25.214	0.1026	544	1.0647
62.8	25.235	0.1025	522	1.0654

TABLE VII (*continued*)

S (–)	ϑ (°C)	$d\vartheta/dS$ (deg)	n (–)	$dn/d\vartheta$ (deg^{-1})
63.0	25.255	0.1025	1.3339 500	1.0660·10^{-4}
63.2	25.276	0.1024	478	1.0666
63.4	25.296	0.1024	456	1.0672
63.6	25.376	0.1023	435	1.0678
63.8	25.337	0.1022	413	1.0684
64.0	25.357	0.1022	1.3339 391	1.0690
64.2	25.378	0.1021	369	1.0696
64.4	25.398	0.1021	347	1.0703
64.6	25.419	0.1020	325	1.0710
64.8	25.439	0.1019	303	1.0716
65.0	25.459	0.1019	1.3339 282	1.0722
65.2	25.480	0.1018	260	1.0728
65.4	25.500	0.1018	238	1.0734
65.6	25.520	0.1017	216	1.0740
65.8	25.541	0.1016	194	1.0746
66.0	25.561	0.1016	1.3339 172	1.0752
66.2	25.581	0.1015	151	1.0759
66.4	25.602	0.1015	129	1.0766
66.6	25.622	0.1014	107	1.0772
66.8	25.642	0.1013	085	1.0778
67.0	25.663	0.1013	1.3339 063	1.0784
67.2	25.683	0.1012	041	1.0790
67.4	25.703	0.1012	020	1.0796
67.6	25.723	0.1011	1.3338 998	1.0802
67.8	25.743	0.1011	976	1.0808
68.0	25.764	0.1010	1.3338 954	1.0814
68.2	25.784	0.1009	932	1.0820
68.4	25.804	0.1009	910	1.0827
68.6	25.824	0.1008	888	1.0834

TABLE VII (continued)

S	ϑ	$d\vartheta/dS$	n	$dn/d\vartheta$
(−)	(°C)	(deg)	(−)	(deg^{-1})
68.8	25.844	0.1008	1.3338 867	1.0840·10^{-4}
69.0	25.864	0.1007	1.3338 845	1.0846
69.2	25.885	0.1007	823	1.0852
69.4	25.905	0.1006	801	1.0858
69.6	25.925	0.1005	779	1.0864
69.8	25.945	0.1005	757	1.0870
70.0	25.965	0.1004	1.3338 736	1.0876
70.2	25.985	0.1004	714	1.0882
70.4	26.005	0.1003	692	1.0888
70.6	26.025	0.1003	670	1.0894
70.8	26.045	0.1002	648	1.0900
71.0	26.065	0.1001	1.3338 626	1.0906
71.2	26.085	0.1001	604	1.0912
71.4	26.105	0.1000	583	1.0918
71.6	26.125	0.1000	561	1.0924
71.8	26.145	0.0999	539	1.0930
72.0	26.165	0.0999	1.3338 517	1.0936
72.2	26.185	0.0998	495	1.0942
72.4	26.205	0.0998	473	1.0948
72.6	26.225	0.0997	452	1.0954
72.8	26.245	0.9996	430	1.0960
73.0	26.265	0.0996	1.3338 408	1.0966
73.2	26.285	0.0995	386	1.0972
73.4	26.305	0.0995	364	1.0978
73.6	26.325	0.0994	342	1.0984
73.8	26.345	0.0994	321	1.0990
74.0	26.365	0.0993	1.3338 299	1.0996
74.2	26.384	0.0993	277	1.1002

TABLE VII (*continued*)

S (−)	ϑ (°C)	$d\vartheta/dS$ (deg)	n (−)	$dn/d\vartheta$ (deg^{-1})
74.4	26.404	0.0992	1.3338 255	1.1008·10^{-4}
74.6	26.424	0.0992	233	1.1014
74.8	26.444	0.0991	211	1.1020
75.0	26.464	0.0991	1.3338 189	1.1026
75.2	26.484	0.0990	168	1.1032
75.4	26.503	0.0989	146	1.1038
75.6	26.523	0.0989	124	1.1044
75.8	26.543	0.0988	102	1.1050
76.0	26.563	0.0988	1.3338 080	1.1056
76.2	26.582	0.0987	058	1.1062
76.4	26.602	0.0987	037	1.1068
76.6	26.622	0.0986	015	1.1074
76.8	26.642	0.0986	1.3337 993	1.1080
77.0	26.661	0.0985	1.3337 971	1.1086
77.2	26.681	0.0985	949	1.1092
77.4	26.701	0.0984	927	1.1098
77.6	26.720	0.0984	906	1.1104
77.8	26.740	0.0983	884	1.1110
78.0	26.760	0.0983	1.3337 862	1.1116
78.2	26.779	0.0982	840	1.1122
78.4	26.799	0.0982	818	1.1127
78.6	26.819	0.0981	796	1.1132
78.8	26.838	0.0981	774	1.1138
79.0	26.858	0.0980	1.3337 753	1.1144
79.2	26.877	0.0980	731	1.1150
79.4	26.897	0.0979	709	1.1156
79.6	26.917	0.0978	687	1.1162
79.8	26.936	0.0978	665	1.1168

TABLE VII (*continued*)

S (−)	ϑ (°C)	$d\vartheta/dS$ (deg)	n (−)	$dn/d\vartheta$ (deg^{-1})
80.0	26.956	0.0977	1.3337 643	1.1174·10^{-4}
80.2	26.975	0.0977	622	1.1180
80.4	26.995	0.0976	600	1.1186
80.6	27.014	0.0976	578	1.1192
80.8	27.034	0.0975	556	1.1197
81.0	27.053	0.0975	1.3337 534	1.1202
81.2	27.073	0.0974	512	1.1208
81.4	27.092	0.0974	491	1.1214
81.6	27.112	0.0973	469	1.1220
81.8	27.131	0.0973	447	1.1226
82.0	27.151	0.0972	1.3337 425	1.1232
82.2	27.170	0.0972	403	1.1238
82.4	27.190	0.0971	381	1.1244
82.6	27.209	0.0971	359	1.1249
82.8	27.228	0.0970	338	1.1254
83.0	27.248	0.0970	1.3337 316	1.1260
83.2	27.267	0.0969	294	1.1266
83.4	27.287	0.0969	272	1.1272
83.6	27.306	0.0968	250	1.1278
83.8	27.325	0.0968	228	1.1284
84.0	27.345	0.0967	1.3337 207	1.1290
84.2	27.364	0.0967	185	1.1294
84.4	27.383	0.0966	163	1.1300
84.6	27.403	0.0966	141	1.1306
84.8	27.422	0.0965	119	1.1312
85.0	27.441	0.0965	1.3337 097	1.1318
85.2	27.461	0.0965	075	1.1324
85.4	27.480	0.0964	054	1.1330

TABLE VII (*continued*)

S	ϑ	dϑ/dS	n	dn/dϑ
(−)	(°C)	(deg)	(−)	(deg^{-1})
85.6	27.499	0.0964	1.3337 032	$1.1335 \cdot 10^{-4}$
85.8	27.518	0.0963	010	1.1340
86.0	27.538	0.0963	1.3336 988	1.1346
86.2	27.557	0.0962	966	1.1352
86.4	27.576	0.0962	944	1.1358
86.6	27.595	0.0961	923	1.1364
86.8	27.615	0.0961	901	1.1369
87.0	27.634	0.0960	1.3336 879	1.1374
87.2	27.653	0.0960	857	1.1380
87.4	27.672	0.0959	835	1.1386
87.6	27.691	0.0959	813	1.1392
87.8	27.710	0.0958	792	1.1397
88.0	27.730	0.0958	1.3336 770	1.1402
88.2	27.749	0.0957	748	1.1408
88.4	27.768	0.0957	726	1.1414
88.6	27.787	0.0956	704	1.1420
88.8	27.806	0.0956	682	1.1425
89.0	27.825	0.0955	1.3336 660	1.1430
89.2	27.844	0.0955	639	1.1436
89.4	27.864	0.0954	617	1.1442
89.6	27.883	0.0954	595	1.1448
89.8	27.902	0.0953	573	1.1453
90.0	27.921	0.0953	1.3336 551	1.1458
90.2	27.940	0.0953	529	1.1464
90.4	27.959	0.0952	508	1.1470
90.6	27.978	0.0952	486	1.1476
90.8	28.000	0.0951	1.3336 464	1.1480
91.0	28.016	0.0951	1.3336 442	1.1486

TABLE VII (continued)

S	ϑ	$d\vartheta/dS$	n	$dn/d\vartheta$
(−)	(°C)	(deg)	(−)	(deg^{-1})
91.2	28.035	0.0950	1.3336 420	$1.1492 \cdot 10^{-4}$
91.4	28.054	0.0950	398	1.1498
90.6	28.073	0.0949	377	1.1504
91.8	28.092	0.0949	355	1.1509
92.0	28.111	0.0948	1.3336 333	1.1514
92.2	28.130	0.0948	311	1.1520
92.4	28.149	0.0948	289	1.1526
92.6	28.168	0.0947	267	1.1532
92.8	28.187	0.0947	245	1.1537
93.0	28.206	0.0946	1.3336 224	1.1542
93.2	28.225	0.0946	202	1.1548
93.4	28.243	0.0945	180	1.1554
93.6	28.262	0.0945	158	1.1559
93.8	28.281	0.0944	136	1.1564
94.0	28.300	0.0944	1.3336 114	1.1570
94.2	28.319	0.0943	093	1.1575
94.4	28.337	0.0943	071	1.1580
94.6	28.357	0.0943	049	1.1586
94.8	28.376	0.0942	027	1.1592
95.0	28.394	0.0942	1.3336 005	1.1597
95.2	28.413	0.0941	1.3335 983	1.1602
95.4	28.432	0.0941	962	1.1607
95.6	28.451	0.0940	940	1.1612
95.8	28.469	0.0940	918	1.1618
96.0	28.489	0.0940	1.3335 896	1.1622
96.2	28.507	0.0939	874	1.1628
96.4	28.526	0.0939	852	1.1634
96.6	28.545	0.0938	830	1.1639

TABLE VII (*continued*)

S (−)	ϑ (°C)	$d\vartheta/dS$ (deg)	n (−)	$dn/d\vartheta$ (deg^{-1})
96.8	28.563	0.0938	1.3335 809	1.1644·10^{-4}
97.0	28.582	0.0937	1.3335 787	1.1650
97.2	28.601	0.0937	765	1.1656
97.4	28.620	0.0936	743	1.1662
97.6	28.639	0.0936	721	1.1667
97.8	28.657	0.0936	699	1.1672
98.0	28.676	0.0935	1.3335 678	1.1678
98.2	28.695	0.0935	656	1.1684
98.4	28.713	0.0934	634	1.1689
98.6	28.732	0.0934	612	1.1694
98.8	28.751	0.0933	590	1.1700
99.0	28.769	0.0933	1.3335 568	1.1706
99.2	28.788	0.0933	546	1.1711
99.4	28.807	0.0932	525	1.1716
99.6	28.825	0.0932	503	1.1722
99.8	28.844	0.0931	481	1.1728
100.0	28.863	0.0931	1.3335 459	1.1733

VI. Application Examples and Their Evaluation

The following examples treat mainly boundary-layer problems in the region of natural convection and combined convection. They have all been evaluated with interferometric methods (MZI) exclusively. In these applications interference methods are of great importance. In the following illustrations small circles denote interference maxima and black dots denote interference minima. Lengths are specified in real size, where the imaging scale of the exposure is considered.

Further examples of interference experiments in heat transfer can be found in, e.g., (94-105).

A. Models in Gaseous Media

1. *Heat Transfer in a Horizontal Annulus in Air* (*stationary wall with a constant temperature gradient at the wall*)

Heat transfer in a horizontal cylindrical annulus can be described in the form

$$\mathrm{Nu}_{\delta_g} = f(\mathrm{Gr}_{\delta_g}; \delta_g/d_i)$$

Nu_{δ_g} and Gr_{δ_g} are the dimensionless characteristic values describing heat transfer in natural convection, as formed with the width of the annulus $\delta_g = (d_0 - d_i)/2$. The parameter δ_g/d_i reflects the fact that cylindrical annuli are, in general, not geometrically equivalent. In the limit $\delta_g/d_i \to 0$, i.e., for a constant width δ_g and an infinitely large inner diameter d_i, the annulus is equivalent to a parallel slit. In the limit $\delta_g/d_i \to \infty$, i.e., an increasing value of δ_g and a constant inner diameter d_i, the case of a horizontal cylinder in an undisturbed medium evolves. For example, $\delta_g/d_i \approx 50$ describes the case of a horizontal cylinder, where the walls of the laboratory are the outer boundary of the "annulus." For small and medium Gr_{δ_g} numbers ($320 \leqslant \mathrm{Gr}_{\delta_g} \leqslant 7 \cdot 10^5$) heat transfer was determined interferometrically for several parameters δ_g/d_i (87).

a. Test Model. The temperature of the wall is to be kept constant in these experiments. The inner cylinder a (see Fig. 64) consists of an electrically heated thick-walled brass tube. The isothermal surface is attained by heat conduction in the wall. The annulus is formed by an outer brass tube b, equipped with two copper coils to provide a countercurrent flow of temperature-controlled water, and the inner brass tube a. Each end of the annulus is closed with a spring-mounted, quick-opening closure of insulating material, which is opened to take photographs. Using these covers, the interaction between the ambient air and that in

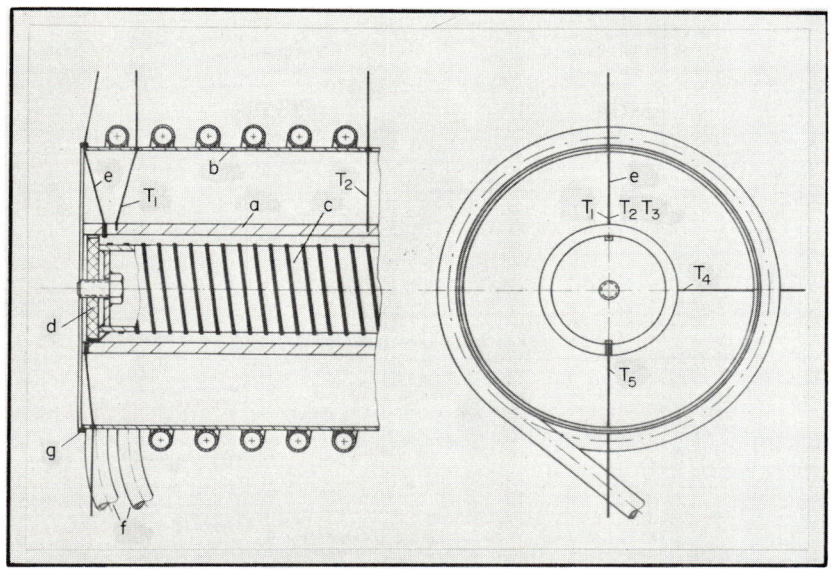

FIG. 64. Cross section of a circular annulus: *a*, inner cylinder; *b*, outer cylinder; *c*, electric heating element; *d*, disk of insulating material; *e*, suspension; *f*, inlet for coolant; *g*, quick-opening closure; $T_1, ..., T_5$, thermocouples.

the annulus is prevented during the short period ($\frac{1}{10}$ to $\frac{2}{10}$ sec) required for the pictures. This method is preferred over using stationary glass plates of interferometric quality since the temperature of the inner cylinder is relatively high. The temperatures of the inner and the outer cylinder and the ambient air in the reference beam of the interferometer are measured by nine thermocouples connected to an indicating millivolt-meter. The thermocouple T_1 serves to check the temperature drop at the end.

b. Interferogram and Evaluation. For example, the interferogram shown in Fig. 65 will be evaluated for the point 30° at the inner cylinder. For a tabulation of the measured values, see Table VIII.

Evaluation procedure: The origin ($y = 0$), i.e., the wall of the model, is not known exactly for the function $S(y)$ as derived from the interferogram since the contour of the wall is blurred because of diffraction effects. The interference fringe nearest to the wall is superposed with the diffraction pattern, so that its position cannot be determined precisely (enlarged portion of Fig. 65). This problem can be circumvented by extrapolating the temperature distribution $\vartheta(y)$ derived from the interference pattern. Because the temperature of the wall ϑ_w is known by thermocouple mea-

TABLE VIII

DATA DERIVED FROM CUTOUT REGION IN FIG. 65

S	y [mm]	b [mm]	b [mm]	ΔS_1 [—]	S_{ideal} [—]	S_{tab} [—]	ϑ [°C]	$\Delta\vartheta$ [deg]	$\Delta\vartheta$ [deg]	y_{corr} [mm]
(22)			(0.308)	(0.24)	(22.24)	(25.90)	(53.65)		(1.55)	
		(0.308)						1.54		
21	0.533		0.308	0.24	21.24	24.90	52.11		1.53	0.02
		0.308						1.52		
20	0.841		0.308	0.24	20.24	23.90	50.59		1.52	0.326
		0.308						1.51		
19	1.149		0.308	0.24	19.24	22.90	49.08		1.50	0.634
		0.308						1.49		
18	1.456		0.310	0.24	18.24	21.90	47.59		1.49	0.941
		0.312						1.48		
17	1.769		0.317	0.23	17.23	20.89	46.11		1.47	1.254
		0.322						1.46		
16	2.091		0.327	0.21	16.21	19.87	44.65			1.576
		0.332								
15	2.422									1.907

Data: $l = 0.5$ m $\quad \delta_g = 29$ mm $\quad \vartheta_w = 52.2$°C $\quad S_{\text{tab}} = 3.66$ ($\triangleq \vartheta_\infty$)
$\lambda = 0.5461 \cdot 10^{-6}$ m $\quad d_i = 40$ mm $\quad \vartheta_\infty = 22.5$°C $\quad S_w = 21.30$
$p = 720$ Torr $\quad\quad\quad\quad\quad\quad\quad\quad\quad\quad\quad\quad \Delta\vartheta_\infty = 29.7$ deg (ideal interferometry)

surements, the origin ($y = 0$) of the temperature distribution can then be defined. In Table VIII the temperature distribution is derived for the seven interference fringes nearest to the wall. The interference order $S = 22$ given in parentheses is fictitious; this point is already inside of the wall contour. The fringe widths $b = y_{i+1} - y_i$ are determined and are used to derive mean values of b in the respective intervals for the corresponding values of S. The interference order of ideal interferometry $S_{\text{ideal}} = S + \Delta S_1$ can be calculated by

$$\Delta S_1 = \frac{n_\infty \cdot \lambda \cdot l}{12 \cdot b^2}$$

[cf. Eq. (92)]. The fringe temperatures can then be determined from Table VI, beginning with $S_{\text{tab}} = 3.66$. This value results for $p = 720$ Torr and $\vartheta_\infty = 22.5$°C (Table VI) and must be added as $\Delta S_{\text{tab}} = 3.66$ to the values S_{ideal} ($S_{\text{tab}} = S_{\text{ideal}} + \Delta S_{\text{tab}}$). Temperature values corresponding to the values of S_{tab} are taken from Table VIII and are extrapolated for $S = 22$. The interference order $S_{\text{w ideal}} = 21.30$ (i.e., $S_w = 21.07$ in the interferogram) can be determined by these methods for $\vartheta_w = 52.2$°C from mean thermocouple measurements, as is shown in Fig. 66 for the specified temperature distribution. Simul-

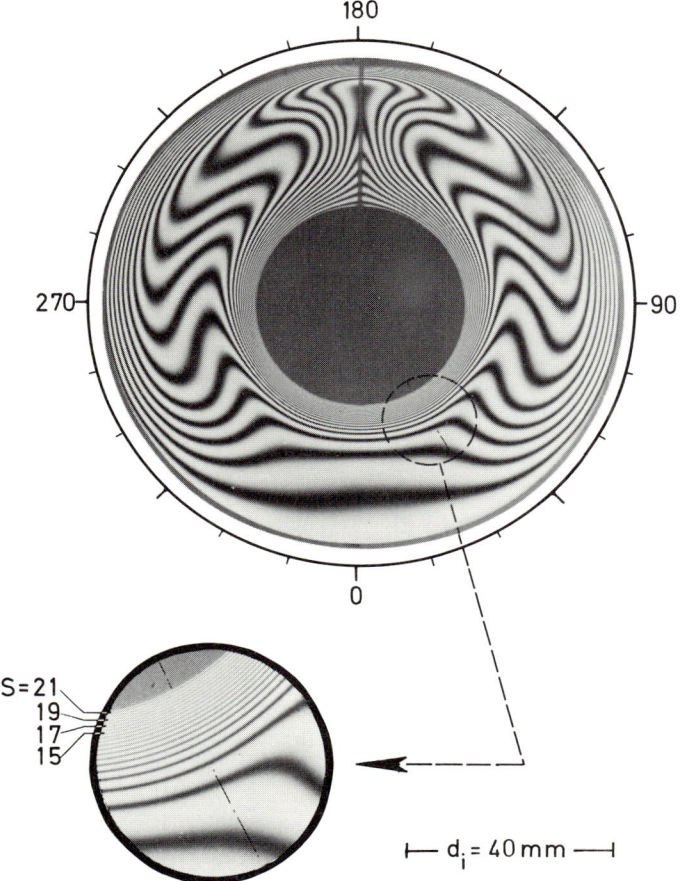

FIG. 65. Interferogram of a horizontal cylindrical annulus in an infinite fringe field: $d_i = 40$ mm, $d_a = 98$ mm, $\delta_g = 29$ mm, $\delta_g/d_i = 0.73$. Position 30°: $\mathrm{Nu}_{\delta_g} = 4.82$, $\mathrm{Gr}_{\delta_g} = 7.28 \cdot 10^4$, $\Delta\vartheta_\infty = 29.7$ deg. (Photograph by W. Hauf.)

taneously, the origin ($y = 0$), i.e., the model wall, is defined for the temperature distribution.

For heat transfer, however, the temperature gradient at the wall is of interest:

$$(d\vartheta/dy)_w \approx (\Delta\vartheta/\Delta y)_w = (\Delta\vartheta/b)_w = (1.53 \text{ deg}/0.310 \text{ mm}) = 4.94 \quad \text{deg/mm}$$

With δ_g and $\Delta\vartheta_\infty = \vartheta_w - \vartheta_\infty$, the dimensionless $\mathrm{Nu}_{\delta_g\infty}$ can be determined from the temperature gradient:

$$\mathrm{Nu}_{\delta_g} = (d\vartheta/dy)_w \cdot (\delta_g/\Delta\vartheta) = 4.82$$

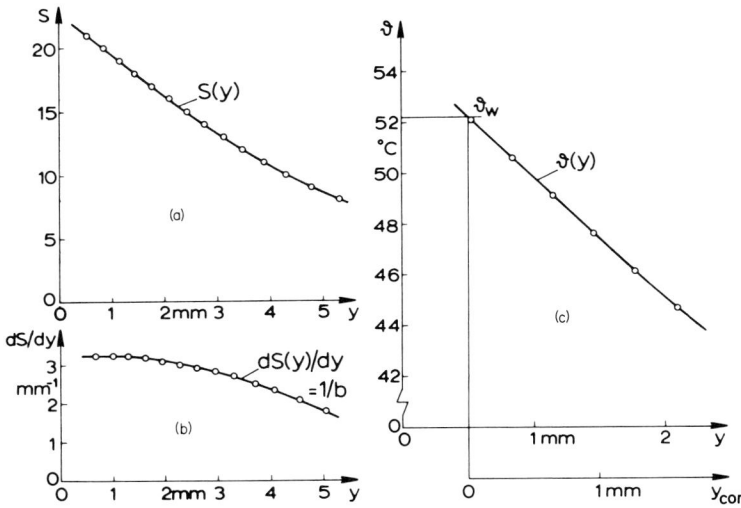

FIG. 66. Evaluation curves of the annulus model for the position 30° in Fig. 65: (a) $S(y)$ phase difference as derived from the interferogram; (b) $1/b(y)$ fringe density, constant at the wall; (c) $\vartheta(y)$ temperature distribution.

The local Nu_{δ_g} numbers are determined for several positions and the mean Nu_{δ_g} number of the annulus can be obtained by integrating along the perimeter. The range of the corresponding Gr_{δ_g} numbers is given by the range of $\varDelta\vartheta_\infty$ which can be realized in interference models and covers about one order of magnitude ($6°C \leqslant \varDelta\vartheta_\infty \leqslant 60°C$). This comparatively small measuring range can be extended by a factor of eight downwards (upwards) by halving (doubling) the dimension of the annulus since the Grashof number is a function of $\delta_g{}^3$.

It is not advantageous to halve the model length l in order to cover twice the temperature range with the same number of fringes because end effects then become more objectionable. The gain (factor of two) is small.

c. Measuring Accuracy. In the course of the evaluation only the phase difference S was corrected. Both the fringe displacement $\varDelta\eta$ for "correct" focusing and end effects were negligible. The closing disks g (Fig. 64) are also heated by the inner cylinder a and function as heating guards and also as heat insulation against the surroundings. The length of the boundary layer at the inner cylinder can be assumed to be about the same as the model length.

The experimental results agree with calorimetric measurements to $\pm 5\%$ (87). In heat transfer measurements an accuracy of about 2–4%

is obtainable for medium fringe numbers ($S = 30$); for small and for very large fringe numbers, accuracies of about 8–10% can be expected. This refers to boundary layers in natural convection.

2. *Heat Transfer in Air at a Rotating, Horizontal Cylinder* (*moving wall where the temperature gradient is not constant*)

a. Test Model. The rotating horizontal cylinder in air is well suited for the observation of separation processes in the boundary layer in the regime of combined convection and incipient turbulence (*88*). The experimental configuration can be seen in Fig. 67. Here this configuration

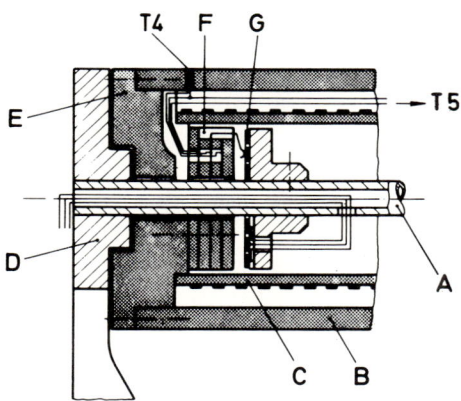

FIG. 67. Cross section of the heated rotating cylinder: A, stationary axis with the connecting wires of the heating element and the thermocouples; B, thick-walled cylinder of brass which can be rotated by means of a string drive; C, tube of insulating material used as a form for the heating wire; D, flange (8 mm wide); E, end disks with bearings; F, G, slip-ring contacts for the heater supply voltage and to the connections of the thermocouples; ($T4$, $T5$) thermocouples at the surface of the cylinder.

shall serve as an example for a variable temperature gradient $(d\vartheta/dy)_w$ at a moving wall (see Section III,C). This can be seen best in Fig. 68 at a point of the perimeter at 135° which is near the separation point of the boundary layer.

The evaluation shows (Fig. 69) that the distribution of the fringe density (practically proportional to the temperature gradient) exhibits only a slight curvature.

b. Interference Pattern and Its Evaluation. The evaluation is performed in the same manner as in the previous example. In the enlarged

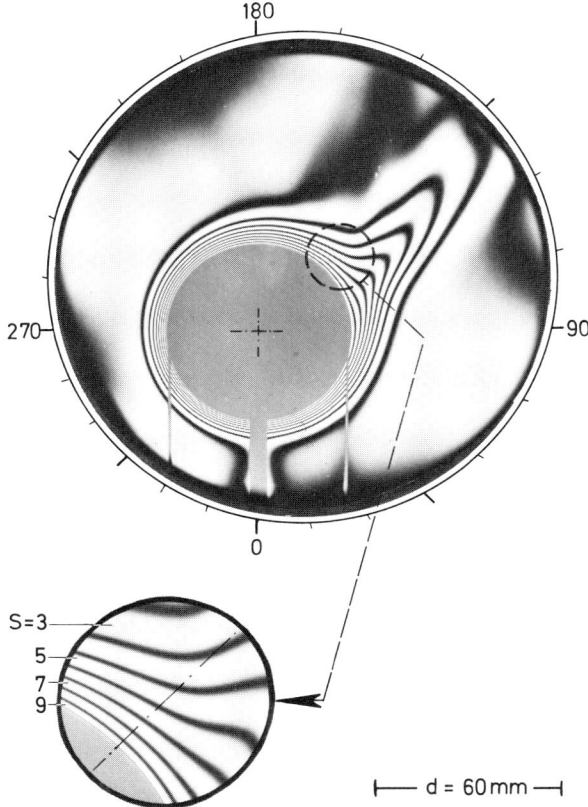

FIG. 68. Interference pattern in air of a rotating heated cylinder (infinite fringe field). The region of interest (135°) is enlarged: $d = 60$ mm, $l = 0.42$ m, $v_\mu = 9.2$ cm/sec, Nu $= 4.91$, Gr/Re$^2 = 2.98$, Re $= 350$. (Photograph by M. Reimann).

segment of Fig. 68 the interference fringes at the wall, which are not yet fully developed, are distorted by diffraction. This region is therefore not suitable for evaluation. Because of the small number of interference fringes in Fig. 69, the distribution can be defined better by evaluating the maxima (S even) *and* the minima (S odd).

Furthermore, it is probable that the model is no longer strictly two dimensional in the separation region. Excluding the laminar sublayer it is probably better to consider this as a motion of turbulence packets evenly distributed along the length of the model. The measuring ray integrates along the length l and the mean value of the local variations of the density (refractive index) is constant in time as is the resulting

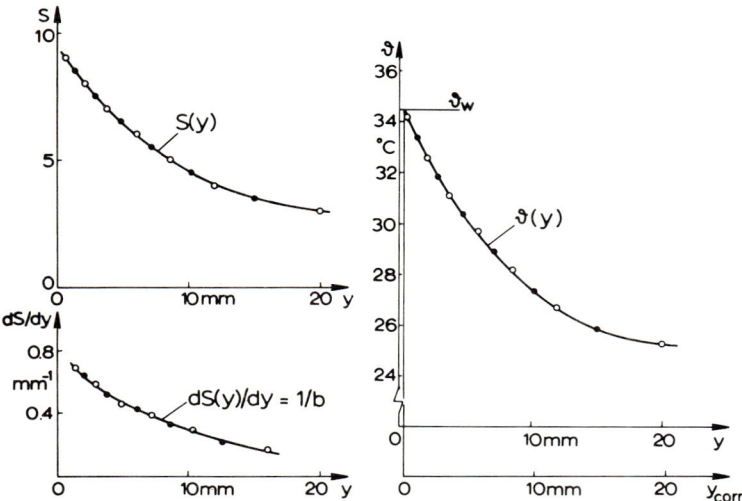

FIG. 69. Distribution of the phase difference $S(y)$, the fringe density $1/b$, and the temperature at the position 135° in Fig. 68 of the rotating cylinder.

phase difference. The resulting temperatures $\vartheta(y)$ are mean temperatures. If it can be assumed—as in this case—that the measuring ray does not deviate from straight-line propagation (as in ideal interferometry), within the limits of measuring accuracy, the correct mean temperature values at the points y_0 in the boundary layer result. This assumption, however, is only valid for moderate temperature gradients in not too large local turbulence packets so that the effects of beam deflection can be neglected. In Table IX the calculated correction terms [Eq. (92)] are listed which are an indication of the deviation from ideal interferometry. Because the model length $l^* = 0.45$ m in this example differs from the model length $l = 0.5$ m on which the tabulated fringe temperatures (Table VI) are based, the values listed in the Table VI must be transformed as $S_{\text{tab}}^* = (0.45/0.5) S_{\text{tab}}$.

c. *Measuring Accuracy.* More pronounced end effects than in the case of the annulus discussed above are to be expected. These can be corrected with the aid of Eq. (104). The quantity Δl in Eq. (104) corresponds approximately to the width of the flange (8 mm). This incomplete correction of the end effects is commensurate with the accuracy of the boundary-layer measurements. The gradient at the wall can only be determined to about $\pm 4\%$, which is a consequence of the nonlinear temperature distribution. The estimated measuring accuracy is about 6–8%.

TABLE IX

DATA DERIVED FROM CUTOUT REGION IN FIG. 68

	Maxima		Minima							
S [—]	y [mm]	b [mm]	y [mm]	b [mm]	ΔS_i	S_{ideal} [—]	S^*_{tab} [—]	S_{tab} [—]	ϑ [°C]	y_{corr} [mm]
9	0.613		(1.398)		$1.04 \cdot 10^{-2}$	9.01	11.22	12.47	34.2	0.088
8.5		1.448	1.367		0.98	8.51	10.72	11.91	33.4	0.923
8	2.061			1.553	0.85	8.0	10.21	11.34	32.6	1.536
7.5		1.709	2.920		0.70	7.5	9.71	10.79	31.9	2.395
7	3.770			1.923	0.55	7	9.21	10.23	31.1	3.245
6.5		2.266	4.843		0.40	6.5	8.71	9.68	30.4	4.318
6	6.036			2.340	0.37	6	8.21	9.21	29.7	5.511
5.5		2.554	7.183		0.31	5.5	7.71	8.57	28.9	6.658
5	8.590			3.03	0.22	5	7.21	8.01	28.2	8.065
4.5		3.37	10.21		0.18	4.5	6.71	7.46	27.4	9.685
4	11.96			4.78	0.09	4	6.21	6.90	26.7	11.44
3.5		6.00	14.99		0.06	3.5	5.71	6.34	25.9	14.77
3	19.96					3	5.21	5.79	25.3	19.44

Data: $l = 0.45$ m $\vartheta_w = 34.5°C$ $S_{tab} = 2.46$ $y_{corr} = 0.53$ mm
$\lambda = 0.5461 \cdot 10^{-6}$ m $\vartheta_\infty = 21.0°C$ $S^*_{tab} = (0.45/0.5)S_{tab}$
$d = 60$ mm $\Delta\vartheta_\infty = 13.5$ deg $\Delta S^*_{tab} = 2.21$

3. *Electric Discharge in Air* (*irregular model, schliere*)

In the following example (the interference pattern of a schliere) the refractive index varies in three dimensions, in contrast to the previous examples where it was considered to be constant in the direction of wave propagation. Here the distribution of the refractive index or the density, respectively, cannot be derived theoretically. In the following the region of heated air in the vicinity of the discharge path in a spark gap is observed for a certain time after the termination of the discharge. Experiments of this kind have been conducted by Hannes (*89*), to whom the following example is due.

a. Experimental Configuration. The spherical electrode *A* is connected to a charged capacitor. The connections exhibit no inductance and negligible resistance. The charge voltage (26 kV) is somewhat smaller than the breakdown voltage of the spark gap in air at standard pressure. The discharge is triggered by a low-energy spark: The triggering electrode is embedded in the grounded plane electrode *B* and is insulated from it. The energy released, which can be calculated from the voltage before and after the discharge and the capacitance, appears in the following forms:

(a) electromagnetic radiation (light) emanating from the plasma in the discharge path;

(b) shock wave emanating from the discharge path;

(c) thermal energy of the plasma which is transferred to the surrounding air as internal and kinetic energy of a rapid eddy movement; a small proportion is dissipated as heat in the electrodes;

(d) losses due to resistance of the connections.

The proportion Q of the thermal energy corresponding to (c) is equal to the total enthalpy of the air schliere and is to be determined from the interference pattern in Fig. 70. The photograph was taken about 30 msec

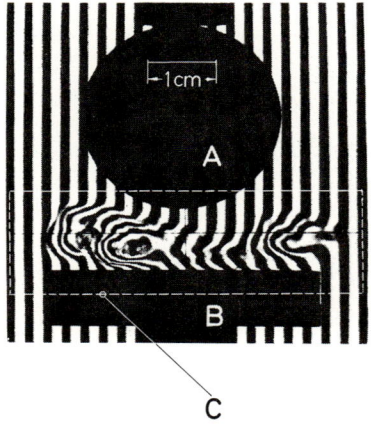

Fig. 70. Interference pattern of a spark gap in a fringe field: A, spherical electrode; B, plane electrode, grounded; C, schlieren region to be evaluated. Total energy of the discharge: 2.1 W-sec at 26 kV maximum voltage and a capacitance of approximately 0.01 μF. Duration of the discharge: $<10^{-5}$ sec. Exposure taken about 30 msec after the discharge. (Photograph by H. Hannes.)

after the discharge. The light emission of the spark and the shock wave have decayed at this point so that constant pressure can be assumed in the schliere. The heated gas particles are now involved in a severe eddy movement, propagating the schliere and intermixing the heated particles with the surrounding air.

The mixing process begins about 20 msec and ends about 200 msec after the discharge. The flow itself, however, is slow enough so that no detectable changes in the density (refractive index) result. The fringe shifts can, therefore, be attributed to temperature changes alone. Photographs taken at different times during the mixing process exhibit schlieren of different size. However, the evaluation of these exposures results in approximately the same energy within the limits of the measuring accuracy, confirming the above assumption [cf. (*89*)].

b. The Interference Pattern and Its Interpretation. In order to evaluate the interference pattern, the interference lines in Region C (Fig. 70) are traced from an enlarged photograph (Fig. 71b). The center of the com-

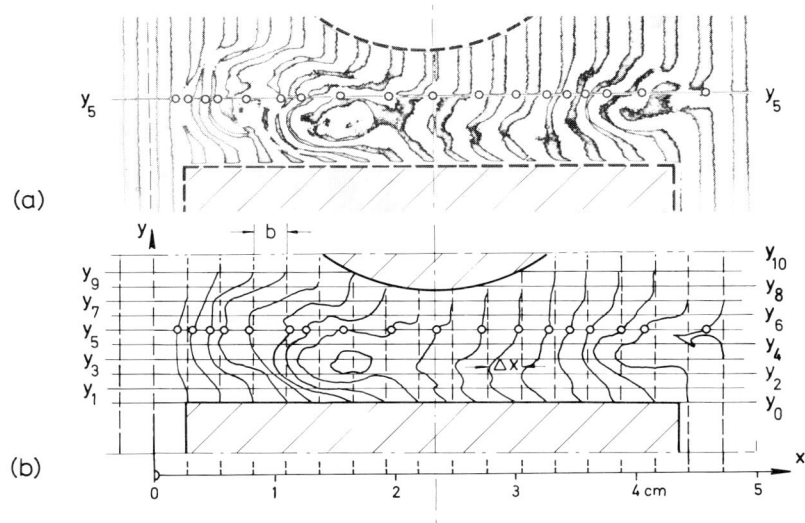

FIG. 71. Evaluation of the interferogram of the schliere in Fig. 70: (a) Equidensometric pattern of the interferogram in Fig. 70; (b) interference lines with x, y raster segments.

paratively broad interference lines can often be determined more accurately by equidensometric methods. An equidensometric pattern is shown in Fig. 71a. The interference lines appear as double lines, whose central trace can be determined more accurately. These double lines are the borders of the fringes, corresponding to a constant degree of darkening in the primary exposure. This technique is described in Section V,E,1 [cf. also Krug and Lau (*82*)].

The original interference fringes, i.e., the pattern before the discharge, are shown in Fig. 71b (dashed lines) and the outlines of the electrodes are indicated. In order to evaluate the pattern, integration over all points of the exposure according to Eq. (111a) is required:

$$H = -\frac{2}{3}\frac{c_p}{\bar{r}} \cdot T_\infty \cdot \lambda \iint S(x, y)\, dx\, dy$$

The integral is now written as a sum of finite steps in the directions of the x and y axes, respectively:

$$H = -\frac{2}{3}\frac{c_p}{\bar{r}} \cdot T_\infty \cdot \lambda \sum_{y=0}^{y=n}\left[\varDelta y \cdot \sum_{x=0}^{x=n}(S \cdot \varDelta x)\right]$$

The path difference S is negative since the density (the refractive index) decreases with increasing temperature. Because of the negative sign in Eq. (111a), the enthalpy H is positive. The schlieren region is now partitioned by a raster, corresponding to finite increments $\varDelta x_i = b$ (vertical original fringes) and $\varDelta y_i$ (Fig. 71b).

Evaluation of the pattern will be demonstrated for the line y_5. The intersection of the interference fringes with the line y_5, denoted by small

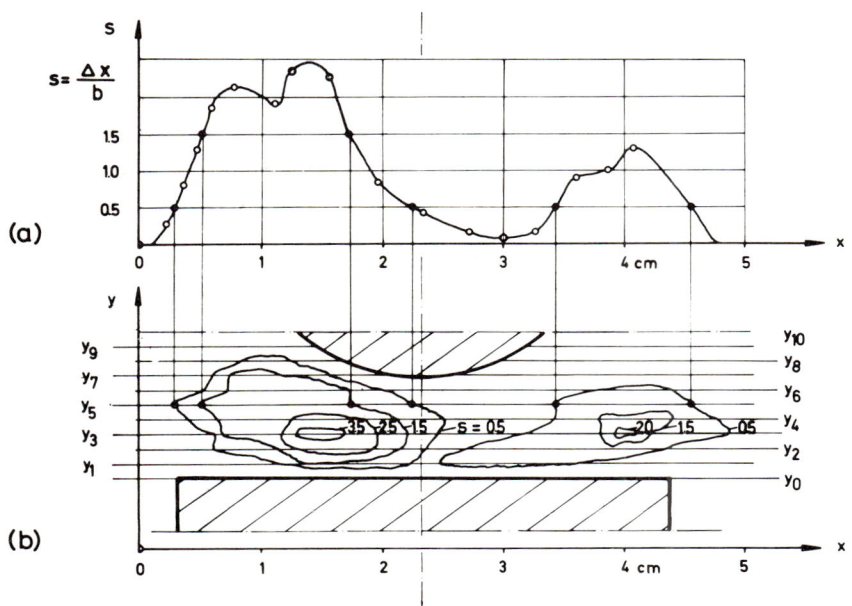

FIG. 72. Evaluation of the interferogram (Fig. 71 continued): (a) Distribution of the phase difference S for y_5; (b) topographical diagram of the eikonal of the schliere (corresponds to a interferogram in the infinite fringe field).

circles, results in measuring points whose corresponding path difference $S = \Delta x/b$ is determined (cf. Fig. 72a); Δx is the deflection of the fringe from its original position and b is the original fringe width. The function $S(x)$ can now be recorded for $y_5 = $ const (Fig. 72a). The phase difference as a function of x can be determined in a similar fashion for other values $y = $ const. [cf. Section V, E, 1, b].

A complete set of these curves represents a topographical diagram of the phase difference $S(x, y)$ (eikonal of the schliere), as is shown in Fig. 72b. The contours $S = 0.5, 1.5, 2.5$, and 3.5 are indicated. These contours correspond to the interference pattern produced in an infinite fringe field. A pattern of this kind is, however, not suitable for quantitative evaluation since the information contained in the small number of interference lines is insufficient.

In the next step, the integral $I(y) = \int S(x)\, dx$ is determined planimetrically for the various intersection planes $y_n = $ const. The values $I(y)$ are shown in Fig. 73. Integration of this curve results in the value of the integral $\iint S(x, y)\, dx\, dy$, which is proportional to the enthalpy.

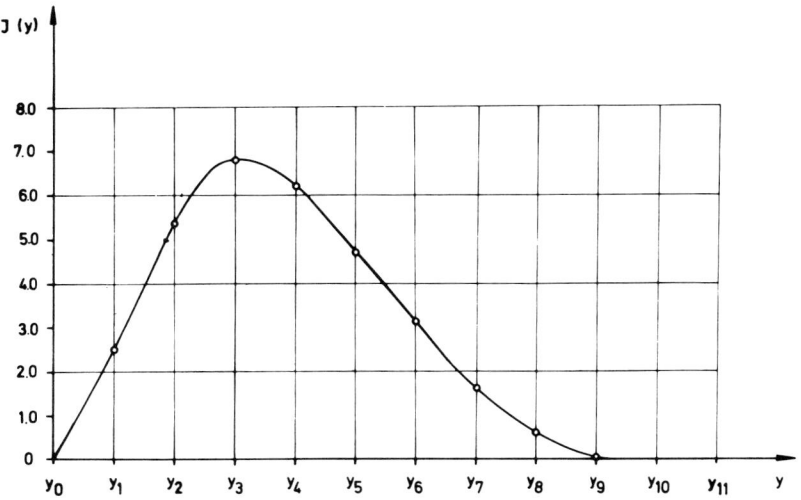

FIG. 73. $I(y) = \int S(x)\, dx$ derived from horizontal intersections $y_0, ..., y_{10}$.

Numerical values:

Horizontal intersection y_5 (Fig. 71b and Fig. 72a) (The values of x are of the same magnitude as in the experimental configuration; this also applies to the raster segments in the y direction: $\Delta y_i = 1.25$ mm.):

TABLE X
Phase Difference in Intersection y_5 (Fig. 72a)

x (mm)	S (—)	x (mm)	S (—)	x (mm)	S (—)
2.0	0.27	12.6	2.36	32.8	0.18
3.2	0.82	15.6	2.28	34.5	0.55
4.7	1.27	19.7	0.82	36.1	0.91
5.8	1.86	23.4	0.45	38.9	1.00
7.9	2.14	27.2	0.18	40.7	1.32
11.3	1.91	30.3	0.09	45.8	0.50

$$(S = \Delta x/b)$$

Numerical values of the integral:

$$y_0 \cdots y_{10}: \quad I(y_n) = \int S(x)\, dx$$

y	y_0	y_1	y_2	y_3	y_4	y_5
$I \cdot 10^2$ m	0	2.653	5.320	6.867	6.188	4.795

y	y_6	y_7	y_8	y_9	y_{10}
$I \cdot 10^2$ m	3.087	1.575	0.462	0	0

$$I_{\text{tot}} = \int_y \left(\int_x S(x)\, dx \right) dy \quad \text{(Fig.73)}$$

$$I_{\text{tot}} = 3.78 \cdot 10^{-4} \quad \text{m}^2$$

Constant of proportionality:

$$k_n \cdot \lambda = \tfrac{2}{3}(c_p/\bar{r})\, T_0 \cdot \lambda$$

$$c_p = 1.008 \text{ kJ/kg-}^\circ\text{K}$$
$$\bar{r} = 0.1505 \cdot 10^{-3} \text{ m}^3/\text{kg}$$
$$T_0 = 293^\circ\text{K}$$
$$\lambda = 0.546 \cdot 10^{-6} \text{ m}$$
$$k_n \cdot \lambda = 0.707 \text{ kJ/m}^3$$

The total enthalpy results to

$$H_{\text{tot}} = k_n \cdot \lambda \cdot I_{\text{tot}} = 0.707 \cdot 3.78 \cdot 10^{-4} \text{kJ} = 0.267 \text{J}.$$

Result: The electric energy of the spark was calculated to be 2.1 J, so that 12.7% of the total energy is dissipated in the schliere.

c. Measuring Accuracy. If it is assumed that the path difference $S(x, y) = \Delta x/b$ is known to an accuracy of $\Delta S = 0.1$ throughout the schliere, the deviation is propagated as follows:

$$H = k_n \cdot \lambda \cdot \iint S(x, y) \, dx \, dy \tag{111}$$

Since $\Delta S = 0.1$,

$$\Delta H = k_n \cdot \lambda \cdot \Delta S \iint dx \, dy$$

where $\iint dx \, dy$ is the area of the schliere in the interference pattern:

$$A = \iint dx \, dy \approx 4.7 \quad \text{cm}^2$$

$$\Delta H = 0.707 \quad \text{kJ/m}^2 \cdot 0.1 \cdot 4.7 \cdot 10^{-4} \quad \text{m}^2 = 0.03 \quad \text{J}$$

Therefore,

$$\Delta H/H = 0.11$$

The error ΔS, which was assumed to be rather large in this example, is comprised of the measuring errors of Δx and b. It is essential that b be determined very accurately. According to the dimensions of the apparatus and the maximum path difference S, an optimum value of b results. In this example it results to $b = 2.78$ mm. Since the local gradient of the refractive index is very small, the error pertaining to ray curvature is negligible compared to the evaluation error of approximately 11%.

4. *Temperature Distribution in a Hydrogen Flame (cylindrical model)*

This example was chosen only to demonstrate the evaluation of an interferogram of a cylindrical phase object.

a. Test Model. The laminar burning flame (Fig. 74a) is evaluated in the cross section t–t, assuming concentrical density distribution. The turbulent burning flame (Fig. 74b) fluctuates periodically about the flame axis (about 15 Hz). It is not possible to evaluate this figure because neither the local flue gas composition is known nor can sure assumptions of a realistic temperature distribution be made. The cusp-shaped distortion of the interference fringes in the flame axis above the burner is caused by hydrogen gas which has not yet burned.

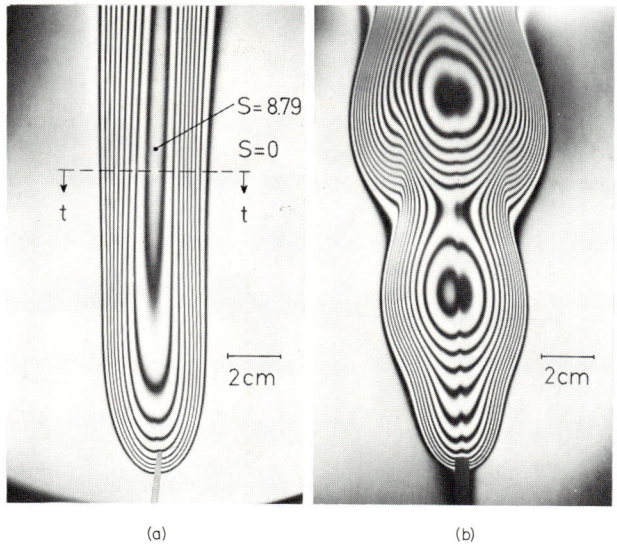

FIG. 74. Interferograms of hydrogen flames with (a) laminar and (b) turbulent flue gas streamer (1/1000 sec). (Photograph by W. Hauf.)

b. The Interference Pattern and Its Evaluation. For the evaluation, with the aid of the method described in Section V,D,1,*a* (step function), the following assumptions on the cross section *t–t*, which is situated nearly 2–3 cm above the cone of the flame, are made:

(1) The hydrogen is burned completely. Along the flame axis there is flue gas of stoichiometric composition, described by the equation

$$H_2 + \tfrac{1}{2}O_2 + \tfrac{1}{2}(0.79/0.21)N_2 \rightarrow H_2O + \tfrac{1}{2}(0.79/0.21)N_2$$

(2) The radial distribution of the concentration would have to be determined in an additional experiment.

In this case, corresponding approximately to the circumstances in a diffusion flame, an exponentially decreasing distribution of the refractive index from $\bar{r}_{\text{flue gas}}$ in the center to a value near to $\bar{r}_{\text{air}}$ at the boundary of the flame (Fig. 75) is assumed. The evaluation with the aid of the function $S(r)$, derived from the interferogram, is carried out first in six and then in eighteen steps to demonstrate the influence of the step width Δr (Fig. 75). In addition, the flue gas is assumed to consist of heated air in order to demonstrate the influence of the varying composition of the flue gas on the temperature profile (indicated by the dash–dotted lines in Fig. 75). The profiles indicated by the solid lines correspond to the exponential

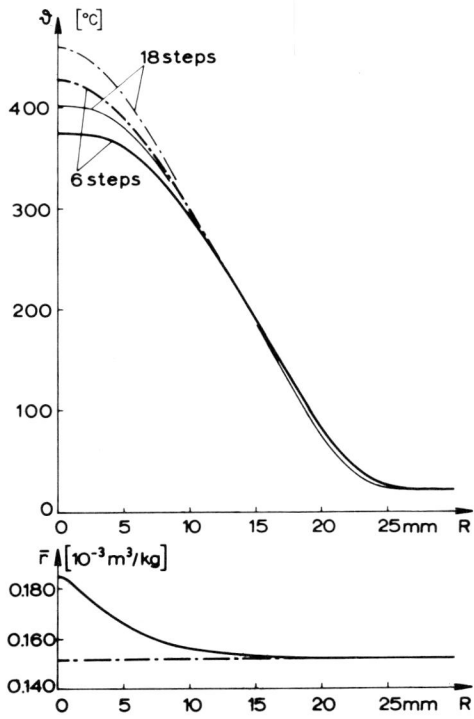

Fig. 75. Temperature distribution in the cross section t–t of the flame shown in Fig. 74. Radius of the flue gas column $R = 25.2$ mm.

distribution of the refractive index in the lower figure (mixture of flue gas and air). The profiles indicated by the dash–dotted lines are calculated for a "flue gas stream" consisting of air.

TABLE XIA

STEP FUNCTION $A(i, k)$ USED IN TABLE XIB

$i =$	5	4	3	2	1	0
$k = 5$	3.3166	1.4721	1.1962	1.0743	1.0171	1
4		3.0000	1.3542	1.1185	1.0260	1
3			2.6458	1.2280	1.0446	1
2				2.2361	1.0964	1
1					1.7321	1
0						1

With equation (107), $S_i \cdot \lambda = 2 \Delta r \sum_{k=i}^{N-1} \Delta n_k \cdot A(i, k)$, for $N = 6$, $\Delta r = 4.2$ mm ($R = 25.2$ mm):

TABLE XIB

Data Derived from Cross Section t–t in Fig. 74

	$i =$	5	4	3	2	1	0
$\Delta n_k \cdot A(i, k)$,	$k = 5$		0.2020	0.1641	0.1474	0.1395	$0.1372 \cdot 10^{-4}$
	4			0.7539	0.6228	0.5713	$0.5568 \cdot 10^{-4}$
	3				1.1671	0.9928	$0.9504 \cdot 10^{-4}$
	2					1.3325	$1.2153 \cdot 10^{-4}$
	1						$1.3850 \cdot 10^{-4}$
$\sum_{k=i+1}^{N-1} \Delta n_k \cdot A(i, k)$		0	0.2020	0.9180	1.9373	3.0361	$4.2447 \cdot 10^{-4}$
S_i (interferogram)		0.70	2.88	5.28	7.16	8.36	8.79
$S_i \lambda / 2 \Delta r - \sum \Delta n_k A$		0.4551	1.6703	2.5146	2.7175	2.3989	$1.4698 \cdot 10^{-4}$
Δn_i		0.1372	0.5568	0.9504	1.2153	1.3850	$1.4698 \cdot 10^{-4}$
$n_i - 1$		243.07	201.11	161.75	135.26	118.29	$109.81 \cdot 10^{-6}$
$\bar{r}_i$ (flue gas)		0.1522	0.1524	0.1532	0.1559	0.1624	$0.1770 \cdot 10^{-3}$
ϑ_i [°C]		38.7	103.6	194.2	280.9	347.5	372.4
$\bar{r}_i$ (air)		0.1509	0.1509	0.1509	0.1509	0.1509	$0.1509 \cdot 10^{-3}$
ϑ_i [°C]		38.8	103.9	195.6	287.3	367.7	417.2

Evaluation (6 steps):

Light wavelength	$\lambda = 0.5461 \cdot 10^{-6}$ m
Specific refractivity (air, Table III)	$\bar{r} = 0.1509 \cdot 10^{-3}$ m³/kg
Gas constant of air	$R_0 = 287.0$ N-m/kg-deg
Atmospheric pressure	$p = 718$ Torr $= 0.95725 \cdot 10^5$ N/m²
Ambient temperature	$T_\infty = 294°$K
Refractive index of the ambient air	$n - 1 = 256.79 \cdot 10^{-6}$
Specific refractivity of the flue gas in the center of the flame calculated from Eq. (117) (Gladstone–Dale)	$\bar{r} = 0.1770$ m³/kg

Corresponding to Table II:

$$A(i, k) = [(k + 1)^2 - i^2]^{1/2} - [k^2 - i^2]^{1/2}$$

c. Accuracy of Measurements. The method of evaluation (Section V,D,1,a) is usually regarded to have a mean error of about 5%. It can be seen from Fig. 75 that for a small number of steps (6 steps) an error of about 10% results, which is of the same order of magnitude as the error caused by neglecting the effect of concentration.

B. MODELS IN LIQUID MEDIA

1. *Nonstationary Heat Transfer in a Horizontal Tube* (*temperature gradient at the wall not constant*)

   a. Test Model. The model of the tube is the cuvette shown in Fig. 76. The glass plates C must be aligned exactly in the parallel ray path according to the procedure detailed in Section V,C,4,*a*. Note that the glass

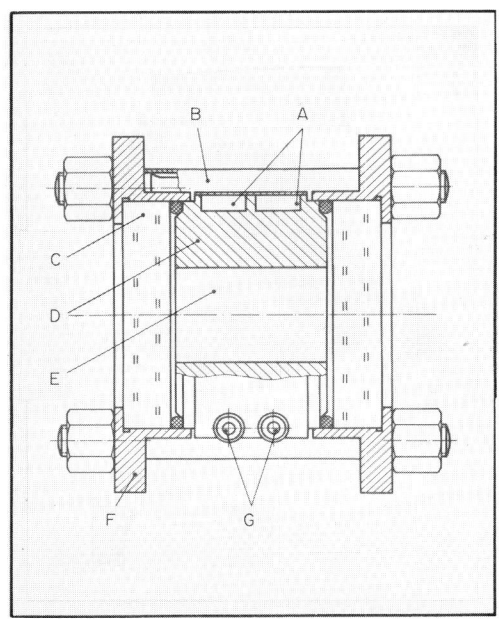

FIG. 76. Cross section of the test chamber: A, channels for temperature-controlled water; B, spaces of precisely equal length (accuracy 0.01 mm) with compression bolts; C, window of high-quality glass suitable for interferometry (ground to an accuracy of $\lambda/10$); D, solid copper jacket with a bore hole representing the tube (model); E, test fluid (water); F, flanges supporting the glass plates; G, intake for the heating fluid.

plates, which reflect slightly, assume a similar role as the auxiliary mirror in Fig. 36(1). The fact that they are mutually parallel can be checked, since each plate produces an individual image of the light source in the focal plane of the objective $L1$ of the MZI. By means of the compression bolts, the spacers can be compressed slightly, changing the relative position of the glass plates C. The two windows are exactly parallel when the individual images of the light source coincide. The "tube" D in Fig. 76 is heated by the thermally controlled fluid passing through the heating channels E. The heating process is begun at $t = 0$. The temperature difference of the thermally controlled fluid with respect to the initial temperature $\vartheta_\infty = 21.53°C$ of the test fluid is $\Delta\vartheta_{tot} = 1.59$ deg. The inner surface of the "tube" C can be assumed to be an isotherm (copper jacket). The following convection processes occur in the test fluid (water) during the heating process (90):

(1) Shortly after the start, the temperature distribution near the wall originates solely by heat conduction. The interference lines (isotherms) in the stationary fluid are concentric.

(2) In the vertical zones (90°, 270°) in Fig. 77 an upward-directed convection current is initiated in the heated layer near the wall. At the uppermost position (180°) the convection currents meet and spread out toward the inner part of the "tube." The interference pattern in Fig. 77 shows this phase. The convection currents in the boundary layers on

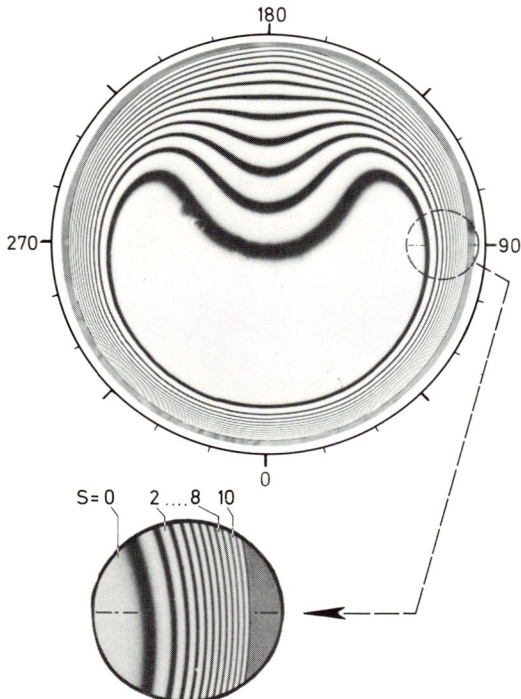

FIG. 77. Interference pattern of nonstationary heat transfer in a horizontal tube (at $t = 30$ sec): $d = 31$ mm, $\vartheta_\infty = 21.53°C$ (undisturbed inner region), $\Delta\vartheta_{tot} = 1.59$ deg (total temperature difference). (Photograph by W. Hauf.)

TABLE XII

DATA DERIVED FROM CUTOUT REGION IN FIG. 77

S [—]	Maxima y [mm]	Maxima b [mm]	Minima y [mm]	Minima b [mm]	ΔS_i [—]	S_{ideal} [—]	S_{tab} [—]	ϑ [°C]
11.5			0.149		(0.126)	11.63	40.23	22.833
11	0.243			0.169	0.126	11.13	39.73	22.783
10.5		0.170	0.318		0.126	10.63	39.23	22.723
10	0.413			0.172	0.125	10.13	38.73	22.673
9.5		0.178	0.490		0.114	9.61	38.21	22.611
9	0.591			0.188	0.094	9.20	37.80	22.570
8.5		0.199	0.678		0.093	8.59	37.19	22.499
8	0.790			0.218	0.091	8.09	36.69	22.449
7.5			0.896					

Data: $l = 0.05$ m $\Delta \vartheta_{90°} = 1.30$ deg $t = 30$ sec $\Delta \vartheta_{tot} = 1.59$ deg
$\lambda = 0.5461 \cdot 10^{-6}$ m $n = 1.334$ (after heating)   (total temperature
$(\vartheta_w = 22.83°C)$ $d = 31$ mm $\Delta S_{tab} = 28.60$ difference)
$\vartheta_\infty = 21.53°C$ $(\triangleq \vartheta_\infty)$

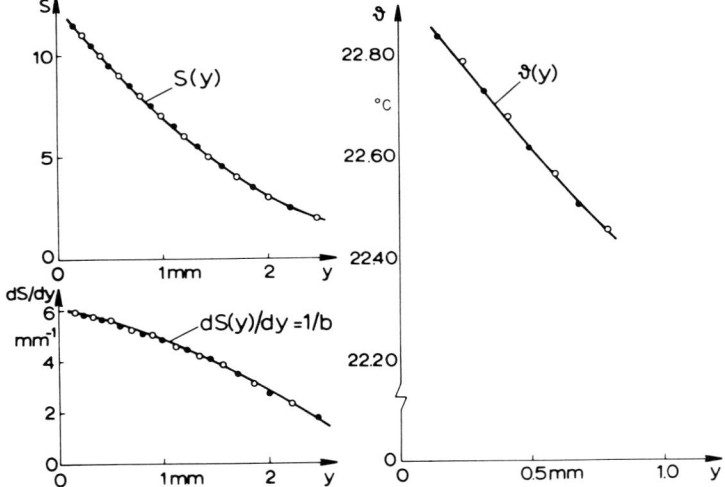

FIG. 78. Phase difference $S(y)$ and the distribution of the fringe density $1/b$ and the temperature $\vartheta(y)$ at the wall corresponding to the position 90° in Fig. 77.

both sides are supplied from the cooler fluid in the inner part of the "tube." The total flow corresponds to two eddy currents, symmetric to the vertical axis, flowing in opposite senses.

(3) During the final phase the convection current diminishes more and more, ceasing altogether in the final state ($\vartheta_\infty + \Delta\vartheta$). An interference pattern corresponding to this final phase is shown in Fig. 1, where the lowest temperature is in the bottom third of the cross section. In this phase one closed interference line after another vanishes as the heating process progresses. In the final state a homogenic field remains.

b. *The Interference Pattern and Its Evaluation.* The local rate of heat transfer at the 90° position can be determined in the same manner as in the example given in Section VI,A,1, although the surface temperature was not measured here. For the evaluation of the function (Sy) in Fig. 77, the maxima and minima are measured. Here it is not necessary to extrapolate the temperature distribution to the exact value at the wall in order to determine the temperature gradient $(d\vartheta/dT)_w$; since dn/dT is practically constant near the wall, it is sufficient to determine the temperature gradient in this region. The fringe temperatures were taken from Table VII (test fluid, water). The results for the 90° position are recorded in Fig. 78. It can be seen from the distribution of the fringe density $1/b = dS/dy$, that the change in the temperature gradient at the wall is relatively small. A local Nusselt number and the local heat flux q_w, at the wall, can be calculated from the temperature gradient at this position, using the value $\delta\vartheta/\delta S$ given in Table VII corresponding to $\vartheta_w = 22.83°$ C and values at the wall derived in Table XII:

$$(\delta\vartheta/\delta y)_w = (\delta\vartheta/\delta S) \cdot (\delta S/\delta y)_w = 0.1103 \text{ deg}/0.168 \text{ mm} = 0.657 \text{ deg/mm}$$

$$\text{Nu}_{90°} = (\delta\vartheta/\delta y)_w (d/\Delta\vartheta_{tot}) = (0.657 \text{ deg/mm}) \cdot (31 \text{ mm}/1.59 \text{ deg}) = 12.8$$

$$q_w = -k_c(\delta\vartheta/\delta y)_w = (0.60 \text{ W/m deg}) \cdot (657 \text{ deg/m}) = 394 \text{ W/m}^2$$

2. *Determination of the Thermal Diffusivity from a Nonstationary Temperature Field*

In order to determine the thermal conductivity of solids, fluids, or gases, stationary methods are customary. A measurable specific thermal flow in a geometrical configuration (usually plane, cylindrical, or spherical gaps) which can easily be treated mathematically is compared with the temperature difference $\Delta\vartheta$ between the two boundaries of the gap. The thermal conductivity k_c is defined by the following equation:

$$q = -k_c \, \partial\vartheta/\partial y \tag{118}$$

or

$$q = -(k_c/\delta_g) \, \Delta\vartheta$$

for a plane parallel gap, when y is the coordinate normal to the wall and δ_g is the width. The experimental result must be corrected because of heat radiation between the walls. In measurements of stationary processes conducted over a long period of time in fluids and gases, it is practically inevitable that convection currents occur, according to the influence of the boundaries of the gap. The measuring errors arising from these effects can be minimized by selecting small gap widths δ_g, since the value of the Raleigh number,

$$\mathrm{Ra} = \mathrm{Gr} \cdot \mathrm{Pr} = \frac{g \cdot \beta \cdot \Delta\vartheta \cdot \delta_g{}^3 \cdot c_p \cdot \rho}{\nu \cdot k_c}$$

which determines the onset of convection, is small. This, however, requires precise knowledge of $\Delta\vartheta$.

Near the critical point of a substance, however, the specific heat c_p is very large and the viscosity is very small so that small values of Ra, i.e., no convection, are hard to attain.

Measurements of k_c by nonstationary processes avoid this disadvantage. These are based on the propagation in isotropic solids of temperature fields emanating from heat sources with a simple geometry (plate, cylinder, sphere, semi-infinite body), which can be described exactly mathematically.

The basic concept of the application of the nonstationary method in stationary fluids and gases of uniform temperature is that the thermal energy emitted suddenly from the surface of the heat source propagates as in a solid by conduction alone (neglecting heat radiation). The temperature field which then develops leads to changes in density and later to slow convection processes. The measurement is then already concluded, however.

The accuracy of conventional nonstationary methods is inferior to that obtainable by stationary methods since the changing temperature field is altered by the heat capacity of the sensors (i.e., thermocouples). This method has been applied particularly using a hot wire strung in the medium to be investigated (Fourier line source). The emitted heat energy can be determined by measuring the voltage across the wire and the current through it. The temperature change in time results from the change in the wire's resistance. Along with other correction terms, the finite heat capacity of the wire must, of course, be considered.

Optical measuring methods (especially interference methods) lend themselves especially well to these applications since the measurements are taken without inertia and without disturbing the test object. The following example has been taken from Bach (*91*).

a. The Mathematical Model of the Nonstationary Method. In order to derive mathematical expressions for the nonstationary method, it is convenient to consider a semi-infinite plate as a heat source surrounded by the medium to be investigated. The plate is electrically heated at the time $t = 0$; the heat flux q_w at the surface is constant.

The definition of k_c [Eq. (118)], $q = -k_c \, \partial \vartheta/\partial y$, satisfies Fourier's differential equation for nonstationary heat conduction:

$$\partial q/\partial t = a \, \partial^2 q/\partial y^2 \tag{119}$$

where $a = k_c/c_p \cdot \rho$ is the thermal diffusivity (containing the desired quantity), c_p the specific heat at constant pressure, ρ the density, t the time, $\Delta\vartheta$ the temperature difference relative to surroundings ($\Delta\vartheta = 0$ for $t = 0$), and y a coordinate normal to the plate. For the boundary condition $q_w = \text{const}$ at $t > 0$ the heat flux q results as a solution of the differential equation (119):

$$q = q_w \, \text{erfc}[y/2(at)^{1/2}] \tag{120}$$

Integration of the above results in

$$\Delta\vartheta = \frac{q_w}{k_c} \int_y^\infty \text{erfc}\left(\frac{y}{2(at)^{1/2}}\right) dy$$

$$\Delta\vartheta = \frac{2q_w}{k_c}\left[\left(\frac{at}{\pi}\right)^{1/2} \cdot \exp\left(-\frac{y^2}{4at}\right) - \frac{y}{2}\,\text{erfc}\left(\frac{y}{2(at)^{1/2}}\right)\right] \tag{121}$$

or, in abbreviated notation,

$$\Delta\vartheta = (2q_w/k_c)(at)^{1/2} \cdot \text{ierfc}[y/2(at)^{1/2}]$$

where $\text{ierfc}(y)$ is the integrated function $\text{erfc}(y)$,

$$\text{ierfc}(y) = \frac{1}{\sqrt{\pi}} \exp(-y^2) - y \cdot \text{erfc}(y) = \int_y^\infty \text{erfc}(\xi)\, d\xi$$

The function $\text{erfc}(y)$ is connected with the Gaussian error function $\text{erf}(y)$ by

$$\text{erfc}(y) = 1 - \text{erf}(y) = 1 - \frac{2}{\sqrt{\pi}} \int_0^y \exp(-\xi^2)\, d\xi$$

The surface temperature $\Delta\vartheta_w$ for $y = 0$ results from Eq. (121):

$$\Delta\vartheta_w = (2q_w/k_c)(at/\pi)^{1/2} \tag{121a}$$

The temperature gradient at this point ($y = 0$) is

$$(\partial\vartheta/\partial y)_w = -q_w/k_c$$

b. Apparatus. A measuring apparatus which meets the theoretical requirements to a large extent usually contains an electrically heated, tautly suspended piece of platinum foil [see also (*91*)]. The medium is separated by the foil into two adjacent "semi-infinite regions." Heat energy is transferred in equal parts to both of these, where the heat capacity of the foil can be neglected. However, the foil is not sufficiently flat for interferometric measurements.

In this case a thin layer of chromium (thickness 0.1 μm) vaporized on a plane glass plate is used as a heater element. The heat flux at the boundaries of the two differing semi-infinite regions is now distributed according to the relation $(\rho_1 k_{c1} c_{p1}/\rho_2 k_{c2} c_{p2})^{1/2}$ which remains constant.

The condition $q_w = $ const is not violated in the medium. In order to suppress convection, the glass plate (suspended in the fluid; in this case water at standard pressure) is so oriented that the heating layer B faces downward. The y axis, the heat flux q, and gravity have the same direction. Convection does not occur because of the stable stratification of the thermal layers. Since the glass plate C is of finite thickness δ, it cannot be considered to be a semi-infinite object for long measuring times. The measurement cannot be extended beyond the point where the temperature field begins to form in Region D of the fluid, i.e., when the first interference fringe becomes visible. Constructional details of the test chamber can be seen in Fig. 79.

The complete apparatus is placed in a MZI and well isolated thermally.

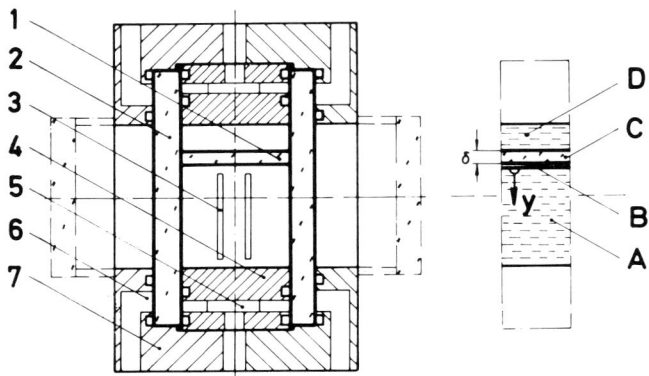

FIG. 79. Test Chamber: 1, glass plate with the heating layer and a scale etched on the end plane; 2, windows, consisting of plane parallel ground-glass plates of high optical quality; 3, cylindrical inner chamber ($d = 56$ mm) containing the test fluid and the thermocouples; 4, nickel-plated copper jacket; 5, circulating channel for the heating fluid; 6, heating channel for the flange; 7, flanges with additional windows (shown schematically), to prevent thermal leakage from the ends.

A compensating chamber in the reference beam is omitted in this case so that the optical path difference (about $135{,}000 \cdot \lambda$, where $\lambda = 0.6328 \cdot 10^{-6}$ m) must be compensated, e.g., with a highly coherent laser light source.

In the exposure in Fig. 80 (exposure time 1/500 sec) interference lines

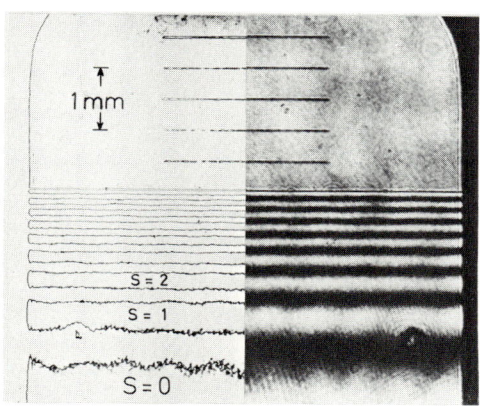

FIG. 80. Cutout of an interference pattern of a nonstationary temperature field in water (infinite fringe field): right half, interference lines $\vartheta =$ const; left half, pattern of equidensometric lines, corresponding to the mirror image of the right half. Photograph by J. Bach.

are superposed on the desired pattern. Furthermore, diffraction patterns stemming from dust particles are distinctly visible. These effects are characteristic of highly coherent laser illumination and difficult to avoid. The evaluation of this somewhat indistinct interference pattern (Fig. 80, right) can often be facilitated by equidensometric procedures. The center of the interference fringes can be determined more precisely by this means.

The extremly narrow interference line nearest to the wall is influenced by diffraction at the wall. This line is not considered in the evaluation of the pattern. In the top half of the picture the glass plate supporting the layer and the etched scale can be seen. The scale serves to determine both the exact position of the wall and the magnification scale of the exposure.

c. Evaluation. Pertinent data for an example are compiled in Table XIII. In this case the test fluid was water at room temperature; the coordinates y of the temperature field were determined photometrically.

The above values for y are given in their original magnitude. The temperatures $\Delta \vartheta$ are derived from the path difference of the interference

TABLE XIIIA

EXPERIMENTAL DATA FOR FIG. 80

$\vartheta_0 = 20.173 \pm 0.003°C$	Temperature of the fluid at the beginning
$t = 8.83$ sec	Elapsed time after switching on the heater
$l = 39.4$ mm	Length of the heating layer in the direction of the light ray
$\lambda = 0.6328 \cdot 10^{-6}$ m	Light wavelength

TABLE XIIIB

TEMPERATURE FIELD (FIG. 80)

S (—)	y (mm)	y (mm)	Δy (mm)	$\Delta\vartheta$ (°C)
3	1.053	1.060	−0.007	0.538
3.5	0.913	0.911	+0.002	0.627
4	0.780	0.778	+0.002	0.715
4.5	0.658	0.657	+0.002	0.803
5	0.543	0.545	−0.002	0.891
5.5	0.440	0.441	−0.001	0.979
6	0.350	0.343	+0.007	1.066
6.5	0.252	0.251	+0.001	1.153
7	0.160	0.164	−0.004	1.239

fringes, considering the temperature dependence of the refractive index of water. (The table VII is not applicable in this case because another wavelength is used.)

During the total measuring time (10 sec maximum), the temperature field is photographed at intervals of 0.2–0.3 sec. The heat flux rates are of the magnitude 0.03–0.12 W/cm², corresponding to a total heating power of 1–2 W. The maximum temperature difference is 1.5 deg so that the physical properties of water, a, c_p, ρ, and k_c, are assumed to be constant.

It would suffice, however, to derive the constants in Eq. (121) from a single photograph:

$$\Delta\vartheta = (2q_w/k_c)(at_n)^{1/2} \text{ierfc}[y/2(at_n)^{1/2}]$$

The thermal diffusivity $a = k_c/c_p\rho$ is the desired quantity. Evaluation of Table XIII results in corresponding values of $\Delta\vartheta$ and y at a specific time $t_n = $ const. The heat flux at the wall q_w, which also appears in Eq. (121), remains constant. It is not necessary to determine q_w by measuring the heating power and calculating the portion transferred to the water. As already mentioned, this is an advantage of optical methods. In Table XIIIB, the measured temperature distribution is shown (first column of y-values). The temperature distribution calculated from the determined value of the thermal diffusivity is also shown for comparison (second column of y-values). The difference Δy of these two values is tabulated to give an idea of the accuracy required in this measurement.

The constants $A = q_w/k_c$ and $B = 2(at)^{1/2}$ are determined by a least-square curve fitting procedure:

$$\Delta\vartheta_i - \Delta\vartheta_{in} = AB \text{ ierfc}(y_i/B) - \Delta\vartheta_{in}$$

$\Delta\vartheta_i$ and y_i are corresponding quantities of the optimized function; $\Delta\vartheta_{in}$ and y_{in} are measured values. The deviation f is to be minimum:

$$f = \sum_{i=1}^{m} [AB \text{ ierfc}(y_i/B) - \Delta\vartheta_{in}]^2$$

By differentiating with respect to A and B and eliminating A the following results:

$$\sum_{i=1}^{m} \Delta\vartheta_{in} \exp[-(y_i/B)^2]\Big/B \cdot \sum_{i=1}^{m} \text{ierfc}(y_i/B) \cdot \exp[-(y_i/B)^2]$$

$$- \sum_{i=1}^{m} \Delta\vartheta_{in} \text{ ierfc}(y_i/B)\Big/B \cdot \sum_{i=1}^{m} [\text{ierfc}(y_i/B)]^2 = 0 \qquad (122)$$

Evaluation of Eq. (122) with the aid of a computer results in

$$A = 1110.861 \quad \text{deg/m}$$

$$B = 0.0022570 \quad \text{m}$$

Since $B = 2(at)^{1/2}$, the thermal diffusivity $a = 1.442 \cdot 10^{-7}$ m²/sec. The reference temperature 20.98°C is the arithmetic mean of the initial temperature and the temperature of the wall.

Other evaluation methods besides the method discussed above are conceivable, e.g., evaluation along an isotherm [Fig. (81)]. Corresponding values of y, t_n and y, $t_n + 1$ are derived from separate exposures. The desired quantity a results from Eq. (121) by optimizing the following in an iterative process:

$$t_n^{1/2} \text{ ierfc}(y_n/2(at_n)^{1/2}) = t_{n+1}^{1/2} \text{ ierfc}[y_{n+1}/2(at_{n+1})^{1/2}]$$

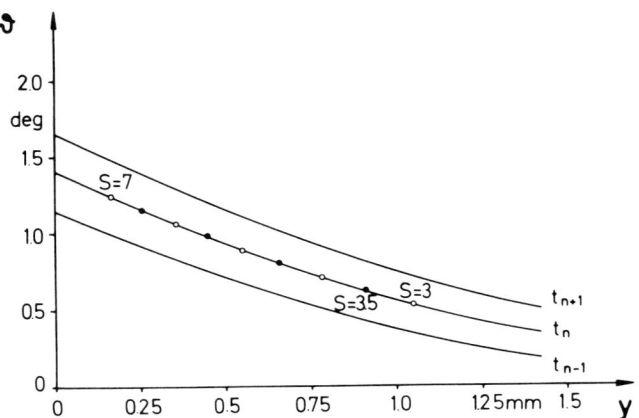

FIG. 81. Nonstationary temperature field. The curve $t_n = 8.83$ sec corresponds to the example discussed in the text. $t_{n+1} = 12.0$ sec, $t_n = 8.83$ sec (meas.), $t_{n-1} = 4.5$ sec.

This method is quite sensitive regarding measuring errors and is inferior in this respect to the first. On the other hand, it can effectively be checked if the heat transfer and the physical properties remain constant during the measurement.

d. *Measuring Accuracy.* Values k_c, obtained by the nonstationary method described above, coincide to an accuracy of $\pm 1\%$ with values of calorimetric measurements. Due to the weak temperature gradient, corrections of the interferogram are very small.

3. *Heat Radiation in Horizontal Liquid Layers*

a. *Test Model.* In order to investigate the contribution of radiation to heat transfer through horizontal liquid layers, a test chamber was used which was, to a high degree free of convection (85, 92). It consists of a glass frame of interferometric quality which is closed from above by a heated and from below by a cooled solid silver plate forming the gap δ_g for the liquid layer.

The temperature difference between both plates was measured by means of thermocouples in connection with a Diesselhorst compensator to a precision of ± 0.004 deg.

Several possible causes of convection in the liquid layer have been investigated, both experimentally and by calculation of temperature fields and streamline fields in the liquid layer. The effect of inclination of the test chamber by, e.g., 0.0025 radiants with respect to the exact horizontal position, can be observed by a slight deformation of the interference fringes. This is caused by weak rollerlike convection in the test chamber. However, an increase in the heat flow rate between both plates can be detected only for inclinations about ten times greater. The angular adjustment by autocollimation methods can be performed to a precision of 0.0002 radiants.

b. The Interference Pattern and Its Evaluation. If conduction alone occurs in the test chamber, the interferogram of the resulting linear temperature distribution would consist of equidistant, parallel interference fringes in the infinite fringe position of the MZI, provided that *dn/dT* is constant.

In order to explain tentatively the interferograms of Fig. 82, such a linear temperature profile forming a virtual wedge is assumed instead of the real nonlinear distribution. This virtual wedge can be compensated for by another virtual wedge of the same size but opposed inclination, as produced in a fringe field position of the MZI (cf. Sections V,B,3 and V,E,1).

Thus, in the case of the linear temperature distribution, an "infinite fringe" field results without fringes. In the real case departures from the linear temperature distribution are now seen as "differences," i.e., a few interference fringes. We can now go a step further and superpose an additional fringe field (vertical fringes). Any departure from a linear distribution of the refractive index (temperature) now leads to a deflection of the vertical fringes, as shown in Fig. 82. This is an indication of the contribution of radiation to heat transfer in the fluid. Fluids with a high coefficient of absorption, such as water, methanol, ethanol, or propanol, do not exhibit this effect: The fringes remain vertical, provided the gradient of the refractive index *dn/dT* is constant. The superposed fringe field was chosen in order to clarify the qualitative form of the temperature distribution. For quantitative evaluation, interferograms in the infinite fringe field without compensation were taken.

Evaluation: The proportion of heat transfer due to radiation versus the proportion due to conduction is of interest. The proportion due to radiation is usually dependent on the infrared absorption spectrum of the liquid, the width of the layer δ_g, and the emissivities ϵ_{rad} of the bounding walls.

The total rate of heat flux due to both conduction and radiation is constant in any cross section of the layer:

$$q_{tot} = q_c + q_{rad} = \text{const}$$

The conductive heat flux q_c is proportional to the local temperature

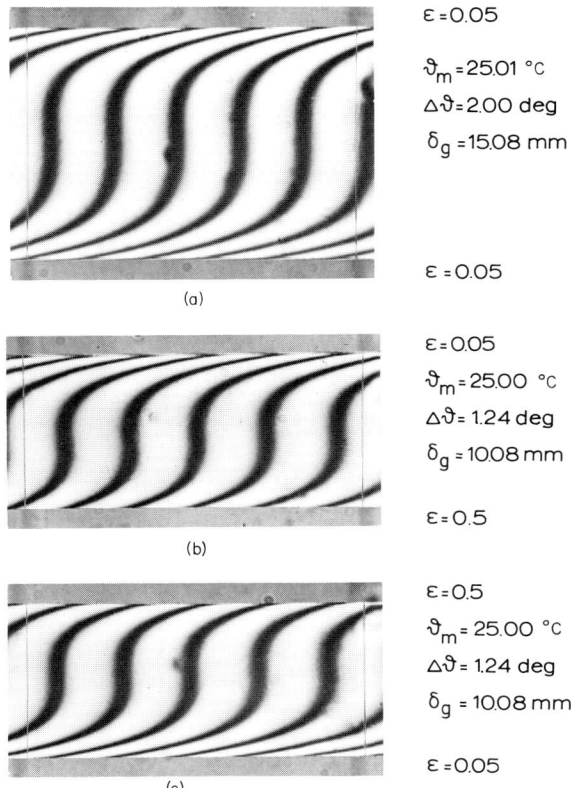

FIG. 82. Interferogram of the temperature field in a layer of carbon tetrachloride (fringe field with a compensating virtual wedge): (a) the nonlinear temperature distribution $\vartheta(y)$, qualitatively indicated by the shape of the interference fringes, is caused by heat conduction and radiation; (b) the influence of different emissivities ($\epsilon_{rad} = 0.5$, $\epsilon_{rad} = 0.05$) of the heated and the cooled plate at the top and the bottom of the liquid layer is shown ($\delta_g = 10$ mm); (c) this interferogram corresponds to (b) except that the cuvette is turned upside down. Photograph by G. Schödel.

gradient $d\vartheta/dy$, which is practically proportional to the local fringe density $1/b$:

$$q_c = -k_c(d\vartheta/dy)$$

$$\frac{d\vartheta}{dy} = \frac{d\vartheta}{dS} \cdot \frac{dS}{dy} = \frac{\lambda}{l} \cdot \frac{dT}{dn} \cdot \frac{1}{b} \quad \left(\frac{dS}{dy} = \frac{1}{b}\right)$$

Because only the quotient q_{rad}/q_c is of interest, the interferogram is evaluated only with respect to the smallest value of conduction

$q_{cm} \sim 1/b_m$, which occurs in the middle (here, subscript m denotes the middle of the layer):

$$q_c/q_{cm} = b_m/b$$

since dn/dT is constant in the region of measurement ($\Delta\vartheta$) for this case. The evaluation of a carbon tetrachloride layer is shown in Fig. 83.

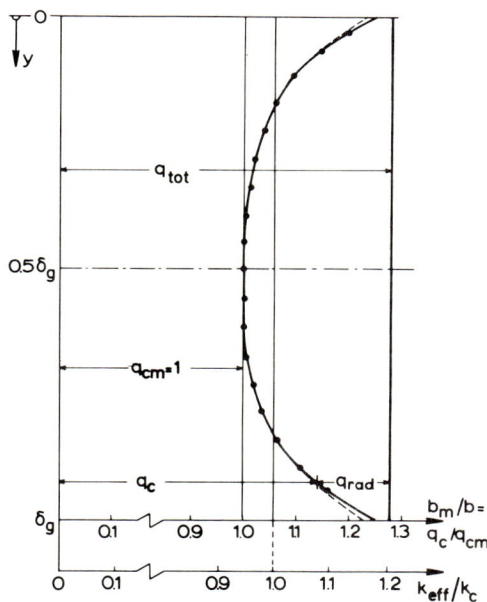

FIG. 83. Distribution of the radiation fraction q_{rad} relative to the fraction due to convection as a function of the coordinate in the test chamber ($0 \leqslant y \leqslant \delta_g$).

The height of the layer is 15 mm and the emissivity of both walls is the same ($\epsilon_{rad} = 0.05$). The quotient of the proportions of conduction $q_c/q_{cm} = b_m/b$ in the middle is defined to be unity and attains its maximum value at the walls. (In this example the quotient $q_{cw}/q_{cm} \approx 1.27$.) The form of the curves q_c/q_{cm} does not depend on the temperature difference between the walls as long as this difference is very small compared to the absolute temperature. The proportion of radiation at the wall is estimated to be 1 % of the total heat flux (85) and can be plotted in the diagram (line q_{tot}). Finally, in order to determine the value $k_{eff}/k_c = (q_c + q_{rad})/q_c = q_{tot}/q_c$, the integral mean value of the flux due to conduction in the liquid layer in Fig. 83 can be assumed to be unity (equal areas). The result is $k_{eff}/k_c = 1.21$.

c. *Accuracy of Measurements.* In the evaluation, the error of the value q_{cw} occurring at the wall, which can only be found by extrapolation, contributes decisively to the error of the quotient k_{eff}/k_c. If we consider the dotted curve in Fig. 83, the result of k_{eff}/k_c is 3% higher. In (*85*), the influence of radiation in carbon tetrachloride was not only determined interferometrically but also with the aid of absorption measurements and caloric measurements. For the quotient k_{eff}/k_c in the case of small emissivities of the walls ($\epsilon_{rad} = 0.05$), the following values, dependent on the height of the layer, have been found:

TABLE XIV

k_{eff}/k_c OF CARBON TETRACHLORIDE

Height of the layer (mm)	Absorption spectrum	Interferogram	Caloric measurements
5	1.09	1.07	1.12
10	1.16	1.15	
15	1.23	1.21	1.25
20	1.28	1.23	1.29
100	1.75		
500	2.16		
∞	2.33		

4. *Thermal Diffusion in a Mixture of Two Fluids*

The test chamber described in the above example was used in order to demonstrate thermal diffusion in a mixture of equal parts of n-hexane and c-hexane, Fig. 84.

a. t = 0. The homogenic mixture is still in an isothermal state. Since this experiment is conducted over a long period of time a weak, virtual wedge (fringe field; four interference fringes) is used in order to be able to check the adjustment of the MZI at the determination of the experiment.

b. t = 20 min. A constant temperature difference $\Delta\vartheta = 1.90°C$ is applied at $t = 0$ which develops a profile similar to the above example.

The number of interference fringes corresponds to the temperature difference $\Delta\vartheta$. The fact that the fringe density is greater at the wall is due to heat radiation within the liquid.

c–e. t = 100 min; t = 150 min; t = 390 min. Due to the influence of the temperature gradient, the components separate: The concentra-

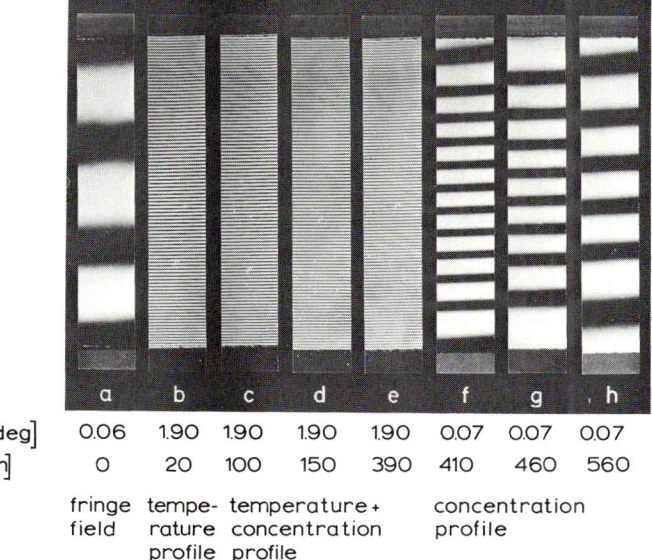

FIG. 84. Thermal diffusion in a horizontal liquid layer in stable stratification. Exposures taken at several times showing small cutout of the rectangular slit. (Photograph by G. Schödel.)

tion of the heavier n-hexane increases near the cold wall, while the concentration of the lighter c-hexane increases at the heated wall. The profile becomes less distinct (the fringe width being approximately uniform in *e*). The deviations from the linearity in the temperature distribution which are due to heat radiation within the liquid are equalized in part by the concentration profile.

f. t = 390 min. The temperature difference is eliminated and isothermal conditions are reached again after about 10 min. The remaining interference fringes stem from the concentration profile and from the virtual wedge adjusted at the start. Afterwards, the concentration profile in the isothermal state of the fluid vanishes, so that finally only the weak virtual wedge remains.

5. *Convection Current in a Horizontal Rectangular Slit*

In this case the same cuvette as in the two previous examples (Sections VI,B,3 and VI,B,4) is also utilized in an experiment to compare the convection in a rectangular slit with the results of a finite step calculation performed by Churchill (93). The temperature fields shown in Fig. 85 are quite similar although the dimensions of the slit in the experiment and in the calculation differ somewhat.

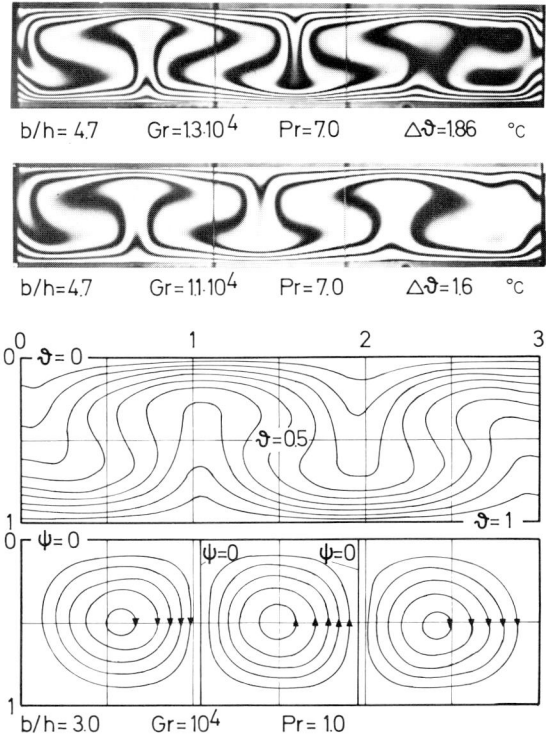

FIG. 85. Rollerlike convection current in a horizontal rectangular slit. Upper half: interferogram (infinite fringe field) of a rectangular slit heated from below and cooled from above; temperature field at a rollerlike convection current (two-dimensional problem); fluid, water. Lower half: temperature field ϑ and streamline field ψ of a strictly two-dimensional chamber for comparison. [Numerical calculation according to Churchill (93).] b_g/h_g: ratio of width to height of the slit. (Photograph by W. Hauf.)

While the Grashof numbers are approximately equal, the Pr numbers and especially the quotient b_g/h_g differ. The size of the convection "rollers" is highly dependent on the height of the slit h_g so that a remarkable agreement between the experiment and the calculation exists.

Further experimental deviations from the calculated model are as follows: There are nonadiabatic side walls of the slit where a linear temperature drop exists. The isotherms (interference lines) are not perpendicular to the wall. The problem is not two dimensional because of the finite length of the model ($l = 0.05$ m).

6. Separation of a Boundary Layer from a Horizontal Abruptly Heated Wire (Fig. 86)

Top: A cylindrical boundary layer is formed at the abruptly heated wire. The first signs of turbulence can already be seen.

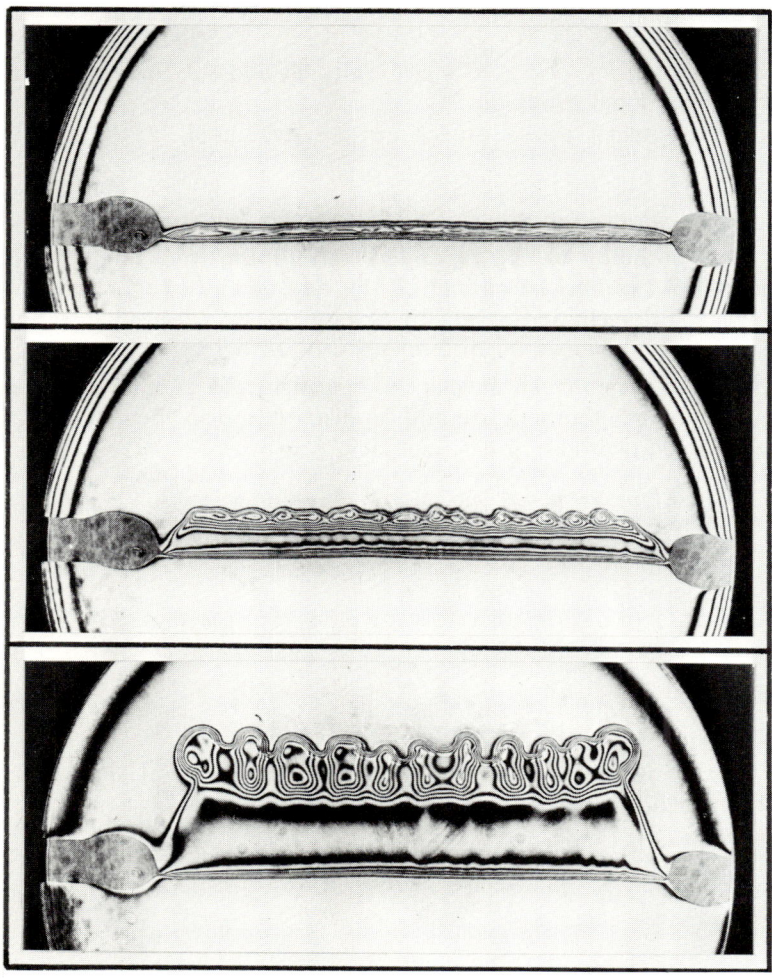

Fig. 86. Phases of the separation process of a boundary layer in the region of transient convection (infinite fringe field): diameter of the cuvette, $d = 31$ mm; heated wire, platinum, 0.1 mm; heat flux at the surface of the wire, 12 W/cm²; test fluid, Freon (CFCl$_3$) near the saturation point ($\vartheta_{\text{sat}} = 23.8°\text{C}$; $P_{\text{sat}} = 760$ Torr); ambient condition, $\vartheta_\infty = 22°\text{C}$, $P = 720$ Torr; time difference between individual exposures, $t \simeq 0.1$ sec (from top to bottom). (Photograph by W. Hauf.)

Center: After attaining a certain size, the boundary layer separates from the wire and drifts off toward the top. The already existing convection cells increase in size. This boundary region is further supplied by heated fluid which flows upward in a thin filmlike zone. The wire is surrounded by a stable boundary layer since the fluid does not reach its boiling point at this heat flux.

Bottom: The convection cells are fully developed. Their rotational motion slows down gradually.

This "irregular" phase object can be evaluated in the following manner:

(1) Mean periodic spacings of the convection cells can be specified.

(2) The enthalpy of the separated part of the transient convection boundary layer can be determined. If the volume of the separated boundary layer is determined—e.g., by means of an additional schlieren exposure from above—a mean temperature can be given [cf. Sect. V,D].

(3) Parts of the convection cells can be considered as parts of spherical phase objects; the temperature and its distribution in the cells can then be determined, without great accuracy, however.

VII. Appendix

A. Mathematical Formulation of the Conditions for a Basic Position of the Mach–Zehnder interferometer (Section V,B)

1. *Rotation Matrix*

(a) According to Eq. (64a), the respective matrices of the reference and the measuring paths must be equal.

$$\bar{D} = (d_{ij}) = (\bar{E} - 2\bar{M}_2)(\bar{E} - 2\bar{M}_1)$$

signifies a rotation matrix corresponding to a double-mirror reflection:

$$\bar{D} = \bar{D}'$$

or

$$(\bar{E} - 2\bar{M}_2)(\bar{E} - 2\bar{M}_1) = (\bar{E} - 2\bar{M}_2')(\bar{E} - 2\bar{M}_1')$$

(b) $\bar{E}$ is the unit matrix. According to Eq. (61a), $\bar{M}$ characterizes the mirror reflection of an arbitrary position vector **v** at a plane with the unit normal vector $\mathbf{m}_i = (u_i\,;\,v_i\,;\,w_i)$:

$$\bar{E} = \begin{pmatrix} 1 & 0 & 0 \\ 0 & 1 & 0 \\ 0 & 0 & 1 \end{pmatrix} \qquad \bar{M} = \begin{pmatrix} u^2 & uv & uw \\ vu & v^2 & vw \\ wu & wv & w^2 \end{pmatrix}$$

$$(\bar{E} - 2\bar{M}) = \begin{pmatrix} 1 - 2u^2 & -2uv & -2uw \\ -2vu & 1 - 2v^2 & -2vw \\ -2wu & -2wv & 1 - 2w^2 \end{pmatrix}$$

(c) According to the rules of matrix multiplication ($c_{ik} = \sum_n a_{in}b_{nk}$), the individual matrix elements of the rotation matrix result:

$$\bar{D} = (d_{ij}) = (\bar{E} - 2\bar{M}_2)(\bar{E} - 2\bar{M}_1):$$

$d_{11} = (1 - 2u_2^2)(1 - 2u_1^2) + 4u_2v_2u_1v_1 + 4w_2u_2w_1u_1$

$d_{12} = -2(1 - 2u_2^2)u_1v_1 - 2u_2v_2(1 - 2v_1^2) + 4w_2u_2w_1v_1$

$d_{13} = -2(1 - 2u_2^2)u_1w_1 + 4u_2v_2v_1w_1 - 2w_2u_2(1 - 2w_1^2)$

$d_{21} = -2v_2u_2(1 - 2u_1^2) - 2(1 - 2v_2^2)u_1v_1 + 4v_2w_2w_1u_1$

$d_{22} = +4v_2u_2u_1v_1 + (1 - 2v_2^2)(1 - 2v_1^2) + 4v_2w_2w_1v_1$

$d_{23} = +4v_2u_2u_1w_1 - 2(1 - 2v_2^2)v_1w_1 - 2v_2w_2(1 - 2w_1^2)$

$d_{31} = -2w_2u_2(1 - 2u_1^2) + 4w_2v_2u_1v_1 - 2(1 - 2w_2^2)w_1u_1$

$d_{32} = +4w_2u_2u_1v_1 - 2w_2v_2(1 - 2v_1^2) - 2(1 - 2w_2^2)w_1v_1$

$d_{33} = -4w_2u_2w_1u_1 + 4w_2v_2v_1w_1 + (1 - 2w_2^2)(1 - 2w_1^2)$

After inserting the components of the mirror normal vectors $\mathbf{m}_i$ [Eq. (60d)] the elements of $\bar{D}$ and $\bar{D}'$ result:

$$u_i = -\psi_i\,; \qquad v_i = -\cos\theta + \varphi_i\sin\theta; \qquad w_i = \sin\theta + \varphi_i\cos\theta$$

Terms of second and higher order are neglected.

(d) $\bar{D} = (\bar{E} - 2\bar{M}_2)(\bar{E} - 2\bar{M}_1):$ $\qquad \bar{D}' = (\bar{E} - 2\bar{M}_2')(\bar{E} - 2\bar{M}_1'):$

$d_{11} = 1 \qquad\qquad\qquad\qquad\qquad\qquad d'_{11} = 1$

$d_{12} = 2\cos\theta\,(\psi_2 - \psi_1) \qquad\qquad\quad d'_{12} = 2\cos\theta\,(\psi_2' - \psi_1')$

$d_{13} = -2\sin\theta\,(\psi_2 - \psi_1) \qquad\qquad d'_{13} = -2\sin\theta\,(\psi_2' - \psi_1')$

$d_{21} = -2\cos\theta\,(\psi_2 - \psi_1) \qquad\qquad d'_{21} = -2\cos\theta\,(\psi_2' - \psi_1')$

$d_{22} = 1 \qquad\qquad\qquad\qquad\qquad\qquad d'_{22} = 1$

$$d_{23} = 2(\varphi_2 - \varphi_1) \qquad d'_{23} = 2(\varphi_2' - \varphi_1')$$
$$d_{31} = 2\sin\theta\,(\psi_2 - \psi_1) \qquad d'_{31} = 2\sin\theta\,(\psi_2' - \psi_1')$$
$$d_{32} = -2(\varphi_2 - \varphi_1) \qquad d'_{32} = -2(\varphi_2' - \varphi_1')$$
$$d_{33} = 1 \qquad d'_{33} = 1$$

Comparison of d_{ij} and d'_{ij} yields Eqs. (65a) and (65b):

$$(\varphi_1 - \varphi_2) = (\varphi_1' - \varphi_2') \quad \text{and} \quad (\psi_1 - \psi_2) = (\psi_1' - \psi_2')$$

2. Displacement Vector

(a) As a consequence of Eq. (64b), the displacement vectors $\mathbf{r} = (r_1, r_2, r_3)$ and $\mathbf{r}'$ must be equal:

$$\mathbf{r} = \mathbf{r}'$$
$$2(\bar{E} - \bar{M}_2)\,\bar{M}_1 \mathbf{q}_1 + 2\bar{M}_2 \mathbf{q}_2 = 2(\bar{E} - \bar{M}_2')\,\bar{M}_1 \mathbf{q}_1' + 2\bar{M}_2' \mathbf{q}_2'$$

(b) If $\mathbf{m}_i = (u_i, v_i, w_i)$ signifies the mirror normal vector, $\bar{M}$ can again be written as

$$\bar{M}_i = \begin{pmatrix} u_i^2 & u_i v_i & u_i w_i \\ v_i u_i & v_i^2 & v_i w_i \\ w_i u_i & w_i v_i & w_i^2 \end{pmatrix}$$

$$(\bar{E} - 2\bar{M}_2)\bar{M}_1 = \bar{M}_1 - 2\bar{M}_2 \bar{M}_1 = \bar{B} = (b_{ij})$$
$$\bar{M}_2 \bar{M}_1 = \bar{A} = (a_{ij})$$

(c) The elements of the matrix $\bar{A} = (a_{ij})$ are

$$a_{11} = u_1^2 u_2^2 + u_2 v_2 u_1 v_1 + u_2 w_2 u_1 w_1$$
$$\qquad = u_2 u_1(u_1 u_2 + v_1 v_2 + w_1 w_2)$$
$$a_{12} = u_2 v_1(u_1 u_2 + v_1 v_2 + w_1 w_2)$$
$$a_{13} = u_2 w_1(u_1 u_2 + v_1 v_2 + w_1 w_2)$$
$$a_{22} = v_2 v_1(u_1 u_2 + v_1 v_2 + w_1 w_2)$$
$$a_{23} = v_2 w_1(u_1 u_2 + v_1 v_2 + w_1 w_2)$$
$$a_{33} = w_2 w_1(u_1 u_2 + v_1 v_2 + w_1 w_2)$$

The elements of the matrix $\bar{B} = (b_{ij})$ are

$$B = \begin{pmatrix} u_1^2 & u_1 v_1 & u_1 w_1 \\ v_1 u_1 & v_1^2 & v_1 w_1 \\ w_1 u_1 & w_1 v_1 & w_1^2 \end{pmatrix} - 2(u_1 u_2 + v_1 v_2 + w_1 w_2)\begin{pmatrix} u_1 u_2 & v_1 u_2 & w_1 u_2 \\ u_1 v_2 & v_1 v_2 & w_1 v_2 \\ u_1 w_2 & v_1 w_2 & w_1 w_2 \end{pmatrix}$$

(d) Inserting the components of the mirror normal vector and neglecting higher-order terms yields

$$u_i = -\psi_i; \quad v_i = -\cos\theta + \varphi_i \sin\theta; \quad w_i = \sin\theta + \varphi_i \cos\theta$$

$$(u_1 u_2 + v_1 v_2 + w_1 w_2) = 1$$

$\bar{B} = \bar{M}_1 - 2\bar{M}_2 \bar{M}_1 \qquad\qquad \bar{B}' = \bar{M}_1' - 2\bar{M}_2' \bar{M}_1'$

$b_{11} = \psi_1(\psi_1 - 2\psi_2) \qquad\qquad b_{11}' = \psi_1'(\psi_1' - 2\psi_2')$

$b_{12} = \cos\theta\,(\psi_1 - 2\psi_2) \qquad\qquad b_{12}' = \cos\theta\,(\psi_1' - 2\psi_2')$

$b_{13} = -\sin\theta\,(\psi_1 - 2\psi_2) \qquad\qquad b_{13}' = -\sin\theta\,(\psi_1' - 2\psi_2')$

$b_{21} = -\psi_1 \cos\theta \qquad\qquad b_{21}' = -\psi_1' \cos\theta$

$b_{22} = -\cos^2\theta + 2\varphi_2 \sin\theta \cos\theta \qquad b_{22}' = -\cos^2\theta + 2\varphi_2' \sin\theta \cos\theta$

$b_{23} = \sin\theta \cos\theta + \varphi_1 - 2\varphi_2 \sin^2\theta \qquad b_{23}' = \sin\theta \cos\theta + \varphi_1' - 2\varphi_2' \sin^2\theta$

$b_{31} = \psi_1 \cdot \sin\theta \qquad\qquad b_{31}' = \psi_1' \sin\theta$

$b_{32} = \sin\theta \cos\theta - \varphi_1 + 2\varphi_2 \cos^2\theta \qquad b_{32}' = \sin\theta \cos\theta - \varphi_1' + 2\varphi_2' \cos^2\theta$

$b_{33} = -\sin^2\theta - 2\varphi_2 \sin\theta \cdot \cos\theta \qquad b_{33}' = -\sin^2\theta - 2\varphi_2' \sin\theta \cos\theta$

(e) Calculation of $\mathbf{d} = \bar{B} \cdot \mathbf{q}_1$, where $\mathbf{q}_1$ and $\mathbf{q}_1'$ are the position vectors of the mirror mid-points [Eq. (60)] follows:

$$\mathbf{q}_1 = (0;\ a \cdot \sin 2\theta - e_1 \cos\theta;\ q_2 + a\cos 2\theta + e_1 \sin\theta)$$

$\mathbf{d} = (\bar{E} - \bar{M}_2)\,\bar{M}_1 \mathbf{q}_1 \qquad\qquad \mathbf{d}' = (\bar{E} - \bar{M}_2')\,\bar{M}_1' \mathbf{q}_1'$

$d_1 = (\psi_1 - 2\psi_2)(a - q_2)\cdot \sin\theta \qquad d_1' = (\psi_1' - 2\psi_2')(a - q_2')\sin\theta$

$d_2 = 2a\varphi_2 \sin^2\theta + e_1 \cos\theta \qquad d_2' = 2a\varphi_2' \sin^2\theta + e_1' \cos\theta - a\sin\theta$

$\quad - a\sin\theta\cos\theta \qquad\qquad\qquad \cdot \cos\theta + q_2'(\cos\theta \sin\theta + \varphi_1'$

$\quad + q_2(\cos\theta \sin\theta + \varphi_1 \qquad\qquad - 2\varphi_2' \sin\theta) + a\varphi_1'(\cos^2\theta - \sin^2\theta)$

$\quad - 2\varphi_2 \sin\theta) + a\varphi_1$

$\quad \cdot (\cos^2\theta - \sin^2\theta)$

$d_3 = a\sin^2\theta - e_1 \sin\theta \qquad d_3' = a\sin^2\theta - e_1' \sin\theta + 2a\sin\theta \cos\theta$

$\quad + 2a\sin\theta \cos\theta\,(\varphi_2 - \varphi_1) \qquad\qquad \cdot (\varphi_2' - \varphi_1') - q_2'(\sin^2\theta$

$\quad - q_2(\sin^2\theta + 2\varphi_2 \cos\theta \sin\theta) \qquad\qquad + 2\varphi_2' \cos\theta \sin\theta)$

(f) Calculation of $\mathbf{f} = \overline{M}_2 \cdot \mathbf{q}_2$ using $\mathbf{q}_2$ and $\mathbf{q}_2'$, respectively follows:

$$\mathbf{q}_2(0;\ -e_2 \cos \theta;\ q_2 + e_2 \sin \theta)$$

$\mathbf{f} = \overline{M}_2 \mathbf{q}_2$ $\qquad\qquad\qquad\qquad$ $\mathbf{f}' = \overline{M}_2 \mathbf{q}_2'$

$f_1 = -\psi_1 q_2 \sin \theta$ $\qquad\qquad$ $f_1' = -\psi_1' q_2' \sin \theta$

$f_2 = \varphi_2 q_2 (\sin^2 \theta - \cos^2 \theta)$ $\qquad$ $f_2' = \varphi_2' q_2' (\sin^2 \theta - \cos^2 \theta)$

$\qquad - q_2 \sin \theta \cos \theta - e_2 \cos \theta$ $\qquad - q_2' \sin \theta \cos \theta - e_2' \cos \theta$

$f_3 = e_2 \sin \theta + q_2 \sin^2 \theta$ $\qquad\quad$ $f_3' = e_2' \sin \theta + q_2' \sin^2 \theta$

$\qquad + 2\varphi_2 q_2 \sin \theta \cos \theta$ $\qquad\qquad + 2\varphi_2' q_2' \sin \theta \cos \theta$

(g) Combining **d** and **f** results in the displacement vector **r**:

$\mathbf{r} = 2(\mathbf{d} + \mathbf{f})$ $\qquad\qquad\qquad$ $\mathbf{r}' = 2(\mathbf{d}' + \mathbf{f}')$

$r_1 = 2 \sin \theta [(\psi_1 - 2\psi_2) a$ $\qquad$ $r_1' = 2 \sin \theta [(\psi_1' - 2\psi_2') a$

$\qquad + (\psi_2 - \psi_1) q_2]$ $\qquad\qquad + (\psi_2' - \psi_1') q_2']$

$r_2 = 2[\cos \theta (e_1 - e_2) + q_2$ $\qquad$ $r_2' = 2[\cos \theta (e_1' - e_2')$

$\qquad \cdot (\varphi_1 - \varphi_2) + 2a\varphi_2 \sin^2 \theta$ $\qquad + q_2'(\varphi_1' - \varphi_2') + 2a\varphi_2' \sin^2 \theta$

$\qquad + a\varphi_1 (\cos^2 \theta - \sin^2 \theta)$ $\qquad + a\varphi_1' (\cos^2 \theta - \sin^2 \theta)$

$\qquad - a \sin \theta \cos \theta]$ $\qquad\qquad - a \sin \theta \cos \theta]$

$r_3 = 2[a \sin^2 \theta + (e_2 - e_1) \sin \theta$ $\quad$ $r_3' = 2[a \sin^2 \theta + (e_2' - e_1') \sin \theta$

$\qquad + 2a \sin \theta \cos \theta (\varphi_2 - \varphi_1)]$ $\qquad + 2a \sin \theta \cos \theta (\varphi_2' - \varphi_1')]$

(h) Comparison of the components of **r** and **r**' ($r_i = r_i'$) and consideration of Eqs. (65a), (65b), and (60e) yields Eqs. (65c), (65d), and (65e):

$$q_1 - q_2 = q_1' - q_2' = 2a \tag{60e}$$

(Fig. 33: Lengths are measured in the direction of the optical axis.)

$$(\varphi_1 - \varphi_2) = (\varphi_1' - \varphi_2') \tag{65a}$$

$$(\psi_1 - \psi_2) = (\psi_1' - \psi_2') \tag{65b}$$

$$(e_1 - e_2) = (e_1' - e_2') \tag{65c}$$

Insertion of specific values corresponding to a parallelogram configuration with $\theta = 60°$ ($\sin \theta = \frac{1}{2}\sqrt{3}$, $\cos \theta = \frac{1}{2}$) results in

$$2\varphi_1 - \varphi_2 - \varphi_2' = 0 \tag{65d}$$

$$2\varphi_2' - \varphi_1' - \varphi_1 = 0 \tag{65e}$$

B. Calculation of the Ray Path $\eta = f(\bar{z})$ of the Displacement Error $\Delta\eta$ and of the Path Difference $S \cdot \lambda$ in Model Boundary Layers

As in Section V,C,2 all lengths are divided by the thickness δ of the boundary layer to obtain nondimensional quantities.

1. Profiles of Refractive Indices

$$n = n(\eta) = n_w + \Delta n \cdot f(\eta)$$

Common boundary conditions are

$$(dn/d\eta)_{\eta=0} = 2\,\Delta n; \qquad (dn/d\eta)_{\eta=1} = 0$$

(except for exponential profile)

$$\Delta n = n_\infty - n_w$$

(a) *Profile of refractive index of thermal boundary layer*:

$$n = n_w + \Delta n\,(2\eta - 2\eta^3 + \eta^4) \tag{93a}$$

$$dn/d\eta = 2\,\Delta n\,(1 - 3\eta^2 + 2\eta^3) \tag{93b}$$

(b) *Quadratic profile of refractive index*:

$$n = n_w - \Delta n\,(2\eta - \eta^2) \tag{94a}$$

$$dn/d\eta = 2\,\Delta n\,(1 - \eta) \tag{94b}$$

(c) *Exponential profile of refractive index*:

$$n = n_w + \Delta n\,[1 - \exp(-2\eta)] \tag{95a}$$

$$dn/d\eta = 2\,\Delta n\,\exp(-2\eta) \tag{95b}$$

2. Path $\eta = f(\bar{z})$ of the Measuring Beam

(a) *Differential equation* [Eq. (10b)]:

$$\frac{d^2\eta}{d\bar{z}^2} = \frac{1}{n} \cdot \frac{dn}{d\eta} \cdot \sin\alpha = \frac{1}{n} \cdot \frac{dn}{d\eta} \cdot \frac{1}{(1 + \tan^2\epsilon)^{1/2}}$$

where $\alpha = \pi/2 - \epsilon$ and $\tan^2\epsilon = d\eta/d\bar{z} \ll 1$,

$$\frac{d^2\eta}{d\bar{z}^2} = \frac{1}{n} \cdot \frac{dn}{d\eta} \approx \frac{1}{n_\infty} \cdot \frac{dn}{d\eta} \tag{10c}$$

with $n_w \leqslant n \leqslant n_\infty$ and $\Delta n = n_\infty - n_w < 10^{-2}$, assuming $n \approx r_\infty$.
Inserting Eq. (93b), (94b), and (95b) in Eq. (10c) yields

$$d^2\eta/d\bar{z}^2 = 2(\Delta n/n_\infty)(1 - 3\eta + 2\eta^3) \tag{10d}$$

$$d^2\eta/d\bar{z}^2 = 2(\Delta n/n_\infty)(1 - \eta) \tag{10e}$$

$$d^2\eta/d\bar{z}^2 = 2(\Delta n/n_\infty) \exp(-2\eta) \tag{10f}$$

In general,

$$d^2\eta/d\bar{z}^2 = (\Delta n/n_\infty) f'(\eta) \tag{10g}$$

(b) *Boundary conditions of the ray path*:

$$d\eta/d\bar{z} = 0; \quad \text{for} \quad \bar{z} = 0 \tag{10h}$$

$$\eta = \eta_0; \quad \text{for} \quad \bar{z} = 0 \quad (\text{coordinate } \eta_0; \text{ entrance of light ray}) \tag{10i}$$

(c) *General solution of the differential equation* (10g):

$$\bar{z} = \int \left\{ d\eta \Big/ \left[C_1 + 2 \frac{\Delta n}{n_\infty} \left(\int f'(\eta) \, d\eta \right) \right]^{1/2} \right\} + C_2 \tag{10j}$$

$$\frac{d\eta}{d\bar{z}} = \left[C_1 + 2 \frac{\Delta n}{n_\infty} \cdot \int f'(\eta) \, d\eta \right]^{1/2} \tag{10k}$$

Inserting the boundary conditions [Eqs. (10h) and (10i)] yields

$$C_1 = - \left(2 \frac{\Delta n}{n_\infty} \right)^{1/2} \left[\int f'(\eta_0) \, d\eta \right]^{1/2}$$

and

$$\left(2 \frac{\Delta n}{n_\infty} \right)^{1/2} \cdot \bar{z} = \int \left\{ d\eta \Big/ \left[\int f'(\eta) \, d\eta - \int f'(\eta_0) \, d\eta \right]^{1/2} \right\} + C_2 \tag{10l}$$

(d) *Thermal boundary layer* [Eq. (10d) inserted in Eq. (10l)]:

$$\left(2 \frac{\Delta n}{n_\infty} \right)^{1/2} \bar{z} = \int \{ d\eta / [(\eta^4 - \eta_0^4) - 2(\eta^3 - \eta_0^3) + 2(\eta - \eta_0)]^{1/2} \} \tag{97a}$$

[Elliptic integral, numerical evaluation shown in Fig. 54 ($C_2 = 0$).]

(e) *Quadratic profile of the refractive index* [Eq. (10e) inserted in Eq. (10l)]:

$$\left(2 \frac{\Delta n}{n_\infty} \right)^{1/2} \cdot \bar{z} = \int \{ d\eta / [(2\eta - \eta^2) - (2\eta_0 - \eta_0^2)]^{1/2} \} + C_2$$

$$= \int \{ d\eta / [(1 - \eta)^2 - (1 - \eta_0)^2]^{1/2} \} + C_2$$

Integration yields

$$(2 \Delta n/n_\infty)^{1/2} \bar{z} = \arcsin[(1-\eta)/(1-\eta_0)] + C_2$$

Inserting Eq. (10i) yields

$$C_2 = -\arcsin 1 = -\pi/2 \pm 2k\pi$$

and

$$\frac{1-\eta}{1-\eta_0} = \sin\left[\frac{\pi}{2} + \left(2\frac{\Delta n}{n_\infty}\right)^{1/2}\bar{z}\right] = \cos\left[\left(2\frac{\Delta n}{n_\infty}\right)^{1/2}\bar{z}\right]$$

Then the equation of the light ray is

$$\eta - \eta_0 = (1-\eta_0)\{1 - \cos[(2\Delta n/n_\infty)^{1/2}\bar{z}]\} \tag{98a}$$

$$d\eta/d\bar{z} = (1-\eta_0)(2\Delta n/n_\infty)^{1/2} \sin[(2\Delta n/n_\infty)^{1/2}\bar{z}] \tag{98b}$$

(f) *Exponential profile of the refractive index* [Eq. (10f) inserted in Eq. (10l)]:

$$\left(2\frac{\Delta n}{n_\infty}\right)^{1/2}\bar{z} = \int \{d\eta/[\exp(-2\eta_0) - \exp(-2\eta)]^{1/2}\} + C_2$$

$$= \int \{d\eta/[\exp(-2\eta_0)(1 - \exp(-2(\eta-\eta_0)))]^{1/2}\} + C_2$$

Integration yields

$$\left(2\frac{\Delta n}{n_\infty}\right)^{1/2}\bar{z} = \exp(\eta_0) \cdot \operatorname{arctanh}^2\{1 - \exp[-2(\eta-\eta_0)]\}^{1/2} + C_2$$

Inserting Eq. (10i) yields $C_2 = 0$. The equation of the light ray is

$$\eta - \eta_0 = -\tfrac{1}{2}\ln\{1 - \tanh^2[\exp(-2\eta_0)(2\Delta n/n_\infty)^{1/2}\bar{z}]\}$$

$$\eta - \eta_0 = \ln\cosh[\exp(-2\eta_0)(2\Delta n/n_\infty)^{1/2}\bar{z}] \tag{99a}$$

$$\frac{d\eta}{d\bar{z}} = \exp(-2\eta_0)\left(2\frac{\Delta n}{n_\infty}\right)^{1/2}\tanh\left[\exp(-2\eta_0)\left(2\frac{\Delta n}{n_\infty}\right)^{1/2}\bar{z}\right] \tag{99b}$$

3. Displacement of the Light Ray $\Delta\eta = f(\eta)$

(a) *Position of the focusing plane with coordinate* $\bar{z}_{mw}$ (according to Eq. (82) with $\eta_0 = 0$ at the wall):

$$\bar{z}_{mw} = \bar{z}_l - \eta_{lw}/\tan\epsilon_{lw} = \bar{z}_l - \eta_{lw}/(d\eta/d\bar{z})_{lw}$$

(b) *Displacement of the light ray in the focusing plane* ($\bar{z}_{\mathrm{mw}}$) (according to Eq. (84)):

$$\Delta \eta = (\eta_l - \eta_0) - (\tan \epsilon_l / \tan \epsilon_{lw}) \cdot \eta_{lw}$$

or

$$\Delta \eta = (\eta_l - \eta_0) - [(d\eta/d\bar{z})_l / (d\eta/d\bar{z})_{lw}] \cdot \eta_{lw}$$

(c) *Thermal boundary layer*: The coordinate of the focusing plane $\bar{z}_{\mathrm{mw}}$ and the displacement of the light ray $\Delta\eta$ (Fig. 55) are calculated with the numerically computed values of the light ray path (Fig. 54).

(d) *Quadratic profile of the refractive index with coordinate* $\bar{z}_{\mathrm{mw}}$: Inserting Eq. (98) in Eq. (82) and setting $\eta_{0w} = 0$ yields

$$\bar{z}_{\mathrm{mw}} = \bar{z}_l - \frac{1 - \cos[(2\,\Delta n/n_\infty)^{1/2}\bar{z}_l]}{(2\,\Delta n/n_\infty)^{1/2} \sin[(2\,\Delta n/n_\infty)^{1/2}\,\bar{z}_l]}$$

of (after trigonometric transformation)

$$\bar{z}_{\mathrm{mw}} = \bar{z}_l - \frac{1}{(2\,\Delta n/n_\infty)^{1/2}} \cdot \tan\left[\left(2\,\frac{\Delta n}{n_\infty}\right)^{1/2} \cdot \frac{\bar{z}_l}{2}\right] \quad (98c)$$

If $(2\,\Delta n/n_\infty)^{1/2}\,\bar{z}_l < \pi$, Eq. (98c) can be rewritten as

$$\bar{z}_{\mathrm{mw}} \approx \tfrac{1}{2}\bar{z}_l - \tfrac{1}{12}(\Delta n/n_\infty)\bar{z}_l^{\,3}$$

Displacement error $\Delta\eta$ can be found as follows: Inserting Eq. (98a, b) in Eq. (84) yields (for $\eta_{0w} = 0$)

$$\Delta\eta = (1 - \eta_0)\{1 - \cos[(2\,\Delta n/n_\infty)^{1/2}\bar{z}_l]\}$$

$$- \frac{(1 - \eta_0)(2\,\Delta n/n_\infty)^{1/2} \sin[(2\,\Delta n/n_\infty)^{1/2}\bar{z}_l]}{(1 - \eta_{0w})(2\,\Delta n/n_\infty)^{1/2} \sin[(2\,\Delta n/n_\infty)^{1/2}\bar{z}_l]}$$

$$\cdot \{1 - (1 - \eta_{0w}) \cos[(2\,\Delta n/n_\infty)^{1/2}\bar{z}_l]\}$$

$$= (1 - \eta_0)\{1 - \cos[(2\,\Delta n/n_\infty)^{1/2}\bar{z}_l]\}$$

$$- \frac{1 - \eta_0}{1 - \eta_{0w}} \cdot \left\{1 - (1 - \eta_{0w}) \cdot \cos\left[\left(2\,\frac{\Delta n}{n_\infty}\right)^{1/2}\bar{z}_l\right]\right\} \quad (98d)$$

$\Delta\eta = 0$ in the focusing plane $\bar{z}_{\mathrm{mw}}$, since $\eta_{0w} = 0$.

(e) *Exponential profile of the refractive index with coordinate* $\bar{z}_{\mathrm{mw}}$: Inserting Eq. (99a, b) in Eq. (82) yields ($\eta_{0w} = 0$)

$$\bar{z}_{\mathrm{mw}} = \bar{z}_l - \frac{\ln \cosh[(2\,\Delta n/n_\infty)^{1/2}\bar{z}_l]}{(2\,\Delta n/n_\infty)^{1/2} \tanh[(2\,\Delta n/n_\infty)^{1/2}\bar{z}_l]} \quad (99c)$$

Displacement of light ray $\Delta\eta$: Inserting Eq. (99a, b) in Eq. (84) yields ($\eta_{0w} = 0$)

$$\Delta\eta = \ln \cosh \left[\exp(-2\eta_0)(2\,\Delta n/n_\infty)^{1/2}\bar{z}_l\right]$$

$$-\exp(-2\eta_0) \frac{\tanh[\exp(-2\eta_0)(2\,\Delta n/n_\infty)^{1/2}\bar{z}_l]}{\tanh[(2\,\Delta n/n_\infty)^{1/2}\bar{z}_l]}$$

$$\cdot \ln \cosh[(2\,\Delta n/n_\infty)^{1/2}\bar{z}_l] \qquad (99d)$$

4. Path Difference $S \cdot \bar{\lambda}$

(a) *Expression for the path difference* $S \cdot \bar{\lambda}$ (according to Eq. (87), where $(d\eta/d\bar{z})^2 \ll 1$ is assumed):

$$S \cdot \bar{\lambda} = n_\infty \bar{z}_l - \int_0^{\bar{z}_l} n(\bar{z})\, d\bar{z} \qquad (87a)$$

(b) *Thermal boundary layer*: Inserting the numerically calculated values of the light path $\eta = f(\bar{z})$ in Eq. (93a), the integral can also be computed numerically.

(c) *Quadratic profile of the refractive index with refractive index function* $n(\bar{z})$: Inserting Eq. (98a) in Eq. (94a) yields

$$n(\bar{z}) = n_w + \Delta n \{1 - (1 - \eta_0)^2 \cos^2[(2\,\Delta n/n_\infty)^{1/2}\bar{z}]\}$$

Path difference is found by integrating Eq. (87a) which yields

$$S\bar{\lambda} = \Delta n\,(1 - \eta_0)^2 \left\{ \frac{\bar{z}_l}{2} + \frac{1}{4(2\,\Delta n/n_\infty)^{1/2}} \cdot \sin\left[2\left(2\frac{\Delta n}{n_\infty}\right)^{1/2}\bar{z}_l\right]\right\} \qquad (87b)$$

(d) *Exponential profile of the refractive index with refractive index function* $n(\bar{z})$: Inserting Eq. (99a) in Eq. (95a) yields

$$n(\bar{z}) = n_w + \Delta n - \Delta n\,\exp\{-2(\eta_0 + \cosh[\exp(-2\eta_0)(2\,\Delta n/n_\infty)^{1/2}\bar{z}])\} \qquad (87c)$$

Path difference is found by integrating Eq. (87a) which yields

$$S\bar{\lambda} = \Delta n \frac{\exp(-2\eta_0)}{(2\,\Delta n/n_\infty)^{1/2}} \tanh\left[\exp(-\eta_0)\left(2\frac{\Delta n}{n_\infty}\right)^{1/2}\bar{z}_l\right] \qquad (87d)$$

(e) *Correction parabola*: In the region $\eta_l - \eta_0$ the profile of the refractive index is replaced by a linear profile (virtual wedge) [Eq. (18)]:

$$n = n_0 + (dn/d\eta)_0 (\eta - \eta_0) \qquad (18a)$$

With $(d\eta/d\bar{z})_0 = 0$, the resulting parabolic ray path [Eq. (89)] is

$$\eta - \eta_0 = (dn/d\eta)_0 \cdot \bar{z}^2/2n_0 \qquad (21a)$$

$$d\eta/d\bar{z} = (dn/d\eta)_0 \cdot \bar{z}/n_0 \qquad (22a)$$

Inserting Eq. (21a) in Eq. (18a) yields

$$n(\bar{z}) = n_0 + (dn/d\eta)_0^2 \cdot \bar{z}^2/2n_0$$

Inserting this equation in the simplified interferometer equation $(d\eta/d\bar{z})^2 \ll 1$, the path difference $S \cdot \bar{\lambda}$ results:

$$S\bar{\lambda} = n_\infty \cdot \bar{z}_l - \int_0^{\bar{z}_l} n(\bar{z})\, d\bar{z}$$

$$S\bar{\lambda} = (n_\infty - n_0)\bar{z}_l - (dn/d\eta)_0^2 \cdot \bar{z}_l^3/6n_0$$

or, substituting $(d\eta/d\bar{z})_l$ with Eq. (22a),

$$S\bar{\lambda} = n \cdot \bar{z}_l - (d\eta/d\bar{z})_l^2 \cdot \bar{z}_l \cdot n_0/6$$

C. CALCULATION OF THE RAY PATH $\eta = f[\bar{z}, (d\eta/d\bar{z})_0]$ AND OF THE PATH DIFFERENCE $S \cdot \bar{\lambda}$, WHEN THE SLOPE OF THE LIGHT RAY AT THE WALL ($\eta_0 = 0$) IS NOT EQUAL TO ZERO AT THE BEGINNING OF THE TEST SECTION

1. *Differential Equation of the Ray Path [Eq. (14)] with Nondimensional Coordinates* $\eta = y/\delta$, $\bar{z} = z/\delta$

$$\frac{d^2\eta/d\bar{z}^2}{[1 + (d\eta/d\bar{z})^2]^{3/2}} = \frac{dn/d\eta}{n}[1 + (d\eta/d\bar{z})^2]^{1/2} \qquad (14a)$$

or integrated,

$$\bar{z} = \int_{\eta_0}^{\eta} \frac{d\eta}{\{(n/n_0)^2[1 + (d\eta/d\bar{z})^2] - 1\}^{1/2}} \qquad (20a)$$

2. *Profile of the Refractive Index (Virtual Wedge) Valid for the Measuring Beam at the Wall* ($\eta_0 = 0$)

$$(dn/d\eta)_w = 2\,\Delta n = 2(n_\infty - n_w) \qquad (32d)$$

$$n = n_w + 2\,\Delta n \cdot \eta$$

3. Equation of the Ray Path

Inserting Eq. (32d) in Eq. (20a) and integrating between the limits $\eta = 0$ and $\eta = \eta$ yields, if $(n/n_w)^2 \approx 1 + (2\,\Delta n/n_w) \cdot \eta$,

$$\eta = (d\eta/d\bar{z})_w \bar{z} + (2\,\Delta n/n_w)[1 + (d\eta/d\bar{z})_w^2]\bar{z}^2$$

or, generally, for the point $\eta_0[(dn/d\eta)_0 = \text{const}]$,

$$\eta - \eta_0 = (d\eta/d\bar{z})_0 \bar{z} + [(dn/d\eta)_0/2n_0][1 + (d\eta/d\bar{z})_0^2]\bar{z}^2 \tag{89a}$$

(correction parabola).

4. Path Difference $S \cdot \bar{\lambda}$

Assuming a parabolic ray path, the general interferometer equation yields (coordinate of the focusing plane $z_{mw} = z_l/2$)

$$S\bar{\lambda} = n_\infty \cdot \bar{z}_l/2\{1 + [1 + (d\eta/d\bar{z})_l^2]^{1/2}\}$$
$$- \int_0^{\bar{z}_l} n(\bar{z})[1 + (d\eta/d\bar{z})^2]^{1/2}\,d\bar{z} \tag{87a}$$

Neglecting terms of higher order and inserting Eqs. (22) and (23), Eq. (87a) can be rewritten

$$S\bar{\lambda} = n_\infty \bar{z}_l + [(dn/d\eta)_0^2/n_\infty^2]\bar{z}_l^3/4$$
$$- n_0 \cdot \bar{z}_l - [(dn/d\eta)_0^2/n_w]\bar{z}_l^3/3 \tag{90c}$$

For the ray nearest to the wall ($\eta_0 = 0$) and with $(dn/d\eta)_w = 2\,\Delta n$,

$$S\bar{\lambda} = \Delta n\,\bar{z}_l - \tfrac{1}{3}(\Delta n^2/n_\infty)\,\bar{z}_l^3[1 + 4\,\Delta n/n_\infty + 4(\Delta n/n_\infty)^2 + \cdots] \tag{90}$$

Acknowledgment

The authors wish to extend their thanks to the following: We are greatly indebted to H. Becker for his assistance in editing the manuscript and especially for his critical comments and valuable suggestions. A part of Section IV is contributed by him. M. Reimann and S. Bloss took care of the illustrations and performed numerical calculations and together with B. Brand assisted in the final corrections. Dr. G. Raithby and H. Spieler translated our article from the German. Finally we wish to express our appreciation to Dr. J. Bach (Augsburg), Dr. H. Hannes (Leverkusen), F. Killermann (Munich), M. Reimann (Erlangen), and Dr. G. Schödel (Burghausen), to whom a number of photos and examples are due.

Nomenclature

a	thermal diffusivity	$g = S \cdot \lambda$	path difference
a	distance between diaphragm slit and screen	g	gravity
		Gr	Grashof number
a	module of parallelogram configuration of the MZI	h	axis of spatial interference
		H	enthalpy per volume
		H_{tot}	enthalpy
A	amplitude	i	subscript integer value
$A_{l,k}$	function of indices of cylindrical models	I	integral, e.g., of the phase difference of a schliere
b	half slit width	I	intensity
b	fringe width, distance between two interference fringes	J_ν	BESSEL function of νth order
$\bar{b} = b/\delta$	dimensionless fringe width	k	$2\pi/\lambda$; wave number
		k_c	thermal conductivity
b_w	fringe width at the wall of the model	k_{rad}	apparent conductivity caused by radiation
b_f	distorted fringe width, caused by fringe displacement Δy	K	modulation of interference contrast
		$\|\mathbf{K}\| = 1/R$	curvature
c	light velocity	l	model length
c_0	light velocity in vacuo	L	distance between model and screen
c_p	specific heat		
C	concentration	L^*	luminous intensity
$C(w)$	real component of FRESNEL integral	m	parameter
		m	subscript for measuring beam
C	axis of a virtual wedge	m_i	normal vector of mirror surface
d	diameter (model)		
$d\mathbf{s}$	path element	$\bar{M}$	projection matrix
$d\mathbf{f}$	area element	MZI	MACH ZEHNDER interferometer
d_p	thickness of a glassplate (e.g., beamsplitter)		
		n	refractive index
$d\mathbf{r}$	line element	n_0	refractive index at the point, where a light ray enters the model
D	differential operator		
D	diffusivity		
e	amount of deflection of a light ray	n'	dn/dy
		n_∞	ambient refractive index
e_i	displacement of a mirror in the MZI	n_w	refractive index at the wall
		n_{gl}	refractive index of glass
erf	error function	N	$\dfrac{(n - n_w)}{(n_\infty - n_w)}$ specific refractive index of a model boundary layer
erfc(x)	$1 - \text{erf}(x)$		
exp	exponential function		
E	eikonal, distorted wavefront		
		Nu	Nusselt number
E^*	specific illumination	0	zero point of coordinate system
$\bar{E}$	unity matrix		
f	focal lenght	$\mathbf{p}_0$	position vector of a point light source
$F(w)$	FRESNEL integral		

p	position vector of the image of the light source	x	cartesian coordinate
P	axis of spatial interference, perpendicular to the interference fringes	x_0	coordinate, where a ray enters the test section
		y	cartesian coordinate
		y'	dy/dz
q	heat flux	y_0	coordinate, where a ray enters the test section
q_w	heat flux at the wall of the model	y_l	coordinate, where the ray leaves the test section
$\mathbf{q}_i$	position vector of a mirror of the MZI	y_1	screen coordinate
r	radius	$\tilde{y}$	displacement of measuring and reference beam in a differential interferometer
r	subscript for reference beam		
$\bar{r}$	specific refractivity		
R	radius of curvature	z	cartesian coordinate (optical axis)
R_0	gas constant		
Ra	Raleigh number	$\tilde{z} = \text{const} \cdot h$	transformed axis of spatial interference
Re	Reynolds number		
s	unity normal vector of a wave front	$\bar{z} = z/\delta$	dimensionless coordinate
		$\bar{z}_l = l/\delta$	dimensionless model length
$s_{x,y,z}$	components of the normal vector of the wave front	$\bar{z}_{mw}$	dimensionless coordinate of the focussing plane
s	coordinate along the ray path	$\bar{z}_m$	dimensionless coordinate of the general position of the focussing plane
$\bar{s} = s/\delta$	dimensionless coordinate of the ray path		
S	interference order, (phase difference: $2\pi S$)	α	angle between light ray and refractive index gradient
$S(w)$	complex component of FRESNEL integral	β	expansion coefficient
		$\gamma = \cos \chi$	cosine of direction
S_w	interference order at the wall	δ	boundary layer thickness
		δ_1	deflection of the ray nearest to the wall
t	time coordinate		
t-t	indication for a intersection plane	δ_g	gap width
		ϵ	angle between the ray and one axis, usually the z axis
t_m-t_m	focusing plane in the measuring beam		
t_r-t_r	focusing plane in the reference beam	ϵ_0	angle at the entrance of the test section
t_f-t_f	focal plane	ϵ_l	angle at the end of the test section
t_i-t_i	image plane		
T	temperature	ζ	$(2\,\Delta n/n_\infty)^{1/2} \cdot \bar{z}$ transformed, dimensionless coordinate of a model boundary layer
T_∞	ambient temperature		
T_w	temperature at the wall		
TS	test section		
u	light amplitude	ζ_l	transformed model length
$U_{1,2}$	LOMMEL functions	ζ_{mw}	transformed coordinate of the focussing plane
v	velocity		
v_u	peripheral velocity	$\eta = y/\delta$	dimensionless boundary layer coordinate
$w_{x,y}$	velocities in x or y direction		

$\eta_0 = y_0/\delta$	ray entrance of the boundary layer	Δn	$n_\infty - n_w$
$\eta_l = y_l/\delta$	ray exit of the boundary layer	θ	angle of parallelogram of the MZI
ϑ	temperature (°C)	$\Delta\vartheta$	temperature difference, e.g., between two isotherms (interference lines)
$\bar{\vartheta}$	$(T - T_\infty)/(T_w - T_\infty)$ dimensionless temperature of a model boundary layer		
		$\Delta\vartheta_\infty$	$\vartheta_w - \vartheta_\infty$
$\bar{\vartheta}'$	$d\bar{\vartheta}/d\eta$	Δt	coherence time
ϑ_w	wall temperature	Δl	coherence length
ϑ_∞	ambient temperature	$\Delta\nu$	band width
λ	wave length	Ω	solid angle of the aperture of the circular diaphragm
λ_0	wave length in vacuo		
λ_m	mean wave length		
$\bar{\lambda} = \lambda/\delta$	dimensionless wavelength in a model boundary layer	ϕ	phase modelution of spatial interference
		Δy	displacement error
μ_{0m}^2, μ_{0r}^2	angular intensity in the measuring and in the reference beam	$\Delta\eta = \Delta y/\delta$	dimensionless displacement
		$\Delta\eta$	displacement by incorrect focussing
μ	subscript	Δr	zone width of circular phase object
ν	frequency of light		
ν_m	mean frequency	ΔS_l	correction of phase difference due to end effects
ν	viscosity		
ρ	density	ΔS_1	correction of phase difference in real interference
τ	period time		
φ	angle of the virtual wedge	$\Delta\eta_{max}$	maximum displacement (correct focussing)
φ_i	rotation angle of a mirror (MZI)		
ψ_i	rotation angle of a mirror (MZI)	ΔS_{tab}	correction of phase difference due to different ambient temperature as in Tables XII, XIV
$\omega = 2\pi\nu$	angular velocity		
$\omega = r/f$	aperture of a circular diaphragm (radius r)	Δx	fringe shift in an interferogram (fringe field)
$\tilde{\omega} = \text{const} \cdot p$	transformed axis of a spatial interference system		

REFERENCES

1. M. Born and E. Wolf, "Principles of Optics." Pergamon Press, Oxford, 1959.
2. H. Schardin, Die Schlierenverfahren und ihre Anwendungen, in "Ergebnisse der exakten Naturwissenschaften," Vol. 20, pp. 303–439. Springer, Berlin, 1942.
3. F. J. Weinberg, "Optics of Flames." Butterworths, London and Washington, D.C., 1963.
4. H. Wolter, Schlieren-, Phasenkontrast- und Lichtschnittverfahren, in "Handbuch der Physik," Vol. 24, pp. 555–641. Springer, Berlin, 1956.
5. E. J. Weyl, Analysis of optical methods, in "Physical Measurements in Gas Dynamics and Combustion," Vol. A1, pp. 1–25. Princeton Univ. Press, Princeton, New Jersey, 1954.

6. S. Tolansky, "An Introduction to Interferometry." McGraw-Hill, New York, 1955.
7. W. Krug, J. Rienitz and G. Schulz, "Beiträge zur Interferenzmikroskopie." Akademie Verlag, Berlin, 1961.
8. ISO Recommendations, R 31, obtainable through the National Standards Organizations.
9. A. Sommerfeld, "Vorlesungen über theoretische Physik, Band IV, Optik," 2nd ed. Akademische Verlagsgesellschaft, Leipzig, 1959.
10. A. Sommerfeld and J. Runge, Anwendung der Vektorrechnung auf die Grundlagen der geometrischen Optik, *Ann. der Phys.* **35**, 277–298 (1911).
11. U. Grigull, Einige optische Eigenschaften thermischer Grenzschichten, *Intern. J. Heat Mass Transfer* **6**, 669–679 (1963).
12. O. Wiener, Darstellung gekrümmter Lichtstrahlen und Verwertung derselben zur Untersuchung von Diffusion und Wärmeleitung, *Wiedemanns Ann.* **49**, 105–149 (1893).
13. J. St. L. Philpot, Direct photography of ultracentrifuge sedimentation curves, *Nature* **141**, 283 (1938).
14. H. Svensson, Direkte photographische aufnahme von elektrophorese-diagrammen, *Z. Kolloid* **87**, 181–186 (1939).
15. M. Françon, Interférences, diffraction et polarisation, "Handbuch der Physik," Vol. 24, pp. 171-460. Springer, Berlin, 1956.
16. H. Wolter, Verbesserung der abbildenden Schlierenverfahren durch Minimumstrahlkennzeichnung, *Ann. Phys.* 6. Folge, **7**, 182–192 (1950).
17. J. Sperling, Das Temperaturfeld im freien Kohlebogen, *Z. Phys.* **128**, 269–278 (1950).
18. H. Wolter, Die Minimalstrahlkennzeichnung als Mittel zur Genauigkeitssteigerung optischer Messungen und als methodisches Hilfsmittel zum Ersatz des Strahlbegriffs, *Ann. Phys.* 6. Folge **7**, 341–368 (1950).
19. F. Killermann, Der Einfluß von begrenzenden Wänden auf den Wärmeübergang bei beheizten, waagrechten Rohren. Inst. Bericht, Thermodynamik A, TH München (1969).
20. E. Schmidt, Schlierenaufnahmen des Temperaturfeldes in der Nähe wärmeabgebender Körper, *VDI (Ver. Deut. Ing.)-Forschungsh.* **3**, 181–189 (1932).
21. G. Smeets, Aufnahmen mit dem Differentialinterferometer und ihre Auswertung, *Proc. Intern. Cong. High-Speed Photography 8th, Stockholm, 1968.* Paper No. 94.
22. R. Ladenburg, Interferometry, *in* "Physical Measurements in Gas Dynamics and Combustion," Vol. A3. Princeton Univ. Press, Princeton, New Jersey, 1954.
23. D. W. Holder and R. J. North, Optical Methods for Examining the Flow in High Speed Wind Tunnels, *NATO Advisory Group Aeron. Res. Develop.* (1965).
24. M. Françon, "Modern Applications of Physical Optics," Tracts of Phys. and Astron. Wiley (Interscience) New York, 1963.
25. L. Mach, Über einen Interferenzrefraktor, *Z. Instrumentenk.* **12**, 89–93 (1892).
26. L. Zehnder, Ein neuer Interferenzrefraktor, *Z. Instrumentenk.* **11**, 275–285 (1891).
27. W. Kinder, Ein Mach-Zehnder-Interferometer für Laserbeleuchtung mit einer vier Meter langen Meßstrecke, Zeiss-Nachrichten, No. 63 (Jan 1967).
28. J. Jamin, Neuer Interferential-Refractor, *Pogg. Ann.* **98**, 345–349 (1856).
29. A. A. Michelson, Interference phenomena in a new form of refractometer, *Phil. Mag*, 236–242 (1882).
30. A. A. Michelson and J. R. Benoit, Détermination expérimentale de la valeur du mètre en longueurs d'ondes lumineuses, *Trav. et Mem. Int. Bur. Poids et Mes.* **11**, 1–85 (1895).
31. A. A. Michelson and E. W. Morley, *Phil. Mag.* **24**, No. 5, 449 (1887).

32. F. Twyman and A. Green, British Patent No. 103832 (1916).
33. W. J. Bates, A wavefront shearing interferometer, *Proc. Phys. Soc.* **59**, 940–950 (1947).
34. J. W. S. Rayleigh, On some physical properties of argon and helium, *Proc. Roy. Soc.* **59**, 201–206 (1896).
35. F. Haber and F. Löwe, Ein Interferometer für Chemiker nach Rayleighschem Prinzip, *Z. Angew. Chem.* **23**, 1393 (1910).
36. H. Kuhn, New techniques in optical interferometry, *Rept. Progr. Phys.* **14**, 64–94 (1951).
37. M. Françon, "Progress in Microscopy." Pergamon Press, Oxford, 1961.
38. F. H. Smith, "Modern Methods of Microscopy." Butterworths, London and Washington, D. C., 1956.
39. R. Kraushaar, A diffraction grating interferometer, *J. Opt. Soc. Am.* **44**, 480–481 (1950).
40. V. Ronchi, *Atti. Fond. Giorgio Ronchi Contrib. Ist. Naz. Ottica* **17**, 240 (1962).
41. D. Gabor, A new microscopic principle, *Nature* **161**, 777–778 (1948).
42. E. N. Leith and J. Upatnieks, Reconstructed wavefronts and communication theory, *J. Opt. Soc. Am.* **52**, 1132 (1962).
43. E. N. Leith and J. Upatnieks, Wavefront reconstruction with continuous tone objects, *J. Opt. Soc. Am.* **53**, 1377 (1963).
44. E. N. Leith and J. Upatnieks, Wavefront reconstruction with diffused illumination and three-dimensional objects, *J. Opt. Soc. Am.* **54**, 1295 (1964).
45. K. A. Haines and B. P. Hildebrand, Surface deformation measurement using the wavefront reconstruction technique, *Appl. Opt.* **5**, 595 (1966).
46. J. M. Burch, J. W. Gates, R. G. N. Hall, and L. H. Tanner, Holography with a scatter plate as beam splitter and a pulsed ruby laser as light source, *Nature* **212**, 1347–1348 (1966).
47. F. Zernike, Diffraction theory of the knife edge test and its improved form, the phase contrast method, *Roy. Astron. Soc.* **94**, 377 (1934).
48. S. F. Erdmann, Ein neues, sehr einfaches Interferometer zum Erhalt quantitativ auswertbarer Strömungsbilder, *J. Appl. Sci. Res.* **B2**, 149–198 (1952).
49. E. L. Gayhart and R. Prescott, Interference phenomena in the schlieren system, *J. Opt. Soc. Am.* **39**, 546 (1949).
50. E. B. Temple, Quantitative measurements of gas density by means of light interference in a schlieren system, *J. Opt. Soc. Am.* **47**, 91 (1957).
51. H. Rottenkolber, Neue einfache Interferenzverfahren und ihre Anwendung auf thermische Grenzschichten, *VDI (Ver. Deut. Ing.) Z.* **6**, No. 8, (1965).
52. F. D. Bennett and G. D. Kahl, A generalized vector theory for the Mach-Zehnder-Interferometer, *J. Opt. Soc. Am.* **43**, 71–78 (1953).
53. L. Silberstein, "Simplified Method of Tracing Rays through Any Optical System of Lenses, Prisms and Mirrors." Longmans, Green, New York, 1918.
54. H. Hannes, Zur Grundstellung des Mach-Zehnder Interferometers, *Z. Opt.* **12**, 17–22 (1955).
55. E. Lamla, Über die Justierung des Mach-Zehnderschen Interferometers, *Z. Instrumentenk.* **62**, 337–346 (1942).
56. W. Kinder, Theorie des Mach-Zehnder-Interferometers und Beschreibung eines Geräts mit Einspiegeleinstellung, *Z. Opt.* **1**, 413–448 (1946).
57. E. R. G. Eckert, R. M. Drake, and E. Soehngen, Manufacture of a Zehnder-Mach Interferometer, Tech. Rept. 5721, Air Material Command, Wright Patterson Air Force Base (1948).

58. B. Gebhart and C. P. Knowles, Design and adjustment of a 20 cm Mach-Zehnder interferometer, *Rev. Sci. Instr.* **37**, 12–15 (1966).
59. R. K. M. Johnstone and W. Smith, A design for a six inch field Mach-Zehnder interferometer, *J. Sci. Instr.* **42**, 231–235 (1965).
60. G. Hansen, Über die Ausrichtung der Spiegel bei einem Interferometer nach Zehnder-Mach, *Z. Instrumentenk.* **60**, 325–329 (1940).
61. R. J. Clark, C. D. Hause, and G. S. Bennett, Adjustment of a Mach-Zehnder interferometer for white light fringes, *J. Opt. Soc. Am.* **43**, 408 (1953).
62. E. W. Price, Initial adjustment of the Mach-Zehnder interferometer, *Rev. Sci. Instr.* **23**, 162 (1952).
63. F. D. Bennett, Effect of size and spectral purity of source on fringe pattern of the Mach-Zehnder interferometer, *J. Appl. Phys.* **22**, 776–779 (1951).
64. R. J. Goldstein, Interferometer for aerodynamic and heat transfer measurements, *Rev. Sci. Instr.* **36**, 1408–1410 (1965).
65. U. Grigull and H. Rottenkolber, Two beam interferometer using a laser, *J. Opt. Soc. Am.* **57**, 149–155 (1967).
66. G. D. Kahl and F. D. Bennett, Experimental verification of source size theory for the Mach-Zehnder Interferometer, *J. Appl. Phys.* **23**, 763–767 (1952).
67. H. G. Stamm, Neuere Arbeiten über die Theorie des Mach-Zehnder Interferometers und neue Anwendungen auf gasdynamische Probleme, Diss. TH München (1943).
68. G. Schulz, Zweistrahlinterferenz in Planspiegelanordnungen I. Lichtquellenbild-Transformation und räumliche Intensitätsverteilung bei beliebiger Spiegelstellung, *Opt. Acta* **11**, 44–60 (1964).
69. G. Schulz, Zweistrahleninterferenz im Planspiegelanordnungen. II. Charakteristische Erscheinungen bei Schraubung zwischen den Lichtquellenbildern, *Opt. Acta* **11**, 132–143 (1964).
70. G. Minkwitz and G. Schulz, Der räumliche Interferenzstreifenverlauf der Keilinterferenzen, *Opt. Acta* **11**, 90–99 (1964).
71. E. Lommel, *Abhandl. Math. Phys. Kgl. Bayer. Akad. Wiss.* **15**, 229 (1886).
72. R. E. Blue, Interferometer Corrections and Measurements of Laminar Boundary Layers in Supersonic Stream, NACA TN 2110, Washington, D. C. (1950).
73. G. P. Wachtell, Refraction Effect in Interferometry of Boundary Layer of Supersonic Flow along Flat Plate, Princeton Univ., Princeton, New Jersey.
74. J. Winckler, The Mach-Zehnder interferometer applied to studying an axially symmetric supersonic air jet, *Rev. Sci. Instr.* **19**, 307 (1948).
75. D. Bershader, An interferometric study of supersonic channel flow, *Rev. Sci. Instr.* **20**, 260–275 (1949).
76. W. Hauf, Interferogramme einiger Modellgrenzschichten, Institutsbericht, Institut A für Thermodynamik, Technische Hochschule München (1969).
77. R. B. Kennard, An optical method for measuring temperature distributions and convective heat transfer, *Bur. Std. J. Res.* **8**, 787–805 (1932).
78. R. Ladenburg, J. Winckler, and C. C. Van Voorhis, Interferometric study of faster than sound phenomena, Part I, *Phys. Rev.* **73**, 1359–1377 (1948).
79. A. Weise and G. Hahn, Praktische Erfahrungen mit dem Mach-Zehnder Interferometer bei gasdynamischen Untersuchungen, Fachausschuß Kurzzeitphysik d. Dtsch. Physikal. Gesellschaft Ernst-Mach-Institut Freiburg (1962).
80. K. Steegmaier, Ein Mach-Zehnder Interferometer für die Strömungsforschung, Bericht Ernst-Mach-Institut, Freiburg (1962).
81. H. Hannes, Neue Möglichkeiten zur interferometrischen Messung bei der Wärme- und Stoffübertragung, *Forsch. Ing. Wes.* **29**, 159–163 (1963).
82. E. Lau and W. Krug, "Die Äquidensitometrie. Grundlagen, Verfahren und Anwendungsbeispiel." Akademie-Verlag, Berlin, 1957.

83. Landolt-Börnstein, "Physikalisch-chemische Tabellen." Springer, Berlin, 1923.
84. "International Critical Tables." McGraw-Hill, New York, 1933.
85. G. Schödel, Kombinierte Wärmeleitung und Wärmestrahlung in konvektionsfreien Flüssigkeitsschichten, Diss. Techn. Hochschule München, Institut A für Thermodynamik (1969).
86. L. W. Tilton and J. K. Taylor, Refractive index and dispersion of destilled water for visible radiation, at temperature 0 to 60 °C. *J. Res. Natl. Bur. Std.* **20** (1938).
87. U. Grigull and W. Hauf, Natural convection in horizontal cylindrical annuli, *Proc. Intern. Heat Transfer Conf. 3rd, Chicago, 1966*, Vol. II, pp. 182–195.
88. M. Reimann, Strömungsverhältnisse und örtlicher Wärmeübergang am rotierenden Zylinder, Dipl. Arbeit, Institut A für Thermodynamik, Techn. Hochschule München (1965).
89. H. Hannes, Interferometrische Messung der thermischen Energie von elektrischen Funken, *Forsch. Geb. Ing. Wes.* **29**, 169–175 (1963).
90. W. Hauf and U. Grigull, Instationärer Wärmeübergang durch freie Konvektion in horizontalen Zylindern, *Inter. Konf. Wärmeübertragung, 4, Versailles, 1970*.
91. J. Bach, Instationäre Messung der Wärmeleitfähigkeit mit optischer Registrierung, Diss. Techn. Hochschule München, Institut A für Thermodynamik (1969).
92. G. Schödel and U. Grigull, Einfluß der Wärmestrahlung auf die effektive Wärmeleitfähigkeit von Flüssigkeiten, *Inter. Konf. Wärmeübertragung, 4, Versailles, 1970*.
93. S. W. Churchill, The prediction of natural convection (invited lecture), *Intern. Heat Transfer Conf. 3rd Chicago (1966)*.
94. E. E. Soehngen, Interferometric studies on heat transfer, *Proc. Intern. Congr. Appl. Mech. 9th Brussels, 1956*.
95. E. R. G. Eckert and E. Soehngen, Distribution of heat transfer coefficients around circular cylinders in crossflow at Reynolds numbers from 20 to 500, *J. Heat Transfer* **74**, 343–347 (1952).
96. R. J. Goldstein and E. R. G. Eckert, The steady and transient free convection boundary layer on a uniformly heated vertical plate. *Intern. J. Heat Mass Transfer* **1**, 208–218 (1960).
97. E. R. G. Eckert and W. O. Carlson, Natural convection in an air layer enclosed between two vertical plates with different temperatures, *Intern. J. Heat Mass Transfer* **2**, 106–120 (1961).
98. H. A. Simon and E. R. G. Eckert, Laminar free convection in carbon dioxide near its critical point, *Intern. J. Heat Mass Transfer* **6**, 681–690 (1963).
99. K. Brodowicz and W. T. Kierkus, Experimental investigation of free convection flow in air above a horizontal wire with constant heat flux, *Intern. J. Heat Mass Transfer* **9**, 81–95 (1966).
100. P. M. Brdlik and V. A. Močalov, Experimental study of free convection with porous blowing and suction at a vertical surface, *Inzh. Fiz. Zh. Akad. Nauk Belorussk. SSR* **10**, 3–10 (1966).
101. J. A. Adams and P. W. McFadden, Simultaneous heat and mass transfer in free convection with opposing body forces, *A.I.Ch.E. (Am. Inst. Chem. Engrs.) J.* **12**, 642–647 (1966).
102. R. J. O'Brien and A. J. Shine, Some effects of an electric field on heat transfer from a vertical plate in free convection, *J. Heat Transfer* **89**, 114–116 (1967).
103. B. Gebhart, R. P. Dring, and C. E. Polymeropoulos, Natural convection from vertical surfaces, the convection transient regime, *J. Heat. Transfer* **89**, 53–59 (1967).
104. B. Gebhart and R. P. Dring, The leading edge effect in transient natural convection from a vertical plate, *J. Heat Transfer.* **89**, 274–275 (1967).
105. W. E. Mercer, W. M. Pearce, and J. E. Hitchcock, Laminar free convection in the entrance region between parallel plates, *J. Heat Transfer* **89**, 251–257 (1967).

Unsteady Convective Heat Transfer and Hydrodynamics in Channels

E. K. KALININ AND G. A. DREITSER

Moscow Aircraft Institute, Moscow, USSR

Introduction	367
I. Statement of Theoretical and Experimental Investigations of Unsteady Heat Transfer and Hydrodynamics in Channels	370
A. General Statement of the Problem in Viscous Fluid Flow	370
B. Statement of the One-Dimensional Problem	373
II. Hydrodynamics of Unsteady Fluid Flow in Channels	386
A. One-Dimensional Theory	386
B. Quasi-Stationary Theory and Its Qualitative Analysis	388
C. Laminar Flows	396
D. Turbulent Flows	398
III. Unsteady-State Heat Transfer in Channel Flow	411
A. Discussion of Experimental Methods in the Determination of the Heat Transfer Coefficient under Unsteady-State Conditions	411
B. Unsteady-State Heat Transfer in Laminar Flow in Channels	419
C. Unsteady-State Heat Transfer in Turbulent Channel Flow	439
D. Conjugate Problems of Unsteady Heat Transfer	457
IV. Some Aspects of Unsteady Heat Transfer	474
A. Unsteady Convective Heat Transfer for Bodies in External Flow	474
B. Transient Free Convective Heat Transfer	480
Nomenclature	484
References	486

Introduction

The present review is devoted to unsteady convective heat transfer and hydrodynamics. The investigations in this field began developing intensely during recent years, since only in this period has there appeared an urgent need for reliable calculation of unsteady-state processes in heat exchangers of nuclear power plants, in aircraft, and in other branches of engineering.

The necessity to analyze the present state of these investigations is caused both by a very great number of publications and by the immense differences of opinion between various investigators regarding the ways and methods of studying and calculating unsteady processes.

The authors hope that such a review will make it possible for the investigators and specialists in industry to understand the present knowledge of the problem and will encourage the development of fresh ideas on methods and ways of further investigations.

The review covers mainly the analysis of unsteady convective heat transfer and hydrodynamics, including the statement of the problem and the methods of investigation. Unsteady heat transfer and hydrodynamics with free convection and in external flows are not presented in detail. A bibliography on these problems is included. Unsteady heat transfer in two-phase flows is not considered in the present paper.

The last decade is characterized by an ever-increasing interest in the study of unsteady convective heat transfer, hydrodynamics in general, and in channels in particular.

This interest is caused by the demands of engineering developments, especially of energetics (including nuclear power), aircraft, energy engineering, metallurgy, chemical technology, etc.

The increase in the power of engines and other energy systems, the use of automatic control of these systems, and a number of technological processes in industry considerably raised the demand for accuracy and reliability in the calculation of unsteady processes in heat exchangers of energy systems and technological equipment.

In general, the purpose of such calculations consists in determining temperature and velocity fields in a heated flowing fluid and temperature fields and thermal stresses in the solid materials which surround this flow. In principle, these fields may be calculated by solving the conjugated[1] problems. However, the solution of conjugated problems in unsteady conditions encounters severe mathematical difficulties in the overwhelming majority of practically important cases. This is not unexpected since even more simple unsteady problems on the determination of temperature and velocity fields in flows with known boundary conditions often cannot be solved, especially for turbulent flows.

---

[1] In the Soviet heat transfer literature conjugated problems are those in which the energy equation is solved simultaneously in the fluid and the boundary walls.

In practice, it is usually of the greatest interest to determine temperature fields only in the wall. Only a mean mass temperature of the flow and the heat flux at the fluid-wall boundary and hydraulic losses need be found in addition.

Under steady-state conditions such a statement of the problem allows the solution of the problem to be divided.

The bulk temperature of the fluid and the friction factor are found by solving a one-dimensional system of equations with the known boundary conditions and by using the concepts of the coefficients of heat transfer and hydraulic resistance. These coefficients are determined either experimentally or by solving three-dimensional problems, after which the temperature field of a wall is determined by solving the heat conduction equation with the third-kind of boundary condition.[1a]

However, even in the steady-state problems, the heat transfer coefficient depends upon the character of the change in wall temperature along the channel surface.

The application of this method of solution to unsteady-state problems is complicated by the fact that the heat transfer coefficient also depends on changes in time of boundary conditions on the wall temperature, and also on the fluid flow rate and its temperature at the inlet.

A number of recent investigations have shown that the only part of the prehistory of a change in the boundary conditions that influences the coefficient of heat transfer and the friction factor is that of time and space. Fortunately, time and distance, at which the prehistory influences these coefficients are especially small for turbulent flows. In this case, it appears that the coefficient of heat transfer and the friction factor depend upon the law of a change in boundary conditions, and also on the first derivatives in time and space. This allows successful application of the well-used method of calculation of heat exchangers under steady-state to unsteady-state conditions.

The aim of the present paper, based on a review of the literature, is to analyze the available techniques and directions of research when unsteady convective heat transfer and hydraulics are studied, mainly with fluid flow in channels.

[1a] Editors Note. The Soviets refer to three general boundary conditions. In the boundary condition of the first kind the boundary temperatures are known functions of position and time. For the second kind, the surface heat flux is a known function of position and time. For the third kind, the heat flux at the wall is specified through a convective heat transfer coefficient.

I. Statement of Theoretical and Experimental Investigations of Unsteady Heat Transfer and Hydrodynamics in Channels

As has been mentioned in the introduction, the skill of reliable calculation of unsteady heat transfer and hydrodynamics (particularly, in channels) has become an urgent problem in many branches of modern engineering.

Due to essential differences between unsteady and steady problems, there have appeared in the literature different, sometimes incompatible, opinions on the statement and methods of the theoretical and experimental investigations of unsteady problems.

Depending on the statement of the problem, various authors use and recommend different methods of investigation and different calculation methods based on these investigations.

It is natural that the statement of the problem, as well as methods and directions of investigation should be considered of primary significance and should be discussed first.

To avoid misunderstandings regarding different interpretations of the most important concepts, it is advisable to present here a general statement of a problem in hydrodynamics and to show its relation to a one-dimensional way of description widely used in engineering practice.

A. GENERAL STATEMENT OF THE PROBLEM IN VISCOUS FLUID FLOW

As is known, the choice of the method for describing real phenomena leads to a rise of certain concepts having physical meaning dependent on the chosen method of description.

All the works presented in the review use the concept of continuum and, particularly, a model of a viscous fluid.

According to the definition accepted in hydrodynamics the viscous fluid is a continuum satisfying the hypotheses of linearity, uniformity, and isotropism, on the basis of which the linear relation between the components of the stress tensors P_{ij} and deformation velocities $\epsilon_{ij} = \frac{1}{2}(\partial W_i/\partial x_j + \partial W_j/\partial x_i)$ is established:

$$P_{ij} = -P\delta_{ij} + (\zeta_{\text{tot}} - \tfrac{2}{3}\mu)\,\delta_{ij}\,\text{div}\,\mathbf{W} + \mu\left(\frac{\partial W_i}{\partial x_j} + \frac{\partial W_j}{\partial x_i}\right) \qquad (1)$$

Here P is the static pressure (pressure in ideal fluid), ζ_{tot} is the coefficient of volumetric viscosity (which is not taken into account in the following), μ is the dynamic viscosity coefficient, δ_{ij} is the Kronecker symbol ($\delta_{ij} = 0$ when $i \neq j$ and $\delta_{ij} = 1$ when $i = j$), $\mathbf{W}$ is the velocity vector,

W_i is its component at the axis x_i, and i, j are the indices, according to which there takes place summation at their repetition ($i, j = 1, 2, 3$).

Note that, for example, the concept of viscosity is defined and has physical meaning, namely, within the framework of a viscous fluid which is used for describing a definite range of real phenomena. This concept has no physical meaning and should not be used for describing the same phenomena by the methods of statistical mechanics, although when establishing the relation with a viscous fluid, μ may be calculated on the basis of its definition and explained from the point of view of kinetic theory.

Within the framework of a viscous fluid any phenomena satisfying Eq. (1) may be uniquely mathematically described; in particular, by means of a closed system of differential equations and boundary conditions.

Usually this system covers (for a one-component medium):

(1) *Navier–Stokes equations of motion:*

$$\rho \frac{dW_i}{d\tau} = F_i \rho - \frac{\partial P}{\partial x_i} + \frac{\partial}{\partial x_i} [(\zeta_{\text{tot}} - \tfrac{2}{3}\mu) \operatorname{div} \mathbf{W}] + \frac{\partial}{\partial x_i} \mu \left(\frac{\partial W_i}{\partial x_j} + \frac{\partial W_j}{\partial x_i} \right)$$
$$(i = 1, 2, 3) \qquad (2)$$

where F_i is the projection of mass forces on the x_i axis and ρ is the density of a medium.

(2) *Equation of continuity:*

$$d\rho/d\tau + \rho \operatorname{div} \mathbf{W} = 0 \quad \text{or} \quad \partial\rho/\partial\tau + \operatorname{div} \rho\mathbf{W} = 0 \qquad (3)$$

(3) *Energy equation:*

$$\frac{di}{d\tau} = -\frac{\operatorname{div} \mathbf{q}}{\rho} + \frac{1}{\rho}\frac{dP}{d\tau} + \frac{\phi_i}{\rho} + \frac{q_v}{\rho} \qquad (4)$$

where i is the specific enthalpy (per unit mass); q_v is the density of distribution of heat sources per unit volume;

$$\phi_1 = \frac{\mu}{2} \left(\frac{\partial W_i}{\partial x_j} + \frac{\partial W_j}{\partial x_i} \right)^2 + (\zeta_{\text{tot}} - \tfrac{2}{3}\mu)(\operatorname{div} \mathbf{W})^2 = \phi + \zeta_{\text{tot}}(\operatorname{div} \mathbf{W})^2 \qquad (5)$$

ϕ is the dissipative Rayleigh function; $\mathbf{q}$ is the heat flux density per unit surface per unit time. In the simplest case of a one-component medium when the heat flux to a medium element is determined only by heat conduction, $\mathbf{q}$ is calculated by the Fourier equation

$$\mathbf{q} = -\lambda \operatorname{grad} T.$$

Here T is the medium temperature and λ is the thermal conductivity (the concept appearing when describing phenomena by a continuum model).

In a general case of a multicomponent medium, the expressions for flows of heat (as well as of mass) may be found, e.g., by the methods of thermodynamics for irreversible processes.

To complete the system of Eqs. (2)–(5), it is necessary to use the additional information on properties and physical laws of a specific problem under investigation with regard to the assumptions adopted. Usually these are equations such as

$$\mathbf{F} = \mathbf{g} = \text{const}; \quad \lambda = \lambda(T, P) \approx \lambda(T); \quad \mu = \mu(T, P) \approx \mu(T)$$

$$\xi_{\text{tot}} = \xi_{\text{tot}}(T, P); \quad F(P, T, \rho) = 0; \quad i = i(\rho, T)$$

and frequently $di = C_p\, dT$; $q_v = q_v(x_1, x_2, x_3, \tau)$. The boundary conditions include initial (time) and boundary (space) conditions. For unsteady processes in a viscous fluid, the prescription of the initial and boundary conditions for the velocity does not usually cause the difficulties that appear when prescribing temperature boundary conditions, e.g., a temperature at wall surfaces $T_w(x_{1w}, x_{2w}, x_{3w}, \tau)$ restricting viscous fluid flow at any time instant $\tau > \tau_0$.

The surface distribution T_w and the character of its change in time with unsteady heat transfer between flow and a wall constraining it depend not only upon hydrodynamic and thermophysical properties of the flow but also upon sizes, configuration, and thermophysical properties of this wall. However, if all boundary conditions (including $T_w = T_w(x_{1w}, x_{2w}, x_{3w}, \tau)$ are prescribed, then the problem is solved uniquely. Its solution allows the fields of all parameters in the flow to be found as functions of space coordinates and time.

It should be noted that the walls of the channels (their sizes, configuration, thermophysical properties, intensity and distribution of heat sources) may influence convective heat transfer by directly influencing only the boundary conditions. Here, we do not consider the effect of electromagnetic fields, which are usually considered to be prescribed and are taken into account by an additional equation of the form $\mathbf{F} = \mathbf{F}(x_1, x_2, x_3, \tau)$.

Therefore, the prescription of the temperature boundary condition in the form of a known function may be replaced by introducing the energy equation for the wall into the system (equation of heat conduction); i.e.,

$$\rho_w C_w\, \partial T/\partial \tau = \text{div}(\lambda_w\, \text{grad}\, T) + q_v \quad (6)$$

or, at $\lambda_w = \text{const}$,

$$\partial T/\partial \tau = a_w \nabla^2 T + q_v/\rho_w C_w, \quad (7)$$

where $a_w = \lambda_w/\rho_w C_w$ is the thermal diffusivity.

Moreover, the conditions are prescribed that the temperature of a wall and fluid as well as heat fluxes on both sides of the body–wall interface are equal.

In this case, the problem is frequently called a conjugated one, i.e., formulated for heat and wall simultaneously. Its solution allows flow and temperature fields and the temperature field in the wall surrounding it to be found at once. If the temperature boundary condition $T_w = T_w(x_{1w}, x_{2w}, x_{3w}, \tau)$ is prescribed, then this may be done independently and separately for flow and the wall surrounding it. The latter is more simple; however, generally, it is impossible to prescribe temperature boundary conditions beforehand. As will be shown, this difficulty can be overcome.

Now consider the difficulties of the description of unsteady turbulent flows. Usually, turbulent flows are described by averaged quantities. After averaging, the equations of Navier–Stokes, energy and continuity, which are considered to be true for instantaneous (true) values of the parameters in turbulent flow, are written down in terms of the averaged quantities. But in the equations there appear terms which are the moments of second and even third order between pulsation components of the parameters. These terms (turbulent stresses in the motion equation $-\rho \overline{W_i' W_j'}$ and turbulent heat flux $-C_p \rho \overline{T' W_1'}$ in the energy equation) appear as a result of the fact that the complex pulsating motion of the continuum is described by averaged values. These terms are new unknown quantities, and new equations should be introduced to close the system of equations. For steady-state flows this problem is being solved in a number of cases with the help of semiempirical turbulence theory or the distribution of the turbulence parameters found experimentally for simple flows. For unsteady flows these ways are still being developed (see Section II). Moreover, for unsteady flows there appears still another problem which is related to admissibility of the averaging of turbulent flows in time.

Regarding the prescription of the boundary conditions and, in particular, the effect of the channel walls surrounding the flow upon a temperature boundary condition, the previous remarks about laminar flows apply.

B. Statement of the One-Dimensional Problem

The solution of the above system of equations for flow and of the conjugated problem, even under steady-state conditions, is very complex and has not been obtained in many interesting, practical cases.

At the same time, in engineering practice the flow and temperature fields in the fluid are not of the greatest interest but only the flow rate, mean temperature, heat flux, wall temperature, and in some cases, temperature fields in the channel walls (i.e., the solution of the flow problem is interesting only from the point of view of determining boundary conditions for a wall).

Therefore, in engineering practice the one-dimensional way of describing heat transfer processes in channels (also in a boundary layer) is widely used as the method of calculation. In this method, flow in a channel is considered to occur with velocity and temperature constant over the channel cross section, which may vary only in one direction along the channel length x. Usually, the mean flow rate velocity

$$W = G/\rho f \tag{8}$$

is taken as a velocity. Here G is the mass flow rate, and f is the area of the cross section of a channel. A mean mass (bulk) temperature in this cross section of the channel

$$T_b = \int_0^{r_0} \rho W \tilde{C}_p T r \, dr \bigg/ \int_0^{r_0} \rho W \tilde{C}_p r \, dr \tag{9}$$

where

$$\tilde{C}_p = \frac{1}{T(r) - T(0)} \int_{T(0)}^{T(r)} C_p \, dT$$

is taken as a temperature. The relationship between the mean mass temperature of a heat agent T_b, specific heat flux through a unit surface of a wall q_w, and wall temperature is established by the relation

$$q_w = \alpha(T_w - T_b) \quad \text{or} \quad \alpha = q_w/(T_w - T_b) \tag{10}$$

where α is the dimensional quantity called the local heat transfer coefficient.

This coefficient takes into account how the real processes in three-dimentional flow determine heat transfer with a wall in a one-dimensional description of these processes.

In the one-dimensional theory the basic equations of motion, continuity, and energy are simplified and assume, for example, the following forms:

(1) *Equation of motion:*

$$\frac{G}{W}\frac{\partial W}{\partial \tau} + G\frac{\partial W}{\partial x} = f\bar{\rho}F_x - (1-\delta)f\frac{\partial P}{\partial x} \tag{11}$$

or

$$\frac{G}{W}\frac{\partial W}{\partial \tau} + G\frac{\partial W}{\partial x} = f\bar{\rho}F_x - f\frac{\partial P}{\partial x} - \xi\frac{\bar{\rho}W^2}{2d_e}f \qquad (12)$$

where F_x is the projection of the density of mass forces on the axis x, ξ is the local friction factor

$$\xi = -\delta\frac{\partial P}{\partial x}\bigg/\frac{\bar{\rho}W^2}{2d_e}, \qquad (13)$$

δ is the fraction of the longitudinal pressure gradient spent for friction and formation of a velocity profile, d_e is the equivalent diameter of a channel

$$d_e = 4f/U, \qquad (14)$$

U is the channel perimeter, $\bar{\rho}$ is the mean fluid density attributed to T_b and P in a cross section x.

(2) *Equation of continuity:*

$$(\partial\bar{\rho}/\partial\tau)f + \partial G/\partial x = 0 \qquad (15)$$

(3) *Equation of energy:*

$$\frac{di_b}{d\tau}f\bar{\rho} = Uq_w + f\frac{\partial}{\partial x}\left(\lambda_{\text{eff}}\frac{\partial T_b}{\partial x}\right) + f\frac{dP}{d\tau} + fT_b\sigma, \qquad (16)$$

where

$$i_b = \int_f \rho iW\, df/G, \qquad (17)$$

λ_{eff} is the mean effective heat conduction over a cross section in the x direction, and σ is the production of entropy per unit volume due to viscous friction.

In common with many heat transfer problems, the last three terms on the right-hand side of Eq. (16) are much less than the first term and often $di_b = C_p\, dT_b$ (a process close to an isobaric one; i.e., $\partial P/\partial x$ are small or the fluid is an ideal gas, $P = \rho RT$). Therefore, the energy equation (16) with the aid of Eq. (10) is simplified:

$$\frac{GC_p}{W}\frac{\partial T_b}{\partial \tau} + GC_p\frac{\partial T_b}{\partial x} = U\alpha(T_w - T_b) \qquad (18)$$

Thus, considerable mathematical simplification of the problem is achieved in the case of a one-dimensional flow model by introducing the coefficient of heat transfer and the friction factor, which are related in a complex way to real three-dimensional flow and cannot, in principle, be determined within the framework of the one-dimensional flow theory. They are found either from experiment or from the solution of the above three-dimensional system of equations with the help of their definitions, Eqs. (10) and (13).

However, if the predetermined conditions α and ξ are known, then the solution of a one-dimensional flow appears much more simple than a three-dimensional one. When α and T_b are known, it is not difficult to find a temperature field in the channel walls by solving the heat conductions equations (6) or (7) under the so-called boundary condition of the third kind obtained from Eq. (10):

$$T_w = T_b + \frac{q_w}{\alpha} = T_b - \frac{\lambda_w \, \partial T/\partial n}{\alpha} \tag{19}$$

where **n** is the normal to the wall surface at a given point.

The above one-dimensional way of describing heat transfer processes and the concept on the heat transfer coefficient resulting from this method proved successful in engineering practice for solving steady-state problems. Moreover, sometimes this concept may be successfully applied even in case α is variable along the channel perimeter, for example, in the noncircular channel. When the distribution of α along the perimeter and the channel length (for some distribution of T_w along the perimeter) is known and its weak dependence upon the character of the distribution of T_w along the perimeter is established (or when this dependence is known), then it is possible to solve the heat conduction problem, Eqs. (7) and (19), and to find a temperature field in the channel walls (*1*).

Nevertheless, in a number of works (*2–8*) there are some objections to the application of a one-dimensional way of description and, in particular, to the use of the concept of the heat transfer coefficient when a temperature boundary condition is unknown beforehand, and even when it is known but $T_w(x) \neq$ const. This problem is presented in more detail in the work by Luikov *et al.* (*5*), which is especially devoted to describing the ways of solution of conjugated problems in those cases when, following the opinion of the authors of Ref. (*5*), the concept of α becomes inapplicable.

Luikov *et al.* (*5*) consider that the local value of α has a physical meaning only when $T_w =$ const. With a change in the wall temperature, they contend the heat transfer coefficient is already not a physical characteristic of a heat transfer process, but is only some quantity dependent on many variables, including also a temperature boundary condition. To illustrate this it is shown that for the steady case under different surface conditions (T_w = const, q = const, etc.) α and Nu have different values for laminar flow. It should also be noted that with heating and a gradual decrease in a wall temperature along the channel length α may become negative. According to the opinion of the authors of Ref. (5) this illustrates, to what absurdities the use of the concept of α may lead. In other works by Luikov and Perelman (3, 4) devoted to the solution of an unsteady conjugated problem [See Section III,D] the inapplicability of the heat transfer coefficient and boundary conditions of the third kind of studying and solving unsteady heat transfer problems is also emphasized.

It is clear that a one-dimensional way of description is not a generalized one. The formulation of unsteady heat transfer problems in the form of three-dimensional conjugated ones is a more general approach.

However, the wide and successful application of a one-dimensional way, the concept of α, and boundary conditions of the third kind in engineering practice in steady problems under different laws of a change in a wall temperature along the channel length leads us to thoroughly analyze the possibility of their application in unsteady conditions (especially for turbulent flows). The necessity of a serious analysis of the possibilities of a one-dimensional description of unsteady problems is conditioned by the fact that despite some known advances (5), the conjugated problems are still very complex for theoretical analysis and give little perspective for an experimental study with modeling due to the necessity of ensuring the similarity with respect to a great number of criteria.

Let us analyze the degree of dependence of the local α(Nu) upon the character of a change in the boundary conditions.

According to the definitions in Eqs. (10) and (9),

$$\alpha = \left(\lambda \frac{\partial T}{\partial y}\right)_{y=0} \bigg/ T_w - \frac{\int_f \rho W_x \tilde{C}_p T \, df}{\int_f \rho W_x \tilde{C}_p \, df}$$

the local value of α in this cross section will depend upon the prehistory of $T_w(x)$ upwards at a section not longer than the length, at which in the channel there occurs a meeting of the thermal boundary layers with prescribed hydrodynamics. For a tube in laminar flow

$$x_{\text{lam}} = \frac{2}{d} \int_0^{d/2} \frac{y^2 \, W_x(y)}{a} \, dy \tag{20}$$

and in turbulent flow

$$x_{\text{tur}} = \frac{2}{d} \int_0^{d/2} \frac{y^2 \, W_x(y)}{a + \epsilon_q(y)} \, dy \tag{21}$$

Applying the concrete law of change of $T_w(x)$ and with constant hydrodynamic conditions, x_{lam} and x_{tur} approximately coincide with a thermal stabilization section. For example, for $T_w = $ const, after substituting $W_x = 8W(y/d - y^2/d^2)$ into Eq. (20) we find $x_{\text{lam}}/d = 0.075\,\text{Pe}$. From the condition that $(\text{Nu} - \text{Nu}_\infty)/\text{Nu}_\infty = 0.01$, the length of the thermal section (9) is $l_T/d = 0.055\,\text{Pe}$.

For a variable law of a change in $T_w(x)$, the thermal stabilization section will be larger, as for example, at $q_w = $ const. In this case, after establishing the law of a change in $T_w(x)$ close to a linear one, it is necessary to have a distance of the order of x_{lam} to completely stabilize a temperature profile and Nu.

For turbulent flow, the distance at which the prehistory of the influence of T_w upon α or Nu plays an essential role is considerably less than x_{tur} since, as is known, the changes of a temperature profile in the core of turbulent flow slightly influence α and Nu. If in this case the power velocity profile with an exponent of $\frac{1}{7}$ is substituted into Eq. (21) and the mean value $(a + \epsilon_q)/a = 40$ is taken at a section $0.1d/2$ from the wall, then upon integration of Eq. (21) when $\text{Pe} = 10^4$, we have $x_{\text{tur}}/d \approx 0.1$. It is clear that under real conditions any laws of change in T_w at such a distance (and even to a greater order) may be successfully approximated by one linear law at this section. It is therefore known from experiments that in turbulent flows heat transfer is only slightly sensitive to different laws governing changes in T_w along the tube length.

Also, it should be noted that in laminar flow, the effect of the influence of T_w upon Nu_∞ is not so great: at $T_w = $ const, $\text{Nu}_\infty = 3.66$ and at $T_w = ax + b$ ($q_w = $ const), $\text{Nu}_\infty = 4.36$ (1.19 times greater). According to Vilenskii (10) (see Section III) $T_w = Ae^x$ gives a further increase in Nu_∞ of less than 4%. In all the laws of change in $T_w(x)$, starting from a linear one, for which

$$\lim_{x \to \infty} \frac{1}{T_w} \frac{\partial T_w}{\partial x} = 0$$

$\text{Nu}_\infty = 4.36$, and only if

$$\lim_{x \to \infty} \frac{1}{T_w} \frac{\partial T_w}{\partial x} \to \infty$$

(e.g., $T_w = Ae^{bx^n}$ at $n \geqslant 2$, which is improbable), Nu is stabilized nowhere. It is true that at the stabilization section, $\text{Nu} > \text{Nu}_\infty$; but for the above practically important laws

$$\left(\lim_{x \to \infty} \frac{1}{T_w} \frac{\partial T_w}{\partial x} = 0 \right)$$

the logarithmic derivatives also differ slightly at the stabilization section (at finite x) and, consequently, Nu will differ slightly. If in turbulent flow $T_w(x)$ is prescribed at a considerable change in T_w at the section Δx (where the prehistory is essential), then Nu increases (heating of a fluid $\partial T_w/\partial x > 0$). But due to a small value of Δx the practically possible laws are successfully approximated by the linear ones (expanding into Taylor's series when only one term is taken into account):

$$\frac{T_w(X)}{T_w(X_0)} = 1 + \frac{(X-X_0)}{1!\, T_w(X_0)} \frac{\partial T_w(X_0)}{\partial X} + \cdots + \frac{(X-X_0)^n}{n!\, T_w(X_0)} \frac{\partial^n T_w(X_0)}{\partial X^n} + \cdots \quad (22)$$

Here

$$X = \frac{1}{\text{Pe}} \frac{x}{d}; \qquad [0, X_0]_{\text{tur}} = \frac{x_{\text{tur}}}{\text{Pe}\, d}$$

In this case, $[0, X_0]_{\text{tur}} = 10^{-5}$. Consequently, it may be expected that in the majority of the practically possible values of $\partial T_w/\partial x$ the change in the local Nu will not depend upon the law of change in $T_w(x)$ itself but only upon its first derivative at a given point. It follows that it is quite possible to establish, even for laminar flows, the dependence of local Nu upon $T_w(x)$ which is general for many laws of change in $T_w(x)$. For turbulent flows, if necessary, this would lead to the experimental or theoretical study of the dependence of Nu upon a new criterion of the form

$$\frac{\partial T_w}{\partial x} \frac{d}{T_w} \quad \text{or} \quad \frac{1}{T(X)} \frac{\partial T_w(X)}{\partial X}$$

which is simpler than solving conjugated problems.

This is well confirmed by an interesting work of Leontiev et al. (11) which is particularly devoted to an experimental study of the effect of a boundary condition $T_w(x)$ upon air-heat transfer in a turbulent boundary layer. On a flat plate, L in length, the following laws of change in temperature boundary conditions along the plate length were investigated:

(1) $T_w = \text{const}$
(2) $T_w - T_0 = b - a(x/L)$
(3) $T_w - T_0 = b + a(x/L)$
(4) $q_w = q_0 \exp[a(x/L)]$
(5) $T_w - T_0 = (T_w - T_0)_0 \exp[a(x/L)]$

The results of the experiments in the coordinates

$$\text{St} = \frac{q_w}{\gamma_0 W_0 C_{p0}(T_w - T_0)} \left(\frac{T_w}{T_0}\right)^{0.5}, \qquad \text{Re}_x = \frac{\rho_0 W_0 x}{\mu_0}$$

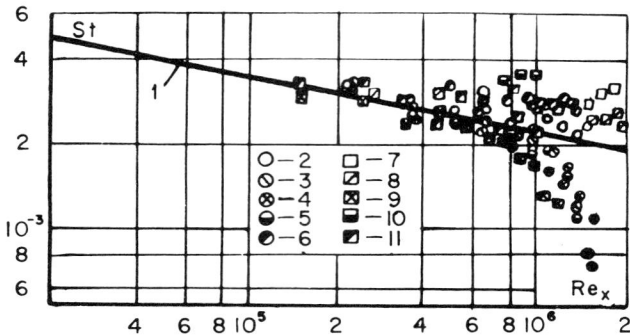

FIG. 1. St versus Re_x for various laws of wall temperature variation:
(1) $St_0 = 0.0288/Re_x^{0.2} Pr^{0.6}$;
(2) $T_w - T_0 = 158.6°C$;
(3) $T_w - T_0 = (222 - 150X)°C$;
(4) $T_w - T_0 = (137 - 81X)°C$;
(5) $T_w - T_0 = (204 - 140X)°C$;
(6) $T_w - T_0 = (44 - 170X)°C$;
(7) $T_w - T_0 = 0.19 \exp(7.0X)°C$;
(8) $T_w - T_0 = (34 + 159X)°C$;
(9) $T_w - T_0 = (159 - 100X)°C$;
(10) $q_w = 3.4 \exp(5.8X)$ W/m²;
(11) $T_w - T_0 = (38 + 110X)°C$.

are presented in Fig. 1, from which it is seen how heat transfer is essentially influenced by $\partial T_w/\partial x$ and by the thickness of a boundary layer δ_t, which increases with Re_x, i.e., with the criterion

$$\frac{\partial T_w}{\partial x} \frac{\delta_t}{T_w - T_b} \quad \left(\text{or} \quad \frac{\partial T_w}{\partial x} \frac{\delta_t}{T_w}\right)$$

In addition, at $\partial T_w/\partial x > 0$, the experimental points lie considerably above those at $T_w = $ const and, at $\partial T_w/\partial x < 0$, considerably lower.

The distributions of velocities and temperature with respect to the thickness of a boundary layer are, respectively, given in Fig. 2a and b. The velocity profile is not sensitive to a change in $\partial T_w/\partial x$, but the temperature profile essentially depends upon it. For $\partial T_w/\partial x > 0$ the temperature profile is more full than at $\partial T_w/\partial x = 0$, and at $\partial T_w/\partial x < 0$ it is less full.

The authors of Ref. (11), analyzing the energy equation

$$\frac{d \, Re_T^{**}}{dX} + \frac{Re_T^{**}}{\Delta T} \frac{d\Delta T}{dX} = St \, Re_L$$

have chosen as a criterion the complex

$$\Gamma_T = \frac{Re_T^{**}}{St_0 \, Re_L} \frac{1}{\Delta T} \frac{d\Delta T}{dX} = \left(\int_0^x q_w \, dx/q_w L\right) \frac{1}{\Delta T} \frac{d\Delta T}{dX}$$

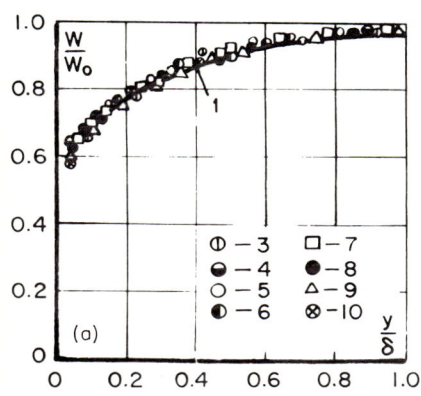
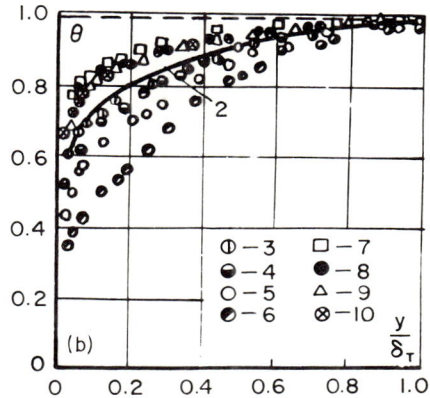

Fig. 2. Change of relative velocity (a) over boundary-layer thickness and of dimensionless temperature (b) over thermal boundary-layer thickness with different laws of wall temperature variation:

(1) $(W/W_0) = (y/\delta)^{1/7}$;
(2) $\theta = (T_w - T)/(T_w - T_0) = (y/\delta_T)^{1/7}$;
(3) $\Gamma_T = -0.57$;
(4) $\Gamma_T = -1.37$;
(5) $\Gamma_T = -3.22$;
(6) $\Gamma_T = -10.1$;
(7) $\Gamma_T = 0.62$;
(8) $\Gamma_T = 0.6$;
(9) $\Gamma_T = 0.52$;
(10) $\Gamma_T = 0.5$.

i.e.,

$$\text{Re}_T^{**} = \delta_T^{**} \rho_0 W_0/\mu_0 ; \qquad X = x/L; \qquad \Delta T = T_w - T_0$$

$$\delta_T^{**} = \int_0^x q_w \, dx/\gamma_0 W_0 C_{p0} \Delta T$$

St/St$_0$ versus this criterion is given in Fig. 3.

Probably the criterion $(d\Delta T/dx) \delta_T/\Delta T$, taking into account the thickness of a thermal boundary layer as a scale for thermal signal propagation, would better satisfy the physics of a process. The criterion Γ_T regards for the prehistory of q_w along the whole length x (at $q_w = $ const,

$$\Gamma_T = \frac{X}{\Delta T} \frac{d\Delta T}{dX} = \frac{x}{\Delta T} \frac{dT_w}{dx}$$

which can influence Nu in terms of δ_T. It may be expected from the analysis of Fig. 1 that the relation

$$\frac{\text{St}}{\text{St}_0} = f\left(\frac{d\Delta T}{dx} \frac{\delta_T}{\Delta T}\right)$$

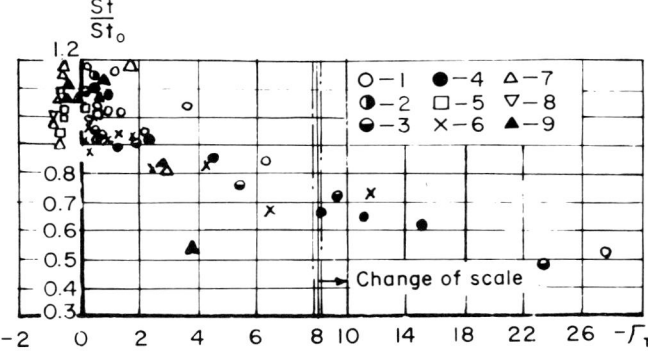

FIG. 3. St/St_0 versus Γ_T:
(1) $T_w - T_0 = (204 - 140X)°C$;
(2) $T_w - T_0 = (44 + 170X)°C$;
(3) $T_w - T_0 = (137 - 81X)°C$;
(4) $T_w - T_0 = (222 - 150X)°C$;
(5) $T_w - T_0 = (38 + 110X)°C$;
(6) $T_w - T_0 = (159 - 100X)°C$;
(7) $T_w - T_0 = (34 + 159X)°C$;
(8) $T_w - T_0 = 0.19 \exp(7.0X)°C$;
(9) $q_w = 3.4 \exp(5.8X)$ W/m².

would better generalize experimental points, and Fig. 3 confirms that, for turbulent flows, heat transfer depends not upon the law of change in $T_w(x)$ but, due to the "short memory" of flow, only upon $\partial T_w/\partial x$. For unsteady heat transfer, this is extended to a change of T_w in time. For laminar flow, the characteristic time $\Delta \tau$, during which Nu at a moment τ_0 may depend upon the prehistory of a change of $T_w(\tau)$ in the interval $(\tau_0, \tau_0 - \Delta \tau)$, will be $\Delta \tau_{\text{lam}} = d^2/4a$. For turbulent flow this characteristic time is

$$\Delta \tau_{\text{tur}} \approx \frac{2}{0.1d} \int_0^{0.1d/2} \frac{y^2\, dy}{a + \epsilon_q(y)} \tag{23}$$

Assuming, as in the above example, the mean integral value $(a + \epsilon_q)/a = 40$ for a layer with a thickness of 10% radius, we find, from Eq. (23) $\Delta \tau_{\text{tur}} \simeq d^2/a \cdot 48{,}000 = \Delta \tau_{\text{lam}}/12{,}000$, that at $d = 0.01$ m and air at 1 atm ($a = \frac{1}{6} \cdot 10^{-4}$ m²/sec) gives $\Delta \tau_{\text{lam}} = 1.5$ sec and $\Delta \tau_{\text{tur}} = 1.25 \cdot 10^{-4}$ sec. For water, $\Delta \tau_{\text{lam}} = 150$ sec and $\Delta \tau_{\text{tur}} = 1.25 \cdot 10^{-2}$ sec.

It is obvious that for turbulent flow in such a tube any practically possible laws may be successfully approximated by one term of the Taylor expansion with respect to $\Delta \tau_{\text{tur}} a/d^2 = [0, \text{Fo}_1]$:

$$\frac{T_w(\text{Fo})}{T_w(\text{Fo}_1)} = 1 + \frac{(\text{Fo} - \text{Fo}_1)}{1!\, T_w(\text{Fo}_1)} \frac{\partial T_w(\text{Fo}_1)}{\partial \text{Fo}} + \cdots + \frac{(\text{Fo} - \text{Fo}_1)}{n!\, T_w(\text{Fo}_1)} \frac{\partial^n T_w(\text{Fo}_1)}{\partial \text{Fo}^n} + \cdots \tag{24}$$

Then again, it is possible to find theoretically or experimentally the dependence of $\alpha(\tau)$ on $\partial T_w(\tau_0)/\partial \tau$, (and not upon the law of change in T_w which is valid during $[\tau_0 - \Delta\tau_{tur}, \tau_0]$) and the dependence of Nu upon the additional criterion of the form $(\partial T_w/\partial \text{Fo})(1/T_w)$. In (12) for a laminar boundary layer the series expansion with respect to the parameters

$$\frac{1}{T_w}\frac{\partial T_w}{\partial \tau}\left(\frac{\delta^2}{a}\right), \quad \frac{1}{T_w}\frac{\partial^2 T_w}{\partial \tau^2}\left(\frac{\delta^2}{a}\right)^2, \quad \frac{1}{T_w}\frac{\partial^n T_w}{\partial \tau^n}\left(\frac{\delta^2}{a}\right)^n$$

is assumed.

In the general case, when $T_w = T_w(x, \tau)$, the Taylor expansion will be

$$\frac{T_w(X, \text{Fo})}{T_w(X_0, \text{Fo}_1)} = 1 + \frac{(X - X_0)}{T_w(X_0, \text{Fo}_1)}\frac{\partial T_w(X_0, \text{Fo}_1)}{\partial X} + \frac{(\text{Fo} - \text{Fo}_1)}{T_w(X_0, \text{Fo}_1)}$$
$$\cdot \frac{\partial T_w(X_0, \text{Fo}_1)}{\partial \text{Fo}} + \cdots + \frac{1}{n!}\frac{d^n T_w(X_0, \text{Fo}_1)}{T_w(X_0, \text{Fo}_1)} \quad (25)$$

From the above estimates it is seen that for turbulent flow the lows of change in T_w realized in practice may be almost always successfully approximated by linear ones within the intervals $[0, X_0]$ and $[0, \text{Fo}_1]$. Moreover, it is always possible to estimate the intervals and to make sure whether it is enough to take only the linear terms in expansion (25). Therefore, it is necessary only to study the dependence of Nu upon

$$\frac{1}{T_w(X, \text{Fo})}\frac{\partial T_w(X, \text{Fo})}{\partial X} \quad \text{and} \quad \frac{1}{T_w(X, \text{Fo})}\frac{\partial T_w(X, \text{Fo})}{\partial \text{Fo}}$$

$$\left(\text{or} \quad \frac{d}{T_w}\frac{\partial T_w}{\partial x} \quad \text{and} \quad \frac{d^2}{aT_w}\frac{\partial T_w}{\partial \tau}\right) \quad (26)$$

Similarly, we may take into account the influence of changes of other boundary conditions in time upon heat transfer: $T_{in} = T_{in}(\tau)$ and $G = G(\tau)$. We use the Zhukovsky number $\text{Zh} = \tau v/d^2$ as a dimensionless time for $G = G(\tau)$. Thus, for turbulent flow with regard to small time $\Delta\tau_{tur}$ or $[0, \text{Fo}_1]_{tur}$ the effect of the prehistory of $T_{in}(\tau)$ and $G(\tau)$ upon Nu may be taken into account by such dimensionless parameters, respectively:

$$\frac{1}{T_{in}(\text{Fo})}\frac{dT_{in}(\text{Fo})}{d\text{Fo}} \quad \text{and} \quad \frac{1}{G(\text{Zh})}\frac{\partial G(\text{Zh})}{\partial \text{Zh}} \quad (27a)$$

or

$$\frac{d^2}{aT_{in}}\frac{dT_{in}(\tau)}{d\tau} \quad \text{and} \quad \frac{d^2}{v\, G(\tau)}\frac{dG(\tau)}{d\tau} \quad (27b)$$

In the majority of cases, these will be accompanied by dimensionless parameters of the form (26), because unsteady changes in T_{in} and G will be accompanied by the appropriate changes of T_w.

The friction factor ξ may be influenced by the boundary conditions $T_{in}(\tau)$ and $T_w(x, \tau)$ only by means of a change in the physical properties that influence the velocity profile and turbulent production. For this reason a change in the physical properties will also influence the Nu number. The friction factor depends upon $G(\tau)$ owing to the change in the velocity profile and the turbulent flow structure.

Under unsteady-state conditions, the form of functional relations for Nu and ξ with regards to the above in the general case of the cylindrical channel will be

$$\mathrm{Nu}_b = f\left(\frac{x}{d}, \mathrm{Re}_b, \mathrm{Pr}_b, \frac{\mu_w}{\mu_b}, \frac{\lambda_w}{\lambda_b}, \frac{\rho_w}{\rho_b}, \frac{C_{pw}}{C_{pb}}, \frac{d}{T_w}\frac{\partial T_w}{\partial x},\right.$$

$$\left.\frac{d^2}{a_b T_w}\frac{\partial T_w}{\partial \tau}, \frac{d^2}{a_b T_{in}}\frac{dT_{in}}{d\tau}, \frac{d^2}{\nu_b G}\frac{dG}{d\tau}\right) \tag{28}$$

$$\xi_b = f\left(\frac{x}{d}, \mathrm{Re}_b, \frac{\mu_w}{\mu_b}, \frac{\rho_w}{\rho_b}, \frac{d^2}{\nu_b G}\frac{dG}{d\tau}\right) \tag{29}$$

It is important that relations (28) and (29) do not include explicit dimensionless time in the form $\mathrm{Ho} = w\tau/d$, $\mathrm{Fo} = a\tau/d^2$, $\mathrm{Zh} = \nu\tau/d^2$, etc. Since the values of the parameters are taken all together, the time scale does not play an essential role either when modeling or calculating according to Eqs. (28) and (29).

However, we should bear in mind that for unsteady heat transfer there exist two regimes: (1) pure heat conduction, i.e., unsteady heating of the flow of a fluid which acts as a solid with variable heat conductivity from the onset of an unsteady process up to the moment when the fluid being at the inlet into a heat transfer section penetrates into the adjacent region of a given cross section, and (2) unsteady convective heat transfer when unsteady heating of a fluid is superimposed on steady convective heat transfer.

In a turbulent flow the second regime starts at a time close to $x/W = (xd/\nu)(1/\mathrm{Re})$.

The search for the general dependences of $\mathrm{Nu}(\tau)$ upon unsteady temperature boundary conditions is also quite useful for the classes of $T_w(\tau)$ interesting in practice and may be, to a considerable extent, solved theoretically.

Note that Vilenskii (13) (see in detail, Section III) has shown that if, for example,

$$\lim_{\tau \to \infty} \frac{d^2}{4a} \frac{1}{T_w(\tau)} \frac{\partial T_w(\tau)}{\partial \tau} \to 0 \tag{30}$$

then at sufficiently great τ (or Fo $= 4a\tau/d^2 = \tau/\Delta\tau$) there begins stabilization of Nu for both the first (similarly to the onset of a regular regime for a solid) and the second regions.

In the second region Nu_∞ will be equal to a steady value. Since the logarithmic derivatives for the laws obeying condition (30) (there are many such practically important laws) are also close to one another during stabilization, then at the sections of time stabilization Nu should be expected to change approximately in the same way.

Thus, if from the solution of three-dimensional problems the common dependence of the Nu number upon the rate of a change in temperature boundary conditions [for example, upon the parameter of the form $(\partial T_w/\partial \tau)(d^2/(T_w - T_b)a]$ will be established theoretically or experimentally, for some class of laws $T_w(\tau)$, then the application of the one-dimensional theory for engineering calculations of unsteady heat transfer will also be effective as it is for steady heat transfer. In this case, for example, the solution of the unsteady heat conduction problem, Eq. (7), under boundary conditions of the third kind, Eq. (19), by the method of successive approximations does not cause any principal difficulties (see Section III,A).

This reasoning allows a statement of future investigations of unsteady heat transfer in channels.

1. *Theoretical Investigations*

(a) Theoretical analysis of steady and unsteady laminar flows with different laws of change in temperature boundary conditions;

(b) Improvement of the methods for solving unsteady heat conduction problems, Eq. (6), for complex configurations with the third-kind of boundary conditions;

(c) Generalization of the semiempirical turbulence theories for unsteady turbulent flows;

(d) Theoretical analysis of steady and unsteady turbulent flows with different laws of change in temperature boundary conditions and fluid flow rate; establishment of general dependences of the Nu number upon these laws;

(e) Improvement of the methods for solving conjugated problems and theoretical analysis of such problems.

2. *Experimental Investigations*

(a) Development and improvement of the methods for experimental study of unsteady heat transfer and friction as well as the structure of unsteady turbulent flows;

(b) Development and making of apparatus for studying unsteady heat transfer, hydrodynamics, and the structure of turbulent flows;

(c) Experimental study of a structure of unsteady turbulent flows;

(d) Experimental investigation of unsteady heat transfer and pressure drop in viscous, viscogravitational, and turbulent flows and their dependences upon unsteady temperature and velocity boundary conditions.

The validity of such a statement of problems for further investigations is confirmed by the state of modern investigations on unsteady heat transfer which is described in the following sections.

II. Hydrodynamics of Unsteady Fluid Flow in Channels

Besides the direct application of the development of the methods for calculating pressure losses, the study of the hydrodynamics of unsteady flows in channels has some other, no less important, purposes. In unsteady turbulent flows the study of hydrodynamics and flow structure, on the one hand, allows correct generalization of an experiment and calculation of unsteady heat transfer and, on the other hand, essensially helps in understanding of the nature of turbulent flow itself.

A. One-Dimensional Theory

As is shown in Section I, in a one-dimensional description of real three-dimensional flows all the peculiarities of real flow in the case of its simplified one-dimensional description may be expressed by the hydraulic resistance coefficient ξ.

In this case the equations of motion and continuity are of the form, respectively,

$$\frac{G}{W}\frac{\partial W}{\partial \tau} + G\frac{\partial W}{\partial x} = f\bar{\rho}F_x - f\frac{\partial P}{\partial x} - \xi\frac{PW^2}{2d}f \tag{12}$$

$$\frac{\partial \rho}{\partial \tau}f + \frac{\partial G}{\partial x} = 0 \tag{15}$$

For incompressible fluid and a variable cross section channel, from the continuity equation ($\partial G/\partial x = 0$) we have

$$G\, \partial W/\partial x = -\rho W^2\, \partial f/\partial x \tag{31}$$

Then, the expression for the coefficient of hydraulic losses is

$$\xi = \frac{-(\partial P/\partial x)d}{\rho W^2/2} - \frac{2d}{W^2}\left[\frac{\partial W}{\partial \tau} - F_x\right] + \frac{2d}{f}\frac{\partial f}{\partial x} \quad (32)$$

(Note that in the gravitational field $F_x = g_x$.) The mean integral value of the coefficient of hydraulic losses along the length at $f =$ const will be

$$\bar{\xi} = \frac{1}{l}\int_0^l \xi(x,\tau)\,dx = \frac{P_0 - P_l}{(\rho W^2/2)(l/d)} - \frac{2d}{W^2}\left[\frac{\partial W}{\partial \tau} - g_x\right] \quad (33)$$

For a compressible ideal gas ($P = \rho RT$) with regard for heat transfer and dissipation of mechanical energy at a constant cross section of a channel, the coefficient of hydraulic losses may be presented as *(14)*

$$\xi = -(1 - M^2K)\frac{(\partial P/\partial x)d}{(\rho W^2/2)}$$

$$-\frac{2d}{W^2}\left[\frac{\partial W}{\partial \tau} - g_x - \frac{MK}{\rho(KRT)^{1/2}}\frac{\partial P}{\partial \tau} + W\frac{\partial \ln T}{\partial \tau} + W^2\frac{\partial \ln T}{\partial x}\right] \quad (34)$$

When deriving Eq. (34), in addition to the equations of motion (12) and continuity (15), the following are used: energy equation for an ideal gas

$$C_p\frac{dT}{d\tau} = T\frac{dS}{d\tau} + \frac{1}{\rho}\frac{dP}{d\tau}$$

where

$$T\frac{dS}{d\tau} = -\frac{\operatorname{div}\mathbf{q}}{\rho} + \frac{\phi}{\rho}$$

the expression for the sound velocity

$$W_{SV}^2 = KRT = \frac{1}{(\partial \rho/\partial P)_S}$$

and the assumption that $\rho = \rho(P, S)$, $P = P(\tau, x)$, and $S = S(\tau, x)$.

To determine the mean integral value of the friction coefficient along the channel length Eq. (34) may be integrated by using the mean value theorem:

$$\bar{\xi} = (1 - K\bar{M}^2)\frac{P_0 - P_l}{(\bar{\rho}\bar{W}^2/2)(l/d)}$$

$$-\frac{2d}{\bar{W}^2}\left[\frac{\partial \bar{W}}{\partial \tau} - g_x - \frac{\bar{M}K\,\partial\bar{P}/\partial\tau}{\bar{\rho}(KRT)^{1/2}} + \bar{W}\frac{\partial \ln T}{\partial \tau} + \frac{\bar{W}^2}{l_h}\ln\frac{T_{lh}}{T_0}\right] \quad (35)$$

where l_h is the length of a heating section ($l_h \leqslant l$).

It is important to note that when calculating flows at $\partial W/\partial \tau \neq 0$ or $\partial W/\partial x \neq 0$ it is necessary to use ξ determined experimentally or theoretically under the same conditions.

Pressure losses due to friction and the redistribution of the velocity profile especially in turbulent flows will differ from steady isothermal flow due, at least, to difference in profiles of tangential stresses. These differences will be analyzed in the next section.

B. Quasi-Stationary Theory and Its Qualitative Analysis

The quasi-stationary theory, as the method of calculation is usually called, uses the basic assumption that at each given moment the real characteristics of the flow in a channel are replaced by steady characteristics, i.e., by such characteristics as would exist in steady and, generally speaking, in isothermal flow with regime parameters (flow rate, temperature, pressure, boundary conditions) equal to the instantaneous values of unsteady flow.

For unsteady turbulent axisymmetric flow of an incompressible fluid in a tube, the equation of motion after comparative estimation of its terms may be written down as

$$\rho \frac{\partial W_x}{\partial \tau} + \rho W_x \frac{\partial W_x}{\partial x} + \rho W_r \frac{\partial W_x}{\partial r} = -\frac{\partial P}{\partial x} + \frac{T}{r} + \frac{\partial T}{\partial r} \tag{36}$$

$$\rho \frac{\partial W_2}{\partial \tau} + \rho W_r \frac{\partial W_r}{\partial r} = -\frac{\partial P}{\partial r} - \frac{1}{2}\rho \frac{\partial \overline{W_r'^2}}{\partial r} \tag{37}$$

and the continuity equation

$$\frac{\partial W_x}{\partial x} + \frac{1}{r}\frac{\partial}{\partial r}(rW_r) = 0 \tag{38}$$

Here

$$T = (\mu + \mu_T)\frac{\partial W_x}{\partial r} \tag{39}$$

In a quasi-stationary case these equations will be

$$\left(\frac{\partial P}{\partial x}\right)_K = \frac{1}{r}\frac{\partial}{\partial r}(rT_K)$$
$$W_r = 0; \quad \partial W_x/\partial x = 0 \tag{40}$$

The subscript K means a quasi-stationary value of a quantity.

Heat Transfer and Hydrodynamics in Channels

For quasi-stationary flow, the pressure gradient is determined from a one-dimensional motion equation (12), assuming

$$-\xi \rho W^2/2d = (\partial P/\partial x)_K \tag{41}$$

Neglecting mass forces, we find from Eq. (12) with regard for Eq. (41)

$$-(\partial P/\partial x)_K = -\partial P/\partial x - \rho\, \partial W/\partial \tau - \rho W\, \partial W/\partial x \tag{42}$$

From Eq. (42) it follows that $(\partial P/\partial x)_K$ is constant with respect to a radius. Then from Eq. (40) we obtain a linear distribution of tangential stress with respect to a radius

$$T = (r/2)(\partial P/\partial x)_K \tag{43}$$

Designating, in Eq. (36), the fraction of the pressure gradient spent for overcoming tangential stresses

$$\left(\frac{\partial P}{\partial x}\right)_T = \frac{T}{r} + \frac{\partial T}{\partial r} = \frac{1}{r}\frac{\partial}{\partial r}(rT) \tag{44}$$

we find

$$-\left(\frac{\partial P}{\partial x}\right)_T = -\frac{\partial P}{\partial x} - \rho\frac{\partial W_x}{\partial \tau} - \rho W_x \frac{\partial W_x}{\partial x} - \rho W_r \frac{\partial W_x}{dr} \tag{45}$$

Obviously $-(\partial P/\partial x)_T = f(r)$ at $x = $ const.

The effect of flow acceleration in time $(\partial W_x/\partial \tau > 0)$ and along the length $(\partial W_x/\partial x > 0)$ upon the character of the dependence $-(\partial P/\partial x)_T$ with respect to a radius is qualitatively the same (as well as in the flow deceleration). The term with the radial velocity will play a secondary role which will be considered below.

With flow acceleration, $-(\partial P/\partial x)_T$ will monotonically decrease from $-\partial P/\partial x$ at the wall to a minimum, $-(\partial P/\partial x)_{T\min}$, at the tube axis. In this case from comparing Eqs. (42) and (45) it follows that

$$|(\partial P/\partial x)_{T\min}| < |(\partial P/\partial x)_K| < |\partial P/\partial x| \tag{46}$$

It is clear that the character of a growth in tangential stress modulus with respect to a radius from the axis to a wall

$$|T(r)| = \left|\frac{1}{r}\int_0^r -r\left(\frac{\partial P}{\partial x}\right)_T dr\right| \tag{47}$$

will be described by a curve with gradient greater than for curve describing $|-\partial P/\partial x|_T$.

However, if

$$-\frac{1}{\pi r_w^2}\int_0^{r_w} 2\pi r \left(\frac{\partial P}{\partial x}\right)_T dr = -\left(\frac{\partial P}{\partial x}\right)_K \tag{48}$$

then the tangential stress at a wall in real and quasi-stationary flows will be the same.

Equation (48) will be valid if, when averaging the right-hand side of Eq. (45) with respect to a cross section, it will convert into the right-hand side of Eq. (42). For this it is necessary that the integral

$$\int_0^{r_w} \rho W_r \frac{\partial W_x}{\partial r} 2\pi r\, dr = 0 \tag{49}$$

or with regard to the continuity equation (38)

$$-\int_0^{r_w} \rho \frac{\partial W_x}{\partial r}\left(\frac{\partial W_r}{\partial r} + \frac{\partial W_x}{\partial x}\right) 2\pi r^2\, dr = 0 \tag{50}$$

These conditions are not obvious although it is clear that deformation of a velocity profile with acceleration should be accompanied by two regions $W_r > 0$ and $W_r < 0$ at least. Moreover, three-dimensional flows may also appear. If $\partial P/\partial r$ and $\partial P/\partial \theta$ are not equal to zero, then Eq. (48) breaks down. From the aforesaid it follows that for accelerated flow ($\partial W_x/\partial \tau > 0$ or $\partial W_x/\partial x > 0$) the tangential stress in the flow core is less than a quasistationary value. This is possible only with a decrease in $-\rho \overline{W_x' W_r'} = \mu_T\, \partial W_x/\partial r$, i.e., with a decrease in the core of intensity of turbulent pulsations or, at least, with that in their correlations. As is known from the experiments in convergent channels ($\partial W_x/\partial x > 0$) and accelerated flows (15–19), this decrease in turbulence may be not only less than the quasi-stationary one but also less than the initial flow before acceleration up to flow laminarization in the core.

Consequently, in the case of accelerating flow ($\partial W_x/\partial x > 0$ or $\partial W_x/\partial \tau > 0$), if the intensity of turbulent pulsations becomes less than the initial value, there should exist a mechanism with the help of which kinetic energy of turbulent pulsations either partially converts into energy of averaged flow or varies and redistributes among different pulsations in such a way that $\overline{W_x' W_r'}$ decreases. With deceleration flow it is the other way around. For qualitative estimates of changes in velocity pulsation and turbulent characteristics in the flow core with acceleration it is possible to use the simple scheme of Prandtl (20) which he developed for estimating a change in these parameters in the case of accelerating flow caused by convergence of a channel and then modified by Uberoi (21) with regard for a change in density.

If we make the following assumptions:

(1) Velocity pulsations W_x' in the axial direction x are primarily determined by eddies, whose axes coincide with the tube radius; W_r' and W_θ' are determined by eddies, whose axes are parallel to the x-axis.

(2) The Helmholz theorem on constancy of circulation with respect to any contour, surrounding an eddy tube and lying on it, is valid for an eddy. Consequently, the product of the angular velocity and the area of a cross section of an eddy is constant, that requires dissipation in an eddy to be neglected.

(3) The density is a single-valued function of pressure.

(4) Assumptions (1) and (2) are similarly valid for eddies of all scales.

Then, with longitudinal acceleration of flow from a velocity W_{x0} to W_x the eddy, whose axis is parallel to the x-axis, elongates $C = W_x/W_{x0}$ times. The area of its cross section decreases by $\pi r^2/\pi r_0^2 = W_{x0}\rho_0/W_x\rho = \rho_0/C\rho$ and the radius will be $r = r_0(\rho_0/C\rho)^{1/2}$. Proceeding from the condition on circulation constancy, the angular eddy velocity ω_x should increase $C\rho/\rho_0$ times. Then, velocity pulsations in radial and tangential directions increase

$$(\overline{W_r'^2}/\overline{W_{r_0}'^2})^{1/2} = \omega_x r/\omega_{x_0} r_0 = (C\rho/\rho_0)^{1/2}$$

times. Relative transverse pulsations $(\overline{W_r'^2})^{1/2}/W_x$ decrease $(\rho/C\rho_0)^{1/2}$ times.

With acceleration in time, the eddy located along the flow axis does not elongate. Its cross section varies by ρ_0/ρ, the angular velocity by ρ/ρ_0, and the radius by $(\rho_0/\rho)^{1/2}$. Consequently, the radial and tangential pulsations vary by (ρ/ρ_0), and when these are related to W_x, they vary by $(1/C)(\rho/\rho_0)^{1/2}$.

A change in velocity along the length dx or during time $d\tau$ will be respectively,

$$C_x = 1 + \frac{\partial W_x}{\partial x} \frac{dx}{W_{x_0}} \tag{51}$$

$$C_\tau = 1 + \frac{\partial W_x}{\partial \tau} \frac{d\tau}{W_{x_0}} \tag{52}$$

Integrally, on the basis of Eqs. (51) and (52), the effect of C_x and C_τ upon turbulent parameters in the core may be taken into account by dimensionless complexes

$$\frac{d}{W}\frac{\partial W}{\partial x} \quad \text{and} \quad \frac{d}{W^2}\frac{\partial W}{\partial \tau}$$

The former is equivalent to the last term in Eq. (32) and the three last terms in Eq. (34); the second coincides with the second term in Eqs. (32) and (34). Consider longitudinal velocity pulsations. For longitudinal flow acceleration due to a density change (heat transfer) in a cylindrical tube, the cross section of eddies located along the channel radius is ellipse-shaped. The major axis of this ellipse b is parallel to the x-axis and the minor one a is normal to it.

Assuming $\partial W_x/\partial x = $ const within the cross section of an eddy with initial radius a_0, we get from Eq. (51),

$$\frac{\partial W_x}{\partial x} = \frac{(C_x - 1) W_{x0}}{\Delta x} = \frac{W_{x0}}{a_0}$$

or $1 + dx/a_0 = C_x$ and where dx/a_0 is the relative elongation of the major ellipse axis during time a_0/W_{x0}.

Consequently, $a = a_0/C_x$ and $b = a_0 \cdot C_x$. Taking into account the density change, we find $a = (a_0/C_x)(\rho_0/\rho)^{1/2}$. The angular eddy velocity ω_r varies by ρ/ρ_0. Then, the longitudinal velocity pulsations $W_x' = \omega_r \cdot a$ vary by $(1/C_x)(\rho/\rho_0)^{1/2}$ and when related to W_x the change is $(1/C_x^2)(\rho/\rho_0)^{1/2}$.

For flow accelerations due to channel convergence the decrease in longitudinal and relative longitudinal velocity pulsations will not depend upon density change and is $1/C_x$ and $1/C_x^2$, respectively (21).

For flow accelerations in time, the sizes of eddies which are parallel

TABLE I

Designation of quantity	Type of flow		
	Acceleration in time $\frac{\partial W_x}{\partial \tau} \neq 0$	$\frac{\partial W_x}{\partial x} \neq 0$ $r = $ const	$\frac{\partial W_x}{\partial x} \neq 0$ $r = f(x)$
$\overline{(W_x'^2)^{1/2}}/\overline{(W_x'^2)_0^{1/2}}$	$(\rho/\rho_0)^{1/2}$	$(1/C_x)(\rho/\rho_0)^{1/2}$	$1/C_x$
$\dfrac{\overline{(W_x'^2)^{1/2}}/W_x}{(\overline{(W_x'^2)^{1/2}}/W_{x0})_0}$	$(1/C_\tau)(\rho/\rho_0)^{1/2}$	$(1/C_x^2)(\rho/\rho_0)^{1/2}$	$1/C_x^2$
$\overline{(W_r'^2)^{1/2}}/\overline{(W_r'^2)_0^{1/2}}$	$(\rho/\rho_0)^{1/2}$	$(C_x\rho/\rho_0)^{1/2}$	$(C_x\rho/\rho_0)^{1/2}$
$\dfrac{\overline{(W_r'^2)^{1/2}}/W_x}{(\overline{(W_r'^2)^{1/2}}/W_{x0})_0}$	$(1/C_\tau)(\rho/\rho_0)^{1/2}$	$(\rho/C_x\rho_0)^{1/2}$	$(\rho/C_x\rho_0)^{1/2}$

to the tube axis and radius may vary only due to a change in medium density. In this case the longitudinal and transverse pulsations vary by $(\rho/\rho_0)^{1/2}$, and their relative values by $(1/C_\tau)(\rho/\rho_0)^{1/2}$. The estimates obtained are valid both for accelerated and decelerated flows. It is natural that they do not take into account diffusion, onset, dissipation, or energy transfer between pulsations which will be considered below.

The estimates found are given in Table I.

As is seen from Table I, in all cases with flow accelerations in time or along the length, the relative intensity of turbulent pulsations in the flow core appears less than the initial one (for flow deceleration it is higher) and, consequently, it is surely less than the quasi-stationary one.

From the balance of energy of turbulence in the core of steady flow in a tube obtained by Laufer and presented in Fig. 4 (22) it is seen that its increment in the tube core occurs partially due to generation of $-\rho \overline{W_x' W_r'}(\partial W_x / \partial r)$ and mainly due to diffusion

$$\frac{1}{u_\tau} \frac{1}{1-R} \frac{d}{d(1-R)} [(1-R)\, \overline{W_r^1 E_T^2}]$$

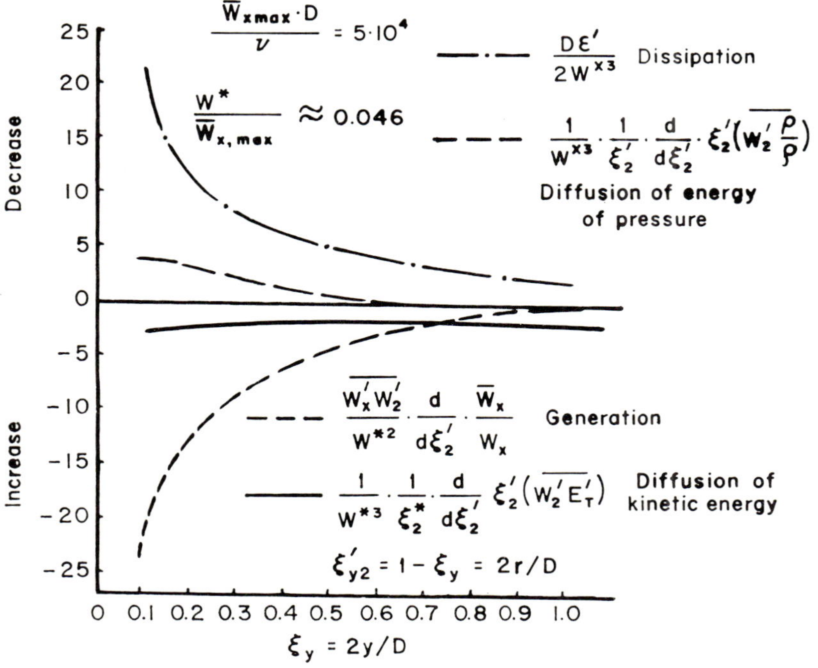

FIG. 4. Energy balance in flow core in tube: $W^* = (T/\rho)^{1/2}$, dynamic velocity; $\epsilon' = \nu(\partial W_i/\partial x_j)(\partial W_j'/\partial x_i)$.

of kinetic energy of turbulent pulsations E_T from the flow region near a wall. Therefore, the delay of intensity of turbulent pulsations in the flow core behind quasistationary values may be at last caught up only due to more intense (than quasistationary) diffusion $E_T = \rho \overline{W_i' W_i'}/2$ from the region adjacent to a wall. For this it is necessary that the generation of E_T near a wall $-\rho \overline{W_x' W_r'}(\partial W_x/\partial r)$ should exceed a quasistationary value. From the balance E_T in the region near a wall also obtained by Laufer [see Fig. 5 (22)] and from the analysis of the equations

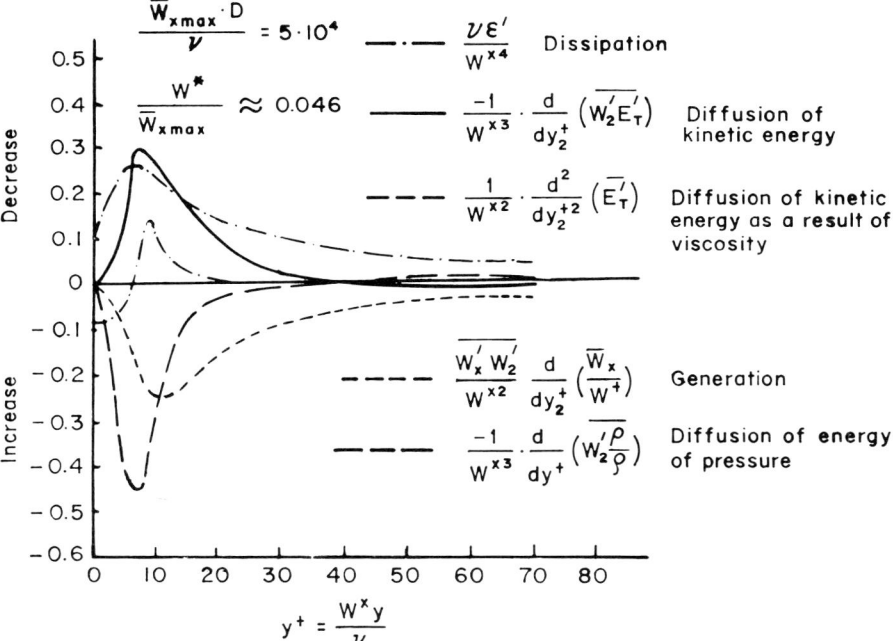

FIG. 5. Energy balance within near-wall region of tube; the same notation as in Fig. 4.

of balance E_T and energy of separate velocity components $E_i = \tfrac{1}{2}\rho \overline{W_i'^2}$, it follows that the increase of generation and diffusion of kinetic energy of pulsations from the region near a wall is possible if near a wall $(\overline{W_r'^2})^{1/2}$, $\rho \overline{W_x' W_r'}$ and $\partial W_x/\partial r$ increase simultaneously. But in this case it is necessary that near a wall the velocity gradient and tangential stress be greater than those in a quasi-stationary case. Note that for a compressible fluid the generation of E_T may also occur (23) due to the term $-\rho \overline{W_x'^2}(\partial W_x/\partial \tau)$. It is known (22) that

$$\nu_T = (\overline{W_r'^2})^{1/2} \Lambda_L = \overline{W_r'^2} \mathcal{T}_L \sim (E_T)^{1/2} \Lambda_L \tag{53}$$

where $\mathcal{T}_L$ is the Lagrangian time scale and $\Lambda_L = (\overline{W_r'^2})^{1/2} \mathcal{T}_L$ is the Lagrangian length scale.

Then it is possible to assume that ν_T as well as $\overline{W_r'^2}$ quantitatively varies in the flow core preserving the distribution with respect to a radius, which is similar to the quasi-stationary one but with some increase near a wall due to the above increase in E_T (and, consequently, $(\overline{W_r'^2})^{1/2}$) and $\mathcal{T}_L$.

Bearing in mind the above character of the distribution of the tangential stress in accelerated (decelerated) flow, it is not difficult to understand qualitative differences of velocity profiles from the quasi-stationary ones in accelerated or decelerated flow. In Fig. 6 are plotted qualitative

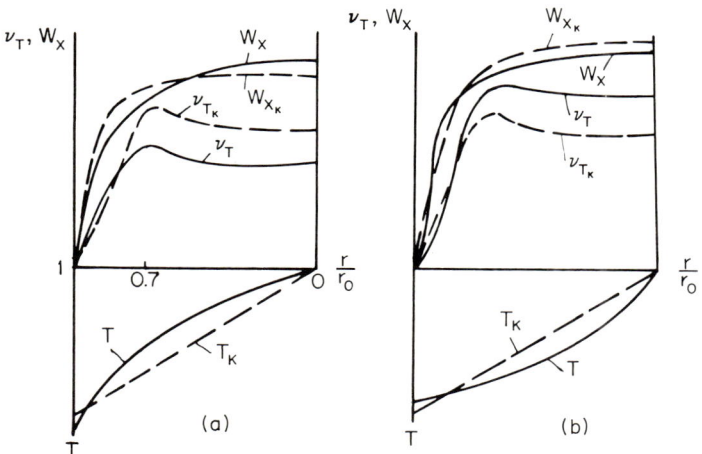

FIG. 6. Qualitative change in profiles of shearing stress T, turbulent viscosity ν_T, and velocity W_x for accelerated (a) and decelerated (b), with respect to time, flows in tube; T_K, W_{xK}, ν_{TK}, quasi-stationary values exponential.

changes in profiles T, ν_T, W_x for accelerated and decelerated flows in time. The same qualitative pattern will take place for longitudinal accelerations.

On the basis of the above analysis it may be expected that at flow accelerations the coefficient of hydraulic losses will be somewhat greater than the quasi-stationary one, since $T_w > T_{wK}$.[2] Heat transfer also increases due to an increase in turbulent pulsations (and consequently λ_T) near a wall. With deceleration, a reverse process takes place if there are no inverse flows. The onset of inverse flows is possible when decelerating flow. At their boundary and even at the boundary of small eddies there are high velocity gradients $\partial W_x/\partial r$ and tangential stresses $\rho \overline{W_x' W_r'}$. In these regions there occurs considerable generation of turbulent pulsations and, consequently, the friction factor increases and, with diffusion of these pulsations to a wall, heat transfer may increase.

At very great accelerations, essential three-dimensional flows may appear since $\partial P/\partial r$ and $\partial P/\partial \theta$ may noticeably differ from zero. In this case it may appear that in the flow core the terms of the form $-\rho \overline{W_i' W_j'}(\partial W_i/\partial x_j)$ will be positive and achieve a large value. Then the energy of turbulent pulsations E_T may again convert into that of averaged flow, which, under definite conditions, will ensure laminarization of the flow core or a decrease of turbulence in it that will be less than the initial one.

C. Laminar Flows

The solution of the motion equation for unsteady laminar flow in channels is in principle not difficult. For a circular cylindrical tube far from the inlet the equation was already solved in 1882 by Gromeka (24) for any initial conditions and the prescribed law of change in the pressure gradient with time.

The review of such works on flat and circular tubes and the solutions for step and periodic change of the pressure gradient in time are presented in the book by Petukhov (9).

A great number of the works are devoted to a theoretical study of an unsteady boundary layer. The review of these works carried out before 1959 is presented in the work by Stewartson (26).

Struminskii's work (27) deals with the theory of a laminar unsteady boundary layer at arbitrary profiles and bodies of revolution.

The most recent works are (28–39). In the work by Yang and Ou (40), the expressions for velocity profiles and tangential stress at a wall in the entrance lengths of circular and flat tubes are found by using

---

[2] In the next paragraphs it will be shown in detail that the onset of a new secondary boundary layer is a reaction of a viscous flow (including a turbulent one) to a change of a velocity of flow. This layer increases in time from a wall to the flow core. That region of flow which is not yet achieved by this boundary layer behaves as an ideal fluid with respect to unsteady conditions. Consequently, from this point of view, with acceleration $T_w > T_{wK}$.

computers when an arbitrary change in the inlet velocity was applied. The analysis of the above works shows a definite likeness between a change of a velocity profile in a channel at the entrance region (9, 41) in steady flow and in time with accelerated flow. In this and other cases the velocity profile appears to be fuller than the stabilized and quasi-stationary ones, respectively. At the starting length of the tube the redistribution of a velocity profile occurs due to a growth of a boundary layer. The thickness of a boundary layer increases with a certain velocity proportional to ν. Therefore, the place where the boundary layers appearing at different walls "close" up and the hydrodynamic entrance length depend only upon the mean velocity of fluid at prescribed ν and channel sizes.

From the exact analytical solution (27) of the case of flow past an infinite plate when the flow velocity changes by a jump from 0 to W_0 it follows that at any time instant $\tau_0 \neq 0$ there are two sections: (1) $0 \leqslant x \leqslant x_0$, where the thickness of a boundary layer $\delta = 5.92(\nu x/W_0)^{1/2}$ exactly coincides with a steady value of the velocity W_0; (2) $x_0 \leqslant x < \infty$, where $\delta = 3.65(\nu \tau_0)^{1/2}$ and does not depend upon x. In this region the thickness of a boundary layer increases with the velocity, which is the same for all x, unless it achieves a value which is steady for each x. The boundary of the regions is $x_0 = 0.392 W_0 \tau_0$.

If flow accelerates not jumpwise but gradually, then inside the existing boundary layer, at a moment τ_0, there appears a new secondary boundary layer corresponding to a new flow velocity. For each new change in the velocity the "old" boundary layer behaves like an ideal fluid unless the "new" boundary layer appearing from a wall achieves this point of flow. The "new" boundary layer is a reaction of a viscous fluid to unsteady-state conditions. Therefore, at flow acceleration near a plate, as at the starting length, the velocity profile is fuller than the quasi-stationary one, and at deceleration the opposite is true. This is well seen in Fig. 7, where are plotted velocity profiles W_x/W_∞ on a plate with acceleration and deceleration of flow at $W_\infty = W_0 e^{\beta \tau}$.

It is natural that with acceleration the tangential stress at a wall will be greater (as at the starting length) and with deceleration it will be less than a quasi-stationary value.

For accelerated motion in the entrance region in channels (40), the tangential stress and the friction factor are greater than the quasi-stationary case and for decelerated motion, they are less. The length of the entrance region varies just as the flow velocity.

For the fully developed region of flow in channels (9) also with acceleration, the friction factor and tangential stresses are greater and with deceleration are less than the quasi-stationary values. Here inside

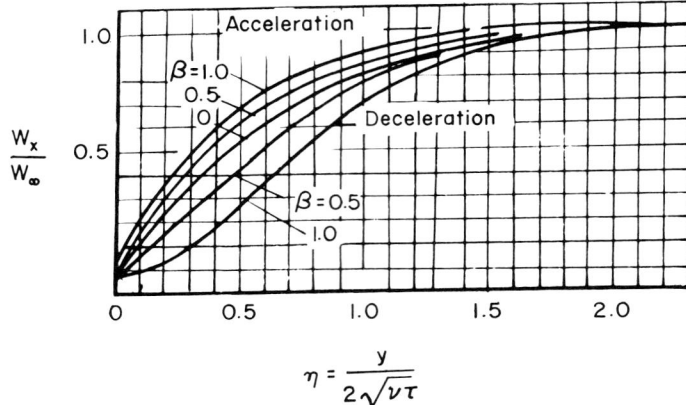

FIG. 7. Velocity profile on plate W_x/W_∞ for exponential flow acceleration and deceleration $W_\infty(x, \tau) = W_0 e^{\rho \Sigma}$.

the "old" boundary layer there appears the "new" one, and the velocity profile begins varying from a wall. The part of the core still not subjected to a change behaves like an ideal fluid with respect to a "new" boundary layer.

These results of the theoretical investigations are not in doubt since for laminar flows the theory is usually in good agreement with experiment. In particular, these are experimentally confirmed for a plate (*42*).

D. TURBULENT FLOWS

The theoretical analysis of unsteady turbulent flows is complicated mainly by the lack of data on the character of the change in turbulent parameters under unsteady conditions. If, when calculating heat transfer in channels it is enough to make an hypothesis on a change in these parameters near a wall making rough assumptions on their values in the flow core, but it is not enough when calculating velocity profiles. Therefore, the works on the investigation of a turbulent structure of unsteady flows and on making hypotheses which allow a system of equations to be closed are of special interest.

Since, in unsteady flows, the turbulent structure varies, the quantities characterizing the structure of turbulence should be introduced into the system of differential equations. According to the semiempirical theory proposed by Kolmogorov and Prandtl and later developed in the works by Monin (*43*), Rotta (*44*), and Glushko (*45*) to obtain a closed system of equations, the equation of balance of kinetic energy of turbulence E_T should be added to the Reynolds equation and to the continuity equation.

Eremenko (46) uses this method for calculating kinematic characteristics of parallel-flat turbulent flow, smoothly varying in time due to a change in channel slope in a gravitational field.

After evaluating the order of the terms for flow under consideration, the Reynolds equation and the balance equation for turbulence energy assume the form, respectively,

$$\frac{\partial W_x}{\partial \tau} + W_x \frac{\partial W_x}{\partial x} + W_y \frac{\partial W_x}{\partial y} = gJ + \frac{\partial}{\partial y}\left(\nu \frac{\partial W_x}{\partial y} - \overline{W_x' W_y'}\right) \tag{54}$$

$$\frac{\partial E_T}{\partial \tau} + W_x \frac{\partial E_T}{\partial x} + W_y \frac{\partial E_T}{\partial y} = -\frac{\partial}{\partial y}\left(W_y' \frac{P}{\rho}\right) - \frac{\partial}{\partial y}(\overline{W_y' E_T'})$$

$$- \overline{W_x' W_y'} \frac{\partial W_x}{\partial y} + \nu \frac{\partial^2 E_T}{\partial y^2} - \overline{W_x'^2} \frac{\partial W_x}{\partial x}$$

$$- \overline{W_y'^2} \frac{\partial W_y}{\partial x} - \nu \sum_i \sum_K \left(\frac{\partial W_i'}{\partial x_K}\right)^2 \tag{55}$$

Here $E_T' = \frac{1}{2}(W_x'^2 + W_y'^2 + W_z'^2)$, $E_T = \overline{E_T'}$, J is the piezometric slope, x, y or x_1, x_2 are the coordinates in longitudinal and transverse directions, and W_x, W_y are the longitudinal and transverse projections of the average velocity vector.

With regard to the continuity equation, the terms

$$-\overline{W_x'^2} \frac{\partial W_x}{\partial x} - \overline{W_y'^2} \frac{\partial W_y}{\partial y} = (-\overline{W_x'^2} + \overline{W_y'^2}) \frac{\partial W_x}{\partial x}$$

may be neglected[3] since $\overline{W_x'^2}$ and $\overline{W_y'^2}$ are of the same order.

Further, separate terms of Eq. (55) are approximated:

(1) Turbulence generation:

$$-\overline{W_x' W_y'} \, \partial W_x/\partial y = \nu_T (\partial W_x/\partial y)^2 \tag{56}$$

(2) Dissipation of turbulent energy according to Rotta's formula (44) using the coefficient $H(R_E)$ introduced by Glushko (45):

$$\nu \sum_i \sum_K (\partial W_i'/\partial x_K)^2 = (C\nu \, E_T/L^2) + (C_1 \, H(R_E) \, E_T^{3/2}/L) \tag{57}$$

---

[3] Note that we omit the terms which determine a decrease in turbulence energy with flow acceleration along the length $\partial W_x/\partial x > 0$ (if $-\overline{W_x'^2} + \overline{W_y'^2} > 0$) or with an increase in turbulence energy at flow deceleration $\partial W_x/\partial x < 0$.

R_E is the turbulence Reynolds number, $H(R_E)$ takes into account the viscosity effect. Near the wall where R_E is small, $H(R_E) < 1$, and at large R_E it is equal to 1.

(3) Diffusion of kinetic energy of turbulence:

$$-\frac{\partial}{\partial y}(\overline{W_y' E_T^1}) = -\frac{\partial}{\partial y}\gamma\nu_T\frac{\partial E_T}{\partial y} \quad (58)$$

(4) Diffusion of energy of pressure:

$$\frac{1}{\rho}\frac{\partial}{\partial y}\left(\overline{\frac{P}{\rho} W_y^1}\right) = aE_T h\left(0.0073 - H(R_E)\frac{\Lambda_L}{h}\right)\frac{\partial^2 W_x}{\partial y^2} \quad (59)$$

This expression is obtained by solving the Poisson equation for pulsation pressure in a limit space after simplifying and using the experimental data of Laufer (22).

Further, the assumption is made that the integral scale of the length Λ_L does not depend upon a coordinate x and time. Then, from $T/\rho = V_*^2(1 - y/h) = (\nu + \nu_T)\partial W_x/\partial y$ and $\nu_T = \Lambda_L\sqrt{E_T}$, we have

$$\Lambda_L = \frac{V_*^2(1 - y/h) - \nu\,\partial W_x/\partial y}{(E_T)^{1/2}\,\partial W_x/\partial y} \quad (60)$$

Here $V_* = (T_0/\rho)^{1/2}$ is the dynamic velocity; h is the distance from the symmetry axis to the wall.

The velocity distribution W_x in Eq. (60) and E_T are found by Nikitin's formulas (47) for steady flow:

$$\frac{W_x}{V_*} = 6.45\log\frac{y}{h}\frac{h}{\delta} + 5.6 + \frac{[(y/h)(h/\delta) - 1]}{(y/h)(h/\delta)} \quad (61)$$

$$E_T = \frac{V_*^2}{2}\left\{\left[0.53 + \frac{0.44h/\delta - 0.44}{(h/\delta)(0.28 + y/h) - 1.28}\right]^2\right.$$

$$\left.+ 2\left[\frac{0.7y/h}{y/h + 0.036}\left(\frac{y^2}{h^2} - \frac{2y}{h} + 2\right)\right]^2\right\} \quad (62)$$

where δ is the thickness of a viscous sublayer determined by the relation $V_*\delta/\nu = 5.6$.

At $\bar{y}/h > 0.55$ the velocity W_x is determined by the formula given in (48).

It has been shown that with an increase in h/δ the function $\Lambda_L/h = f(y/h)$ ceases to be dependent on the Reynolds number. This allows the function for large Reynolds numbers to be plotted. This plot has been used in calculations.

Thus, a closed system of equations has been obtained, assuming that approximating expressions (56)–(59) are valid, that Λ_v/h may be obtained according to the plot described above, and that the constants C, C_1, γ, and a are universal for steady and unsteady flows.

After neglecting the convective terms in Eqs. (54) and (55) the system of equations obtained was numerically solved on an electronic computer by using the local uniform method of Samarsky and the factorization method.

The calculation results with $C = 3.93$; $C/C_1 = 0.19$; $\gamma = 0.9$; $a = 0.13$, and $J = 0.00084 \pm 0.1 \sin(2\pi\tau/10)$ are given in Fig. 8, from which it is seen that E_T/W^2 with acceleration near a wall becomes greater, and with deceleration it becomes less. For deceleration with $\tau = 1.5$ sec the velocity profile with inverse flow near a wall is obtained.

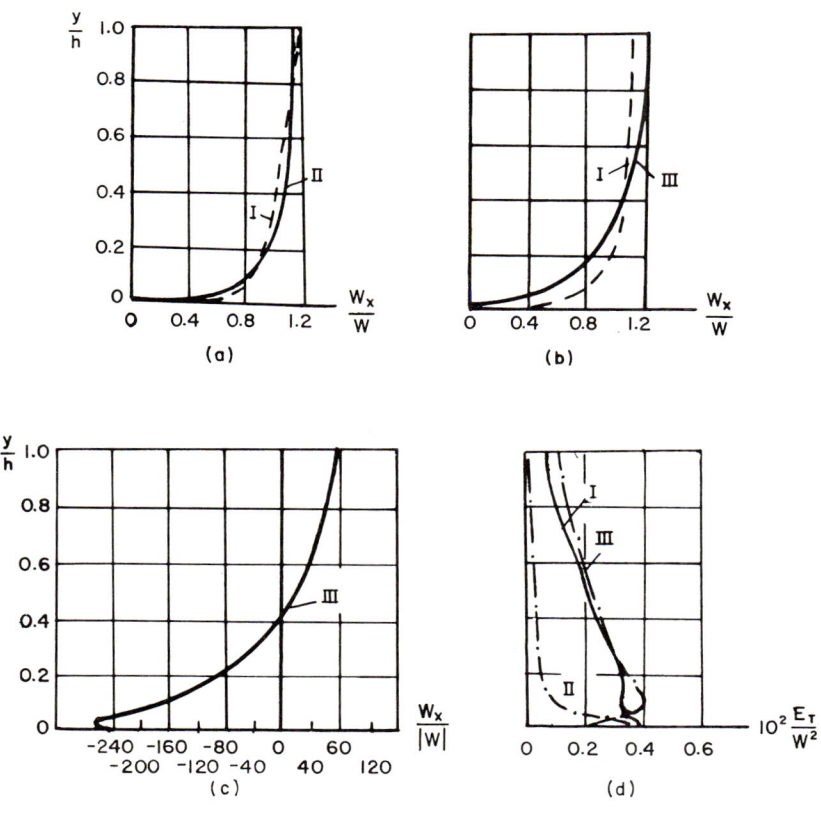

FIG. 8. Results of Eremenko's calculations (46) for kinematic characteristics of transient motion. I, uniform motion; II, accelerated motion; III, decelerated motion; (a) $\tau = 0$ sec; (b) $\tau = 1$ sec; (c) $\tau = 1.5$ sec; (d) $\tau = 1.0$ sec.

The character of distribution of W_x and E_T/W^2 found theoretically by Eremenko is in qualitative agreement with the experimental data of Markov (*18*). Markov conducted his experiments in a rectangular water channel, located horizontally, with cross section 100×380 mm and length 6.5 m. The kinematic characteristics were obtained by statistical treatment of photographs of moving particles of degreased aluminum powder of size about 10 μm. The photographs were made in the fully developed region and covered the region adjacent to the bottom of the channel up to 0.088 cm.

In Fig. 9 the velocity profiles and $(\overline{W_x'^2})^{1/2}/W$ are shown from this work for accelerating and decelerating flows. It is noteworthy that in an accelerating flow the value of $(\overline{W_x'^2})^{1/2}/W$ near the wall is higher and in a decelerating flow is lower than the stationary distribution, while in the measured region the value $(\overline{W_y'^2})^{1/2}/W$ has the reverse relation. Unfortunately both in the theoretical work of Eremenko and in the experiments of Markov the velocity profiles were not brought to the wall and the shear stresses and the friction factors were not determined. These data together with the distribution of $\overline{W_x' W_y'}$ could give highly valuable information as to the peculiarities of the structure of unsteady flows.

Vasiliev and Kvon (*49*) made a theoretical study of a planar, open, turbulent liquid flow in an inclined channel with a flat bottom. The system of equations they obtained, together with the relations of the semiempirical theory, made it possible to take into approximate account the unsteady kinetic and dynamic characteristics of the flow at high Reynolds numbers.

In particular, the shear stresses and velocity profiles were calculated. It was determined that (a) in an accelerating flow a shift in the velocity maximum down from the free surface is possible; (b) with uniform flow velocity (averaged over the depth) the friction stress at the bottom of the accelerating flow is higher, and of the decelerating flow is lower, than in a uniform flow. In the experimental work of Shabrin (*50–52*), made on the flow models with a free surface, it is noted that in the region near the bottom, where the measurements were made, the longitudinal oscillations and their correlation with the cross oscillations are much lower than the quasi-steady values in an accelerating flow.

As was pointed out in Section II,B, the peculiarities of the redistribution of the turbulent flow structures suffering accelerations in the longitudinal direction (divergent–convergent flows) and in time have common features. Therefore, to understand the nature of turbulent structure redistribution in unsteady flow it seems advisable to use the data on the structure of the divergent–convergent turbulent flows.

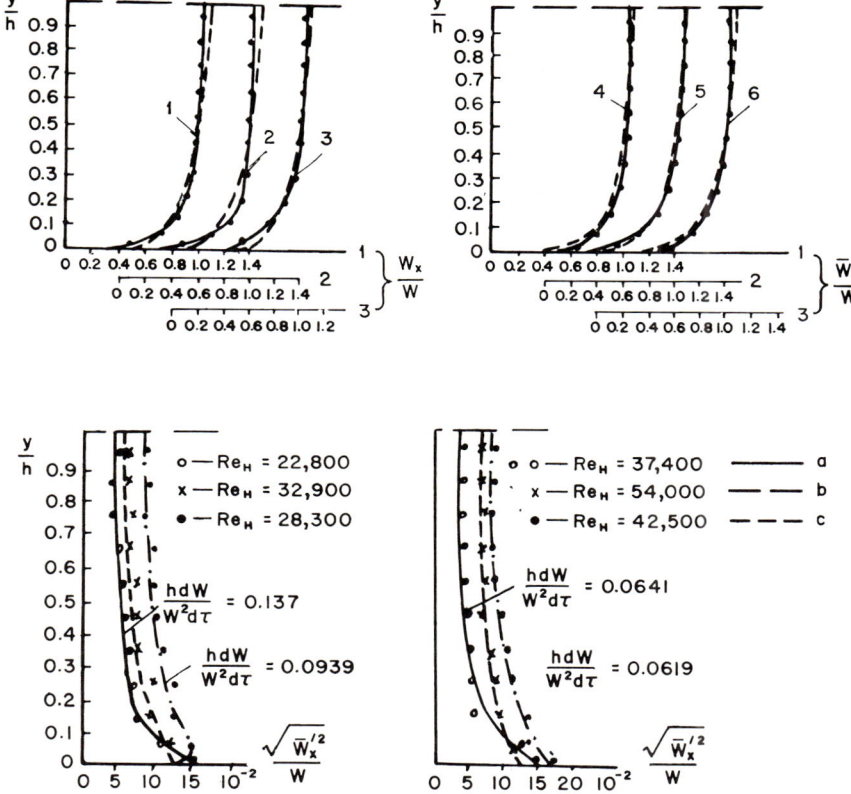

Fig. 9. Profiles of velocity W_x/W and mean-square values of pulsating velocities $(\overline{W_x'^2})^{1/2}/W$ for acceleration and deceleration of flow from Markov's experiments (18):

(1) Re = 22,800, $(h/W^2)(dW/d\tau) = 0.137$;
(2) Re = 37,800, $(h/W^2)(dW/d\tau) = 0.0501$;
(3) Re = 60,200, $(h/W^2)(dW/d\tau) = 0.0204$;
(4) Re = 43,100, $(h/W^2)(dW/d\tau) = -0.0316$;
(5) Re = 28,300, $(h/W^2)(dW/d\tau) = -0.0939$;
(6) Re = 16,150, $(h/W^2)(dW/d\tau) = -0.310$.

Dashed line shows quasi-stationary profiles according to Nikuradze. a, accelerated motion; b, decelerated motion; c, uniform motion.

Zaric (53) made a rather detailed analysis of the turbulent flow structure in a four-sectional convergent–divergent rectangular channel with a cone angle of 16° and one plane wall. The measurements in the last difuser section showed a considerable increase in turbulent oscillations near the curved wall as compared with the opposite smooth wall due to the formation of small swirling zones on it. Graham and Deissler (54) analyzing the effect of flow acceleration on the transverse velocity oscillations gave an explanation on why the turbulent heat transfer from the gas to the wall in the critical cross section of the nozzle is considerably lower than the calculated values where acceleration is not taken into account. The estimation is based on the simplest assumptions. The turbulent heat flux through the boundary layer of the thickness δ in the nozzle is

$$q_T \approx \rho C_p v_T \, dT/dy$$

Assuming $dT/dy \approx \Delta T/\delta$, and, by (53), $v_T \approx (\overline{W_r'^2})^{1/2} \Lambda_L$ where $\Lambda_L \sim \delta$, we shall obtain the heat transfer coefficient

$$\alpha = \frac{q_T}{\Delta T} \approx \rho C_p (\overline{W_r'^2})^{1/2} \quad \text{or} \quad \text{St} = \frac{\alpha}{\rho W C_p} \approx \frac{(\overline{W_r'^2})^{1/2}}{W}$$

taking an average v_T over the thickness of the boundary layer.

The change of $(\overline{W_r'^2})^{1/2}/W$ during acceleration in the nozzle is estimated in two ways:

(a) $(\overline{W_r'^2})^{1/2}$ does not change in the process of acceleration, as was observed by Bradshaw and Ferris (55).

(b) $(\overline{W_r'^2})^{1/2}$ changes in the same way as was found by Deissler (56) for the local structure of the uniform turbulence in a noncompressible accelerating flow.

The results of calculations in terms of the Stanton number by method (a) were somewhat lower, while those by method (b) were somewhat higher than those obtained by Moretti and Kays (57). Note that the simplest estimation of the change in $(\overline{W_r'^2})^{1/2}/W$ presented in Table I, gives results very close to the estimation made by Deissler (56). Good coincidence of such simple estimates with experiment is evidently possible only for thin boundary layers in an entrance region flow in the interior of the thermal boundary layer. In the developed flow region and at relatively high values of δ (as compared to δ_T) the picture may even be the reverse. In these cases, q decreases essentially over the thickness of the layer or over the channel radius. For the process of heat transfer the changes in $(\overline{W_r'^2})^{1/2}/W$ near the wall are most essential, and here (see Section II,B) they may exceed the quasi-stationary ones, though in the core of the flow they will be lower.

Large convergence angles of nozzles may cause essential transverse and secondary flows. Thus, for example, the experiments of Boldmen et al. (58) demonstrated that heat transfer in the nozzle throat with the convergence half-angle of 60° is 40% higher than in a 30° nozzle, though in both cases they are lower than the predicted values where the flow acceleration was not accounted for.

According to Talmor (*16*) and Gukhman (*15, 19*) large accelerations $[(\nu/W^2)(dW/dx) \geqslant 3.5 \cdot 10^{-6}]$ may cause reverse transition of turbulent flow into laminar flow in the nozzle throat. This is favored by the cooling of the flow and the decrease in the Reynolds numbers. Apparently this is connected with a considerable decrease in $(\overline{W_r'^2})^{1/2}/W$ and particularly in $(\overline{W_x'^2})^{1/2}/W$ in the core of the flow due to its acceleration. Thus, with the 16-fold increase of velocity in the nozzle ($C_x = 16$) we find from Table I that $(\overline{W_r'^2})^{1/2}/W$ decreases 4 times, and $(\overline{W_x'^2})^{1/2}/W$ 256 times, as compared to their values at the nozzle inlet. Apparently, laminarization occurs in the core of the flow or in that portion of the boundary layer which is far from the wall (*19*), for near the wall turbulent oscillations, and especially longitudinal ones, continue to generate. This has been proved by Markovin (*28a*).

Panchurin (*59*) made an attempt to theoretically calculate the velocity profiles in an unsteady turbulent flow in tubes. He solved the Navier–Stokes equation with an exponential change of pressure gradient. Real velocities were replaced by averaged ones, while the kinematic viscosity was replaced by an average over the cross section value of the turbulent viscosity, depending on velocity; W_r was assumed equal to zero. It is natural that with such assumptions not only nonstationary but also the quasi-stationary changes in the turbulent structure of the flow were not taken into account.

Consider now a number of theoretical and experimental works in which the influence of the nonstationary state on the friction factor in channels is studied.

Carstens and Roller (*60*) obtained an expression for the relation between the nonstationary friction factor in the channel ξ and the quasi-stationary ξ_K calculated with the following assumptions:

(1) In a nonsteady flow the velocity profile in the core of the flow is governed by the conventional power law and does not depend on x:

$$\frac{W_x}{W} = \frac{(2n+1)(n+1)}{2n^2}\left(1 - \frac{r}{r_0}\right)^{1/n} = f\left(\frac{r}{r_0}\right) \qquad (63)$$

Here n is a function of Re; $n = 7$ at Re $\leqslant 10^5$ and increases up to $n = 10$ at Re $= 3.2 \cdot 10^6$. Physically this is substantiated by the fact that in the core of the flow higher turbulent viscosity will hinder any essential changes in the velocity profile.

(2) Average integral value of the acceleration along the radius is equal to the average mass flow rate acceleration $\partial W/\partial \tau$.

(3) The gradient of the shear stress on the channel axis is equal to the quasi-steady-state value:

$$(T)_{r=0} = (T_K)_{r=0} = 0$$
$$(\partial T/\partial r)|_{r=0} = (\partial T/\partial r)_K|_{r=0} = T_{wK}/r_0 \tag{64}$$

The second equation of (64) further contradicts the calculations performed by the authors of (60) as they use the average rate acceleration and not the acceleration on the axis of the channel. The final expression has the form:

$$\frac{\xi}{\xi_K} = 1 + \frac{4F_3(1)}{\xi_K W^2}\frac{d}{d\tau}\frac{\partial W}{\partial \tau} \tag{65}$$

where $F_3(r/r_0)$ is determined by the following values:

n	7	8	9	10
$4F_3(1)$	0.449	0.391	0.346	0.31

In Fig. 10 this dependence is represented by a solid line, while the dashed line shows its approximate direction in the case when the second equation of (64) is corrected to bring it into agreement with the average mass rate acceleration. The same figure contains the results of the experiments of (60) and of Daily et al. (61). The experiments in (60) were carried out on water in a horizontal tube 500 diam in length. They began with zero mass flow rate by sudden opening of a value. First, a laminar regime was established and then, with the rise of the mass flow rate in the inlet sections, a turbulent flow arose, developing to the exit and reaching it at $\tau(gh_0/L)^{1/2} \geqslant 1.38$ (h_0 is the piezometric head in a tank; L is the tube length).

Experiments in (61) were also carried out on water but in a vertical tube with a length of 27 diam at Re $\leqslant 5 \cdot 10^5$ and

$$-0.3 < (2d/\bar{\xi}_K W^2)(\partial W/\partial \tau) < 0.3.$$

They are generalized by the formula

$$\bar{\xi}/\bar{\xi}_K = 1 + C_2(2d/\bar{\xi}_K W^2)(\partial W/\partial \tau) \tag{66}$$

where C_2 is the test coefficient, having the value of $2F_3(1)$ in Eq. (65). With acceleration it was equal to 0.01, while with deceleration, to 0.62.

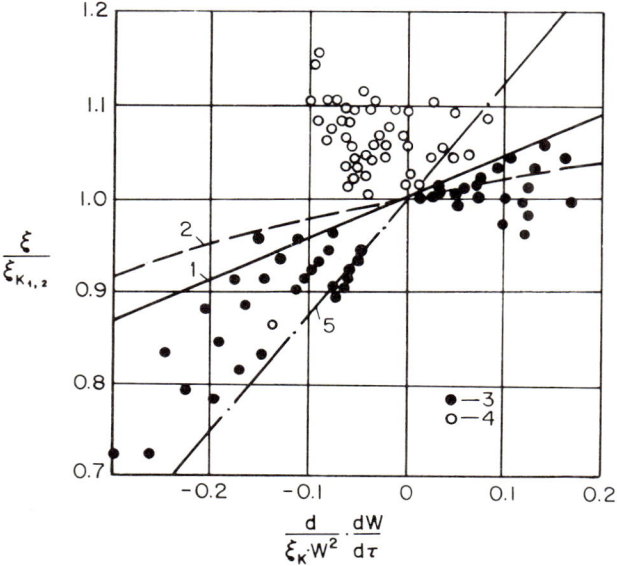

Fig. 10. Friction coefficient for unsteady turbulent flow in tube: (1) Eq. (65); (2) Eq. (65) accounting for error made by authors of (60) in distribution of shearing stresses; (3) experimental data of (61); (4) experimental data of (60); (5) Eq. (67) according to experimental data of (62–65).

Considerable scatter of experimental data is the result of inaccurate measurements and analysis, and in particular, of graphical analysis and of the velocity curve differentiation.

Carstens and Roller on the basis of analysis of their results and of those obtained by Daily *et al.* (61) make a conclusion on the applicability in the first approximation of the quasi-steady-state calculation method for turbulent flows. Their explanation of the phenomena is that in a quasi-steady-state flow the inertia components greatly contribute to the total pressure loss and the relative value of the losses due to friction is small. Moreover, the unknown values of local resistances at unsteady conditions may lead to a greater error in calculations than disregard for the difference between unsteady-state losses due to friction and quasi-steady-state ones.

In the work of Aitsam *et al.* (62) and Liyv (63–65) the friction factor was determined experimentally on water for accelerating and decelerating flows. A tube of 0.075 m in diameter had length of 53 diam. The mass flow rate changed from 0.8 to 16 liters/sec at $\mathrm{Re}_{max} = 16.5 \cdot 10^5$; $(d/W^2)(\partial W/\partial \tau)$ changed from -1.25 to 0.8.

The tests are generalized by the relation

$$\xi/\xi_K = 1 + (1.28 d^2/\xi_K W^2)(\partial W/\partial \tau) \tag{67}$$

which is given in Fig. 10 by a dashed-dotted line.

Contrary to (60), the authors of (62) contend that disregard for the dependence of ξ on the unsteadiness of the flow leads to great errors. In the example given in (62) it is shown that, at the given mass flow rate and with the time for the opening of a value of 5 sec, the error in the mass flow rate for an accelerating flow is 20–50%.

The experiments of Kochenov and Kuznetsov (66, 67) with laminar and turbulent water flow demonstrated a very pronounced effect of unsteadiness on the friction coefficient (several times). This was also proved by their theoretical analysis of the laminar flow. With acceleration of the flow they obtained $\xi > \xi_K$, while with deceleration, $\xi < \xi_K$ and even $\xi < 0$.

Panchurin (68) carried out his experimental investigation on an U-shaped tube 7 m in length and 47 mm in diameter at $Re \leqslant 10^5$ and $\partial W/\partial \tau \leqslant 1.4$ m/sec. The water in this tube oscillated and the results on the water column oscillation were compared with the calculated data obtained from the solution of the equation of oscillation with friction. They substituted the quasi-steady-state value $\bar{\xi}_K = 0.019$ into the equation which corresponds to the self-similar flow region at $Re \geqslant 6 \cdot 10^5$. It follows from experiments that the coefficient of hydraulic resistance is greater than the quasi-steady-state one, both with acceleration and deceleration. This is to be expected, since in the calculation the losses on the formation of the velocity profile and the higher value of ξ_K at small velocities, secondary flows during flow oscillations, and so on, were not taken into account.

Tsetserin (69), proceeding from the motion equation in the form

$$\rho W\, dW = -dP + \frac{\partial P}{\partial \tau} d\tau - \xi_K \frac{\rho W^2}{2d} dx \tag{68}$$

introduces the determination of the unsteady-state friction factor

$$\xi/\xi_K = 1 - (\partial P/\partial \tau)(1/W)/\xi_K \rho W^2/2d \tag{69}$$

which allows us to reduce Eq. (68) to a form similar to the steady-state one,

$$\rho W\, dW = -dp - \xi(\rho W^2/2d)\, dx$$

Then, using the continuity equation at $P/\rho = \text{const}$, in the form $-\partial P/\partial \tau = W_{SV}^2 \partial(\rho W)/\partial x$ we transform Eq. (68) into

$$\frac{\xi}{\xi_K} = 1 + \frac{2}{\xi_K} \frac{1}{M^2} \frac{1}{G} \frac{\partial G}{\partial (M \text{ Ho})}$$

and introducing the average Mach number $\overline{M}$ (for the period of charge or discharge of the gas duct)

$$\frac{\xi}{\xi_K} \cong 1 + \frac{2}{\xi_K} \frac{1}{\overline{M}^3} \frac{\partial \ln \overline{G}}{\partial \text{Ho}} \qquad (70)$$

Here $\text{Ho} = W_{SV} \Delta\tau/d$, $\Delta\tau$ is the averaging time (charge or discharge of a gas duct), $\overline{G} = G_m/G_{24}$, G_m is the average mass flow rate for $\Delta\tau$, and G_{24} is the average mass flow rate for 24 hours.

Then the average (for $\Delta\tau$) friction factor is

$$\bar{\xi} = \frac{1}{\Delta\tau} \int_0^{\Delta\tau} \xi \, d\tau = \xi_K - \frac{2}{\overline{M}^3 \text{ Ho}} \ln \overline{G} \qquad (71)$$

The drawback of this expression is the necessity for the experimental determination of the values $\overline{G}$, Ho, and $\overline{M}$.

Calculations using (71) have been compared with experimental data obtained on a gas duct 0.309 m in diameter and 5424 m in length. Agreement is within 10%. However, it is not clear if $\bar{\xi}$ was found experimentally: If it was determined by Eq. (71) or by Eq. (70), then such agreement is not surprising; however, it proves nothing. It follows that the unsteady-state hydraulic resistance coefficient (69) with the rise of pressure in the gas pipe $\partial P/\partial \tau > 0$ (charge) is always greater than ξ_K, while at the pressure drop (discharge) it is less than ξ_K. If it is assumed that discharge occurs when the mass flow rate on the pipe exit is higher than the rate at the inlet, then in this case the gas flow in it will be accelerating. During the charge there is a reverse picture, i.e., the flow will be decelerating. Thus, there is qualitative discrepancy found in (60–66). Apparently, this discrepancy is the result of assumptions which lead to the motion equation in the form (68) and the unusual way of determining ξ in Eq. (69).

Experimental investigation made by Kalishevsky and Selikhovkin (70) is in qualitative discrepancy with works (60–66). They studied the effect of unsteadiness on the friction factor in a turbulent air flow in a tube with $d = 107$ mm and $l/d = 35$. They measured the shear stress at the wall by a Stanton tube, and the air flow rate by a Pitot tube on the axis and in another point of the flow core.

The authors of (70) found that in the entrance region the measured shear stress at the wall with a scatter of ±20% agrees with the quasi-steady-state value at

$$\left| \frac{\nu}{WV_*^2} \frac{dW}{d\tau} \right| = \left| \frac{8}{\xi_K \operatorname{Re}} \frac{d}{W^2} \frac{dW}{d\tau} \right| < 4 \cdot 10^{-3} \tag{72}$$

Moreover, on the basis of two velocities measured by Pitot tubes they found that the unsteady-state velocity profile in the flow core is governed by the power law. These data together with the qualitative theoretical analysis permitted Kalishevsky and Selikhovkin (70) to conclude that under condition (72) the velocity profile and the friction factor in unsteady-state conditions do not differ appreciably from the quasi-steady-state one.

The work lacks analysis of the reasons of qualitative discrepancy with other investigators, though, for example, in (70) the parameter

$$\left| \frac{d}{\xi_K W^2} \frac{dW}{d\tau} \right| = 21\text{--}250$$

and in (60, 61, 65) did not exceed 0.3 and the essential influence of unsteadiness was found.

Work (70) needs the following comments:

(1) It is difficult to agree with the reliability of measurements of the shear stress by the Stanton tube 0.3 mm in height when the thickness of a viscous sublayer was 0.02–0.2 mm.

(2) The power velocity law in the flow core obtained by two measurements on the tube radius cannot be used as a proof of even quasi-steady-state character of the velocity profile in the flow core since the power laws are valid for a great range of Reynolds numbers.

Moreover, the quasi steadiness of the velocity profile in the core cannot be used as the proof of its quasi steadiness along the whole section. As was mentioned, the power law in the flow was utilized in work (60) to prove unquasi steadiness of the whole velocity profile (see also Section II,B).

In the work the qualitative theoretical analysis is based on the hypothesis of quasi steadiness of the whole flow if the flow is quasi-steady in the wall region. This hypothesis cannot be assumed correct owing to Section II,B, and moreover in (70) the quasi steadiness of the flow in the wall region has not proved.

Daily et al. (61) as well as Panchurin (68) made experimental investigation of the local friction factors ζ of diaphragms in an unsteady-state flow. It was established that with acceleration ζ is lower than the quasi-steady-state value, and with deceleration it is greater. This effect increases with decrease in the ratio of the diameter of diaphragm orifice to the tube diameter.

The turbulent boundary layer on a plate, cylinder, and wing profiles at various unsteady-state conditions was studied in works (*42, 71–74*).

From what has been said above, the following conclusions may be drawn:

(1) Theoretical investigations of the hydrodynamics of unsteady-state turbulent flow are limited, besides mathematical and technical difficulties, due to the absence of data on the character and regularities of the change in the flow structure under these conditions. Extension of the semi-empirical turbulence theory to the case of unsteady-state and convergent–divergent flows is an urgent problem in this respect.

(2) Experimental investigations of the hydrodynamics of unsteady-state flows (both of their structure and integral characteristics) is in the initial stage. Due to considerable methodical and technical difficulties there is great scatter of, and discrepancy in, the data available and their analysis. A number of the most important concepts are determined by different authors in different ways, as, for example, the unsteady-state friction factor.

(3) The investigation carried out shows rather convincingly the nonvalidity of the quasi-steady-state method of calculation of hydraulic losses in the general case. At the same time, they do not permit us yet to establish the boundaries of applicability of the quasi-steady-state methods and do not give reliable recommendations for calculations outside these boundaries.

III. Unsteady-State Heat Transfer in Channel Flow

A. Discussion of Experimental Methods in the Determination of the Heat Transfer Coefficient under Unsteady-State Conditions

As was shown in Section I, the methods of calculating convective heat transfer with the use of heat transfer coefficients as well as the methods of calculating temperature fields within a wall by using the boundary conditions of the third kind, all widely applied in engineering practice for steady-state problems, may without any principal difficulties be generalized to the solution of unsteady-state problems.

For experimental determination of the heat transfer coefficient at unsteady-state heat transfer

$$\alpha(x, \tau) = \frac{q_w(x, \tau)}{T_w(x, \tau) - T_b(x, \tau)} \tag{73}$$

it is necessary to know the variation in time of the average fluid bulk temperature $T_b(x, \tau)$, wall temperature $T_w(x, \tau)$, and heat flux at the wall $q_w(x, \tau)$. Direct measurement of these quantities is impossible in most cases, and therefore indirect methods must be utilized to determine these quantities.

The quantities $T_w(x, \tau)$ and $q_w(x, \tau)$ may be determined through measurements of the temperature of the channel outside wall $T_H(x, \tau)$ and the density of heat sources per unit volume in the wall q_v (the outside wall in insulated or heat losses are known) by solving the reverse heat conduction problem (75). Let us consider the basic methods of indirect determination of $T_w(x, \tau)$ and $q_w(x, \tau)$, the analysis of which is given in (14).

(1) The method of indefinite coefficients developed by Berlin (14) takes into account the heat transfer from the channel wall and the initial conditions; it has no principal restriction as to accuracy. Practically, it permits determination of $T_w(x, \tau)$ and $q_w(x, \tau)$ after the same time Fo* $= a\tau/\delta^2$ (where δ is the wall thickness) from the beginning of the change of T_w starting from which the change in the outside wall temperature is recorded carefully by instruments. The value of Fo* depends on the rate of change of $T_H(x, \tau)$ and the accuracy of instruments and may be estimated in each specific case (usually Fo* < 0.5). This method is a particular case of a more general method of solution of reverse problems developed by Temkin (76, 77). As an example we shall consider the application of the method of indefinite coefficients for a hollow infinite cylinder with heat sources changing in time by a given law, with arbitrary initial conditions and with known temperature and heat flux at any time on the outside nonheated surface ($r = r_H$). The heat conduction equation has the form

$$\frac{\partial^2 T}{\partial R^2} + \frac{1}{R}\frac{\partial T}{\partial R} + \frac{r_H^2}{\lambda} q_V(\text{Fo}) = \frac{\partial T}{\partial \text{Fo}} \tag{74}$$

where $R = r/r_H$; $R_w = r_w/r_H$; Fo $= a\tau/r_H^2$.

Initial condition:

$$T(R\ 0) = T_0(R) \tag{75}$$

Boundary conditions:

$$T(1, \text{Fo}) = T_H(\text{Fo})$$
$$q(1, \text{Fo}) = q_H(\text{Fo}) \tag{76}$$

The solution is obtained in a form which automatically takes into account the initial conditions:

$$T(R, \text{Fo}) = \sum_{K=0}^{\infty} [\alpha_K(R) \, T_H^{(K)}(\text{Fo}) + \beta_K(R) \, q_H^{(K)}(\text{Fo}) + \epsilon_K(R) \, q_V^{(K)}(\text{Fo})] \tag{77}$$

Substituting (77) into Eq. (74) and into boundary conditions (76), equating the coefficients (functions of R) at equal functions of Fo, and solving the differential equations with respect to unknown functions $\alpha_K(R)$, $\beta_K(R)$, and $\epsilon_K(R)$ we obtain

$$\alpha_0(R) = 1; \qquad \beta_0(R) = (r_H/\lambda) \ln R; \qquad \epsilon_0(R) = (r_H^2/\lambda)(\tfrac{1}{2} \ln R + \tfrac{1}{4}(1 - R^2))$$

α_K is calculated successively by the recursion formula:

$$\alpha_K(R) = B(R) - A(1) \ln R - B(1) \tag{78}$$

where

$$A(R) = \int R \, \alpha_{K-1}(R) \, dR; \qquad B(R) = \int \frac{A(R)}{R} \, dR$$

Recursion formulas for $\beta_K(R)$ and $\epsilon_K(R)$ are identical to formula (78).

The temperature field for a hollow cylinder will be described by Eq. (77) after substitution into it of the values obtained for $\alpha_K(R)$, $\beta_K(R)$, and $\epsilon_K(R)$. At $R = R_w$, temperature and heat flux on the inside surface will be determined by the following expressions:

$$T_w(\text{Fo}) = T(R_w, \text{Fo})$$
$$= \sum_{K=0}^{\infty} [\alpha_K(R_w) \, T_H^{(K)}(\text{Fo}) + \beta_K(R_w) \, q_H^{(K)}(\text{Fo}) + \epsilon_K(R_w) \, q_V^{(K)}(\text{Fo})] \tag{79}$$

$$q_w(\text{Fo}) = -\frac{\lambda}{r_H} \frac{\partial T(R_w, \text{Fo})}{\partial R}$$
$$= -\frac{\lambda}{r_H} \sum_{K=0}^{\infty} [\alpha_K^{(1)}(R_w) \, T_H^{(K)}(\text{Fo}) + \beta_K^{(1)}(R_w) \, q_H^{(K)}(\text{Fo}) + \epsilon_K^{(1)}(R_w) \, q_V^{(K)}(\text{Fo})] \tag{80}$$

There the index K designates the Kth derivative.

(2) Hill's method (78), convenient for practical applications, is developed for an infinite plate with zero initial temperature. The idea of this method consists in dividing the time interval we are interested in into a finite number of equal intervals ΔFo with the replacement of the temperature curve by a chord on each time interval.

The temperature on the inner surface of the wall T_{wn} at time $\text{Fo} = n\,\Delta\text{Fo}$ and the mean value of the heat flux $\bar{q}_{wn}$ through the heat exchange surface for the time $(n-1)\,\Delta\text{Fo} \leqslant \text{Fo} \leqslant n\,\text{Fo}$ are determined in this method by the following expressions:

$$T_{wn} = (T_{Hn} + \theta_2 T_{w,n-1} + \theta_3 T_{w,n-2} + \cdots + \theta_n T_{w1})\frac{1}{1-\theta_1} \quad (81)$$

$$\bar{q}_{wn} = \frac{8\lambda}{\pi^2 \delta\,\Delta\text{Fo}}(M_1 T_{w,n} + M_2 T_{w,n-1} + \cdots + M_n T_{w1}) \quad (82)$$

There θ_n and M_n are dimensionless quantities depending on ΔFo and n. Hill gives the table of parameters θ_n and M_n for a whole number of values of ΔFo and n and corresponding formulas for calculation of the values of θ_n and M_n which are not given in the table.

The drawback of this method is the condition restricting the value $\Delta\text{Fo} \gg \text{Fo}^*$; when $\Delta\text{Fo} < \text{Fo}^*$, this method is not applicable. When $\Delta\text{Fo} \simeq \text{Fo}^*$, the error in determination of T_w will be very great at the first steps. Therefore, it is convenient only at small values of Fo^*.

(3) The essence of the method of successive intervals, suggested by Kudryavtsev *et al.* (6) is in the replacement of the real dependence $q_w = q_w(\text{Fo})$ by the step dependence (for each time interval ΔFo_i the heat flux $\bar{q}_{wi}$ is constant and equal to the integral mean value over this time interval).

For each interval ΔFo_i the heat conduction problem is solved with boundary conditions of the second kind

$$q_w|_{Y=0} = 0; \quad q_w|_{Y=1} = \text{const} \quad (83)$$

where $Y = y/\delta$ (δ is the plate thickness).

The temperature distribution at the end of the previous interval is used as the initial condition. The values of T_{wi} and $\bar{q}_{wi}$ are calculated by the following formulas:

$$\bar{q}_{wm}\frac{\delta}{\lambda} = \frac{1}{\Delta\text{Fo}_m - \frac{1}{6}}\left(T_{Hm} - \sum_{K=1}^{m-1}\bar{q}_{wK}\frac{\delta}{\lambda}\Delta\text{Fo}_K\right) \quad (84)$$

$$T_{wm} = \bar{q}_{wm}\frac{\delta}{\lambda}\left(\Delta\text{Fo}_K + \frac{1}{3}\right) + \sum_{K=1}^{m-1}\left(\bar{q}_{wK}\frac{\delta}{\lambda}\Delta\text{Fo}_K\right) \quad (85)$$

The drawback of this method is the restriction of the interval by the condition $\varDelta \text{Fo}_i > 0.5$, even for very small Fo*. The advantage is the absence of the requirement of the equality of time intervals to each other. On first intervals, T_w and q_w are determined with large error, to decrease which it is necessary that $\varDelta \text{Fo}_i \gg \text{Fo}*$.

(4) The authors of (6) also suggested the method of mean temperature based on the condition that there is a point (surface) in the body the temperature of which changes in a way similar to the bulk mean temperature. This method is simple, but it may be applied only at $\text{Fo} > 0.5$ and when thermocouples are correctly placed at the definite point (surface).

(5) The method suggested by Sparrow et al. (79) uses the Laplace transformations with the help of which exact integral formulas are obtained, by which numerical calculation of T_w is made (for an infinite plate)

$$T_w(\text{Fo}) = \frac{1}{2} Z(\text{Fo}) + \frac{1}{4\sqrt{\pi}} \int_0^{\text{Fo}} \frac{T_0(\theta)}{(\text{Fo} - \theta)^{3/2}} e^{-1/4(\text{Fo}-\theta)} \, d\theta \qquad (86)$$

where $T_0(\text{Fo})$ is the given temperature change of the insulated surface, θ is the integration variable, and $Z(\text{Fo})$ is the unknown function determined by numerical solution of the integral equation

$$T_0(\text{Fo}) = \frac{1}{2\sqrt{\pi}} \int_0^{\text{Fo}} \frac{Z(\theta)}{(\text{Fo} - \theta)^{3/2}} e^{-1/4(\text{Fo}-\theta)} \, d\theta \qquad (87)$$

with the initial condition $Z(\theta) = 0$. The value of q_w is determined from the solution of the direct problem with the boundary conditions of the first kind.

This method has the general drawback of all indirect methods of determining T_w and $q_w \mid \varDelta \text{Fo}_i > \text{Fo}* \mid$ and, moreover, the drawback of all step methods, i.e., great error on first intervals. As compared with the method of successive intervals this method is advantageous only at very small Fo*. As compared with the Hill method, the Sparrow method has the advantages that the whole time interval is not divided into equal intervals; but the Sparrow method is quite laborious.

It should be pointed out that the methods considered, although developed under the conditions of constant physical properties, are applicable in case λ depends on temperature but temperature variation along the coordinate is sufficiently small. In this case λ can be considered constant along the coordinate and variable in time. The most convenient method of accounting for the dependence of λ on time is the method of successive intervals, as it permits substitution of required values of λ

on each interval. When Hill's method is used, it is necessary to choose intervals of time in such a way that the intervals of dimensionless time are equal to one another. In cases when intervals of ΔFo are equal, the method of Sparrow *et al.* is similar to the method of Hill, while with nonequal intervals of ΔFo the method of Sparrow *et al.* is similar to the method of successive intervals. When using the method of indefinite coefficients and the method of mean temperature, the time of the process should be divided into intervals such that on each of them the physical properties may be considered constant.

In a number of works (*25, 80–95*) the particular solutions of the reverse problems of unsteady-state heat conduction are considered. Some of them, for example (*80, 86–90*), are solved for conditions of variable thermophysical properties.

To determine the unsteady-state heat transfer coefficient it is necessary to know also the mean mass (bulk) temperature of the fluid $T_b(x, \tau)$. A method of determining $T_b(x, \tau)$ under arbitrary laws of the fluid mass rate variation $G(\tau)$, the inlet flow temperature $T_0(\tau)$, and the specific heat flux at the wall $q_w(x, \tau)$ is given in (*96*). This method is based on the solution of the one-dimensional energy equation

$$\rho C_p \, \partial T_b/\partial \tau + \rho C_p W \, \partial T_b/\partial x = q_v \qquad (88)$$

by the method of characteristics and solution of two Cauchy problems. Here q_V is the heat flux related to the volume unit of the moving fluid $(dV = f \, dx)$; f is the channel cross section area.

In (88) the heat supply due to dissipation is negligibly small as compared to q_V and is not taken into account. The substitution of the variables

$$X = \frac{x}{d}; \qquad \text{Ho} = \int_0^\tau \frac{W(\tau)}{d} \, d\tau; \qquad \theta = T_b/T_0 \qquad (89)$$

leads Eq. (88) to the form

$$\frac{W}{\overline{W}} \frac{\partial \theta}{\partial \text{Ho}} + \frac{\partial \theta}{\partial X} = 4 \, \text{St}_0 \qquad (90)$$

Here T_0 is the characteristic temperature ($T_0 = 273°\text{K}$):

$$\text{St}_0 = f(\text{Ho}, X) = \frac{U d \, q_w(\text{Ho}, X)}{4 T_0 C_p \, G(\text{Ho})} = \frac{\text{Nu}_0}{\text{Re Pr}}; \qquad \text{Nu}_0 = \frac{q_w d}{\lambda T_0}$$

U is the heated perimeter. Eq. (90) is solved with the following assumptions: (1) Heat capacity $C_p = \text{const}$; (2) $\overline{W}/W = 1$. This assumption is not essential, as the length of the channel is divided into several sections for determining local values of α. A uniform equation equivalent to (90) with regard to assumption (2) is

$$\partial F/\partial \text{Ho} + \partial F/\partial X + 4\,\text{St}_0\,\partial F/\partial \theta = 0 \qquad (91)$$

where $F(\text{Ho}, X, \theta) = 0$. The equation of characteristics is

$$d\text{Ho} = dX = d\theta/4\,\text{St}_0(\text{Ho}, X) \qquad (92)$$

The general solution of the equivalent equation (91) is obtained by the method of characteristics with the subsequent solution of the Cauchy problem for which it is necessary to consider two regions of the process.

The first region of the process ($\text{Ho} \leqslant X \leqslant X_e = L/d$) covers those channel sections such that at the given instant Ho, the particles of the fluid at the inlet into the channel has not yet reach the section considered. For this region the problem of unsteady-state heat conduction is solved in which heating of the fluid may be considered as the heating of a solid rod with a variable cross section heat conductivity. In this case, the dimensionless temperature distribution along the channel length is given at the initial moment of the dimensionless time $\text{Ho} = 0$ as $\theta = \varphi(X)$. The solution has the form

$$\theta(\text{Ho}, X) = 4 \int_{X-\text{Ho}}^{X} \text{St}_0(Y + \text{Ho} - X, Y)\,dY + \varphi(X - \text{Ho}) \qquad (93)$$

$$\theta(\text{Ho}, X) = 4 \int_{0}^{\text{Ho}} \text{St}_0(Y, Y - \text{Ho} + X)\,dY + \varphi(X - \text{Ho}) \qquad (94)$$

The second region ($X \leqslant \text{Ho} \leqslant \infty$) is the region of unsteady-state convective heat transfer which covers those sections of the channel and those times for which the initial temperature distribution does not play any role and the process is determined by the inlet conditions. For any section, this part of the process appears at that moment when particles of the fluid, which at the initial moment were at the inlet section of the channel, pass the section considered. In this case the boundary condition is the prescription of flow temperature at the channel inlet $\theta = \psi(\text{Ho})$ at $X = 0$. Solution of the Cauchy problem has the form

$$\theta(\text{Ho}, X) = 4 \int_{0}^{X} \text{St}_0(Y + \text{Ho} - X, Y) + \psi(\text{Ho} - X) \qquad (95)$$

$$\theta(\text{Ho}, X) = 4 \int_{\text{Ho}-X}^{\text{Ho}} \text{St}_0(Y, Y - \text{Ho} + X)\,dY + \psi(\text{Ho} - X) \qquad (96)$$

In Reference (97), this method was extended to the case of unsteady-state heat exchange of chemically active media. Thus, the existing methods of calculation allow us to determine the heat transfer coefficient from an experiment under unsteady-state conditions as well as to take into account its dependence on the rate of change of boundary conditions.

The existing methods also allow the use of the dependences obtained for the heat transfer coefficient on the unsteady-state conditions for calculation of the temperature fields of solid boundaries in unsteady-state. There are a large number of studies (98–152) which deal with the methods of calculation of unsteady-state temperature fields within a wall with boundary conditions of the third kind. Methods are developed which allow us to take into account the variation of the heat transfer coefficient in time and along the length, as well as to account for the variable thermophysical properties. Let us consider, e.g., the method of solving one-dimensional problems of unsteady-state heat conduction with the boundary conditions of the third kind with heat transfer coefficient and medium-temperature variable in time. There is no exact analytical solution (in elementary functions) of the problem of unsteady-state heat conduction with boundary conditions of the third kind and with arbitrary time varying temperature of the medium T_b and heat transfer coefficient α. However, there is a numerical method which permits us to obtain the solution of this problem with any desired accuracy for an infinite flat plate with zero initial temperature if one surface of the plate is insulated while the other transfers heat and the heat transfer coefficient $\alpha(\tau)$ and the medium temperature $T_b(\tau)$ are given in time. This method was suggested by Hill (78). The idea of the method is as follows: The whole time interval is divided into equal intervals ΔFo. The heat transfer surface temperature change on each interval is assumed linear; the average (on the interval) heat flux from the heat medium to the surface is equal to the half-sum of its values at the boundaries of the interval. Hill obtained the following calculation formulas:

$$T_{wm} = \frac{H_m T_{bm} + H_{m-1}(T_{b,m-1} - T_{w,m-1}) - M_2 T_{w,m-1} - \cdots - M_m T_{w1}}{M_1 + H_m} \quad (97)$$

$$\bar{q}_{wm} = \frac{8}{\pi} \frac{\lambda}{\delta} \frac{1}{\Delta \text{Fo}} (M_1 T_{w,m} + M_2 T_{w,m-1} + \cdots + M_m T_{w1}) \quad (98)$$

$$T_{Hm} = T_{wm}(1 - \theta_1) - \theta_2 T_{w,m-1} - \theta_2 T_{w,m-2} - \cdots - \theta_m T_{w1} \quad (99)$$

where T_{wm} and T_{Hm} are the temperatures of the heat exchange and insulated surfaces at the end of the mth interval; $\bar{q}_{wm}$ is the mean value of the heat flux from the fluid to the wall on the mth interval; H_m is the dimensionless heat transfer coefficient at the end of the mth interval, which is calculated by the formula

$$H_m = \frac{\pi^2}{16} B_{im} \varDelta\text{Fo} = \frac{\pi^2}{16} \frac{\alpha_m \delta}{\lambda} \varDelta\text{Fo} \tag{100}$$

M_K and θ_K are the dimensionless coefficients depending only on the value of the interval $\varDelta\text{Fo}$ and the number K.

Expressions (97)–(99) are derived for the case of constant thermophysical properties of the plate. But they are also applicable to the case when these properties depend on temperature. In this case the difference $T_w - T_H$ should be sufficiently small to permit accounting for only time dependence on temperature, and the intervals $\varDelta\text{Fo}$ should be equal. If the heat transfer coefficient is prescribed by the functional equation into which enter the terms containing T_w and $\partial T_w/\partial \tau$, then calculation by formula (97) should be made on each spacing by the method of successive approximations. The accuracy of the method is very high as the value of the interval $\varDelta\text{Fo}$ may be chosen as small as necessary.

Thus, the existing methods make it possible to experimentally determine the unsteady-state heat transfer coefficient and take into account its dependence on the rate of heat transfer surface temperature change, fluid temperature at the inlet, and of flow rate. There are also methods permitting us to apply the heat transfer coefficient in the case of unsteady-state processes accounting for the dependence of the heat transfer coefficient on the factors characterizing the degree of unsteadiness. The heat transfer coefficient in the process of unsteady-state heat transfer does not lose its meaning and knowledge of it permits, in particular, a complicated conjugated "flow-channel" problem to be reduced to a considerably more simple problem of heat conduction with boundary conditions of the third kind.

B. Unsteady-State Heat Transfer in Laminar Flow in Channels

In the majority of theoretical works on unsteady-state heat transfer, a laminar flow regime is considered.

In the book by Petukhov (9) the author gives a systematic treatment of the theory and calculation methods of heat transfer in laminar noncompressible fluid flow in tubes, as well as a detailed treatment of the works on heat transfer with unsteady-state flow regime. A brief review of the works on unsteady-state convective heat transfer in a laminar flow is given also in (153).

In published studies, the analytical solution for the case of unsteady-state heat transfer in a laminar fluid flow in channels is obtained only with the following assumptions:

(1) The fluid is incompressible;
(2) The physical properties are constant;
(3) Heat supply due to energy dissipation is absent;
(4) Heat conduction along the channel axis is equal to zero;
(5) $W_y = 0$; $W_z = 0$.

With these assumptions (for a tube) the problem is reduced to the separate solution of the motion equation,

$$\frac{\partial W_x}{\partial \tau} + W_x \frac{\partial W_x}{\partial x} = -\frac{1}{\rho} \frac{\partial P}{\partial x} + \nu \frac{1}{r} \frac{\partial}{\partial r}\left(r \frac{\partial W_x}{\partial r}\right) \tag{101}$$

and the energy equation,

$$\frac{\partial t}{\partial \tau} + W_x \frac{\partial t}{\partial x} = \frac{\lambda}{\rho C_p} \frac{1}{r} \frac{\partial}{\partial r}\left(r \frac{\partial t}{\partial r}\right) \tag{102}$$

As a rule, the energy equation (102) is solved approximately. When the integral method of calculation is used the distribution of velocity W_x along the channel section is taken into account, but instead of the initial differential equation (102) the energy equation in the integral form is used, which is obtained upon integration of (102) along the tube radius. The solution is sought in the form of the series similar to the steady-state solution but in which the energy terms is multiplied by the function $F_n(x, \tau)$. This function is obtained from the solution of the energy equation in the integral form by the method of characteristics.

The applied method of the "rod" model consists in assuming that the velocity in (102) is constant along the channel section but variable in time and equal to the mean flow rate ($W_x = W$). In contrast with the integral method, in this case the function $F_n(x, \tau)$ is found from the solution of the initial differential equation. Such an approach permits, in a number of cases, an exact mathematical solution to be obtained for different boundary conditions which vividly show their influence.

Both the integral method and the "rod" model do not permit estimation of the deviation from the true values, but permit qualitative and exact description of the basic peculiarities of the unsteady-state heat transfer.

In Siegel's work (*154*) the integral method is used for the solution of Eq. (102) for laminar flow in tubes and in plane-parallel channels. A case is considered where unseady-state heat transfer appears at a stepwise wall temperature change with a constant fluid velocity. At the initial moment, heat transfer is absent and the wall temperature t_w is equal to the temperature of the flow t_0 at the inlet. In the calculation, the heat capacity of the wall is neglected. We may assume that the velocity profile in the channel during the transient period remains constant according to the conditions of the problem. For a fully developed flow in a tube the velocity distribution along the channel section is a parabola:

$$W_x/W = 2[1 - (r/r_0)^2] \tag{103}$$

where r is the tube radius.

Equation (102), with the help of (103), is written in a dimensionless form

$$\frac{\partial T}{\partial \text{Fo}} + (1 - R^2) \frac{\partial T}{\partial X} = \frac{1}{R} \frac{\partial}{\partial R} \left(R \frac{\partial T}{\partial R} \right) \tag{104}$$

where

$$\text{Fo} = \frac{\nu \tau}{r_0^2 \, \text{Pr}}; \quad X = \frac{x/r_0}{\text{Re Pr}}; \quad R = \frac{r}{r_0}; \quad T = \frac{t - t_0}{t_w - t_0}$$

t_0 is the inlet temperature and is solved for the following initial and boundary conditions at the inlet, on the tube axis, and at the wall:

$$\begin{aligned}&(1) \quad T(X, R, 0) = 0 \\ &(2) \quad T(0, R, \text{Fo}) = 0 \\ &(3) \quad (\partial T/\partial R)_{R=0} = 0 \\ &(4) \quad T(X, 1, \text{Fo}) = 1\end{aligned} \tag{105}$$

The solution of the unsteady-state problem is sought in the following form:

$$T = 1 - \sum_{n=0}^{\infty} C_n F_n(X, \text{Fo}) \, \varphi_n(R) \tag{106}$$

where C_n and $\varphi_n(R)$ are determined from the solution of the corresponding steady-state problem (155).

The unknown function $F_n(X, \text{Fo})$ is found from the integral relations

$$\frac{\partial F_n}{\partial \text{Fo}} \int_0^R R \varphi_n \, dR + \frac{\partial F_n}{\partial X} \int_0^1 (R - R^3) \, \varphi_n \, dR - F_n \left(\frac{\partial \varphi_n}{\partial R} \right)_{R=1} = 0 \tag{107}$$

The solution of Eq. (107) is obtained by the method of characteristics and has the form

$$T = 1 - \sum_{n=0}^{\infty} C_n \, \varphi_n(R) \begin{cases} e^{-\gamma_n \text{Fo}}; & \text{Fo} \leqslant a_n X \\ e^{-\beta_n^2 X}; & \text{Fo} \geqslant a_n X \end{cases} \tag{108}$$

where

$$\gamma_n = \frac{\beta_n^2}{a_n}; \quad \beta_n^2 = \frac{\partial \varphi_n/\partial R \,|_{R=1}}{\int_0^1 (R - R^3)\, \varphi_n \, dR}; \quad a_n = \frac{\int_0^1 R \varphi_n \, dR}{\int_0^1 (R - R^3)\, \varphi_n \, dR}$$

β_n^2 and φ_n are the eigenvalue and eigenfunction of the Sturm–Liouville type problem.

With sufficiently large time, solution (108) transforms into the steady-state one.

The change in the heat flux during unstabilized process necessary for supporting the constant wall temperature after the step is found from Eq. (108) and the Fourier equation:

$$\frac{qr_0}{\lambda(t_w - t_0)} = - \sum_{n=0}^{\infty} C_n \frac{d\varphi_n}{dR}\bigg|_{R=1} \begin{cases} e^{-\gamma_n \text{Fo}}; & \text{Fo} \leqslant a_n X \\ e^{-\beta_n^2 X}; & \text{Fo} \geqslant a_n X \end{cases} \quad (109)$$

In Fig. 11, the calculated results are given from Eq. (109) for several sections along the length of the channel. As is seen from the plot, the

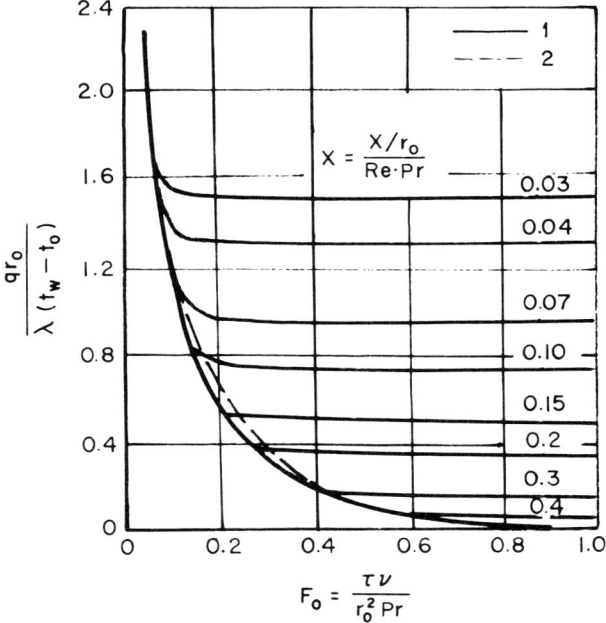

FIG. 11. Change in heat flux for stepwise variation of circular tube wall temperature: (1) integral method (*154*); (2) unsteady heat conduction, Eq. (111).

heat flux at the beginning is greater than the steady-state value. The decrease in the heat flux in time after the stepwise wall temperature change is explained by the heating of the fluid.

At the first moment of time (Fo $< a_n x$) such heating occurs due to unsteady-state heat conduction (since the wall temperature does not change along the length, then the temperature of the fluid also remains constant along the length of the channel, and the convective terms in the energy equation is equal to zero). The heat flux on the wall of the channel may be determined from the solution of the equation for "pure" unsteady-state heat conduction

$$\frac{\partial t}{\partial \tau} = \frac{\lambda}{\rho C_p} \frac{1}{r} \frac{\partial}{\partial r}\left(r \frac{\partial t}{\partial r}\right) \tag{110}$$

with the given boundary and initial conditions, the solution of the equation may be written in the form

$$\frac{q r_0}{\lambda(t_w - t_0)} = 2 \sum_{n=0}^{\infty} e^{-l_n^2 \text{Fo}} \tag{111}$$

where l_n are the roots of the Bessel functions of zero order. In the plot of Fig. 11, solution (111) is shown by a dotted line. As is seen from the plot, the results are in good agreement. The time of establishing the steady-state value of the heat flux depends on the position of the section considered relative to the channel inlet.

From the given approximate solution it follows that the unsteady-state heating of the fluid rod due to heat conduction at a stepwise wall temperature change for the section with coordinate x proceeds for $\tau_s = a_n x/W$. This process in a laminar flow does not depend on the velocity profile. This situation does not taken place in reality since:

(1) The fluid flow velocity is not equal to W and for a laminar flow changes from 0 at the wall to $W_{r=0} = 2W$ on the channel axis. Therefore for the section x, beginning from the instant $\tau_0 = x/W_{r=0}$, the regime of unsteady-state heat conditions will cease in the center of the channel and the solution obtained above will not reflect the real process. For the steady-state process to be established in the section x after the moment τ_0, additional time is needed which is somewhat less than the time of the passage of the heat signal from the axis to the wall. This signal is proportional to $\tau_r = r_0^2/a$. Thus, the time of establishing the steady-state process is $\tau_s \approx a_n x/W \leqslant \tau_0 + \tau_r$ because the velocity is not constant across the cross section.

(2) In real problems owing to thermal inertia the wall temperature and the heat flux cannot be changed stepwise. Therefore, the time of the change of the heat flux and of the wall temperature will be considerably greater than τ_s. Thus, at $\tau < \tau_0$ an unsteady-state conduction or at $\tau > \tau_s$ an unsteady-state heat exchange with time variable t_w or q will take place.

For comparison, in Fig. 12 the dimensionless time $\mathrm{Fo}_s = \nu \tau_s / r_0^2 \Pr$ of establishing steady-state thermal regime is given, determined by the

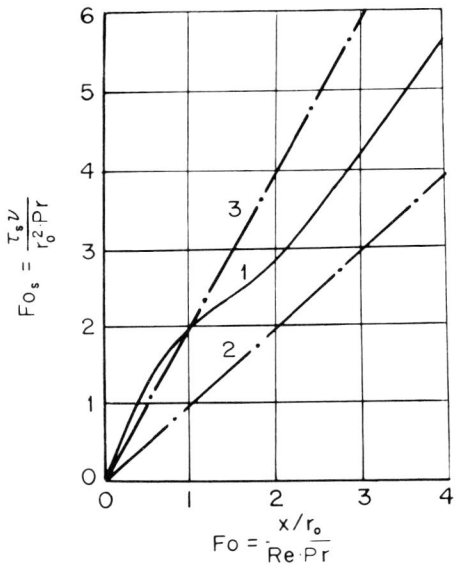

FIG. 12. Time of development of steady regime for stepwise wall temperature variation in circular tube ($\Pr = 0.7$): (1) integral method (*154*); (2) low limit $\mathrm{Fo}_s = X$; (3) "rod" model $\mathrm{Fo}_s = 2X$.

solution obtained with the integral method (*154*) and by using the "rod" model depending on the distance from the inlet $X = (x \mid r_0)/\mathrm{Re}\,\Pr$. As is seen from the plot, the time determined by the integral method is within the bounds $x/W_{r=0} \leqslant \tau_s \leqslant x/W$.

This same problem for the stepwise change of the wall temperature was solved by the integral method in (*156*). The solution was obtained in the form of (108). In Ref. (*157*) a method is given for the solution of the same problem using the Laplace transform. It considers the asymptotic solution for constant wall temperature along the length (the

Graetz problem). A similar result was obtained by Tsoi in (*158*) also using the Laplace transform and solving the boundary problem by the Bubnov–Galerkin method. The relative excess temperature of the flow is reduced to the formula

$$\frac{t(R, X, \text{Fo}) - t_w}{t_0 - t_w} = \begin{cases} \sum_{K=1}^{3} \varphi_K^*(R) \exp(-\beta_K \text{Fo}); & X > \alpha_K \text{Fo} \\ \sum_{K=1}^{3} \varphi_K^{**}(R) \exp(-\gamma_K X); & X < \alpha_K \text{Fo} \end{cases} \quad (112)$$

where

$$X = \frac{1}{\text{Re Pr}} \frac{x}{r_0}; \quad R = \frac{r}{r_0}; \quad \text{Fo} = \frac{\tau \nu}{r_0^2 \text{Pr}} = \frac{a\tau}{r_0^2}$$

The eigenvalues β_K and γ_K, the coefficient α_K, and eigenfunctions $\varphi_K^*(R)$ and $\varphi_K^{**}(R)$ for $K = 1, 2, 3$ are given in the study.

For a plane channel, the solution is obtained in a similar way (*154*). The arbitrary time change of the wall temperature may be given as the sum of the separate step changes. The integral method may also be used for more complex forms of unsteady-state conditions, for example, for a nonstationary laminar flow of a noncompressible fluid in a plane channel (*159*) when the unsteadiness is caused by simultaneous step change in fluid pressure and wall temperature with initially nonheated state and with initial steady heating. The solution is given for the cases when the thermal resistance of the wall is small and when it should be taken into account (see Section III,D). In (*159*) they take into account the velocity change in the transverse section of the channel and in time which is obtained from the solution of the motion equation for an incompressible laminar flow with a stepwise change of the pressure gradient.

First a particular case is considered: Stepwise change of the pressure gradient (initially the pressure gradient is equal to zero; the fluid temperature is equal to the wall temperature and is not equal to the fluid temperature upstream of the channel entrance). The temperature distribution along the cross section of the channel during the transient period is found from the solution of the energy equation under the following initial and boundary conditions:

$$T(X, Y, 0) = T_{s1}; \quad T(0, Y, \text{Fo}) = 0;$$
$$T(X, \pm 1, \text{Fo}) = 1; \quad \left(\frac{\partial T}{\partial Y}\right)_{Y=0} = 0 \quad (113)$$

where

$$X = \frac{8x/r_0}{3\,\mathrm{Re}\,\mathrm{Pr}}; \quad Y = \frac{y}{r_0}; \quad \mathrm{Fo} = \frac{\tau\nu}{r_0^2\,\mathrm{Pr}};$$

$$\mathrm{Re} = \frac{W 4 r_0}{\nu}; \quad \mathrm{Pr} = \frac{C_p \mu}{\lambda}$$

r_0 is the half-distance between the plates; T_{s1} is the temperature in the steady-state regime.

As in the previous case, a solution of the problem may be given in a form similar to the solution for the steady-state case

$$T = 1 - \sum_{n=0}^{\infty} b_{n_0} \psi_n(Y) F_n(X, \mathrm{Fo}) \tag{114}$$

where $F_n(X, \mathrm{Fo})$ is the function sought. It may be determined from the energy equation in the integral form, the solution of which is found by the method of characteristics. The term F_n depends on the relation between X and Fo. At $\mathrm{Fo} < X$, as was pointed out, the heat exchange will be determined by the unsteady-state heat conduction, and at $\mathrm{Fo} > X$ convective heat transfer will take place. In Fig. 13 the time variation of the heat flow on the wall is shown for several values of X. At the beginning of the unsteady-state process, until the liquid reaches the section considered ($\mathrm{Fo} < X$), the heat transfer in this section

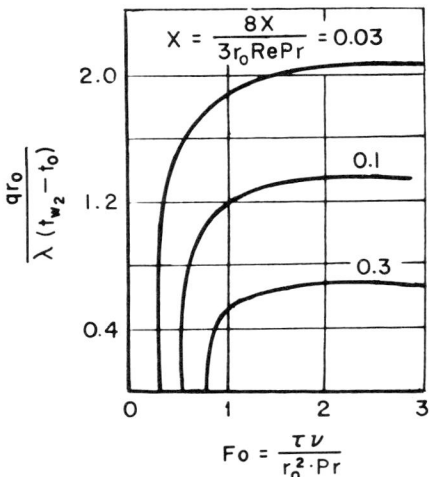

FIG. 13. Change in heat flux in flat channel for stepwise variation of pressure gradient and constant wall temperature Pr = 0.7.

remains equal to 0 since the fluid was not moving and its temperature was equal to the wall temperature and different from the flow temperature at the inlet. During the subsequent period, the heat flux increases rapidly and then acquires a constant value corresponding to the established velocity distribution.

For the case of a simultaneous change of the pressure gradient and the wall temperature (without initial heating), if the dimensionless temperature is determined in the form $T = (t - t_0)/(t_{w_2} - t_0)$, where t_0 is the fluid temperature at the inlet, then boundary conditions (113) apply, while the initial condition will change: $T(X, Y, 0) = 0$. The solution is obtained in a similar way. The change of the heat flux after a stepwise change of the fluid pressure and of the wall temperature is shown in Fig. 14. For each value of X, the heat transfer goes through

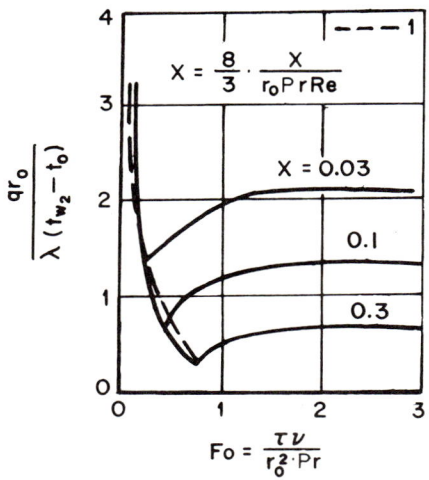

Fig. 14. Change in heat flux in flat channel for variation of pressure gradient and wall temperature, $Pr = 0.7$. Initial moment: $W_x = W_1 = 0$; $t_w = t_{w_1} = t_0$. Final moment: $W_x = W_2$; $t_w = t_{w_2}$. (1) transient heat conduction.

a minimum, and then increases up to a constant value corresponding to the steady-state value. The curve, corresponding to the decrease in the heat flux, reflects the process of unsteady-state heat conduction. The increase in the heat flux corresponds to the time when the temperature of the flow along the axis of the channel starts changing. The convective term in the energy equation has a finite value and under the action of the increasing velocity the heat flux increases until the velocity reaches a finite value W_2. The dotted line shows the solution of the problem of unsteady-state heat conduction in an infinite plate with the stepwise temperature change on the surface.

The same problem is solved by Siegel and Perlmutter in (*160*) with the help of the "rod" model. The expression for the uniform velocity with the assumption of rod flow is obtained by averaging the velocity profiles along the section as given in (*159*). For a stepwise change of the wall temperature and the pressure gradient without initial steady-state heating, the temperature distribution in the transient period is determined by the equation

$$T = 1 - 2 \sum_{n=0}^{\infty} \frac{(-1)^n}{E_n} F_n(\text{Fo}, X) \cos E_n Y \qquad (115)$$

where $E_n = (n + \tfrac{1}{2})\pi$; $F_n(\text{Fo}, X)$ is the function taking into account the unsteady-state condition. At sufficiently large time $F_n(\text{Fo}, X) \to e^{E_n^2 X}$ (i.e., steady-state solution). The results of calculation for this case are given in Fig. 15.

For stepwise change in the pressure gradient with constant wall temperature the solution is obtained in a similar way. In this case the

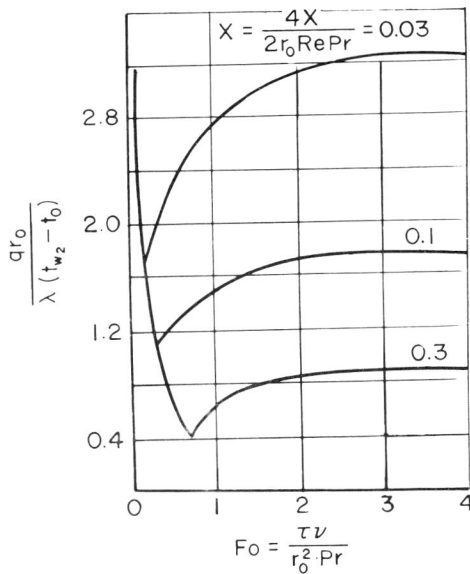

FIG. 15. Change of heat flux in flat channel at stepwise variation of pressure gradient and wall temperature, $\text{Pr} = 0.7$. Initial moment: $W = W_1 = 0$; $t_w = t_{w_1} = t_0$. Final moment: $W = W_2$; $t_w = t_{w_2}$.

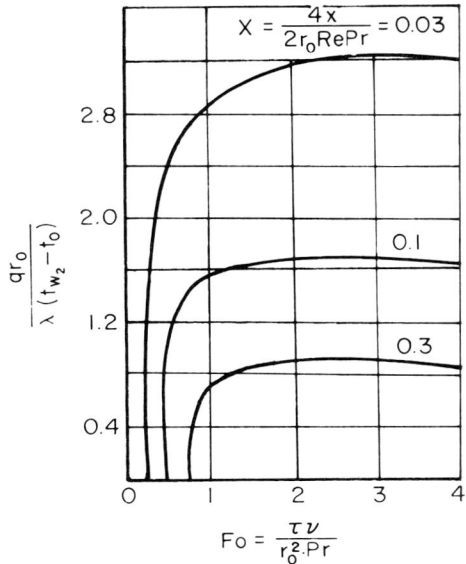

FIG. 16. Change of heat flux in flat channel at stepwise variation of pressure gradient and constant temperature of wall, Pr = 0.7. Initial moment: $W = W_1 = 0$. Final moment: $W = W_2$.

unsteady-state condition is accounted for by the function $G_n(\text{Fo}, X) = e^{-E_n^2 X W_2/W_1}$. The results of the solution are given in Fig. 16.

For a simultaneous stepwise change in wall temperature and pressure gradient with initial steady-state heating, the solution may be obtained by superposition of the two partial solutions obtained earlier. The final result has the form

$$T = \frac{t - t_0}{t_{w_2} - t_0} = \left(\frac{t_{w_2} - t_{w_1}}{t_{w_2} - t_0}\right) [F_n \text{ solution}] + \left(\frac{t_{w_1} - t_0}{t_{w_2} - t_0}\right) [G_n \text{ solution}] \quad (116)$$

where F_n and G_n are solutions obtained for the partial cases of the wall temperature and pressure gradient changes.

In this case, when unsteadiness is caused by the step in the pressure gradient and heat flux, the qualitative picture is the same. In Fig. 17 the results of calculations are given for simultaneous stepwise increase in the pressure gradient and heat flux. For the interval of time $\text{Fo} < X$ the wall temperature increases with time. This is explained by the fact that heat exchange in this regime is caused by unsteady-state heat conduction (convective terms in the energy equation is equal to 0). If

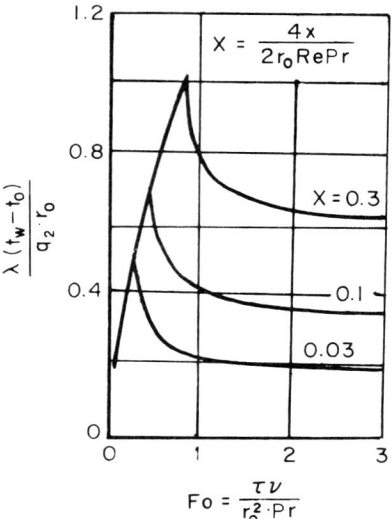

FIG. 17. Wall temperature variation in flat channel at stepwise variation of pressure gradient and heat flux, Pr = 0.7. Initial moment: $W = W_1 = 0$; $q = q_1 = 0$. Final moment: $W = W_2$; $q = q_2$.

the liquid were stationary, then in principle the wall temperature would increase infinitely ($t_\infty \to \infty$), but, owing to the fact that at the moment Fo > X the convective transfer appears, the increase in the wall temperature ceases and the wall temperature starts decreasing approaching the new steady state. In this region convective heat exchange is effective by the time variation of the velocity profile which is not accounted for by the method considered. In the case of decreasing heat flux and pressure gradient (Fig. 18) the wall temperature for the time Fo < X decreases reaching a minimum since the velocity still has the initial value, and at Fo > X starts increasing and reaches the steady-state value. It should be noted that for one and the same unsteady-state process the solutions by the integral method (159, Figs. 13 and 14) and with the use of the "rod" model (160, Figs. 15 and 16) gave qualitatively the same results. The results obtained in (154, 157, 160) pertain to the case when the heat capacity of the channel walls is negligibly small as compared with the fluid heat capacity. It can be expected that the heat capacity of the walls is somehow smoothed by the time variation of the boundary conditions. For walls with very high heat capacity, the flow velocity can reach a new steady-state value up to the moment when heating begins in the channel. In this case, there is no necessity in accounting for the

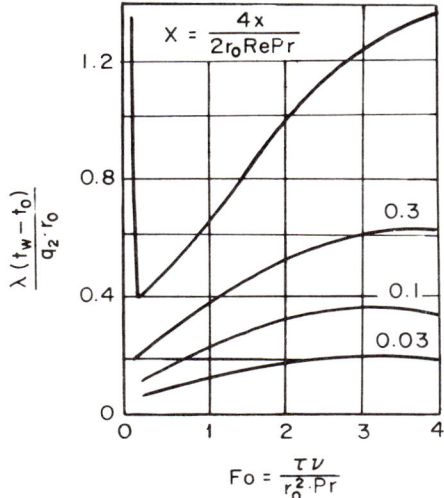

FIG. 18. Wall temperature variation at stepwise variation of pressure gradient and heat flux: $\text{Pr} = 0.7$; $q_1/q_2 = 5$; $W_1/W_2 = 10$.

unsteadiness of the velocity, and the results of works (*154, 155, 161, 162*) may be utilized.

Siegel and Sparrow (*161, 162*) solve the problem of unsteady-state heat transfer for the thermal starting length of the tube and the plane channel. The case considered is the stepwise change of wall temperature (heat exchange is absent at the initial moment). The solution is obtained by the integral method with the application of this approximate method of boundary layer theory. The initial integral relations are obtained by integrating the energy equation along the thickness of the boundary layer (Δ is the thickness of the boundary layer). For the distribution of the velocity in the boundary layer, the Poiseuille law is utilized. As a result of the solution, the dependence (in a dimensionless form) of the boundary-layer thickness $\bar{\Delta} = \Delta/r_0$ on the time $\text{Fo} = a\tau/r_0^2$, the distance from the beginning of the section $(1/\text{Pe})(x/2r_0)(\text{Pe} = W2r_0/a)$ for the tube (Fig. 19), and the time of the onset of the steady-state regime Fo_s in the tube (1) and plane channel (2) (Fig. 20) may be used. The results for the thermal starting length are in good agreement with the results of external flow problems.

In the work of Yang and Ou (*40*), the hydrodynamics and heat transfer in the inlet section of the tube and plane channel are investigated theoretically with an arbitrary law of the flow velocity distribution at the inlet. The medium is considered incompressible with constant physical

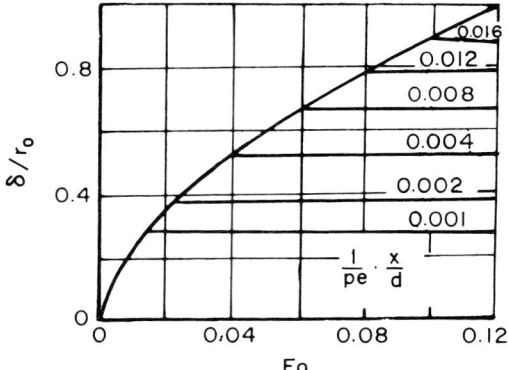

FIG. 19. Change in boundary-layer thickness δ/r_0 for thermal starting length in tube for variation of heat flux q_w.

properties. The cases of constant wall temperature and constant heat flux are considered. Asymptotic solutions are found for very small and large times. The solutions are obtained by series expansion of the dimensionless temperature and flow function with respect to corresponding dimensionless parameters. The coefficient for the Prandtl numbers 0.7 and 5 are obtained on a digital computer. Both solutions have a satisfactory convergence. It is shown that with positive accelerations and small times the shear stress on the wall is higher, and heat transfer lower, than in a quasi-steady-state case (Fig. 21).

Examples are calculated for the stepwise, monotonic, and periodic laws of flow velocity variation. In the latter case, at low-frequency inlet

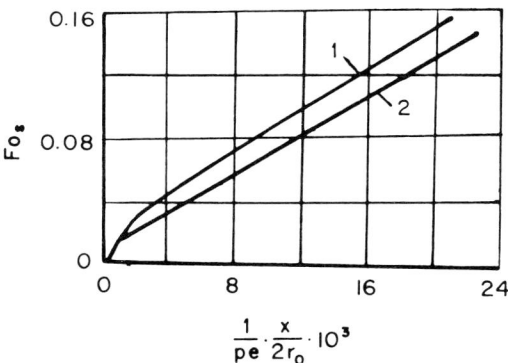

FIG. 20. Time of development of steady regime on thermal starting length of flat channel (1) and tube (2) at stepwise variations of heat flux.

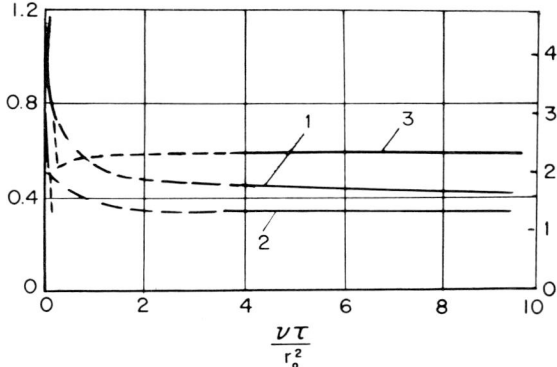

FIG. 21. Shearing stress on wall, $T_w \sqrt{Re}/\rho W_0^2$ (1), heat transfer at given wall temperature $Nu/\sqrt{Re}$ (2), and (3) wall temperature at given heat flux versus time for entrance tube section with flow acceleration, $(T_w - T_0)/(q_w/\lambda)(\nu x/W_0)^{1/2}$ (Pr $= 0.7$; $x/2r_0 = 1$; $U_0(\tau) = \beta\tau$).

velocity pulsations, the steady-state periodic and secondary steady-state solutions are obtained for the functions of flow, velocity, temperature, shear stress on the wall, and the Nusselt number. This points to the appearance of the secondary steady-state flows both in the core and in the laminar boundary layer directed into the side contrary to the main flow. Secondary flows cause a very small rise in the shear stress on the wall and decrease in heat transfer as compared to the quasi-steady-state case.

Calculations of unsteady-state heat transfer in a plane channel with the assumption of a "rod" flow are made also for more complex conditions in (163) where arbitrary changes of the heat flux in time and along the channel are considered with the simultaneous time variation of the velocity. In the example given in (160) the heat flux through the wall is determined in the form

$$q/q_2 = (1 - e^{AX})(1 - e^{BFo}) \tag{117}$$

The calculations are made for various B and A and are compared with the results of calculations where the heat transfer coefficient α is constant. The results of calculation for the case when $B = 0.5$ (Fig. 22a) correspond to smooth variation in time and are in good agreement with the solution where $\alpha = $ const. Inconsistency for small X follows from the fact that in the solution for $\alpha = $ const the coefficient of heat transfer was used for the fully developed flow while the solution of (163) takes into account the change in the local value of α at the heated section.

The results given in Fig. 22b correspond to the sharp change of q up to the time ($B = 100$). In this case, the change in time may be considered close to the step change. For small values of Fo we observed considerable deviation of the solution in (163) from the case when the heat transfer coefficient is constant.

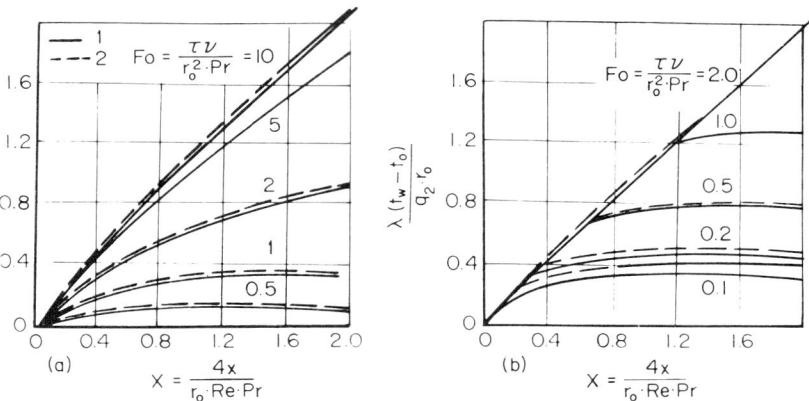

FIG. 22. Wall temperature variation with changing heat flux and constant velocity: (1) solution of (163); (2) solution for constant heat transfer coefficient.

In Ref. (164) the rod model is used for calculation of the temperature field in a circular tube at the prescribed initial and boundary conditions. To account for the dependence of fluid thermal conductivity on temperature it is suggested that the following variable introduced into the energy equation

$$\theta = \frac{1}{\lambda(0)} \int_0^t \lambda(t)\, dt \tag{118}$$

Some general regularities of unsteady-state heat transfer in a laminar flow in channels were stated in the study by Vilenskii (10, 13). The fluid velocity profile is considered fully developed and constant in time, the physical properties of the fluid are considered constant, and the heat flux along the channel axis due to heat conduction is negligibly small. With these assumptions the fluid temperature field is described by the equation

$$\partial T/\partial \text{Fo} + U\, \partial T/\partial X = \partial^2 T/\partial Y^2 + \partial^2 T/\partial Z^2 \tag{119}$$

where $T = (t_b - t^*)/\Delta t^*$ (t_b is the fluid temperature, and t^* and Δt^* are the characteristic temperature and temperature difference); $U = W_x/W$ (W_x, W are, respectively, the axial fluid velocity and its mean value); $\text{Fo} = a\tau/r_r^2$ (r_r is the hydraulic channel radius);

$X = 2x/r_r$ Pe (x is the coordinate along the channel axis, Pe $= 2r_r W/a$); $Y = Y/r_r$; $Z = r/r_r$ (Y, Z are the coordinates in the channel cross section plane).

The unsteady-state heat transfer is considered with time variation of temperature or heat flux at the channel wall.

In the region $X \geqslant U_{\max}$ Fo ($U_{\max} = W_{x\max}/W$ is the maximum velocity value) the problem is reduced to the solution of the heat conduction equation; the fluid temperature field is uniquely defined by the initial temperature field at the channel wall and does not depend on the channel inlet conditions. At $X < U_{\max}$ Fo the fluid temperature field is determined by the channel inlet conditions, channel wall conditions, and the conditions at Fo $= X/U_{\max}$; it is the solution of an ultraparabolic equation. The author obtained the asymptotic solutions of this equation for large Fo with the help of which the limiting behavior of Nu is studied with increase in time for various laws of wall temperature change t_w. It is shown in the study that in both regions the distribution of Nu along the channel length with increase in Fo tends to the limiting value, if t_w does not increases more quickly than $\exp(K$ Fo$)$. The time-limiting properties of the fluid temperature field and the Nusselt number do not depend on the specific form of the function describing the law of boundary-layer change caused unsteady heat transfer and are determined by the limit of its logarithmic derivative at Fo $\to \infty$, i.e.,

$$\lim_{\text{Fo}\to\infty} \frac{d\ln \phi}{d\text{Fo}} = \lim_{\tau\to\infty} \frac{r_T^2}{a} \frac{1}{t_w} \frac{\partial t_w}{\partial \tau}$$

If the limit of the logarithmic derivative is equal to zero, then stabilization of the temperature field and of the Nusselt number starts both in the first and in the second regions. In the second region, Nu_∞ will be equal to the steady-state value. In the first region, stabilization (having the same meaning as the regular regime, for example, in the heating of a solid rod) may not start when x is small and at $\tau = \tau_{\text{stab}}$ the condition $x > W_{x\max}\tau$ will not hold. If this limit is equal to $K_0 > 0$ (but $K_0 < \infty$), then stabilization will be present in both regions, but the values of Nu_∞ in them will be greater than the previous ones.

In a number of studies (*158, 165–173*) some specific problems are considered which refer to the subject discussed here. In (*158, 171*) the heat transfer problem in a plane parallel channel is solved, the heat transfer being caused only by the heat of friction. The energy equation has the form (*158*)

$$\frac{\partial t}{\partial \tau} + W_x(y) \frac{\partial t}{\partial x} = a \frac{\partial^2 t}{\partial y^2} + \frac{\eta}{C\gamma}\left(\frac{dW_x}{dy}\right)^2 \qquad (120)$$

with initial and boundary conditions

$$(x, y, \tau)_{\tau=0} = t_0; \qquad t(x, y, \tau)_{x=0} = \varphi_0(\tau); \qquad t(x, y, \tau)_{y=(-1)^K G} = \varphi_K(x, \tau)$$

Here

$$W_x(y) = \frac{3W_0}{2}(1 - Y^2); \qquad Y = \frac{y}{r_0}; \qquad X = \frac{1}{\text{Re Pr}}\frac{x}{r_0}$$

and r_0 is the half-width of the channel. The temperature distribution inside the channel under the condition of constant wall temperature and initial fluid temperature $[\varphi_K(x, \tau) = t_0]$ in the first approximation is written in the form

$$t(Y, X, \text{Fo})$$
$$= t_0 + \frac{3}{4}\frac{\eta W_0^2}{\lambda}\begin{cases}[1 - \exp(-3.214\,\text{Fo})](1 - Y^4); & X > 1.2W_0\tau \\ [1 - \exp(-2.673X)](1 - Y^4); & X < 1.2W_0\tau\end{cases} \quad (121)$$

At large distances from the starting length and at large times $\tau(\text{Fo} \geq 1)$ the fluid temperature approaches the limiting value

$$t_{\text{stab}}\left(\frac{y}{r_0}\right) = t_0 + \frac{3}{4}\frac{\eta W_0^2}{\lambda}\left[1 - \left(\frac{y}{r_0}\right)^4\right] \quad (122)$$

The heat flux at the channel wall $q^* = -\lambda(\partial t/\partial y)_{y=1}/\eta W_0^2$ (Fig. 23) increases as X becomes larger.

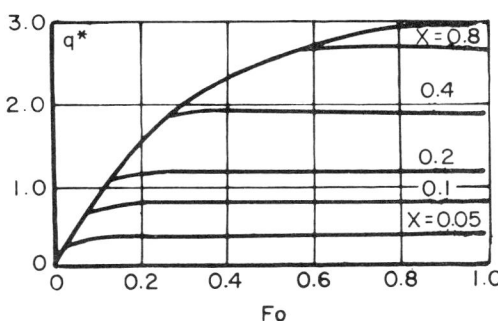

FIG. 23. Change of heat flux on flat channel wall at energy dissipation.

In works (165, 166), for the solution of the unsteady-state heat transfer problem, a new calculation method is suggested based on the approximate change of derivatives entering the energy equation by difference relations. The unsteady-state temperature field in a plane channel with a sudden heating at the channel inlet was found by this method.

In (*167*) the numerical finite-difference method was used to solve the problem of unsteady-state heat transfer in a tube with a stepwise wall temperature change and with heating by the moving flow of a wall of infinitely large thermal conductivity. Storage capacity of the computer limited the number of coordinate nodes considered, and therefore what was studied was the time variation of heat transfer at the starting length of the tube with the length $L = 4x/2r_0 \text{ Pe} = 0.3$ with radius pitches $\Delta R = \Delta r/r_0 = 0.1$ and length pitches $\Delta L = 0.01$. It was assumed that $\beta = \gamma_w C_w \delta_w / C\gamma r_0 = 1000$ (the subscript w refers to the wall; thus, δ_w is the wall thickness). Beginning from Fo $= 0.25$, the solution for both cases coincide and, for points not very close to the inlet ($L > 0.15$), they practically do not differ from the steady solution (deviation is less than 0.5% at Fo > 0.35).

In (*168–170, 173*) unsteady-state heat transfer in a liquid metal laminar flow in a tube is considered. The solution is obtained by the integral method. Axial heat conduction is taken into account.

Kuznetsov (*336, 337*) investigated unsteady heat transfer in laminar (or turbulent) flow in tubes with constant flow rate and fluid properties. The starting point of the analysis is the unsteady energy equation for the fluid

$$\frac{\partial \theta}{\partial \text{Fo}} + U(R) \frac{\partial \theta}{\partial Z} = \frac{1}{R} \frac{\partial}{\partial R}\left[\gamma(R) R \frac{\partial \theta}{\partial R}\right] \tag{A}$$

where

$$\theta = \frac{T - T_{in}}{T_{in}}; \quad Z = \frac{2x}{r_0 \text{ Pe}}; \quad \text{Fo} = \frac{a\tau}{r_0^2}; \quad R = \frac{r}{r_0}$$

$U(R)$ is the known velocity profile. The boundary condition

$$\theta|_{\text{Fo}=0} = \theta_0(R, Z); \quad \theta|_{Z=0} = 0; \quad \theta|_{R=1} = \theta_w = f(\text{Fo}) \tag{A*}$$

is known. Substituting new variables $V = \theta(R, Z, \text{Fo})/f(\text{Fo})$ in the energy equation one can obtain

$$\frac{\partial V}{\partial \text{Fo}} + K_\theta(\text{Fo}) V + U(R) \frac{\partial V}{\partial Z} = \frac{1}{R} \frac{\partial}{\partial R}\left[\gamma(R) R \frac{\partial V}{\partial R}\right] \tag{B}$$

with boundary condition

$$V|_{\text{Fo}=0} = 0; \quad V|_{Z=0} = 0; \quad V|_{R=1} = 1; \quad \partial V/\partial R|_{R=0} = 0 \tag{B*}$$

Here

$$K_\theta(\text{Fo}) = \frac{1}{f(\text{Fo})} \frac{df(\text{Fo})}{d\text{Fo}}$$

In the general case heat transfer is defined by all prehistory of $K_\theta(\text{Fo})$; for large values of Fo only the value of K_θ defines the heat transfer. As was shown by Vilenskii (*13*), if

$$\lim_{\text{Fo}\to\infty} K_\theta(\text{Fo}) = \tilde{K}_\theta < \infty$$

then $V(R, Z, \text{Fo}) \to \tilde{V}(R, Z)$, where $\tilde{V}(R, Z)$ is a solution of

$$\tilde{K}_\theta \tilde{V} + U(R) \frac{\partial \tilde{V}}{\partial Z} = \frac{1}{R} \frac{\partial}{\partial R} \left[\gamma(R) R \frac{\partial \tilde{V}}{\partial R} \right] \quad \text{(C)}$$

$$\tilde{V}\,|_{Z=0} = 0; \quad \tilde{V}\,|_{R=1} = 1; \quad \partial \tilde{V}/\partial R\,|_{R=0} = 0 \quad \text{(C*)}$$

The solution of Eq. (B) when

$$\theta_w = f(\text{Fo}) = \sin \text{Fo } \Omega = e^{i\Omega \text{Fo}}$$

is

$$\tilde{\theta} = A(R, Z, \Omega) \sin[\Omega \text{ Fo} + \varphi(R, Z, \Omega)]$$

where A and φ are the modulus and argument of the complex function $\tilde{V}$. The solution of problem (A)–(A*) is found in the integral form

$$\theta(R, Z, \text{Fo}) = \int_0^{\text{Fo}} G(R, Z, \text{Fo} - \tau) f(\tau) \, d\tau$$

It is shown that

$$G(R, Z, \text{Fo}) = \frac{2}{\pi} \int_0^\infty A(R, Z, \Omega) \cos[\varphi(R, Z, \Omega)] \cos \Omega \text{ Fo } d\Omega$$

The method is illustrated for laminar flow in the tube when $U(R) = 2(1 - R^2)$ and $\gamma(R) = 1$. Solved for were the bulk temperature

$$\theta_b = \int_0^{\text{Fo}} G_\theta(Z, \text{Fo} - \tau) f(\tau) \, d\tau$$

and the heat flux

$$q_w = \int_0^{\text{Fo}} G_q(Z, \text{Fo} - \tau) f(\tau) \, d\tau$$

Here

$$G_\theta = 4 \int_0^1 G(1 - R)^2 R \, dR \quad \text{and} \quad G_q = \left. \frac{\partial G}{\partial R} \right|_{R=1}$$

θ_w, θ_b, and q_w have a shift in phase; therefore the Nusselt number rises with Fo to minus infinity and then returns from plus infinity. The method can be used when θ_w changes with distance.

As a result of the review of the studies dealing with unsteady-state heat transfer in a laminar flow we may draw the following conclusions:

(1) Unsteady-state heat transfer in laminar flow in channels has been studied in detail for step disturbances. Considered was laminar flow of an incompressible fluid with constant physical properties without accounting for viscous dissipation and axial heat conduction.

(2) Theoretical solutions have been obtained for the "rod" model or by the integral method. These techniques do not take into account the redistribution of the velocity profile which occurs in an unsteady-state process. Therefore the solutions are not accurate and are applicable mainly for qualitative analysis of unsteady-state processes.

(3) The time of transient process for a fixed section may be divided into two regions: (a) To the first region pertains those time instants when the fluid which was outside the channel before the beginning of the unsteady-state process did not succeed in coming into the section considered. (b) To the second region refer those time instants when the fluid which was outside the channel before the beginning of the unsteady-state process did reach the section considered.

(4) The solutions obtained give satisfactory agreement with the solutions for "pure" unsteady-state heat conduction in the first region of the transient period.

(5) Solutions for the stepwise wall temperature change (or of heat flux) are valid only for a wall with zero heat capacity. With finite heat capacity, the wall temperature (or heat flux) cannot change stepwise, and therefore the time of unsteady-state heat transfer practically stretches up to the steady-state values of wall temperature (or heat flux).

(6) Unsteady-state heat transfer in laminar flow has not been investigated experimentally.

C. Unsteady-State Heat Transfer in Turbulent Channel Flow

The theoretical analysis of unsteady-state heat transfer in turbulent flow by even approximate methods is much more complex than in laminar flow. Moreover, data on the distribution of turbulent coefficients of momentum and heat transfer along the channel section are not known for unsteady-state flows. Therefore theoretical studies only consider unsteady-state heat transfer with a nonvariable velocity profile and steady-state distribution of turbulent parameters along the flow section.

Sidorov (*174, 175*) made an approximate estimation of the applicability of steady-state convective heat transfer formulas to unsteady-state processes and calculated the possible errors when the formulas are used in internal flows. The problem is solved by the method of successive approximations. Assuming the fluid to be incompressible, the surface to be straight, and neglecting the energy dissipation due to internal friction and the dependence of fluid physical properties on temperature, the author writes down the energy equation for the boundary layer

$$\frac{\partial \Delta T}{\partial \tau} + W_x \frac{\partial \Delta T}{\partial x} + W_y \frac{\partial \Delta T}{\partial y} = a_\Sigma \frac{\partial^2 \Delta T}{\partial y^2} \tag{123}$$

where $\Delta T = T_b - T_w$ is the difference between the flow temperature T_b and invarient temperature of the wall T_w; a_Σ is the summation thermal diffusivity coefficient. When the energy equation (123) is transformed, the convective term is dropped out. The author considers that the presence of convective transfer is taken into account by the boundary conditions

$$y = 0; \quad \Delta T = 0; \quad y = \delta; \quad \Delta T = \Delta T_0 \tag{124}$$

where δ is the boundary-layer thickness; $\Delta T_0 = T_0 - T_w$ is the flow temperature on the edge of the boundary layer. Then from the solution of energy equation (123), taking into account (124), the relation of unsteady q and quasi-steady-state value q_0 is obtained

$$\frac{q}{q_0} = 1 - \frac{1}{6} \frac{\rho C \lambda \, \Delta T_0^2}{q_0^2} \left(\frac{\partial \Delta T_0}{\partial \tau} \frac{1}{\Delta T_0} + m \frac{1}{W_x} \frac{\partial W_x}{\partial \tau} \right) \tag{125}$$

where m is the dimensionless exponent of the dependence of the boundary-layer thickness on velocity (for laminar flow $m = 0.5$; for turbulent flow $m = 0.2$). With flow acceleration the heat transfer coefficient is less than that obtained by quasi-steady calculations, and with flow deceleration it is higher. The calculated transfer of heat from the fluid to the wall under conditions of temperature rise in the flow is less than by quasi-static calculations, and with fluid temperature decrease it is higher. According to Eq. (125) the influence of unsteady conditions on heat transfer is slight. This is perhaps the results of neglecting convective terms. If these terms are taken into account, then the initial temperature profile will not be linear and the wall temperature change should influence heat transfer more strongly and the temperature gradient near the wall will change. The conclusion of Sidorov on the effect of hydrodynamic unsteady conditions seems to be highly approximate since in the solution, the redistribution of the velocity profile is not taken into account, while the momentum equation is not considered at all.

In (176) Sidorov considers a plane unsteady nonisothermal flow of an incompressible fluid under the same assumptions as in (174). By integrating the energy and momentum equations of the boundary layer the author obtained the following expressions to account for the influence of unsteady conditions on the friction factor:

$$C_f = \frac{2T}{\rho W_x^2} = 0.0263 \text{ Re}^{-1/7} \left(1 - H \frac{1}{W_x^2} \frac{\partial W_x}{\partial \tau}\right)^{-1/7} \quad (126)$$

and on heat transfer:

$$\text{St} = 0.0132 \text{ Re}^{-1/7} \text{ Pr}^{-2/3} \left\{\frac{6}{7\phi x}[1 - \exp(-\tfrac{7}{6}\phi x)]\right\}^{-1} \quad (127)$$

where T is the shear stress; W_x is the flow velocity outside the boundary layer; $H = \delta^*/\delta^{**} = \delta_T^*/\delta_T^{**} = (1 - 0.78 \text{ Re}^{-1/14})^{-1}$ is considered constant; $\delta^*, \delta^{**}, \delta_T^*$, and δ_T^{**} are displacement thickness, impulse, temperature displacement losses, and enthalpy losses, respectively;

$$\phi x = \frac{\partial T_w}{\partial \tau} \frac{H}{W_x(T_0 - T_w)} - \frac{\partial T_b}{\partial \tau} \frac{72/7 - H}{W_x(T_0 - T_w)} - \frac{1}{7} \frac{H}{W_x^2} \frac{\partial W_x}{\partial \tau}$$

is the unsteady parameter; and T_0 is the flow temperature on the outer border of the thermal boundary layer.

When obtaining (127), for transition from $\partial \delta_T^{**}/\partial \tau$ to $\partial(T_0 - T_w)/\partial \tau$ and $\partial W_x/\partial \tau$ the form of bond is used from the steady-state flow. From (126) it follows that with flow acceleration C_f is higher than given by quasi-steady-state relation, while with deceleration it is lower.

With small values of the dimensionless parameter ϕx, we take the first terms of the series and obtain

$$\text{Nu} = 0.0132 \text{ Re}^{6/7} \text{ Pr}^{1/3} (1 - \tfrac{7}{6}\phi x)^{-1/7} \quad (128)$$

With an increase in the temperature of the incoming fluid and with heat transfer from the fluid to the wall at constant T_w and W_x the heat transfer will be lower than according to quasi-steady-state calculations. With a decrease in the fluid temperature, $\phi x > 0$, the heat transfer will be higher than given by steady-state calculations. The conclusion is quantitatively the same as in the previous work of Sidorov (174). With heat transfer from the fluid to the wall and with wall temperature rise,

$\phi x < 0$, the heat transfer is less than the quasi-steady-state value. It follows that with the fluid temperature rise and with a wall temperature rise the heat transfer from the fluid to the wall decreases. With a rise in the wall temperature, the fluid temperature gradient near the wall decreases and the heat transfer coefficient is lower than according to quasi-steady-state calculations. With heat transfer from the wall to the fluid, an increase in wall temperature will lead to an increase in heat transfer. With an acceleration of the fluid, the unsteady-state parameter $\phi x < 0$ and the correction to unsteady-state conditions will be less than 1, i.e., according to the work considered, the acceleration of the flow decreases the heat transfer coefficient. Deceleration of the flow increases the heat transfer as compared to the quasi-steady-state calculations. Numerical calculations for real unsteady-state conditions give small differences from quasi-steady-state solutions, and according to (*176*) the deviation is much higher than given by (*174*). Such a conclusion of Sidorov is the result of the fact that in the solutions considered, the hydrodynamic analogy is used, i.e., the similarity of the velocity and temperature fields near the wall. It is obvious that, under unsteady-state conditions with a velocity or fluid temperature change, the velocity and temperature fields are not similar as a rule.

In the work of Sparrow and Siegel (*177*) unsteady-state turbulent heat transfer in a tube is considered with constant mass flow rate and stepwise time variation of the wall temperature. At the initial moment the fluid and wall temperatures are equal and the heat flux is zero. The calculation method is the same as in (*154*). For hydrodynamically stabilized flow the energy equation is

$$\frac{\partial T}{\partial \tau} + W_x \frac{\partial T}{\partial x} = \frac{1}{r} \frac{\partial}{\partial r} \left[r(a + \epsilon_q) \frac{\partial T}{\partial r} \right] \qquad (129)$$

The following assumptions were made when Eq. (129) was written: The axial heat conduction and energy dissaption were neglected; fluid rate and inlet fluid temperature were also considered constant. The wall temperature changed in time but not along the channel length. The dimensionless velocity profile and the coefficient of turbulent thermal conductivity were taken by data known for steady-state flow. The solution of Eq. (129) should satisfy the equation of pure heat conduction at the initial moment since, at the beginning of the process, heat transfer was determined by "pure" heat conduction, and for large times it should approach the solution for a steady-state process.

Sparrow and Siegel obtained the solution of Eq. (129), applicable both to the unsteady-state and to the steady-state cases, both for the tube inlet and for the fully developed section. To obtain the solution, Eq. (129) was integrated with respect to r from 0 to r_0:

$$\frac{1}{(r_0^+)^2} \frac{\partial}{\partial \tau^+} \left[\int_0^{r_0^+} r^+ T^+ \, dr^+ \right] + \frac{1}{2r_0^+} \frac{\partial}{\partial x^+} \left[\int_0^{r_0^+} r^+ W_x^+ T^+ \, dr^+ \right]$$

$$= \frac{r_0^+}{\Pr} \left(\frac{\partial T^+}{\partial r^+} \right)_{r_0^+} \tag{130}$$

where

$$W_x^+ = \frac{W_x^0}{U_*}; \qquad \tau^+ = \frac{\tau \nu}{r_0^2}; \qquad x^+ = \frac{x}{d}; \qquad T^+ = \frac{T - T_w}{T_0 - T_w}$$

T_0 is the inlet flow temperature,

$$r_0^+ = r_0 U_*/\nu; \qquad r^+ = r U_*/\nu,$$

and U_* is the friction velocity. The temperature distribution in the transient period should satisfy the following initial and boundary conditions:

$$T(x, r) = T_0 \quad \text{at} \quad \tau = 0 \quad \text{or} \quad T^+(x^+, r^+) = 1 \quad \text{at} \quad \tau^+ = 0$$

$$T^+(0, r^+) = 1; \qquad T^+(x^+, r_0^+) = 0; \qquad \partial T^+/\partial r^+(x^+, 0) = 0 \tag{131}$$

The solution for the unsteady-state process is sought in the form

$$T^+ = \sum_{n=1}^{\infty} C_n F_n(x^+, \tau^+) \phi_n(r^+) \tag{132}$$

where C_n and ϕ_n are taken from the solution for the steady-state process; F_n is the series of functions dependent on x^+ and τ^+.

Here F_n is found with the help of the integral energy equation (130). Substitution of (132) into (130) gives the partial differential equation which makes it possible to determine T^+ as a function of x^+ and τ^+. This equation is solved by the method of characteristics. As a result the following expression is obtained for F_n:

$$F_n = \exp \left\{ \frac{(r_0^+)^3 \, (d\phi_n/dr^+)_{r_0^+} \, \tau^+}{\Pr \int_0^{r_0^+} r^+ \phi_n \, dr^+} \right\} \quad \text{at} \quad \tau^+ \leqslant a_n x^+$$

$$F_n = \exp \left\{ -\frac{4B_n^2 x^+}{\text{Re}} \right\} \quad \text{at} \quad \tau^+ \geqslant a_n x^+ \tag{133}$$

During the period just after the jump, the heat is transferred only due to one-dimensional heat conduction, both molecular and turbulent

($\tau^+ \leqslant a_n x^+$). For this period the expression for the dimensionless heat transfer $qr_0/\lambda(T_w - T_0)$ obtained on the basis of solution of differential equation (129) agrees well with the expression obtained from the general solution (133).

The solution obtained for the dimensionless heat transfer is given in Figs. 24 and 25 at various x/d and stepwise wall temperature changes.

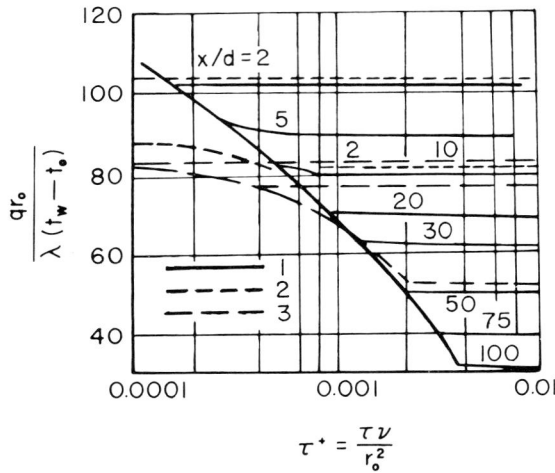

FIG. 24. Heat transfer in circular tube with stepwise variation of wall temperature; $Pr = 0.7$; $Re = 10^5$: (1) solution (177); (2) local heat transfer; (3) heat transfer for developed motion.

In Fig. 24 $Re = 10^5$ and $Pr = 0.7$; in Fig. 25 $Re = 10^5$ and $Pr = 100$. Dependencies of heat transfer on time are qualitatively the same: At the initial instants, heat transfer is determined only by "pure" heat conduction and is represented by an encircling curve, falling with time. Then at certain τ^+, increasing with x/d, the effect of convection increases and the curve deviates from the curve for pure heat conduction, heat transfer then continues to decrease till it reaches the value corresponding to the steady state. It is seen in Figs. 24 and 25 that the dependence obtained from (133) has a comparatively short curvilinear position. This occurs due to the approximate nature of the theoretical solution satisfying the integral energy equation but not the initial differential equation. The higher x/d is, the less considerable are the terms entering into the solution and therefore the extent of the curvilinear portions decreases. Since solution (133) is restricted by seven terms, the curves in Figs. 24 and 25 cannot extend to all times up to $\tau^+ = 0$; rather they

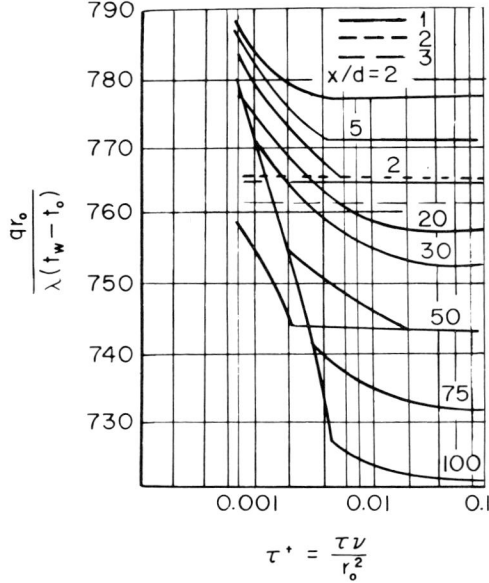

FIG. 25. Heat transfer at stepwise variation of wall temperature in tube; Pr = 100; Re = 10^5: (1) solution of (*177*); (2) local heat transfer; (3) heat transfer for developed motion.

may be applied beginning from some time that is dependent on Re and Pr. It is interesting to evaluate in Figs. 24 and 25 the time of the transient period after the stepwise wall temperature change. Since approximation to the established state is an asymptotic process, the authors evaluate the period of the transient process as the time at which heat transfer deviates by 5% from the established value. These times are given in Fig. 26, depending on x/d, by the solid lines. The dotted line corresponds to the time $\tau_{ss} = x/W_x$ which is approximately equal to that when the convective heat transfer of the fluid coming from the inlet into the tube dominates. In the dimensionless form $\tau_{ss}^+ = (4x/d)/\text{Re}$, i.e., it is the lower, the greater Re is. With the rise of Pr the stabilization time decreases; with the rise in Pr from 0.7 to 100 the time decreases two to three times. The time of the establishment of the steady state agrees well with τ_{ss} at Pr = 0.7, but with the rise in Pr it decreases. The value x/W_x is useful for preliminary estimations.

Further, Sparrow and Siegel consider the case of smooth wall temperature change in time. The linearity of the energy equation allowed the authors to apply the superposition method to use the solution

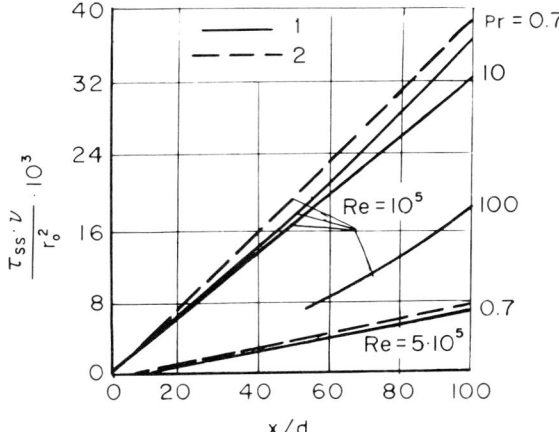

FIG. 26. Time of development of steady regime for turbulent liquid flow in circular tube and stepwise variation of wall temperature: (1) solution of (*177*); (2) $\tau_{ss} = x/W$.

obtained for the stepwise change. For example, Siegel and Sparrow consider the case of linear wall temperature change in time. In Fig. 27 the solution is given for $\mathrm{Pr} = 0.7$, $\mathrm{Re} = 10^5$, $x/d = 5$ and 50.

The results obtained in (*177*) are compared with the solution for constant heat transfer coefficient α. For each x/d the heat flux was calculated twice (at $\alpha = $ const) for the local Nu corresponding to the given x/d and for the fully developed flow. Both solutions are given in Figs. 24, 25, and 27 by dotted lines. The discrepancy between the solutions (*177*) under the condition $\alpha = $ const is noticeable at the inlet section; with the increase of x/d it decreases.

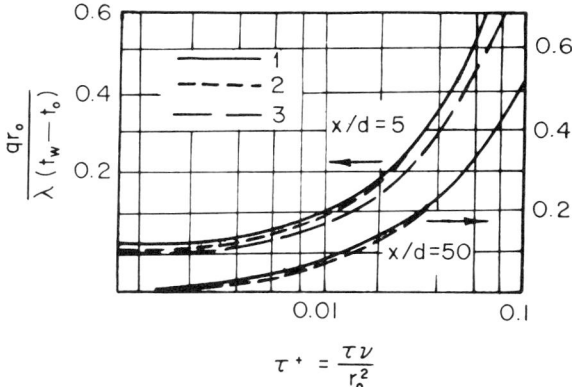

FIG. 27. Heat flux for linear variation of wall temperature with time; $\mathrm{Pr} = 0.7$; $\mathrm{Re} = 10^5$: (1) solution (*177*); (2) local heat transfer; (3) heat transfer for developed motion.

With stepwise wall temperature change, the solutions obtained at $\alpha = $ const differ considerably from the solution of Siegel and Sparrow (*177*) (Figs. 24 and 25), while for smooth wall temperature changes they are in good agreement with the solutions of the authors of (*177*) (Fig. 27) if the local heat transfer is taken for comparison. Consequently, the accuracy of solutions in which the quasi-steady-state value of the heat transfer coefficient is used strongly depends on the way the wall temperature varies.

It should be noted that the quantitative conclusions made by Siegel and Sparrow are applicable only in the range of validity of the integral method as well as of other assumptions, and most of all on the constant thermophysical properties.

In the work of Gill (*178*) unsteady turbulent heat transfer is considered under time varying inlet conditions caused by disturbances in the thermally developing flow. The inlet temperature change is stepwise. The initial condition is a steady temperature distribution. Fluid properties are considered constant; therefore the velocity profile does not depend on the inlet temperature change. The solution is obtained by the integral method (*177*). On the basis of the analysis made the authors consider that quasi-steady-state relations may be applied to conditions of the variable inlet temperature. It should be noted that in this case the quasi-steady-state solution will always coincide with the solution by the integral method since the function F_n [Eq. (133)] is a multiplier independent of r, by which the steady-state solution should be multiplied in order that it become quasi-steady and account for the inlet temperature change.

In (*179*) an approximate method is suggested for the solution of a two-dimensional problem of unsteady heat transfer with the help of Laplace transforms. A circular cylindrical channel is considered. The hydrodynamic characteristics of the flow are considered constant. The solution of this problem depends on the solution of the energy equation for the flow

$$\frac{\partial T(r, x, \tau)}{\partial \tau} + W_x(r) \frac{\partial T(r, x, \tau)}{\partial x} = \frac{1}{r} \frac{\partial}{\partial r} \left[r\, a_\Sigma(r) \frac{\partial T(r, x, \tau)}{\partial r} \right] \quad (134)$$

where a_Σ is the total thermal diffusivity coefficient taking into account the heat transfer in a flow owing to turbulent mixing and molecular heat conduction.

In this work, the cases relate to the changes in the fluid inlet temperature, wall temperature, and heat flux. The solution is obtained in a series form. No concrete calculation results by this method are given in the study (*179*).

A specific solution was obtained by the authors in another work (*180*) with very rough assumptions. The fluid velocity was assumed uniform over the duct cross section, and the total thermal diffusivity coefficient a_Σ was considered uniform along the radius. Introduction of the dimensionless coordinates $R = r/r_0$, $\text{Fo}' = \tau a_\Sigma/r_0^2$, and $X = (x/W)(a_\Sigma/r_0^2)$ reduces Eq. (134) to

$$\frac{\partial T}{\partial \text{Fo}'} = \frac{\partial^2 T}{\partial R^2} + \frac{1}{R}\frac{\partial T}{\partial R} - \frac{\partial T}{\partial X} \tag{135}$$

The case of a discontinuous wall temperature change from 0 to t_w uniform along the length is considered. Laplace transformation gives the solution

$$T = \begin{cases} T_w - \sum_{n=1}^{\infty} \frac{2T_w\, I_0(Rm_n)}{m_n\, I_1(m_n)} e^{-\text{Fo}'m_n^2} & \text{at } \text{Fo}' \leqslant X \\ T_w - \sum_{n=1}^{\infty} \frac{2T_w\, I_0(Rm_n)}{m_n\, I_1(m_n)} e^{-Xm_n^2} & \text{at } \text{Fo}' \geqslant X \end{cases} \tag{136}$$

where m_n is the eigenvalue, I_0 and I_1 are the Bessel functions of the real argument of zero order and first order, respectively. The heat transfer is found in the following way:

$$\begin{aligned} \text{Nu} &= \frac{\alpha d}{\lambda} = \frac{a_\Sigma}{a} \sum_{n=1}^{\infty} e^{-\text{Fo}'m_n^2} \bigg/ \sum_{n=1}^{\infty} \frac{e^{-\text{Fo}'m_n^2}}{m_n^2} & \text{at } 0 \leqslant \text{Fo}' \leqslant X \\ \text{Nu} &= \frac{\alpha d}{\lambda} = \frac{a_\Sigma}{a} \sum_{n=1}^{\infty} e^{-Xm_n^2} \bigg/ \sum_{n=1}^{\infty} \frac{e^{-Xm_n^2}}{m_n^2} & \text{at } \text{Fo}' \geqslant X \end{aligned} \tag{137}$$

Here

$$\alpha = \frac{\lambda_\Sigma (\partial T/\partial R)_{R=1}}{(T_w - T_b)_{r=r_0}}; \qquad T_b = 2\int_0^1 RT\, dR$$

In a steady state ($\text{Fo}' \to \infty$) at a large distance from the entrance

$$\text{Nu}_\infty = (a_\Sigma/a)\, m_1^2$$

from which the averaged equivalent thermal diffusivity coefficient a_Σ is evaluated using the experimental Nusselt number (*180*). In Fig. 28 Nu versus time is plotted for different cross section of the duct. A sharp increase in the heat flux causes an increase in the heat transfer rate.

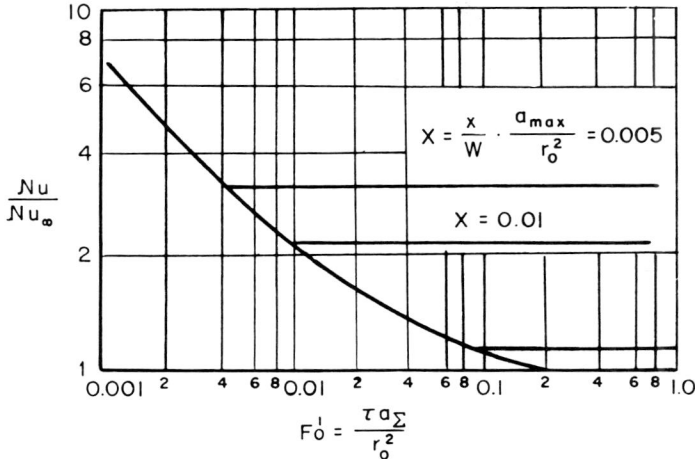

FIG. 28. Heat transfer at stepwise variation of wall temperature.

The assumptions used in this study reduce the problem to that of laminar flow with a fictitious thermal diffusivity coefficient solved by the "rod" model rather than a solution of the problem for turbulent flow.

We may say, thus, that unsteady heat transfer has not been studied adequately by analytical methods. All results available are obtained by the integral method which does not account for the rearrangements of the velocity and temperature distributions in the flow during unsteady process. This method also does not include the effects of an unsteady state on the turbulent properties of the flow which are adopted from the relations for steady state. In all the works reported, variability of thermal properties is also neglected. For problems of practical interest, analytical solutions are mainly qualitative. Therefore experimental studies of unsteady turbulent heat transfer in ducts is as important as further theoretical developments occur.

Works (*181–184, 193*) are devoted to experimental studies of unsteady heat transfer in a turbulent air flow in a circular tube heated by an electric current. The subject of these works is the effect of unsteady boundary conditions on the Nu number.

The authors proceed from the fact that, in the turbulent flows under consideration, the characteristic time $\Delta\tau_T$ (23), within which the prehistory of the boundary conditions may affect $\mathrm{Nu}(x, \tau)$, is small. Therefore, the laws of the wall temperature $T_w(\tau)$ or heat transfer fluid flow rate $G(\tau)$ typical in practice may be approximated within the range $[\tau, \tau - \Delta\tau_T]$ by the first term of the Taylor series (24). Then, assuming

$\partial T_w(\tau)/\partial \tau$ and $\partial G(\tau)/\partial \tau$ to be only slightly changing for time $\Delta\tau_T$ we may analyze the change of Nu with the derivatives of $T_w(\tau)$ or $G(\tau)$ at a certain time rather than the change of Nu with $T_w(\tau)$ or $G(\tau)$ themselves. Thus, it is postulated that for small $\Delta\tau_T$ and the laws $T_w(\tau)$ and $G(\tau)$, which are accurately described by the series expansion of $\Delta\tau_T$ (24), the Nu at a certain moment is determined by the first derivatives $\partial T_w/\partial \tau$ and $\partial G/\partial \tau$. The corresponding dimensionless numbers are

$$K_T = (\partial T_w/\partial \tau)\, d^2/(T_w - T_b)a \qquad (138)$$

$$K_G = (\partial G/\partial \tau)(d^2/G\nu) \qquad (139)$$

From theory and experiments, the relation for unsteady-state heat transfer (Nu) is of the form

$$K = \mathrm{Nu}/\mathrm{Nu}_0 = f(K_T, K_G, T_w/T_b) \qquad (140)$$

$$\mathrm{Nu}_0 = f_0(\mathrm{Re}, \mathrm{Pr}, x/d, T_w/T_b) \qquad (141)$$

where Nu_0 is the steady-state relation. Special studies have confirmed that steady Nu_0 and unsteady Nu depend equally on Re and x/d. The Pr number was unchanged in the experiments. The temperature factor T_w/T_b in the unsteady state has a greater effect on Nu compared to the steady state, since the temperature profiles under steady and unsteady conditions are different.

The following types of boundary conditions were studied:

(1) Increase of the wall temperature ($\partial T_w/\partial \tau > 0$; $K_T > 0$) with discontinuous increase of the heat q_v generated in the wall (due to heating by an electric current) from $q_{v_1} = 0$ to q_{v_2} or from $q_{v_1} > 0$ to q_{v_2}. The fluid flow rate in these experiments was found in advance and it remained unchanged within the run ($\partial G/\partial \tau = 0$; $K_G = 0$).

(2) Decrease of the wall temperature ($\partial T_w/\partial \tau < 0$; $K_T < 0$) with a discontinuous decrease of the heat quantity generated in the wall from q_{v_1} to $q_{v_2} = 0$ or from q_{v_1} to $q_{v_2} > 0$. In these experiments the fluid flow rate was also unchanged ($\partial G/\partial \tau = 0$; $K_G = 0$).

(3) Simultaneous change of the wall temperature and the fluid flow rate in various combinations was also studied.

The experimental facility was an open loop. It consisted of two working tubes with the walls of different thickness made of steel X18H10T. The first working section was 5.39 mm i.d. (d_1), 0.32 mm in thickness (δ_1), and 1086 mm in length (L_1); the dimensions of the second tube were $d_2 = 5.56$ mm, $\delta_2 = 0.225$ mm, $L_2 = 1076$ mm. The tubes were heated by alternating current. Under unsteady conditions, the time changes of the temperatures of the external walls were recorded in seven locations along the tube length. The procedure for the determination of the unsteady heat transfer coefficient was similar to that described in Section III,A and provided highly accurate results.

The first two cases of unsteady conditions were studied within the Reynolds number range $Re = 1.8 \cdot 10^4$ to $4.12 \cdot 10^5$; pressure at the entrance $P = (4-20) \cdot 10^4 \text{ kg/m}^2$; $T_w/T_b = 1-1.65$. The maximum rate of the wall temperature increase was $\partial T_w/\partial \tau = 200$ deg/sec for the tube $\delta_1 = 0.32$ mm that gave $K = 1.3$ and $\partial T_w/\partial \tau = 360$ deg/sec ($K = 1.6$) for the tube $\delta_2 = 0.225$ mm. The maximum rate of the temperature

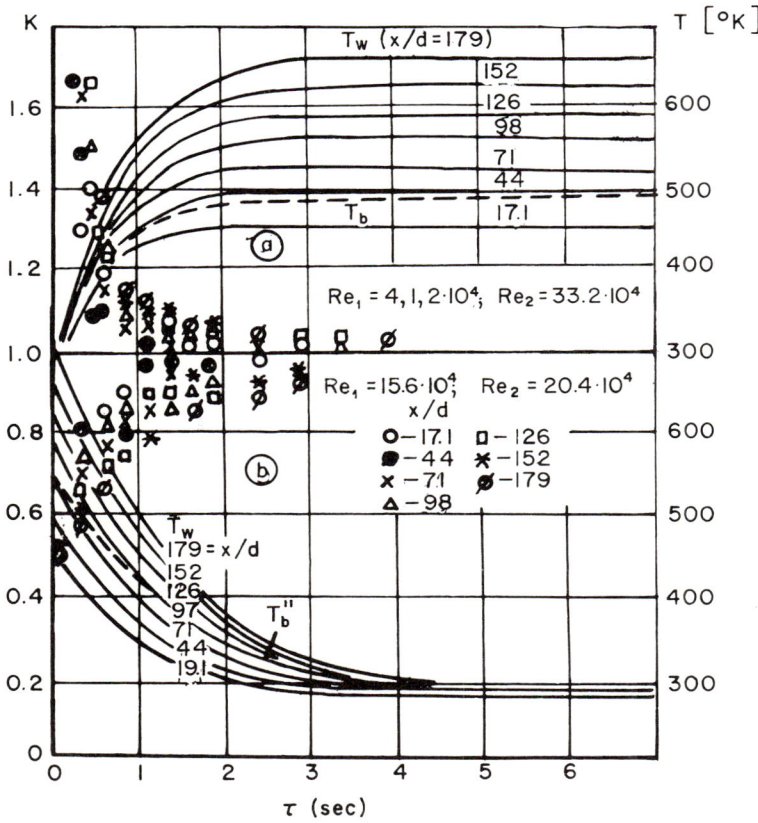

FIG. 29. Change of wall temperature T_w, coefficient K, for seven values of x/d and exit flow temperature at stepwise increase (a) and decrease (b) of heat transfer in tube (tube $\delta_2 = 0.225$ mm).

decrease T_w was $\partial T_w/\partial \tau = -220$ deg/sec ($K = 0.75$) for the first tube and $\partial T_w/\partial \tau = -390$ deg/sec ($K = 0.6$) for the second tube. The typical curves of the wall temperature, outlet flow temperature, and K number versus time are shown in Fig. 29.

Figure 30 illustrates the temperature factor effect on K obtained by

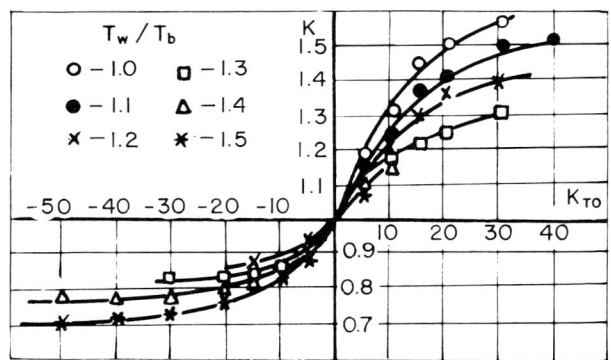

Fig. 30. K versus $K_{T_0} = (\partial T_w/\partial \tau)[d^2/(T_w - T_b)_0 a]$ and temperature factor T_w/T_b at increasing and decreasing wall temperature T_w.

matching points for different ratios T_w/T_b. This effect may be attributed to the fact that at $\partial T_w/\partial \tau > 0$ the temperature gradient $\partial T/\partial r$ near the wall is higher than that in a quasi-steady case and the density is lower. Therefore, the decrease of production of turbulent pulsations $\overline{\rho W_x' W_r'} \, \partial W_x/\partial r$ due to the increase of T_w/T_b is more pronounced than steady-state case, and the decrease of λ_T and the heat transfer coefficient are, respectively, more pronounced. With $\partial T_w/\partial \tau < 0$, the temperature gradient $\partial T/\partial r$ near the wall is less than a quasi-steady value that reduces the temperature factor effect.

The results of unsteady-state experiments covering the first two cases may be correlated by the following relations:

The wall temperature growth ($K_{T_0} = 0$–30; $T_w/T_b = 1$–1.4):

$$K = 1 + (2.12 - 1.12 T_w/T_b)[\exp(0.01913 K_{T_0} - 0.000248 K_{T_0}^2) - 1] \qquad (142)$$

The wall temperature decrease ($K_{T_0} = -30$–0; $T_w/T_b = 1$–1.5):

$$K = 1 + (1.08 T_w/T_b - 0.62)[\exp(0.02015 K_{T_0} - 0.000352 K_{T_0}^2) - 1] \qquad (143)$$

where $K_{T_0} = (\partial T_w/\partial \tau) \, d^2/(T_w - T_b)_0 a$; $(T_w - T_b)_0$ is the temperature driving force after steady-state conditions set in (142) and before the onset in unsteady regime (153). In changing from the quantity q_{v_1} to q_{v_2} in Eqs. (142) and (143), K_{T_0} should be replaced by

$$K^*_{T_0} = \frac{\partial T_w}{\partial \tau} \frac{d^2}{[(T_w - T_b)_1 - (T_w - T_b)_2]a} \quad (144)$$

Figure 31 illustrates fair agreement of Eqs. (142) and (143) obtained for changes from 0 to q_{v_1} and from q_{v_2} to 0 with the data obtained in

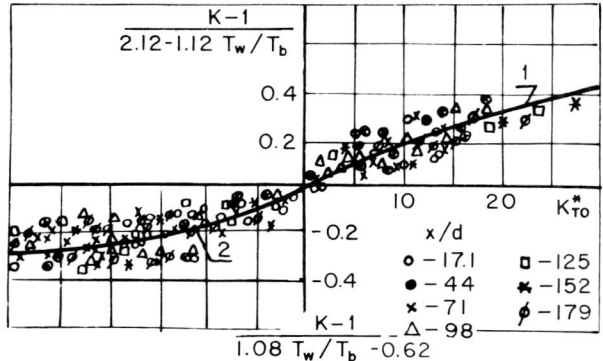

FIG. 31. K versus $K^*_{T_0} = (\partial T_w/\partial \tau)[d^2/|(T_w - T_b) - (T_w - T_b)_2|a]$ at stepwise increase ($q_{v_1} \neq 0$) and decrease ($q_{v_2} \neq 0$) of heat release in tube ($\delta_2 = 0.225$ mm): (1) by formula (142); (2) formula (143).

changes from $q_{v_1} \neq 0$ to $q_{v_2} \neq 0$ using $K^*_{T_0}$ from (144). This has allowed the authors (181–184, 193) to recommend the relations obtained for a wide class of different laws of $T_w(\tau)$ in the range of K_T and T_w/T_b considered.

The third case of unsteady process was studied in the following range of parameters:

(a) *Increase of the mass flow rate:*

$$Re_1 = 2.1 \cdot 10^4 – 9.7 \cdot 10^4; \quad Re_2 = 4.6 \cdot 10^4 – 2.6 \cdot 10^5$$
$$(T_w/T_b)_1 = 1.18–1.45; \quad (T_w/T_b) = 1.15–1.3$$
$$P_1 = (2.5–7.3) \cdot 10^4 \text{ kg/m}^2; \quad P_2 = (6.6–16.5) \cdot 10^4 \text{ kg/m}^2$$
$$G_1/G_2 = 0.24–0.55$$

(b) *Decrease of the mass flow rate:*

$$Re_1 = (4.6–25) \cdot 10^4; \quad Re_2 = (2.1–9.7) \cdot 10^4$$
$$(T_w/T_b)_1 = 1.14–1.3; \quad (T_w/T_b)_2 = 1.2–1.43$$
$$P_1 = (5.6–17) \cdot 10^4 \text{ kg/m}^2; \quad P_2 = (2.8–8) \cdot 10^4 \text{ kg/m}^2$$
$$G_1/G_2 = 1.8–3.66$$

The variations of $G(\tau)$ and the corresponding K studied are shown in Fig. 32. When the mass flow rate increases according to Law (Curve) 1, Fig. 32, it shows that $G(\tau)$ increases very rapidly, passing a maximum

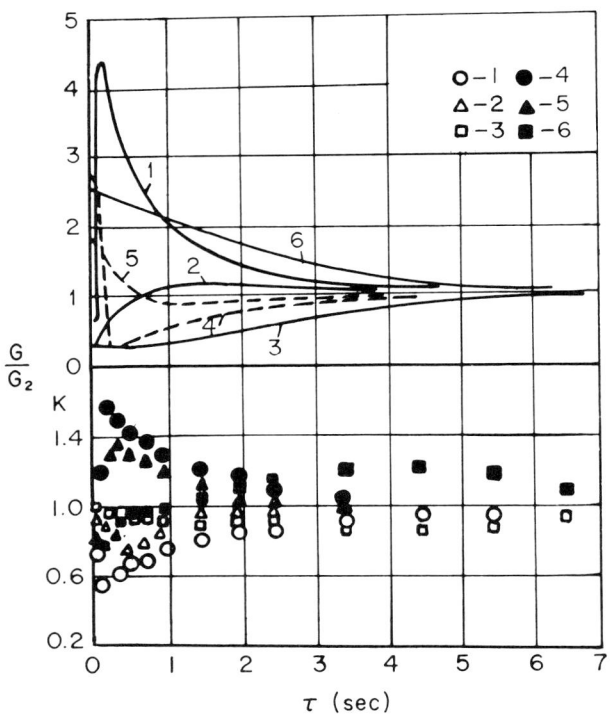

FIG. 32. Change in coefficient K at various laws of flow rate variation for $x/d = 44$.

and falling to the steady-state value $G_2 > G_1$ within the remaining period of time. This is accompanied by a fall of the wall temperature. Consequently, in this case the boundary conditions [$\partial G/\partial \tau < 0$ (from -0.024 to 0 kg/sec^2) and $\partial T_w/\partial \tau < 0$ (from -250 to 0 deg/sec)] have reduced the heat transfer rate compared to the quasi-steady-state which resulted in the change of K from 0.6 to 1.

When the mass flow rate changes smoothly (Law (Curves) 2 and 3 in Fig. 32) $\partial G/\partial \tau > 0$ (from 0.004 to 0 kg/sec^2) the increased heat transfer rate and $\partial T_w/\partial \tau < 0$ (from -70 to 0 deg/sec) reduced K. Since the effect of $\partial T_w/\partial \tau$ dominated, K changed from 0.8 to 1. When the heat flow rate decreased from G_1 to $G_2 < G_1$ the wall temperature increased, thus increasing K. In the case of Law (Curve) 4, the mass flow rate fell rapidly, passed a minimum, and increased within the remaining period of time. Therefore, $\partial G/\partial \tau > 0$ (from 0.007 to 0 kg/sec²) and $\partial T_w/\partial \tau > 0$ (from 150 to 0 deg/sec) had the same effect which yielded the change of K from 1.8 to 1. Under the conditions of a smooth decrease of $G(\tau)$ (Laws Curves 5 and 6) $\partial G/\partial \tau < 0$ (from -0.005 to 0 kg/sec²) and $\partial T_w/\partial \tau > 0$ (from 75 to 0 deg/sec); i.e., the effects of these parameters were opposite, but the effect of $\partial T_w/\partial \tau$ dominated (K changed from 1.2 to 1).

The results of these experiments show that when the mass flow rate in the region adjacent to the wall which controls the whole heat transfer process increases, turbulent thermal conductivity λ_T and, consequently, the turbulence level are higher than those in a quasi-steady case, though these values are lower in the core. The picture is quite the reverse when the mass flow rate decreases. This agrees well with the statements of Section II.

The experimental data were correlated by the selection of experimental points with similar values of K_T. They were used for plotting the relation $K = f(K_G)$. At $K_G = $ const, from these curves the families of the curves $K = f(K_T)$ were obtained. The data for two working sections ($\delta_1 = 0.32$ mm and $\delta_2 = 0.225$ mm) for Laws 1 and 4 are shown in Fig. 33 and are correlated by the following relations:

(a) When the mass flow rate and the wall temperature increase (Law 1) for $K_T = 0\text{--}25$ and $K_G = 0\text{--}15$,

$$K = 1 + 0.1155(K_T)^{0.353} + (0.0213 + 0.000415 K_T) K_G^{0.75} \qquad (145)$$

(b) When the mass flow rate and the wall temperature decrease (Law 4) for $K_T = -3\text{--}32$ and $K_G = -40\text{--}0$,

$$K = \exp[A(K_T) K_T] - C(K_T)(-K_G)^{n(K_T)} \qquad (146)$$

where

$A(K_T) = 0.044/(-K_T)^{0.5}$ at $-6.2 < K_T < -3.2$

$A(K_T) = 0.0892/(-K_T)^{0.915}$ at $-3.2 < K_T < -6.2$

$C(K_T) = 0.132(-K_T)^{0.8}$;

$n(K_T) = 0.424(-K_T)^{0.14}$ at $-14.1 < K_T < -3$

$C(K_T) = 239(-K_T)^{-3.66}$;

$n(K_T) = 0.0669(-K_T)^{0.84}$ at $-32 < K_T < -14.1$

K_T and K_G in (145) and (146) are found from (138) and (139). The experimental results for Laws 2, 3, 5, and 6 are shown in Fig. 34.

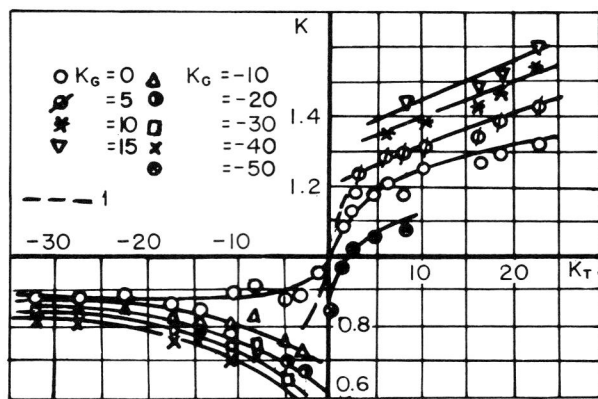

Fig. 33. K versus K_T and K_G for sharp changes in flow rate: (1) smooth variation of flow rate ($K_G \approx 0$).

The comparison of the curves $K = f(K_T)$ with $K_G = 0$ in Figs. 33 and 34 shows good agreement with the curves in Fig. 30 if the correction for the temperature factor from Eqs. (142) and (143) is introduced and K_{T_0} is found from (144).

In the experiments with changing mass flow rates the authors (*181, 183, 184*) have failed in their attempts to determine separately the effect of the temperature factor, which decreases somewhat the generality of Eqs. (145) and (146). The aim of further investigations is to obtain a general relation of the form

$$K = f(K_T, K_G, T_w/T_b)$$

Thus, studies (*181–184*) have shown that under unsteady-state conditions, when the heat flux and mass flow rate of the gas change with time, the heat transfer may differ from a quasi-steady value and depend

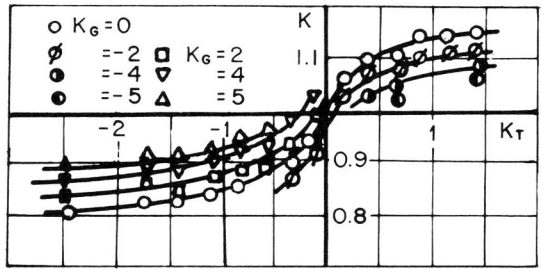

Fig. 34. K versus K_T and K_G for smooth variation of flow rate.

on K_T, K_G, and T_w/T_b. Results in qualitative agreement with those in (*181–184*) have been obtained in the experiments with water described in Kochenev and Kuznetsov's works (*66, 185*). The experiments were carried out in a tube 7.6 mm in diam with a wall thickness of 0.3 mm. Disturbances were introduced by changes in electric power supplied to the tube or by changing the water flow rate. The heat transfer coefficients were found by the method described in (*181–184*). When the heat flux increased, the heat transfer coefficient was larger than the quasi-steady value (Fig. 35) ($1 \leqslant K \leqslant 6$) and it was smaller ($0 < K \leqslant 1$) when

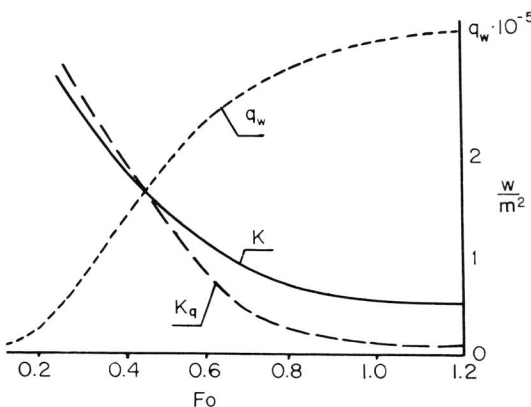

FIG. 35. Change of heat flux q_w, criterion K_q, and reduced coefficient of unsteady-state heat transfer K with increasing heat flux in experiments with water.

the heat flux decreased. The divergence of the unsteady heat transfer coefficient from the quasi-steady value increased with increasing the absolute value of $K_q = (1/q_w)(\partial q_w/\partial \text{Fo})$, where q_w is the heat flux from the wall to the water and $\text{Fo} = a\tau/d^2$. The aim of further investigations is to obtain the correlating equations for the calculation of unsteady-state heat transfer in water under the conditions of strong changes of thermal properties.

D. CONJUGATE PROBLEMS OF UNSTEADY HEAT TRANSFER

As it was stated in Section III,A, in case of unsteady heat transfer the walls around the flow affect the flow, giving rise to specific changes in the boundary conditions, such as the wall temperature or the heat flux. Including the energy (heat conduction) equation for the enclosing walls in the set of equations for the flow allows us to avoid prescription of the boundary conditions for the flow. Here equality of temperatures and heat fluxes at the interface "flow-body" is assumed, and the boundary conditions are prescribed at the external surfaces of the channel walls and at the entrance to the channel. Solution of such a problem in a general case yields temperature fields simultaneously in the walls and in the flow. However, the solution of conjugate problems in general formulations encounters great difficulties. This demands different simplifications, e.g., use of a heat transfer coefficient. In this case this problem is virtually divided into two separate ones. First, the dependence of the heat transfer coefficient on specific laws of the history of the boundary conditions is assumed (or it is assumed that the heat transfer coefficient is independent of these changes); then the heat conduction problem for the wall with boundary conditions of the third kind is solved.

Now, we shall discuss the reported works on conjugate problems. Perlmutter and Siegel's work (159) is a theoretical study of laminar heat transfer in a plane channel with a wall of finite heat capacity with a stepwise change of the pressure gradient and the temperature of the external wall surface. The results show (Fig. 36) that with zero heat conductivity of the wall, the dimensionless heat flux (the Nusselt number) falls abruptly as the heat flux increases and until the fluid which was at the inlet at the initial moment reaches a certain cross section (the period of pure heat conduction or unsteady heating of the fluid flow). Then as the convective heat transfer develops, the dimensionless heat flux approaches the steady-state value. When the thermal conductivity of the wall is not zero, the heat flux into the fluid is zero at the initial moment. As the whole thickness of the wall is heated, the heat flux into the fluid increases and, after passing a maximum (for example, for $X = 0.1$ and $\delta = 0.79$ mm), it falls again since the period of unsteady heat conduction still goes on. When convection sets in, heat flux and Nu increase up to the steady-state values.

For a wall of a greater thickness ($\delta = 1.59$ mm), within the period of pure heat conduction the heat flux increases very slowly and has no time to reach the maximum. After the convection period sets in, heat flux increases to the steady-state value.

Siegel (186) has also studied a laminar flow in a plane channel with walls of finite thermal conductivity. "Rod" flow was assumed (with a velocity uniform across the thickness) and the wall temperature was assumed uniform across the thickness. The effects of different laws of change in time of heat quantities generated initially in the isothermal wall have been studied. The numerical results for a stepwise heat transfer are shown in Fig. 37. With zero thermal conductivity of the

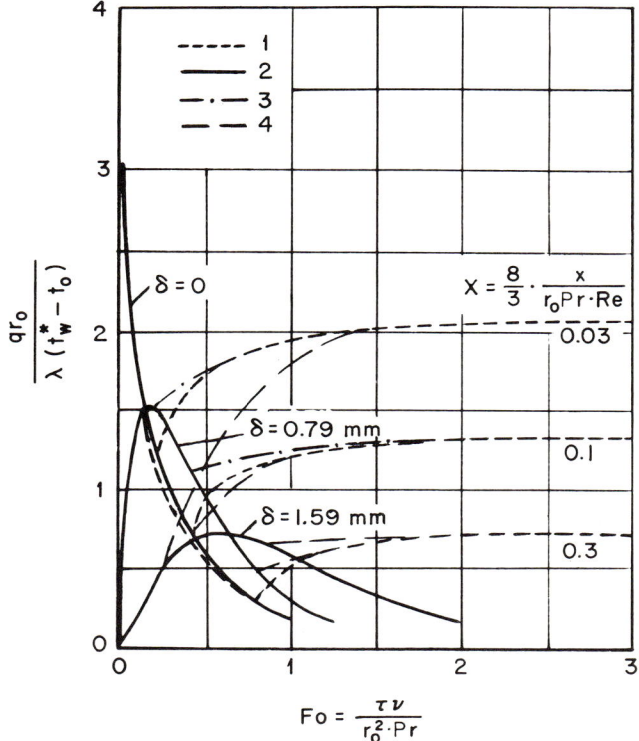

FIG. 36. Effect of channel wall thickness upon unsteady heat flux from wall to liquid at stepwise variation of pressure gradient and outer wall surface temperature; $\text{Pr} = 0.7$; $r_0 = 3.18$ mm. Initial moment: $W = W_1 = 0$; $t_w{}^H = t_{w_1}^H = t_0$. Final moment: $W = W_2$; $t_w{}^H = t_{w_2}^H$. (1) (159), $\delta = 0$; (2) transient heat conduction; (3, 4) transient process for $\delta = 0.79$ mm and $\delta = 1.59$ mm (159) (r_0, half-width of channel; δ, wall thickness).

wall two periods are clearly discerned: (1) The period of unsteady heat conduction $x > W\tau$, which may be well described by heat conduction equations for one-dimensional heating of two bodies in contact with heat generation in either, uniform along the coordinate and stepwise in time; (2) The steady-state period $W\tau > x$.

With the thermal conductivity of the wall different from zero, the heat flux into the fluid is less than that generated in the wall and the rate of the heat flux increase is inversely proportional to the wall heat capacity.

If a smooth change in temperature of such a wall is considered as the extreme case of a stepwise change, an equal but displaced in time period

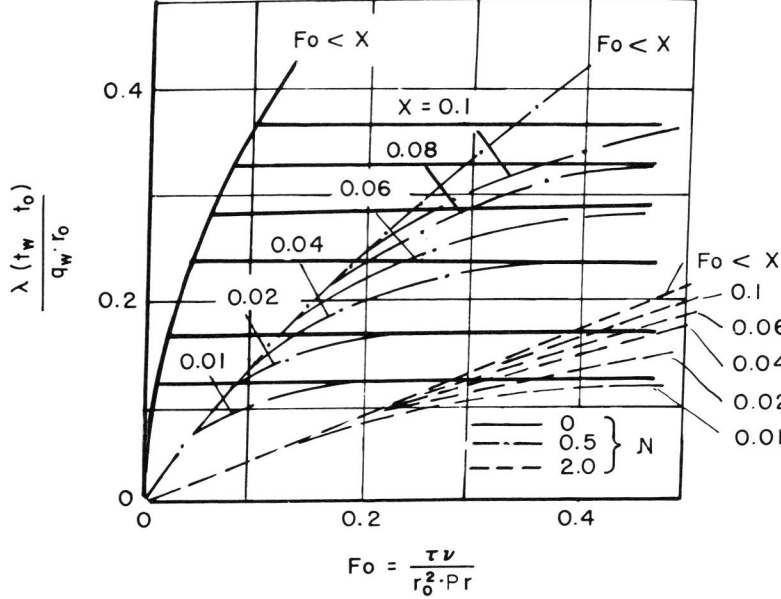

FIG. 37. Wall temperature versus X and Fo at stepwise variation of heat flux in wall and various values of wall heat capacity $N = \delta\rho_w C_w / r_0 \rho C_p$ (r_0, half-width of channel; δ, wall thickness); $X = 4x/r_0 \, \text{Pr Re}$.

of unsteady heat conduction will exist for each step. This period superimposes steady convective heat transfer of previous steps which results in increasing heat transfer in direct proportion with the value of a step, or otherwise proportionally to $dT_w/d\tau$.

Adams and Gebhart (187) have carried out a theoretical and experimental study of heat transfer on a wall in a laminar flow with stepwise heat generation. The theoretical problem has been solved with the use of heat transfer coefficients for a steady flow. The experiments were carried out on a nichrome ribbon, 13 μm in thickness, in a laminar air flow. The surface temperature was measured by an infrared detector. The experimental data are found to be in good agreement with the predicted values which may be attributed to the small heat capacity of the plate and a short period of pure heat conduction.

Chambre (188) has obtained a precise closed solution of a conjugate problem for a laminar plug flow around a plate of a finite heat capacity with the wall temperature uniform over the thickness, heat fluxes across the axis being absent.

Johnson and Chambre (189, 190) have simplified this solution for the determination of local and mean unsteady temperatures of the plate, with heat generation following an exponential law. Soliman (189, 191) has extended this solution to the case of a stepwise heat generation. The workers mentioned have also considered a "rod" laminar flow along a plate of large heat capacity with the temperature uniform over the thickness, and heat fluxes across the axis of the plate with the flow being absent. The physical properties of the fluid were assumed constant and energy dissipation due to viscosity was considered small.

The energy equation for the flow is

$$\partial T/\partial \tau + W\, \partial T/\partial x = \lambda\, \partial^2 T/\partial y^2; \qquad \tau > 0,\ x > 0,\ y > 0 \qquad (147)$$

The boundary conditions are

$$T(0, x, y) = 0; \qquad x > 0,\ y > 0$$

$$T(\tau, 0, y) = 0; \qquad \tau > 0,\ y > 0$$

$$T(\tau, x, \infty) = 0; \qquad \tau > 0,\ x > 0$$

$$-\lambda \frac{\partial T(\tau, x, 0)}{\partial y} = q_0 - \frac{\rho_w C_w \delta}{2} \frac{\partial T(\tau, x, 0)}{\partial \tau}; \qquad \tau > 0,\ x > 0 \quad (148)$$

where λ is the fluid thermal conductivity; q_0 is the heat generated in the plate per unit surface area; δ is the plate thickness; ρ_w and C_w are the plate density and specific heat, respectively; T is the temperature difference between a given point and infinity.

The solution for a stepwise heat generation is the following:

(1) *Local temperature:*

$$\frac{T_w(\theta, X)}{T_{RS}} = \frac{2}{\sqrt{\pi}} \sqrt{\theta} + e^\theta\, \mathrm{erfc}(\sqrt{\theta}) - 1 \quad \text{at}\quad \theta = \frac{\tau}{B^2} \leqslant X = \frac{x}{W_\infty B^2} \tag{149}$$

$$\frac{T_w(\theta, X)}{T_{RS}} = \frac{2}{\sqrt{\pi}} \sqrt{X} + e^\theta\, \mathrm{erfc}\left(\frac{\theta + X}{2\sqrt{X}}\right) - \mathrm{erfc}\left(\frac{\theta - X}{2\sqrt{X}}\right) \quad \text{at}\quad \theta \geqslant X \tag{150}$$

where $T_{RS} = q_0 \sqrt{\pi} \delta/\lambda$ is the characteristic temperature;

$$B = \rho_w C_w \delta \sqrt{a}/\lambda$$

is a parameter; a is a fluid thermal diffusivity.

(2) *Mean temperature along the plate length L:*

$$\frac{\bar{T}(\theta)}{T_{RS}} = \frac{T_w(\theta, X)}{T_{RS}} - \left(\frac{T_w(\theta, X)}{T_{RS}} - \frac{\bar{T}_1}{T_{RS}}\right)\frac{\theta}{X_L} \quad \text{at} \quad \theta \leqslant X_L \quad (151)$$

where

$$\frac{\bar{T}_1}{T_{RS}} = \frac{4\sqrt{\theta}}{3\sqrt{\pi}} + \frac{1}{\theta}\int_0^\theta \left[e^\theta \operatorname{erfc}\left(\frac{\theta + X}{2\sqrt{X}}\right) - \operatorname{erfc}\left(\frac{\theta - X}{2\sqrt{X}}\right)\right] dX$$

and

$$\frac{\bar{T}(\theta)}{T_{RS}} = \frac{4\sqrt{X_L}}{3\sqrt{\pi}} + \frac{1}{X_L}\int_0^{X_L} \left[e^\theta \operatorname{erfc}\left(\frac{\theta + X}{2\sqrt{X}}\right) - \operatorname{erfc}\left(\frac{\theta - X}{2\sqrt{X}}\right)\right] dX$$

$$\text{at} \quad \theta \geqslant X_L \quad (152)$$

(3) *Mean heat transfer coefficient:*

$$\alpha_m = (q_0 - \tfrac{1}{2}\rho_w C_w \delta\, \partial T_w/\partial \tau)/\bar{T} \quad (153)$$

is, respectively, for $\theta \leqslant X_L$,

$$\frac{\alpha_m(\theta)\, a\rho_w C_w \delta}{2\lambda^2}$$

$$= \left[1 - \left(1 - \frac{\theta}{X_L}\right) e^\theta \operatorname{erfc}\sqrt{\theta} - \frac{1}{X_L}\int_0^\theta e^\theta \operatorname{erfc}\left(\frac{\theta + X}{2\sqrt{X}}\right) dX\right]\Big/(\bar{T}/T_{RS}) \quad (154)$$

and, for $\theta \geqslant X_L$,

$$\frac{\alpha_m(\theta)\, a\rho_w C_w \delta}{2\lambda^2} = \left[1 - \frac{1}{X_L}\int_0^{X_L} e^\theta \operatorname{erfc}\left(\frac{\theta + X}{2\sqrt{X}}\right) dX\right]\Big/(\bar{T}/T_{RS}) \quad (155)$$

or

$$\mathrm{Nu}_L = \frac{\alpha_m L}{\lambda} = \frac{2L\rho C}{\delta \rho_w C_w} \cdot \frac{T_{RS}}{\bar{T}}\left[1 - \frac{1}{X_L}\int_0^{X_L} e^\theta \operatorname{erfc}\left(\frac{\theta + X}{2\sqrt{X}}\right) dX\right] \quad (156)$$

The analysis of expressions (154)–(156) shows that, for the case considered, the heat transfer coefficient or the Nusselt number depends not only on the physical properties of the fluid but also on the plate heat capacity and the rate of heat generation inside it. The conclusion follows from this analysis that, with the hydrodynamics of a laminar flow and the temperature difference known, it is the wall heat capacity and heat generated in it that determine the rate of the wall temperature change. The higher the wall heat capacity and the smaller the rate of heat generation, the less the rate of the wall temperature change. Naturally, this fact cannot serve as the basis for the statements of some authors (6–8) that the physical properties of a wall may have a direct effect on the unsteady-state heat transfer coefficient, (rather than through boundary conditions). Such statements imply a doubt as to the existence of the single solution of a closed system of equations with given boundary conditions, this system of equations describing fluid flow in a certain channel with heat transfer.

The above statement that unsteady-state heat transfer coefficients depend on the physical properties of a body (a wall) and its dimensions was originally formulated by Kudryavtsev et al. (6, 7) on the basis of their experimental study of the conjugate problem. They studied heat transfer between insulated rods (only one end-face was not thermally insulated) with the diameter d and the length R and a flow of heated water in a thermostat or in a rectangular duct with Re up to 5000.

The conclusion of these workers (6, 7) on the effects of the material and the dimensions of the cylinders studied on α (which is impossible from the viewpoint of modern hydrodynamics) probably may be attributed to the erroneous treatment of the results. The authors (6, 7) assumed the problem to be one-dimensional along the axis of the samples. In their experiments, one end-face of a cold cylinder was imbedded into the adiabatic wall of a thermostat (unit A) or a rectangular duct (unit B). The cylinder was heated by circulating water. Therefore, at the heated end-face of the cylinder a thermal boundary layer of variable thickness was formed. Consequently, at any time moment the local heat transfer coefficient changed considerably, decreasing along the chords from the front edge of the end-face to the back one. This should result in a nonuniform heat flux over the cylinder surface. Thus, it should be a two-dimensional or even three-dimensional problem for a rod. The value of nonuniformity of the heat flux was proportional to the rod length R and the value of the heat flux (or, more exactly, to the temperature difference and thermal conductivity).

Therefore at large temperature differences $\bar{T}_b - \bar{T}_w$ and large R the heat transfer coefficient α_1 calculated for a one-dimensional problem is found to be considerably higher than the true mean (over the surface) α_m, and when $(\bar{T}_b - \bar{T}_w) \to 0$ and $R \to 0$, $\alpha_1 \to \alpha_m$. It is evident, too, that

$$\bar{\alpha} = \frac{\bar{q}}{\bar{T}_b - \bar{T}_w} \neq \alpha_m = \frac{1}{f} \int_f \alpha \, df$$

Bearing in mind the above, it is easy to explain why the dependence of α_1 on the temperature difference was weaker on unit A than on unit B, the materials and R being the same. The rod diameter was 23 mm in case A, and $d = 16$ mm in case B. The deviation from the one-dimensional case was smaller on unit A. We can also understand the experimental fact found by the authors (6, 7) that with higher R the value of α_1 is also higher, and with sufficiently small R (when the deviation from the one-dimensional problem was not important) further changes in R did not affect the changes in α_1 with $\overline{T}_b - \overline{T}_w$ any more, and so on. For this very reason, experiments on cooling rods in a cold bath without circulation [these are mentioned in (6), but the results are not presented] should have yielded results qualitatively different from the experiments on units A and B.

In works (6, 7) great attention is paid to the experimental facts that under identical conditions of heat transfer of rods: $q(\tau) = idem$ when $\tau/\rho_w C_w R = idem$; this seems quite clear without experiments.

An attempt to explain theoretically the dependence of unsteady heat transfer coefficients not only on the flow properties but also on the dimensions of the body and its material (rather than on the changes of the boundary conditions) is made in Kudryashev and Smirnov's work (8).

Luikov and Perelman (3, 4) have theoretically studied a conjugate problem of heat transfer between the end-face of a rod of diameter d and height R, imbedded into the wall of a rectangular duct with laminar flow (Fig. 38). The remaining surfaces of the rod were thermally insulated; the fluid with constant physical properties was assumed incompressible. The flow was developed. For simplicity, the boundary conditions and equations were averaged along the duct length X by $(0, d)$. Here $\partial T/\partial x$ is found from the linear fluid temperature distribution

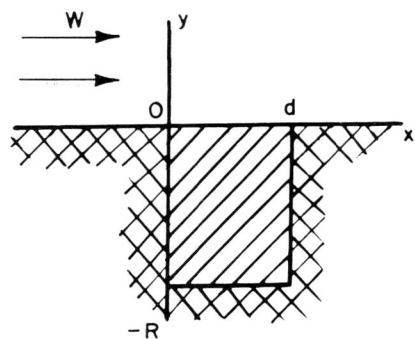

Fig. 38. Scheme to solution of conjugated problem of Luikov and Perelman (3, 4).

over X. Since the velocity distribution is replaced by a certain velocity W_e, we see that a rod flow was adopted here.

Upon simplification, the problem became of the form:

The wall:

$$\partial T(y, \tau)/\partial \tau = a_w \, \partial^2 T(y, \tau)/\partial y^2 \tag{157}$$

The flow:

$$\partial \theta(y, \tau)/\partial \tau + C_1 \, \theta(y, \tau) = \partial^2 \theta(y, \tau)/\partial y^2 \tag{158}$$

The initial conditions:

$$\tau = 0; \quad \theta = 0; \quad T = T_0 \tag{159}$$

The boundary conditions:

$$\begin{aligned} y = \infty; & \quad \theta = 0; \quad y = -R; \quad \frac{\partial T}{\partial y} = 0; \\ y = 0; & \quad \theta(0, \tau) = T(0, \tau); \quad \lambda_w \frac{\partial T}{\partial y} = \lambda \frac{\partial \theta}{\partial y} = q_w(\tau) \end{aligned} \tag{160}$$

Such a statement of the problem refers to the problem of unsteady heat conduction between an infinite plate of a finite thickness and heat capacity with a single face thermally insulated and a rod flow of an infinite thickness with heat generation proportional to the flow temperature rather than to a conjugate heat transfer problem. Such an interpretation of the problem statement may better explain the results obtained. However, the simplifications adopted were useful for the description of the initial, essentially unsteady, portion of the original problem. This follows from a comparison with the work of Baida *et al.* (*192*), which contains a useful numerical solution of the above conjugate problem. At the next stages of the solution, these simplifications have greater effect. Therefore, the solution obtained in (*3, 4*) for

$$\gamma = \frac{C\rho(D \cdot d)^{1/2}}{C_w \rho_w R \sqrt{\text{Pe}}} \ll 1$$

(where D is the equivalent diameter of the duct),

$$\text{Nu}(\xi) = (D\text{Pe}/d)^{1/2} \, (1 - \gamma/2)[1 + e^{-(1-\gamma)\xi/2} \sqrt{\pi} \, \xi^{2/3}] \tag{161}$$

when the dimensionless time $\xi = W_e \tau/d \to \infty$ gives a steady-state value which still depends on γ (i.e., on the heat capacity of a solid body).

According to (9) the steady-state solution for the original statement of the problem should be of the form

$$\text{Nu}_{\xi \to \infty} \sim (D\text{Pe}/d)^{1/3}$$

In the work by Kudryashev and Smirnov (8), theoretical consideration is given to the problem of unsteady convective heat transfer between a plane infinite plate and a laminar flow past two sides of the plate. The thermal properties of the plate material and the flow are constant. However, further into the solution some simplifications are made which reduce the problem first to that of a plate and a rod flow and then to the problem of heat transfer between a plate and fluid at rest. Moreover, heat transfer in the flow alone is considered (external problem) for different cases of prescribing the wall temperature; however, because of the simplifications used, the Nusselt number found for an unsteady process tends to zero as $\tau \to \infty$. Then the heat conduction problem is solved for a plate with boundary conditions of the third kind. Next, a "thin" plate is considered with the temperature uniform across the thickness, and the expression for Nu is used which is found for the "external problem." This is substituted into the simple expression for a heat balance of the wall:

$$C_w \rho_w \delta \, d(T_w - T_b)/d\tau = -\bar{\alpha}(T_w - T_b) \qquad (162)$$

which may be expressed in a dimensionless form:

$$d\theta_w/d\text{Fo}_L = -A \, \overline{\text{Nu}_L} \, \theta_w \qquad (163)$$

where

$$A = \frac{2C_p \rho L}{C_w \rho_w \delta}; \qquad \text{Fo}_L = \frac{4\alpha\tau}{L^2}; \qquad \theta_w = \frac{T_w - T_{b\infty}}{T_{w_0} - T_{b\infty}}$$

T_{w_0} is the initial surface temperature; θ_w is replaced by its mean value for the time Fo_L. Further, after considerable simplifications for small Fo_L, the expression for θ_w is found

$$\theta_w = \frac{1}{(1 + \tfrac{2}{3}A^2 \, \text{Fo}_2)^3} \qquad (164)$$

This is again substituted into (163) whence it is found that

$$\text{Nu}_L = \tfrac{2}{3}A/[(1 + \tfrac{2}{3}A^2 \, \text{Fo})^3 - 1] \qquad (165)$$

Such a very simplified solution cannot be correct and cannot confirm the statement of the authors of (8) that: "From the solution obtained [from Eq. (165)] it is clear that in high-rate cooling of plates, the heat transfer coefficient depends not only on thermal properties of the main flow, but also on thermophysical properties of the wall and the length-to-thickness ratio L/δ. This fact was originally found experimentally by Kudryavtsev et al. (6)."

Sparrow and De Farias (172) investigated the conjugate problem with laminar flow in a parallel plate channel. The fluid inlet temperature is assumed to vary sinusoidally in time with an arbitrary frequency. The velocity of the flow is unchanged with time. The plates are separated by a distance $2L$ and are each of thickness l. The plates are insulated on their external surfaces, and are sufficiently thin so that the temperature across the thickness of the wall is assumed constant, but is a function of time and of axial position.

The energy equations for flow and wall are analyzed simultaneously. The axial conduction terms and property variations are neglected. The flow velocity is assumed constant in any cross section. Numerical results are obtained for the time and space dependence of the wall and bulk temperatures and of the Nusselt number. The wall and bulk temperatures vary sinusoidally with different amplitude and have a phase lag. It follows that the Nusselt number experiences discontinuities, soaring to plus infinity and then returning to minus infinity. The excursions occur when the wall and bulk temperatures are equal, and the heat flux is not zero. It was found that, for a range of operating conditions, the quasi-steady model is able to give accurate performance predictions, especially where it is used in conjunction with spatially varying heat transfer coefficients.

To conclude this section, we consider a theoretical and experimental study of a conjugate problem on unsteady heat transfer of a plane plate of large heat capacity with variable heat generation and a turbulent flow past its two surfaces. This study was carried out by Soliman and Johnson (189) with the following assumptions:

The fluid is incompressible and with constant physical properties. Heat fluxes in the longitudinal direction x are neglected. The temperature changes in the flow are unsteady but of such a sort that allows the use of time-averaged values of T, W_x, and W_y of a turbulent flow. The velocity field, momentum, and heat transfer coefficient for a turbulent process are independent of time. The plate temperature is uniform across its thickness. Then the energy equations for the flow and the boundary conditions become

$$\frac{\partial T}{\partial \tau} + W_x(x,y)\frac{\partial T}{\partial x} + W_y(x,y)\frac{\partial T}{\partial y} = \frac{\partial}{\partial y}\left[(a + \epsilon_q)\frac{\partial T}{\partial y}\right] \quad (166)$$

$$\tau > 0, \quad x > 0, \quad y > 0$$

$$T(\tau, x, \infty) = 0, \quad \tau > 0, \quad x > 0 \tag{167}$$
$$T(0, x, y) = 0, \quad x > 0, \quad y > 0, \quad T(\tau, 0, y) = 0, \quad \tau > 0, \quad y > 0$$

$$\lim_{y \to 0}[-\rho C_p(a + \epsilon_q)\, \partial T/\partial y] = q_0 - \tfrac{1}{2}\rho_w C_w \delta\, \partial T(\tau, x, 0)/\partial \tau \tag{168}$$

where q_0 is the rate of heat generation per unit area of the plate; δ is the plate thickness.

Assuming $\Pr = 1$, $\epsilon_q = \epsilon_\tau$, and tangential stresses across the boundary-layer thickness $T = \text{const}$, we find

$$a + \epsilon_q = \nu + \epsilon_\tau = \frac{T_w/\rho}{\partial W_x/\partial y} = \frac{V_*^2}{\partial W_x/\partial y} \tag{169}$$

The velocity distribution over the boundary-layer thickness δ_0 is assumed constant with the exponent $\tfrac{1}{7}$:

$$W_x/W_\infty = (y/\delta_0)^{1/7} \tag{170}$$

Then the tangential stress at the wall may be expressed by

$$\frac{T_w}{\rho W_\infty^2} = \frac{V_*^2}{W_\infty^2} = \frac{0.0228}{(W_\infty \delta_0/\nu)^{1/4}} = \frac{0.0296}{(W_\infty x/\nu)^{1/5}} \tag{171}$$

From (169)–(171) we find

$$a + \epsilon_q = F(x)\, y^{6/7} \tag{172}$$

where

$$F(x) = 0.179 W_\infty^{21/35} \nu^{8/35} x^{-3/35} \tag{173}$$

It is assumed that for the range of Re covered and with the time of the second reading not less than 1 μsec, the thermal boundary layer has time to pass through a viscous sublayer and enter the region of the power velocity distribution. For time less then 1 μsec, Chambre's solution is recommended (*188*). The aim of the analysis discussed is to find an approximate solution of system (166)–(168) with account for (172) with stepwise heat generation in a plate.

The first step is to find the solution for a plate with zero heat capacity. In this case the boundary conditions (168) are

$$T(\tau, x, \infty) = 0, \quad \tau > 0, \quad x > 0$$
$$\lim_{y \to 0}[-\rho C_p(a + \epsilon_q)\, \partial T/\partial y] = q_0, \quad \tau > 0, \quad x > 0 \tag{174}$$

Two limiting cases are considered:

(1) The case of short times (less than 1 μsec) when convective terms in (166) may be neglected and one-dimensional heat conduction dominates.

With this assumption and with (174) and $T(0, x, y) = 0$, solution (166) is found by Laplace transformation, and for the wall temperature it has the form

$$\frac{T_w(\tau, x)\, \rho C_p W_\infty}{q_0} = 30\, \text{Re}_x^{0.2} \theta_x^{0.125} \tag{175}$$

where $\theta_x = \tau W_\infty / X$.

(2) A steady-state solution when the term $\partial T/\partial \tau$ in Eq. (166) may be neglected. Then solution (166) with (174) and $T(\tau, 0, y) = 0$ from (189) is

$$T_w(x)\, \rho C_p W_\infty / q_0 = 30.3\, \text{Re}_x^{0.2} \tag{176}$$

other assumptions being the same.

From References (187) and (177) it is suggested that, in the case considered, when $\tau \leqslant x/W_\infty$ or $\theta_x \leqslant 1$ the heat transfer is controlled by one-dimensional heat conduction [Eq. (175)] where, for agreement with Eq. (176) the coefficient of 30.3 is taken. Here $\tau = x/W_\infty$ or $\theta_x = 1$ corresponds to the commencement of the steady regime, or, with $\theta_x \geqslant 1$, Eq. (176) is valid. The average over the length L temperature of the wall

$$\bar{T}(\tau) = \frac{1}{L} \int_0^L T_w(\tau, x)\, dx$$

is

$$\bar{T}(\tau) = \frac{q_0 L}{\lambda\, \text{Re}_L^{0.8}} (28.19 \theta_L^{0.125} - 2.94 \theta_L^{1.2}), \qquad \theta_L = \frac{\tau W_\infty}{L} \leqslant 1 \tag{177}$$

$$\bar{T} = 25.25\, \frac{q_0 L}{\lambda\, \text{Re}_L^{0.8}}, \qquad \theta_L \geqslant 1 \tag{178}$$

To account for a finite heat capacity of the wall $q_w = q_0 f(\tau, x)$, or, with averaging over the length,

$$\bar{q}_w = q_0 \bar{f}(\tau) \tag{179}$$

Then (177) and (178) are, respectively,

$$\frac{\bar{T}(\theta_L)}{T_{RS}} = \frac{(28.19 \theta_L^{0.125} - 2.94 \theta_L^{1.2})\, \bar{f}(\theta_L)}{C\, \text{Re}_L^{0.8}}, \qquad \theta_L \leqslant 1 \tag{180}$$

$$\frac{\bar{T}(\theta_L)}{T_{RS}} = \frac{25.25\, \bar{f}(\theta_L)}{C\, \text{Re}_L^{0.8}}, \qquad \theta_L \geqslant 1 \tag{181}$$

where

$$T_{RS} = q_0 \sqrt{\pi}\, \delta/\lambda \quad \text{and} \quad C = \tfrac{1}{2}\rho_w C_w \delta/\rho C_p L$$

however, on the other hand

$$q_w = q_0 - \tfrac{1}{2}\rho_w C_w \delta\, \partial T(\tau, x)/\partial \tau \tag{182}$$

Then, averaging over x with (179)

$$\bar{f}(\theta_L) = 1 - \frac{1}{X_L} \frac{d}{d\theta_L}\left(\frac{\bar{T}}{T_{RS}}\right) \tag{183}$$

where

$$X_L = 4\rho C_p \lambda L / \rho_w^2 C_w^2 \delta^2 W_\infty$$

substitution of (183) into (180) and (181) allows $\bar{T}(\theta_L)/T_{RS}$ and the heat transfer coefficient

$$\bar{\alpha} = (q_0 - \tfrac{1}{2}\rho_w C_w \delta\, \partial \bar{T}/\partial \tau)/\bar{T} \tag{184}$$

to be obtained.

The Nusselt number found by the same procedure is

$$\mathrm{Nu}_L = \bar{\alpha} L/\lambda = \mathrm{Re}_L^{0.8}/(28.19\theta_L^{0.125} - 2.94\theta_L^{1.2}), \quad \theta_L \leqslant 1, \tag{185}$$

$$\mathrm{Nu}_L = 0.0396\, \mathrm{Re}_L^{0.8}, \qquad\qquad\qquad\qquad\qquad \theta_L \geqslant 1 \tag{186}$$

Thus the heat transfer coefficients for the walls with zero and nonzero heat capacities obtained by the method discussed above are the same.

Experiments were carried out under conditions approaching the prescribed ones. The plate was replaced by a ribbon made of 50%-nickel and 50%-iron alloy, 3.175 mm in width, 76.2 mm in length, and 0.35 mm and 0.1 mm in thickness. The ribbon soldered to massive flow guides was placed in the middle of a vertical rectangular duct with cross section 22.2 × 9.5 mm. Two sides of the ribbon were cooled by a turbulent water flow at a temperature of about 132°C and pressure of 56 atm that provided Pr ≈ 1. The mean temperature of the ribbon (for this aim it served as a resistance thermometer) and heat generated in it were measured.

In Figs. 39–41 the experimental and predicted dimensionless wall temperatures are presented for plates 0.1 mm and 0.35 mm in thickness for the largest Reynolds number ($\mathrm{Re}_L = 2 \cdot 10^6$) and for a plate 0.1 mm in thickness at a small Reynolds number ($\mathrm{Re}_L = 1.66 \cdot 10^5$). The solution discussed above agrees well (within the experimental accuracy) with the experimental data for both heat capacities (thicknesses) of the

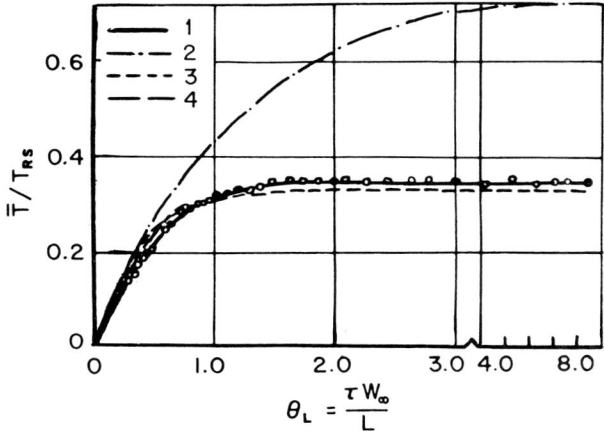

FIG. 39. Change in mean surface temperature at stepwise heat generation: $Re_L = W_\infty L/\nu = 2 \cdot 10^6$; $X_L = 4x\lambda^2/W_\infty \rho_w^2 C_w^2 \delta a^2 = 0.915$; $C = \rho_w C_w \delta/2\rho C_p L = 0.68 \cdot 10^{-3}$. (1) experiment; (2) solution (151) and (152) for rodlike flow; (3) solution (180) and (181) for turbulent flow; (4) solution for constant heat transfer coefficient.

wall within the range $Re_L = 5.3 \cdot 10^5$ to $2 \cdot 10^6$. The comparison of the experimental and predicted values at a constant steady-state value of the heat transfer coefficient shows disagreement (overestimated wall temperatures) during the initial period ($\theta_L \leqslant 1$) and better agreement at a larger heat capacity of the wall. At a rather small Reynolds number

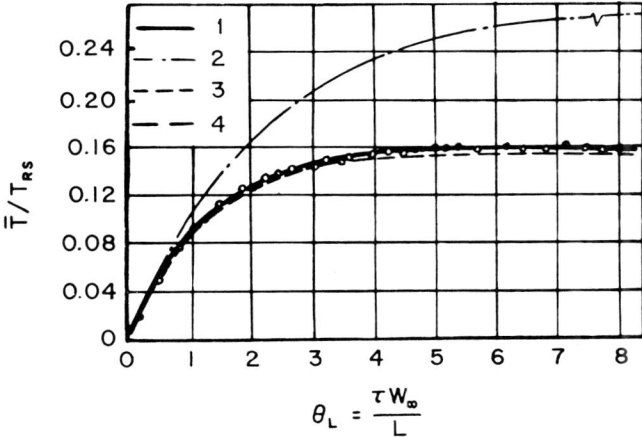

FIG. 40. Change in mean surface temperature at stepwise heat release ($Re = 2 \cdot 10^6$; $X_L = 0.075$; $C = 2.39 \cdot 10^{-3}$); the same notation as in Fig. 39.

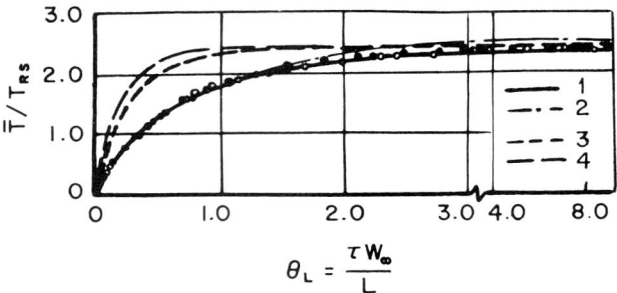

Fig. 41. Change in mean surface temperature at stepwise heat release ($Re_L = 1.66 \cdot 10^5$; $X_L = 12.3$; $C = 0.68 \cdot 10^{-3}$); the same notation as in Fig. 39.

($1.11 \cdot 10^5$), the layer, where the effect of turbulent transfer is small, is sufficiently large. Therefore, initially the thermal boundary layer grows more slowly and the heat transfer process is controlled by one-dimensional molecular heat conduction and the solutions (151) and (152) agree well with the experimental data. In the case of low Reynolds numbers ($1.66 \cdot 10^5$), the time for steady regime onset is considerably larger than the predicted value ($Q_L = 1$), whereas at $Re_L = (5.3–20) \cdot 10^5$ it is larger than the predicted one by 15%.

Figures 42 and 43 show mean experimental heat transfer coefficients for ribbons of various heat capacity: $W = 4.27$ and 2.43 m/sec, respectively ($W = 0.8 W_\infty$). In the same figures are presented the values for

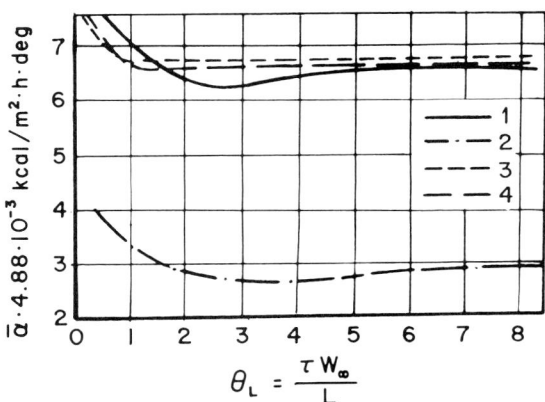

Fig. 42. Change in mean heat transfer coefficient at transitional regime due to stepwise heat release; $\delta = 0.35$ mm; $W_\infty = 4.27$ m/sec: (1) experiment; (2) solution (154) and (155) for rodlike flow; (3) solution (185) and (186) for turbulent flow; (4) experiment for ribbon at $\delta = 0.1$ mm.

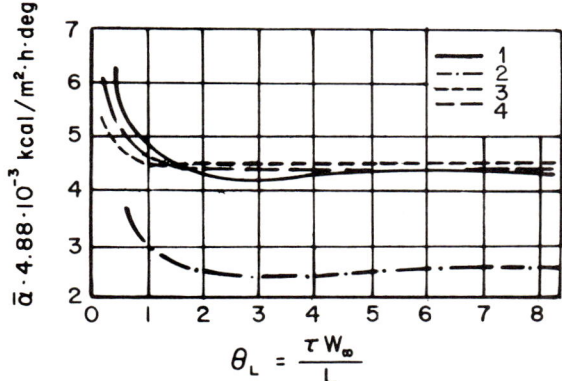

FIG. 43. Change of mean heat transfer coefficient at transitional regime caused by stepwise heat release; $\delta = 0.35$ mm; $W_\infty = 2.43$ mm; the same notation as in Fig. 42.

a turbulent flow predicted by Eqs. (185) and (186) and those predicted by Eqs. (154) and (155) for a rod flow. The best agreement with the experimental data is shown for the ribbon with the lowest heat capacity, which is quite clear from the assumptions made (α for a zero heat capacity). The experimental α pass a minimum, being qualitatively the same as those predicted by (154) and (155) for a thicker ribbon. This minimum may probably be attributed to the fact that, when the conductive contribution to α prevails over the convective one (as an average, the time increases with the plate heat capacity), the heat transfer falls below the steady-state value. It may be noticed that, when the heat capacity of the plate increases, the dimensionless time Θ_L for the commencement of the steady regime also increases (Figs. 39 and 40). Therefore, the time will also increase, during which the heat conduction process will cause superposition of heat transfer due to unsteady heating of the fluid on convective heat transfer. However, since the rate of the wall temperature change falls, the deviation of heat transfer from a convective one decreases.

From the above discussion of the works on conjugate problems, the following conclusions may be made:

(1) As regards the determination of the temperature field in a fluid, the significance of the solution of conjugate problems is in determining the boundary conditions (the surface temperature).

(2) The conjugate problems which allow us in principle to find simultaneously the temperature fields in a body and in the fluid are involved with great mathematical difficulties that make necessary

considerable simplifications; thus the solved problem differs considerably from the original one. Theoretical methods need further development.

(3) Experimental studies of conjugate problems with turbulent flows are probably useful only for verification of specific analytical solutions. Correlation of such experimental data is almost impossible since it involves two or three additional dimensionless numbers of the type Bi, Fo, and $\rho_w C_w \delta / \rho C_p x$.

This indicates that the experiments should be carried out under natural conditions. Therefore, it may be useful to study experimentally and to generalize the dependence of unsteady heat transfer on the typical laws of time development of the boundary conditions and to use the data obtained for the solution of unsteady heat conduction problems for a body with boundary conditions of the third kind.

IV. Some Aspects of Unsteady Heat Transfer

Besides the works discussed above, some other studies on unsteady heat transfer are of particular interest. In Refs. (*12, 30, 31, 194–224*) unsteady convective heat transfer is treated in external flow around bodies. References (*225–264*) are devoted to various aspects of unsteady free convection and Refs. (*236, 237, 263*) deal with unsteady heat transfer in channels with simultaneous free and forced convection. A considerable number of works (*265–319*) are devoted to theoretical or experimental studies of dynamic characteristics of heat exchangers. In the books by Serov and Korolkov (*316*) and by Shevyakov and Yakovleva (*318*) a great number of these problems are analyzed.

Some problems of unsteady radiative heat transfer are considered in (*320–330*), and works (*321, 324, 326, 329*) are devoted to unsteady heat transfer with simultaneous thermal radiation and convection. Problems of unsteady heat transfer in capillary-porous bodies are considered in works (*331–335*).

A. Unsteady Convective Heat Transfer for Bodies in External Flow

A number of works are devoted to the problem of unsteady heat transfer in external flow in a laminar boundary layer with constant physical properties. Sparrow and Gregg (*194*) have estimated the deviation of an unsteady process from quasi-stationary heat transfer by the disturbance method and have presented the results over a wide range of Prandtl numbers.

Cess (*196*) determined the time development of a heat transfer process in laminar flow along a flat plate with unsteady surface temperatures. The velocity and temperature of the undisturbed flow were assumed constant. At a stepwise change of the surface temperature, the time of development of stationary heat transfer is longer the larger the Prandtl number. In the case of a linear increase of the surface temperature with time, the deviation from quasi-stationary heat transfer is larger the shorter the time τ and the larger the Prandtl number. In Goodman's works (*197, 198*) the effect of the wall temperature jump on the heat transfer of an incompressible fluid has been studied. The case is similar to that studied by Cess but analyzed by the integral method. Goodman has assumed that the temperature profile corresponds to a cubic parabola and has taken a linear velocity profile. The results did not agree with the known result for the steady state; therefore the existence of a heat wave causing the transition of unsteady-state conditions to steady-state ones has been assumed. Rozenshtok (*199*) also solved the problem of an unsteady-state thermal laminar boundary layer on a semi-infinite plate in a viscous liquid flow. The solution is obtained by the integral method with the use of approximate velocity and temperature profiles in the boundary layer in the form of a fourth-power parabola. It is shown in the paper that the assumption of a steady-state velocity distribution in the problem on an unsteady-state thermal boundary layer is valid only at $Pr > 1$, but at $Pr < 1$ only the steady-state part of the problem can be solved. In this case the integral method leads to a prediction of the time of transition from unsteady state to steady state, lower by a factor of 2. Adams and Gebhart (*187*) consider unsteady-state forced convection from a flat plate subjected to a stepwise heating from inside with a steady-state flow with a laminar boundary layer. The following cases are considered: a plate of zero heat capacity and a plate of relatively high heat capacity. In the case of a plate with zero heat capacity, the problem is solved by the integral method with the use of various approximations of the velocity profile. For the plate with relatively high heat capacity, a quasi-stationary solution is obtained.

It should be noted that, for a transient external flow with constant properties with a laminar boundary layer, the time of development of the steady-state velocity distribution is about 2.5 times as much as the time required for the fluid at the entrance at $\tau = 0$ to reach a given point. Besides, the time of development of a steady-state temperature distribution depends on Pr. With increasing Pr the time of temperature development increases, proportional to $Pr^{1/3}$. Riley (*202*) included in the Cess solution (*196*) a correction for convective heat transfer in fluids. For small times, the solution to the energy equation is of the form

$$\frac{q}{\lambda(T_w - T_\infty)} = \left[0.564\,\text{Pr}^{1/2}\left(\frac{W_\infty \tau}{x}\right)^{-1/2} - 0.0208\left(\frac{W_\infty \tau}{x}\right)\right.$$
$$\left. - 0.00052\,\text{Pr}^{-3/2}(4.9\,\text{Pr} - 1)\left(\frac{W_\infty \tau}{x}\right)^{5/2} + \cdots\right]\left(\frac{W_\infty x}{\nu}\right)^{1/2}$$
(187)

where the first terms is pure thermal conductivity of the fluid.

Knuth (*12*) derived a general expression for the deviation of the heat flux q from the quasi-stationary value q_0 in the case of a time-variable surface temperature

$$q = q_0\left[1 + 2.67\,\text{Pr}^{1/2}\frac{\theta'(\tau)}{\theta(\tau)}\left(\frac{x}{W_\infty}\right) - 1.00\,\text{Pr}^{2/3}\frac{\theta''(\tau)}{\theta(\tau)}\left(\frac{x}{W_\infty}\right)^2\right.$$
$$\left. - 2.67\frac{Z_2(F)}{Z_2(0)}\,\text{Pr}^{1/2}\frac{\theta'(\tau)}{\theta(\tau)}\left(\frac{x}{W_\infty}\right) + 1.00\frac{Z_4(F)}{Z_4(0)}\,\text{Pr}^{2/3}\frac{\theta''(\tau)}{\theta(\tau)}\left(\frac{x}{W_\infty}\right)^2 + \cdots\right]$$
(188)

where

$$\theta(\tau) = \frac{T_w(\tau)[1 + \text{Pr}[(K-1)/2]\,M^2] - T_r}{T_w(0) - T_r}$$

T_r is the restoration temperature; $Z_2(F)/Z_2(0)$ and $Z_4(F)/Z_4(0)$ are functions quickly decreasing with increase of $F = 0.2\,\text{Pr}^{-1/2}(W_\infty \tau/x)$ (from 1 at $F = 0$ to 0.1 at $F = 0.25$ and to 0.01 at $F = 0.5$). This formula is in agreement with the solutions by Sparrow and Gregg (*194*) at $F > 1$ and by Goodman (*197, 198*). As seen from (188), in most cases of a forced external flow, the heat flux to the surface, the temperature of which is variable in time, differs only slightly from the quasi-stationary one.

In (*204*), Chao and Jeng studied analytically the unsteady-state heat transfer for the case of a laminar flow in a two-dimensional axisymmetric frontal zone with stepwise variations of wall temperature and heat flux. Two asymptotic solutions are obtained for small and large time intervals which are in satisfactory agreement at $\text{Pr} = 0.01$–100. With a stepwise variation of surface temperature, the deviation of the heat flux from the quasi-stationary value is larger the lower the Pr number, where $\theta = K\tau/\text{Pr}$ is a dimensionless time with $K = 2W_\infty/R$ for a cylinder and $K = 3W_\infty/2R$ for a sphere (R is the radius of the body). The stabilization time is inversely proportional to the velocity of the incoming flow W_∞ and directly proportional to $\text{Pr}^{1/4}$. A similar problem is solved in (*30*) and (*31*) for a time-variable flow velocity.

The positive flow acceleration increases the shear stress at the wall, decreases heat transfer at a given wall temperature, or increases the temperature of the wall at a given heat flux (as compated to the quasi-stationary values).

Chao and Cheema (224) investigate the unsteady heat transfer in an incompressible, laminar boundary layer over a semiinfinite flat plate due to a step change in either the wall temperature or the wall flux for Prandtl numbers ranging from 0.01 to 100. When the flow is steady they found that, with the help of a new analytical method valid at all times for a restricted range of Prandtl numbers (0.72–100), the time required for the thermal layer to achieve steadiness varies inversely with the free stream velocity and directly with $\frac{1}{3}$ power of the Prandtl number.

In (189) Soliman and Johnson solved the problem of flow with a turbulent boundary layer with a stepwise variation of heat release. The solution is presented in detail in Section III,D. The heat transfer process for the time $\tau \leqslant x/W_\infty$ is mainly determined by one-dimensional thermal conduction, and at $\tau = x/W_\infty$ a steady-state regime sets up immediately. Experimental data (203) on a plate heated by an exponential time-dependent heat source $q = q_0 \exp(\tau/\tau_0)$ agree with this solution at $0.28 \leqslant L/W_\infty\tau_0 \leqslant 2$ [$\tau_0 = (0.5\text{–}4)L/W_\infty$], and at $5 \leqslant L/W_\infty\tau_0 < \infty$ ($\tau < 0.2L/W_\infty$), with the solution of Chambre (188) obtained for a rodlike flow (L is length of the plate). Thus, the theoretical investigations of transient heat transfer in an external flow have been mainly carried out with the same assumptions as the problem of heat transfer in channels, i.e., neither the effect of unsteadiness upon turbulent flow characteristics nor variability of thermal properties are taken into account. As the above papers show, in the external flow the time of heat transfer stabilization with a stepwise change of the boundary conditions is longer the larger the distance from the starting length of the plate and the larger the Prandtl number.

In a number of studies the effect of thermal unsteadiness upon heat transfer in an external flow has been investigated experimentally. Katsnelson and Timofeeva (211) investigated heat transfer of spheres in a liquid flow by the method of analogy with diffusion. The essence of the method is that particles of NaCl were allowed to fall in a liquid, and the salt concentrations in the solution were determined at different levels. It was found that, during the initial period of particle motion in the liquid, the diffusional Nu increased from the values close to zero at the starting length (constant velocity) to the constant value corresponding to the steady state at given Re. This can be explained by the fact that in the initial period (till the steady state is set up) a fraction of the substance participates in saturating the boundary film and cannot be found in the volume. Therefore, this method cannot investigate unsteady diffusion on the surfaces of particles, and hence not the heat transfer either. In the same paper, the authors describe an experiment where the diffusion Nu has been determined from direct measurements of the amount of salt removed from the particle surfaces. In this test, the diffusional Nu decreases from infinity to the steady value more quickly the larger the Reynolds number. In the papers by Kryukova (*212*), Bitsyutko *et al.* (*213*), and Bitsyutko *et al.* (*214*) the transient heat transfer of a sphere in an air flow was studied. No effect of thermal unsteadiness upon the heat transfer coefficient has been found there (which follows from the regularity of the regimes of cooling the spheres under study). It should be noted that experiments (*213, 214*) have been carried out at comparatively high Re($2.4 \cdot 10^4$ to $2.4 \cdot 10^5$) and low rates of sphere temperature variations [maximum value $| \partial T_w/\partial \tau = 1\text{--}2$ deg/sec], which certainly affected the result obtained.

In the paper by Kudryashev and Smirnov (*215–217*) the effect of thermal unsteadiness upon convective heat transfer in an external incompressible liquid flow around an infinite cylinder and sphere is discussed. It is assumed there that the heat transfer starts at the moment when the cylinder and sphere are immersed in the liquid flow.

Theoretical analysis (*215, 216*) with the use of a semiempirical turbulence theory allowed the integration of the momentum and energy equations, and the solution was obtained

$$\mathrm{Nu}^2/\mathrm{Nu}_k^2 = 1 + \mathrm{const}/\mathrm{Fo}^m\,\mathrm{Re}^n \qquad (189)$$

where Nu_k corresponds to the regime set up. The constants included in (189) are determined experimentally. The method of regular regime has been used. Empirical relationships are obtained for the transient heat transfer coefficient

$$\begin{aligned}\mathrm{Nu}^2/\mathrm{Nu}_k^2 &= 1 + [3.6/(\mathrm{Fo}\,\mathrm{Re}^{0.7})^{0.56}] \quad \text{at} \quad 0 < \mathrm{Fo}\,\mathrm{Re}^{0.7} < 23 \\ \mathrm{Nu}^2/\mathrm{Nu}_k^2 &= 1 + [282/(\mathrm{Fo}\,\mathrm{Re}^{0.7})^2] \quad \text{at} \quad 23 < \mathrm{Fo}\,\mathrm{Re}^{0.7} < 45\end{aligned} \qquad (190)$$

Within the range studied ($1000 < \mathrm{Re} < 5000$) the steady state starts at $\mathrm{Fo}\,\mathrm{Re}^{0.7} \simeq 45$.

From the experiments carried out, the authors conclude the following: At the beginning of the process, $\tau = 0$, the heat transfer coefficient $\alpha \to \infty$, then it decreases to a stationary heat transfer coefficient (as $\tau \to \infty$, $\alpha \to \alpha_k$). The duration of the process of thermal stabilization is very short (relative to the whole process of cooling the body). With increasing Re the period of unsteadiness decreases, and at large Re the effect of unsteadiness upon the heat transfer problem becomes rather slight.

In (205) Grigoriev showed that the solution by Kudryashev and Smirnov (217) for small Re is very approximate and can be performed only for a limited class of functions. In the case of a power variation of the surface temperature of the form $\varphi(\tau) = K\tau^m$ the calculation error increases with m and at $m = 2$ is 55%. At $\varphi(\tau) = A \sin \omega\tau$, the calculation according to (217) yields a wrong result. In this case, according to (205), steady-state heat transfer is impossible, and in (217) $\lim_{\tau \to \infty} \mathrm{Nu} = 1$.

In a paper by Rutsky (218) an experimental study is described of a thermistor in the form of a cylinder in a transverse air flow around at unsteady-state thermal regime and with constant velocity. The thermistor was heated by an electric current. The heated thermistor was suddenly immersed in the air flow. The constant power supplied to the thermistor under transient conditions was provided by a special electric circuit. It was found that the heat transfer coefficient increased with thermistor cooling by 2–4 times and increased with the flow velocity. Such a strange result can be explained by an improper determination of the heat transfer coefficient; i.e., in determining the heat flux the heat released by the cooling thermistor was not taken into account. Hence, the value of the heat transfer coefficient was lower the greater $\partial T_w/\partial \tau$, i.e., the lower τ at constant flow velocity, or the greater flow velocity at $\tau = \mathrm{const}$. As the calculations showed, this correction is commensurate with the heat transfer coefficient determined in (218).

Popov (219) in his study of the local heat transfer behind the step on a flat surface in a supersonic air flow by a transient method (method of a "thin wall") has found that for the stabilization of heat transfer processes in a separation region, a considerably longer time is needed than for the stabilization of the velocity fields. This time is about 50 times larger than $\Delta\tau_a = H/W_\infty$ (H is the height of the step; W_∞ is the velocity of the flow beyond the displacement region), which in (220) is recommended as the characteristic time of heat transfer process stabilization. At the initial moment, the wall temperature derivative $\partial T_w/\partial \tau$ and the heat transfer coefficient are several times as much as the quasi-stationary values.

In the paper by Parnas (221), in contrast to the above papers, the transient heat transfer of a cylinder has been studied when it is heated by air with a sudden change in velocity and temperature of the incoming flow. In this case, the heat transfer coefficient changes from zero at the starting moment to a quasi-stationary value. This can be attributed to the fact that in a transient process the gas layers near the wall are cooled more quickly and the velocity profile is longer. Within the range of Re numbers $Re_b = 760$–2900, the heat transfer stabilizes at $Ho = W\tau/d \simeq 20$ (d is cylinder diameter), i.e., 10–20 times as fast as in the experiments by Kudryashev and Smirnov (215, 216).

In (223) an unsteady transfer process on a plate with a stepwise variation of boundary conditions in a laminar boundary layer has been studied by an electrochemical method. For zero angle of attack, the time necessary for the mass transfer processes to approach the steady-state regime after a stepwise variation of the concentration at the boundary of the body is in fair agreement with Goodman's solution (197, 198). With positive angles of attack, the time of stabilization is shorter, and with negative it is longer than zero angle of attack.

Pelepeichenko and Simbirsky (222) studied the transient heat transfer from wires, $d = 0.02$–0.1 mm, in a transverse air flow with an harmonic velocity variation at frequencies 0.6–18 Hz, homochronous criterion $Ho = 37$–$148 \cdot 10^4$ and $Re = 7$–180, which agreed with the maximum value of the criterion of velocity unsteadiness (139) $K_G = 0.01$. The instantaneous heat transfer was quasi-stationary under such conditions.

Thus, the majority of the experimental studies dealt with heat transfer with time variations of wall temperature. The deviation of transient heat transfer values from quasi-stationary ones were greater, the larger $\partial T_w/\partial \tau$.

B. Transient Free Convective Heat Transfer

The majority of papers (226–229, 231–235, 239–243, 247, 248, 258, 264) deal with the laminar regime of transient free convection which arises near vertical elements immersed in a large volume of liquid at rest with heat release inside the elements. The following peculiarities are typical for the process. Energy generation in an element causes a temperature increase on its surface. During a short time interval, heat is transferred to a motionless liquid only by simple heat conduction. This process can be considered as one-dimensional. The temperature field generated in the liquid causes it to become buoyant. Accelerated motion of the liquid begins. In this process equilibrium will be attained between the buoyancy, inertia, and viscous forces. Also equilibrium is set up between the heat flux supplied to the element surface by conduction, and heat flux removed from the element to the liquid by convection. The temperature of the surface increases with time and tends to an asymptotic value.

Siegel (226) suggested an approximate solution for the cases of stepwise temperature variation and stepwise flux variation on the element surface in a laminar flow. The integral relationship by Karman–Pohlhausen has been used. In the first case, the time of stabilization is

$$\text{Fo}_S = \frac{a\tau_S}{L^2} = \frac{[5.24(0.952 + \text{Pr})^{1/2} + 7.1(0.377 + \text{Pr})^{1/2}]}{2} \left[\frac{x}{4\,\text{Ra}_L\,\text{Pr}}\right]^{1/2} \quad (191)$$

and, in the second,

$$\text{Fo}_S = a\tau_S/L^2 = 4.78(0.8 + \text{Pr})^{2/3}\,(\text{Ra}_L^*\,\text{Pr})^{-2/5}\,(x/L)^{2/5} \quad (192)$$

Here L is the characteristic length; x the distance from the starting length of the plate;

$$\text{Ra}_L = (g\beta\,\Delta T L^3/\nu^2)\,\text{Pr}, \qquad \text{Ra}_L^* = gq\beta L^4\,\text{Pr}/\nu^2\lambda$$

are Rayleigh numbers determined from $\Delta T = T_w - T_\infty$ and q; β the volumetric expansion coefficient; and T_∞ the liquid temperature far from the wall. In the process of stabilization the heat transfer coefficient passes through a minimum.

The results of Gebhart's solution (*231*) for stepwise heat flux variation in terms of generalized variables do not depend on the Prandtl number. During the transient process, the surface temperature increases slowly to its asymptotic value.

According to the solution by Sparrow and Gregg (*228*) the deviation of the transient heat transfer coefficient from the quasi-stationary value α_0 for gases (Pr = 0.72) can be determined by the relationship

$$\frac{\alpha}{\alpha_0} = 1 + 1.53\,\frac{1}{T_w - T_\infty}\,\frac{\partial(T_w - T_\infty)}{\partial\tau}\left[\frac{xT_\infty}{g(T_w - T_\infty)}\right]^{1/2} + \cdots \quad (193)$$

The rise of heat transfer with T_w is attributed to the inertia of the boundary layer. With increasing T_w the increase of the boundary-layer temperature lags behind that of T_w, and during the transitional period the boundary layer is cooler than it was when steady state was attained at the given value of T_w. Therefore, the intensity of heat transfer with an unsteady increase of T_w is greater than at steady-state conditions.

The relationship (193) allows the determination of the time of heat transfer coefficient stabilization at the given law of variation of $(T_w - T_\infty)$. Thus, for

$$T_w - T_\infty = A\tau \quad \text{at} \quad \tau \geqslant 9.75(xT_\infty/gA)^{1/3}$$

heat transfer deviation from the quasi-stationary value does not exceed 5%.

Chung and Anderson (232) considered the time variation of both the wall temperature and the acceleration field. With increasing g, heat transfer decreases as compared to quasi-stationary values, which is attributed to the fact that with an increase of g the average velocity of the medium in the boundary layer remains less than that which may develop in the system after attaining the steady state at some instantaneous value of g.

At a given value of the unsteadiness parameter,

$$\xi_0 = \frac{1}{2}\left(\frac{x}{\beta g}\right)^{1/2} \frac{\partial(\Delta T)}{\partial \tau} \frac{1}{(\Delta T)^{3/2}}$$

for the case of variable wall temperature and

$$\xi_0 = \frac{1}{2}\left(\frac{x}{\beta \Delta T}\right)^{1/2} \frac{\partial g}{\partial \tau} \frac{1}{g^{3/2}}$$

for the case of an unsteady field of acceleration; then effect of the first unsteadiness is considerably greater than that of the second one. In the first case (at $\mathrm{Pr} = 0.72$)

$$\frac{\mathrm{Nu}_L}{(\mathrm{Gr}_L)^{1/4}} = \frac{1}{\sqrt{2}}\left(\frac{x}{L}\right)^{-1/4}(0.5046 + 1.5522\xi_0 + \cdots) \qquad (194)$$

and in the second

$$\frac{\mathrm{Nu}_L}{(\mathrm{Gr}_L)^{1/4}} = \frac{1}{\sqrt{2}}\left(\frac{x}{L}\right)^{-1/4}(0.5046 - 0.03357 + \cdots) \qquad (195)$$

Schetz and Eichhorn (235) and Menold and Tzu (233) obtained solutions for the velocity and temperature profiles at arbitrary Prandtl numbers and for a number of boundary conditions, namely surface temperature and heat flux on the surface. On the basis of velocity and temperature fields along an infinite plate, Goldstein and Briggs (242) obtained a solution which helps to calculate when the effect of the front edge of a semi-infinite vertical plate would distribute over a given distance x_p. In a transient process, the minimum of the heat transfer coefficient is most noticeable with a stepwise variation of the surface temperature.

Gebhart considered transient natural convection from vertical elements with heat capacity at stepwise (240) and linear (239) variations of internal heat generation. At stepwise variations of heat generation, depending on the wall heat capacity, transient processes are divided into three groups: (a) one-dimensional transient heat conduction till the temperature level is set up at

$$Q = C_w/\rho C \delta_0 \; M_0 < 0.1$$

[where C_w is wall heat capacity per unit surface; ρ and C, the density and heat capacity of the fluid; δ_0, the thickness of the thermal boundary layer, $M_0 = f(Pr)$; for $Pr = 0.01$, $M_0 = 1.88$; for $Pr = 1$, $M_0 = 1.79$; for $Pr = 1000$, $M_0 = 1.76$]; (b) transient convection at $Q = 0.1$–1; (c) actually quasi-stationary process at $Q > 1$; with linear variations of heat generation the quasi-stationary region is considerably wider.

Transient free convection on a vertical surface was studied experimentally in (227, 241, 247, 258). Goldstein and Eckert (227) carried out interferometric investigations of transient temperature fields in air and water with stepwise variations of heat release. The boundary layer increases with time gets to its maximum and decreases to its steady value. The heat transfer is in agreement with the solution for unsteady heat conduction and then attains its stationary value. In (258) the period of transient convection ($0.1 < Q < 1$) is studied. No temperature jump has been found.

In (232, 234) a theoretical study of transient free convection near horizontal cylinders is presented. Experimental investigations (225, 238, 246) are carried out on horizontal cylinders, and (261) on spheres.

Transient laminar free convection on horizontal surfaces was studied in (250, 255, 259). In (250) the experimental results are presented. According to (255) the time of free convection development with a stepwise heating of the surface is less the more intensive is the heat generation and the larger is the ratio $K = (\lambda C \rho / \lambda_w C_w \rho_w)$. This solution takes into account only the vertical component of the velocity.

Transient free convection in vertical channels is considered in (243) (tube) and (253) (slot).

As on a vertical plate, at the initial moment heat transfer due to heat conduction dominates. Transit to the steady-state conditions is accompanied by wave processes and wall temperature oscillations.

In a number of papers (244, 245, 249, 251, 252, 254, 256, 257, 260) transient free convection is studied in closed volumes. Polezhaev (250) did numerical solutions to the system of two-dimensional unsteady-state Navier–Stokes equations for a compressible gas in a closed rectangular region between two vertical walls at various temperatures and with heat insulated bases. The temperature dependence of gas thermal properties is taken into account. The solution fields the values of the mean heat transfer and a picture of onset and development of secondary flows at large Grashof numbers.

A method of solution for three-dimensional unsteady-state Navier–Stokes equations is presented in the paper by Aziz and Hellums (*257*). The problem on convection in closed volumes heated from below is solved. In the three-dimensional case, the fluid velocity and temperature deviations from the developed values are less than in two-dimensional cases since the additional vertical walls hamper free motion.

Evans and Stefany (*254*) experimentally studied heat transfer in a cylindrical region with a stepwise wall temperature variation. The cylinder was placed vertically and horizontally, and L/D changed from 0.75 to 2. The heat transfer coefficient remained constant within the greater portion of the transient process. In the paper by Barakat and Clark (*251*) is presented an experimental and theoretical study of transient laminar free convection in a cylindrical tank with a free surface of the fluid. An experimental study of transient turbulent free convection in a tank is carried out by Tatom and Carlson (*252*). The tank was heated both from the side and from below.

Transient heat transfer with simultaneous forced and free convection has been studied in (*236*) for a flat vertical channel, and in (*237*) for a circular vertical tube, with heating of the liquid flowing upwards and with cooling of the liquid flowing downwards. Unsteadiness due to time changes of the wall temperature, pressure gradient, and the power of wall heat source is considred. Time variations of velocity, temperature of the fluid, and heat transfer are of the nature of oscillations with damping amplitudes. The oscillations are most noticeable at $Pr = 1$ and large Gr numbers.

The onset of oscillations is explained as follows: For example, with increasing pressure gradient the velocity and heat transfer increase. At constant wall temperature this leads to an increase in the average temperature of the fluid and decrease of the buoyancy, which in its turn decreases the velocity and heat transfer and hence the average temperature of the fluid. As a result, the buoyancy and velocity rise. The frequency of oscillations is higher the greater the number $Ra = Gr \cdot Pr$.

Nomenclature

a thermal diffusivity
C_p, C_w heat capacity
d internal tube diameter
G flow rate
g acceleration of gravity
P pressure
Q heat flux
q_w specific heat flux through internal surface
q_H specific heat flux through external surface
q_v volumetric heat release

r	radius	ρ	density
t_b, T_b	mean mass (mean calorimetric) flow temperature	τ	time
		ν	kinematic viscosity coefficient
t_H, T_H	temperature of external surface of tube	ξ	hydraulic resistance coefficient
t_w, T_w	temperature of internal surface of tube	T	tangential stress [Section II]
		Bi	$\alpha\delta/\lambda_w$, Biot number
T_0	const, characteristic temperature	K	$\mathrm{Nu}/\mathrm{Nu}_0$, ratio of unsteady Nusselt number to a quasi-stationary value;
$(T_w - T_b)_0$	characteristic temperature difference		$K_T = (\partial T_w/\partial \tau)[d^2/(T_w - T_b)a]$ and
W	mean mass flow velocity in cross section		$K_{T_0} = (\partial T_w/\partial \tau)[d^2/(T_w - T_b)_0 a]$ are dimensionless criteria for unsteady temperature conditions
$\overline{W}$	mean velocity along channel length		
W_x	axial velocity component		
W_x', W_r'	turbulent pulsations of axial and radial components of velocity	K_G	$(\partial G/\partial\tau)(d^2/G\nu)$, dimensionless criterion for unsteady hydrodynamic conditions
x	longitudinal coordinate	Re	$4G/\pi g\, d\mu$, Reynolds number
y	cross-sectional coordinate	Pr	ν/a, Prandtl number
X	x/d, dimensionless distance	Nu	$\alpha d/\lambda$, Nusselt number
α	heat transfer coefficient	Nu_0	Nusselt number calculated by quasi-stationary relation
γ	specific weight		
δ	thickness of a tube wall; thickness of a boundary layer	Fo	$a\tau/r_0^2$, Fourier number
		Ho	$\int_0^\tau [W(\tau)/d]\, d\tau$, homochronous criterion
ϵ_q	coefficient of turbulent thermal diffusion		
ϵ_τ	turbulent viscosity coefficient		
θ	T/T_0, dimensionless temperature	SUBSCRIPTS	
$\varphi(X)$	distribution of dimensionless flow temperature along channel at initial time instant	1, 2	initial and finite steady states
		1, 11	value at inlet and outlet of test section
$\psi(\mathrm{Ho})$	dependence of dimensionless flow temperature at channel inlet upon time	k	quasi-stationary value of a quantity
		b	value determined by flow temperature
λ	thermal conductivity		
λ_T	coefficient of turbulent heat conduction	H	value determined by temperature of an external wall surface
μ	dynamic viscosity coefficient	w	refers to wall conditions
μ_T	turbulent viscosity coefficient	in	inlet

References

1. V. K. Koshkin, E. K. Kalinin, S. A. Yarkho, and A. A. Ter-Mkrtchyan, Heat transfer in a channel with packed arrangement of a bundle of tubes or rods, *Teplofiz. vysokikh Temperatur* **5**, No. 2, 317–321 (1967).
2. T. L. Perelman, On conjugated problems of heat transfer, *Intern. J. Heat Mass Transfer* **3**, No. 4, 293–303 (1961).
3. A. V. Luikov and T. L. Perelman, On unsteady heat transfer between a body and flow around it. Teplo-i Massoobmen tel s Okruzhayushchei Gazovoi Sredoi, Izd. Nauka i tekchnika pp. 3–24. Minsk, 1965.
4. A. V. Luikov and T. L. Perelman, Problems of unsteady heat transfer between a body and a gas flow around it. Teplo-i Massoperenos Vol. 6, pp. 63–85. Izd. Nauka i Tekhnika. Minsk, 1966.
5. A. V. Luikov, T. L. Perelman, and V. B. Ryvkin, On determination of the heat transfer coefficient in simultaneous conductive and convective heat transfer. *Proc. Intern. Heat Transfer Conf. 3rd. Chicago, 1966*, Vol. 2, No. 42, pp. 12–24. AIChE, New York, 1966.
6. E. V. Kudryavtsev, K. N. Chakalev, and N. V. Shumakov, "Unsteady Heat Transfer" (Nestatsionarnyi meploobmen). Izd. Akad. Nauk SSSR, Moscow, 1961.
7. E. V. Kudryavtsev and N. V. Shumakov, The effect of sizes and material of a solid upon the process of unsteady heat transfer, *Inzh. Fiz. Zh. Akad. Nauk Belorussk.* **4**, No. 1 (1961).
8. L. I. Kudryashev and A. A. Smirnov, Conjugated unsteady convective heat transfer of bodies in external flow, *Tr. VNIITsEM Prom.* **5**, 89–110 (1966).
9. B. S. Petukhov, "Heat Transfer and Resistance with Laminar Liquid Flow in Ducts" (Teploobmen i soprotivlenie pri laminarnom techenii zhidkosti v trubakh). Izd. Energiya, Moscow, 1967.
10. V. D. Vilenskii, General laws of stabilization of a heat transfer coefficient with liquid flow in a channel *Teplofiz. Vysokikh Temperatur* **4**, No. 5, 675–682 (1966).
11. A. I. Leontiev, V. A. Mukhin, B. P. Mironov, and V. P. Ivakin, Effect of boundary conditions upon development of thermal boundary layer, *in* "Teplo-i Massoperenos," Vol. I, pp. 125–132. Energia, Moscow, 1968.
12. E. L. Knuth. Forced convection heat transfers with time dependent surface temperatures, *AIAA J.* **1**, No. 5 (1963).
13. V. D. Vilenskii, Some general laws of unsteady heat transfer with laminar liquid flow in channels, *Teplofiz. Vysokikh Temp.* **4**, No. 6, 838–845 (1966).
14. E. K. Kalinin, Unsteady convective heat transfer and hydrodynamics in channels, *Izv. Akad. Nauk BSSR Ser. Fiz.-Tekh. Nauk*, No. 4, 44–55 (1966).
15. A. A. Gukhman, A. F. Gandelsman, V. V. Usanov, and G. N. Shorin, New data on properties of transsound flows *in* "Teplo-imassoperenos." Vol. 1, pp. 10–14. Izd. Nauka i tekhnika, Minsk, 1965.
16. E. Talmor, Effect of pressure gradient on sonic-point heat transfer, *Papers JSME semi-Intern. Symp. Heat Mass Transfer, Tokyo* **1**, 171–180 (1967).
17. E. V. Zalutskii, Study of steady accelerated turbulent flows in hydraulically smooth channel, *in* "Dokladov po Gidrotekhnike" (Institut gidromekhaniki Akad. Nauk USSR Vol. 7. Izd. Energiya, Moscow, 1966.
18. S. B. Markov, Experimental investigation of kinematic characteristics of unsteady pressure flow, Sb. "Gidravlika i gidrotekhnika," Izd. "Naukova dumka," Kiev, No. 8, 1968.
19. A. A. Gukhman, A. F. Gandelsman, G. G. Katsnelson, B. A. Kader, L. P. Naurits,

and V. V. Usanov, Effect of large negative pressure gradients upon the turbulent flow structure, *in* "Teplo-i Massoperenos," Vol. I, pp. 812–824. Izd. Energia, Moscow, 1968.
20. L. Prandtl, Attaining a Steady Air Stream in Wind Tunnels, V.D.I., **77**, No. 5 (1933).
21. M. S. Uberoi, Effect of wind-tunnel contraction on free-stream turbulence, *J. Aeron. Sci.* No. 8, 754–764, 1956.
22. J. O. Hinze, "Turbulence, An Introduction to its Mechanism and Theory." McGraw-Hill, New York, 1959.
23. A. S. Monin and A. M. Yaglom, "Statistical Hydromekhanics" (Statisticheskaya gidromekanika), Part I. Izd. Nauka. Moscow, 1966.
24. I. S. Gromeka, To the theory of liquid motion in narrow cylindrical pipes, Sobranie sochinenii, *Izd. Akad. Nauk SSSR*, 149–171, 1952.
25. G. T. Aldoshin, A. S. Golosov, and V. I. Zhuk, The solution of an inverse problem in transient heat conduction for a plate, *in* "Teplo-i massoperenos," Vol. 8, pp. 186–199. Izd. Nauka, Minsk, 1968.
26. K. Stewartson, The theory of unsteady laminar boundary layers, *Advan. Appl. Mech.* **6**, 1–37 (1960).
27. V. V. Struminskii, Theory of Unsteady Boundary Layer, pp. 230–252. Sb. teoreticheskikh rabot po aerodinamike, Oborongiz, 1957.
28. F. K. Moore and S. Ostrach, Displacement thickness of the unsteady boundary layer. *J. Aeron. Sci.* **24**, No. 1, 77–78 (1957).
28a. M. V. Markovin, private communication. Illinois Institute of Technology 1969.
29. L. Rosenhead, "Laminar Boundary Layers," Chapter 7. Oxford Univ. Press (Clarendon), London and New York, 1963.
30. N. Tohuda and W. J. Yang, Unsteady stagnation point heat transfer due to arbitrary timewise-variant free stream velocity, *Proc. Intern. Heat Transfer Conf. 3rd Chicago 1966*, Vol. 2, 223–232. AIChE, New York, 1966.
31. W. J. Yang, General non-steady laminar boundary layer in two-dimensional and axisymmetric flows, *Papers JSME Semi-Intern. Symp. Heat Mass Transfer Tokyo* **1**, 121–132 (1967).
32. V. Ya. Neiland, On solution of the equations of a laminar boundary layer at arbitrary initial conditions, *PMM 30* **4**, 674–678, 1966.
33. M. D. Duric, A Method for Solution of Unsteady Incompressible Laminar Boundary Layers, Publications de l'Institut Mathématique Nouvelle. Serie tower 6 (20), 29–55 (1966).
34. S. Ishirawa, The unsteady laminar flow between two parallel discs with arbitrary varying gap width, *Bull. JSME* **9**, No. 35, 533–549 (1966).
35. Yang K. T. Unsteady laminar layers over an arbitrary cylinder with heat transfer in an incompressible flow, *Trans. ASME Ser. E. J. Appl. Mech.* **81**, No. 2, 171–178 (1959).
36. G. N. Sarma, A general theory of unsteady compressible boundary layers with and without suction or injection, *Proc. Cambridge Phil. Soc.* **61**, 795–807 (1965).
37. G. N. Sarma, Solutions of unsteady compressible boundary layer equations, *Proc. Cambridge Phil. Soc.* **62**, No. 3, 511–518 (1966).
38. O. A. Oleinik, On the system of the boundary layer equations for unsteady flow of incompressible fluid, *Dokl. Akad. Nauk SSSR* **168**, No. 4, 751–754 (1966); On stability of solutions of the system of the boundary layer equations for unsteady flow of incompressible fluid, *PMM* **30**, 3, 417–423 (1966); To the mathematical theory of a boundary layer for unsteady flow of incompressible fluid, *PMM* **30**, No. 5, 801–821 (1966).

39. M. Kawaguti and P. Jain, Numerical study of a viscous fluid flow past a circular cylinder, *J. Phys. Soc. of Japan* **21**, No. 10, 2055–2062 (1966).
40. W. J. Yang and J. W. Ou, Unsteady forced convection of the entrance region of closed conduits due to arbitrary time-variant inlet velocity. *Papers JSME Semi-Intern. Symp. Heat Mass. Transfer Tokyo* **1**, 133–143 (1967).
41. S. Goldstein, ed., "The Modern Art of Hydroaerodynamics of Viscous Fluid" (Sovremennoe sostoyanie gidroaerodinamiki vyazkoi zhidkosti), Vol. 1. IL., Moscow, 1948.
42. A. D. Yung, Boundary layer, *in* "Sovremennoe sostoyanie aerodinamiki bol'shikh skorostei," Vol. 1. IL, Moscow, 1955.
43. A. S. Monin, Dynamic turbulence in atmosphere. *Izv. AN SSSR, Ser. Geograf. i Geofiz.* **14**, No. 3 (1950).
44. J. Rotta, Statistische theorie nichthomogener Turbulenz, *Leitschrift Phys. Bd.* **129**, 131 (1951).
45. G. S. Glushko, Turbulent boundary layer at a plate in incompressible fluid. *Izv. Akad. Nauk SSSR Mekh.* No. 4, 13–23 (1965).
46. E. V. Eremenko, Calculation of kinematic characteristics of a turbulent flow with unsteady motion, *Proc. Symp. Probl. Turbulent Fows Including Geophys. Appl., Kiev, 1967.* Izd. Naukova dumka, 1967.
47. I. K. Nikitin, "Turbulent Channel Flow and Processes in Bottom Region" (Turbulentnyi ruslovoi potok i protsessy v pridonnoi oblasti). Izd. AN USSR, Kiev, 1963.
48. Pai Shi-I, "Turbulent Flows of Liquids and Gases" (Turbulentnye techeniya zhidkosti i gazov). IL, Moscow, 1962.
49. O. F. Vasiliev and V. I. Kvon, On the velocity distribution with respect to a depth and friction resistance with unsteady motion of an open turbulent fluid flow, *Proc. Symp. Probl. Turbulent Flows Including Geophys. Appl.* Izd. Naukova dumka, Kiev, 1967.
50. A. N. Shabrin, On turbulent characteristics with unsteady motion, *Dokl. Akad. Nauk USSR* No. 8, 1030–1034, 1963.
51. A. N. Shabrin, Velocity structure of open flows with unsteady motion, *Dokl. Akad. Nauk SSSR* No. 11, 1448–1451 (1963).
52. A. N. Shabrin, A study of longitudinal and vertical pulsation components of velocity and moments of correlation between them in unsteady open flow, *in* "Gidrotekhnika i gidromekhanika, pp. 7–10. Izd. Naukova dumka, Kiev, 1964.
53. Z. Zaric, Turbulent heat transfer in a divergent-convergent channel, *Papers JSME Semi-Intern. Symp. Heat Mass Transfer Tokyo* **1**, 161–170 (1967).
54. R. W. Graham and R. G. Deissler, Prediction of flow acceleration effects on turbulent heat transfer, *Trans. ASME, Ser. C, J. Heat Transfer* **89**, No. 4 (1967).
55. P. Bradshaw and D. H. Ferris, The response of a retarded equilibrium turbulent boundary layer to the sudden removal of pressure gradient, Natl. Phys. Lab. (England) Aeron. Report, No. 1145 (Mar. 16, 1965).
56. R. G. Deissler, Weak locally homogeneous turbulence in idealized flow through a cone, NASA TND-3613 (1966).
57. P. M. Moretti and W. M. Kays, Heat transfer to a turbulent boundary layer with varying free-stream velocity and varying surface temperature — an experimental study, *Intern. J. Heat Mass Transfer* **8**, No. 9, 1187–1202 (1965).
58. D. R. Boldmen, J. F. Shmidt, and R. C. Ehlers, Effect of uncooled onlet length and nozzle convergence angle on the turbulent boundary layer and heat transfer in conical nozzles operating with air, *Trans. ASME, Ser. C J. Heat Transfer* **89**, No. 4, 341–350 (1967).

59. N. A. Panchurin, Velocity distribution in some cases of unsteady turbulent flow in tubes, *Tr. LIVT* **46**, 38–44 (1963).
60. M. K. Carstens and J. E. Roller, Boundary-shear stress in unsteady turbulent pipe flow, *J. Hydraulics Div. Proc. Am. Soc. Civil Eng.* 67–81 (Feb. 1959).
61. I. W. Daily, W. L. Hanrew, K. W. Olive, and J. M. Jordan, Resistance coefficients for accelerated and decelerated flows through smooth tubes and orifices, *Trans. ASME* No. 9, 78 (1956).
62. A. M. Aitsam, L. L. Paal, and U. R. Liyv, Calculation of transient head flow of incompressible fluid in rigid cylindrical tubes, *Tr. Tallin Polytekhn. Inst. Ser. A*, No. 223, 3–19 (1965).
63. U. R. Liyv, Head losses in a transient flow of incompressible fluid in rigid head tubes, *Tr. Tallin Polytekhn. Inst. Ser. A*, No. 223, 21–28, 1965.
64. U. R. Liyv, Hydraulic laws in decelerated fluid flows in a head cylindrical pipeline, *Tr. Tallin Polytekhn. Inst. Ser A*, No. 223, 29–42, 1965.
65. U. R. Liyv, Hydraulic laws in accelerated flows in a head cylindrical pipelines, *Tr. Tallin. Polytekhn. Inst. Ser. A* No. 223, 43–50 (1965).
66. I. S. Kochenov and U. N. Kuznetsov, Study of unsteady hydrodynamic and convective heat transfer processes, Report on Soviet-French symposium in Grenoble (1966).
67. I. S. Kochenov and U. N. Kuznetsov, Unsteady flows in tubes, Sb. "Teplo-i massoperenos," Vol. I, 306–314. Izd. Nauka i teknika, Minsk, 1965.
68. N. A. Panchurin, Hydraulic resistance in a transient turbulent flow in tubes, *Tr. LIVT* **13**, 43–56 (1961).
69. V. A. Tsetserin, Determination methods of the effect of unstationary conditions on the hydraulic resistance coefficient, *Tr. Kuybyshev Avia. Inst.* **24**, 137–142, 1967.
70. L. L. Kalishevsky and S. V. Selikhovkin, Some results of the study of an unsteady-state turbulent flow, *Teploenerget.* No. 1, 69–72 (1967).
71. K. K. Fedyaevsky and A. S. Ginevskii, Unsteady state turbulent boundary layer of a wing profile and body of revolution, *Zh. Tekh. Fiz* **7**, 916–923 (1959).
72. A. S. Ginevsky and K. K. Fedyaevsky, Some laws of transient translatory motion of bodies in viscous fluid, *Izv. Akad. Nauk SSSR Otd. Tekhn. Nauk Mekhan. i mashinostr.* No. 3, 207–209 (1959).
73. S. K. F. Karlsson, An unsteady turbulent boundary layer, *J. Fluid Mech.* **5**, No. 4, 622–636 (1959).
74. T. Sarpkaya and C. J. Garrison, Vortex formation and resistance in an unsteady flow, Paper *ASME* No. WA-62 (1962).
75. H. S. Carslow and J. C. Jaeger, Thermal Conductivity of Solids, "Izd. Nauka." Moscow, 1964.
76. A. G. Temkin, Inverse heat conduction problem of a symmetric body, *Inzh. Fiz. Zh.* **4**, No. 9, 45–55 (1961).
77. A. G. Temkin, Inverse heat conduction problem of an asymmetric body, *Inzh. Fiz. Zh.* **4**, No. 10 (1961).
78. P. R. Hill, A method of computing the transient temperature of thick walls from arbitrary variation of adiabatic wall temperature and heat transfer coefficient, NACA Rept. 1372 (1958).
79. E. M. Sparrow, A. Haji-Sheikh, and T. S. Lundgren, The inverse problem in transient heat conduction. ASME Paper, No. 64-APM-10 (1964); *Trans. ASME* **E86**, 369–375 (1964).
80. G. Stolz, Numerical solution to an inverse problem of heat conduction for simple shapes, *Trans. ASME Ser. C J. Heat Transfer* **82**, No. 1, 20–26 (1960).

81. J. V. Beck and H. Wolf, The nonlinear inverse heat conduction problem, Paper ASME No. 65-HT-40 (1965).
82. O. R. Burggraf, An exact solution of the inverse problem in heat conduction theory and applications, *Trans. ASME Ser. C J. Heat Transfer* **86**, No. 3, 373–380 (1964).
83. J. V. Beck, Discussion to the inverse problem in transient heat conduction, *Trans. ASME Ser. E J. Appl. Mech.* **87**, No. 2, 472–473 (1965).
84. J. V. Beck, Analytical determination of optimum transient experiments for measurements of thermal properties, *Proc. Intern. Heat Transfer Conf. 3rd Chicago, 1966*, Vol. 2, pp. 102–119. AICh E, New York, 1966.
85. D. G. Stephenson and G. P. Mitalas, The calculation of surface temperature and heat flux flow subsurface temperature measurements, *Trans. Eng. Inst. Canada* **6**, No. B-4, 26 (1963).
86. A. G. Temkin and G. P. Buynyachenko, The determination of internal heat transfer parameters from the measurements in an asymmetric plane field, *Izv. Akad. Nauk Latv. SSR Ser. Fiz.-Tekn. Nauk* No. 4, 91–98 (1963).
87. M. I. Pak and V. A. Osipova, A quasi-stationary method of combined determination of thermal properties of solids in a wide temperature range, *Teploenerget.* No. 6, 73–76 (1967).
88. M. I. Pak and V. A. Osipova, The determination of heat flux density in high-temperature studies of thermal properties of solids, *Teploenerget.* No. 12, 68–71 (1966).
89. V. A. Koverianov, An inverse problem in transient heat conduction, *Teplofiz. Vysokikh Temperatur* **5**, No. 1, 141–148 (1967).
90. A. G. Temkin, A. G. Gromyko, and R. V. Amitonov, Application of inverse heat conduction methods for determination of transient heat transfer conditions, *Inzh. Fiz. Zh.* **11**, No. 2, 261–263 (1966).
91. Y. A. Polyakov, A method of an exponential point in transient heat transfer, *Teplofiz. Vysokikh Temperatur* **5**, No. 1, 137–140 (1967).
92. L. I. Zhemkov, The determination of boundary conditions by the method of mean temperature of initial stages of a heat transfer process, *Tr. KuAI vyp.* **24**, 109–114, 1967.
93. L. I. Deverall and R. S. Channapragada, A new integral equation for heat flux in inverse heat conduction, *Trans. ASME Ser. C J. Heat Transfer* **88**, No. 3, 327–328 (1966).
94. V. Kmonicek, The determination of unsteady flow to wall from the measurements of surface temperature made with thin film resistance thermometers, *Intern. J. Heat Mass Transfer* **9**. No. 2, 199–213 (1966).
95. S. D. Kovalev and G. A. Pleshchenkov, The determination of temperature at inner channel wall in experimental study of unsteady heat transfer processes, *Izv. Akad. Nauk BSSR Ser. Fiz. Energ. Nauk* No. 1, 68–71 (1968).
96. E. K. Kalinin, The determination of flow temperature and friction coefficient in channels with a transient nonisothermal heat transfer agent flow, *in* "Teplo-i massoperenos," Vol. 1, pp. 228–297. Izd. Nauka i Teknika, Minsk, 1965.
97. A. N. Devoino, S. D. Kovalev, G. A. Pleshchenkov, and B. E. Tverkovkin, The methods of experimental study of transient heat-transfer of chemically reacting media, *Izv. Akad. Nauk BSSR Ser. Fiz-Energ. Nauk*, No. 1, 72–75 (1968).
98. A. N. Gordov, Temperature field of bodies at variable temperature of medium and variable heat transfer, *Tr. VNII Metrol. vyp.* **35** (95), 129–152 (1957).
99. A. N. Gordov. Temperature of an infinite cylinder due to flow with pulsating velocity and temperature, *Prikl. Mekh. Matem.* **19**, 240–243 (1955).

100. H. G. Elrod, New finite difference technique for solution of the heat conduction equation, especially near surfaces with convective heat transfer, *Trans. ASME* **79**, 1519–1526 (1957).
101. G. I. Pavlovsky and R. G. Akman, Determination of the heat transfer coefficient in vapour turbine startings *in* "Convective Heat Transfer," pp. 57–66. Izd. Naukova Dumka, Kiev, 1965.
102. S. Y. Berkovich, P. P. Golovastikov, and A. K. Chentsov, Calculation of unsteady heat transfer from a film to a bed, *Inzh. Fiz. Zh.* **6**, No. 5, 99–105 (1963).
103. A. W. Pratt and E. F. Ball, Transient cooling of a heated enclosure, *Intern. J. Heat Mass Transfer* **6**, No. 8, 703–708 (1963).
104. Y. L. Rozenshtok, Temperature field of an infinite plate in case of time dependence of the surrounding medium and heat transfer coefficient, *Inzh. Fiz. Zh.* **6**, No. 3, 45–50 (1963).
105. P. I. Khristichenko and S. I. Prokopets, Transient temperature field of an open cylindrical envelope, *Inzh. Fiz. Zh.* **7**, No. 11, 90–93 (1964).
106. I. A. Kuznetsov, Unsteady heat transfer in a pipeline, *Inzh. Fiz. Zh.* **7**, No. 11, 16–21 (1964).
107. N. Hayasi, K. Inouye, Transient heat transfer through a thin circular pipe due to unsteady flow in the pipe, *Trans. ASME Ser. C J. Heat Transfer* **87**, No. 4, 513–520 (1965).
108. Yu. V. Vidin, Study of asymmetric heating of a plate at variable heat transfer coefficients, *Izv. Akad. Nauk SSSR Energ. i Transp.* No. 3, 110–114 (1967).
109. N. A. Fridlender, Method of simultaneous simulation of heat transfer processes, *Inzh. Fiz. Zh.* **9**, No. 5, 577–582 (1965).
110. M. M. Sidlyar, Transient temperature field of an infinite cylinder at variable heat transfer coefficient, *Prikl. Mekh. Izd. Akad. Nauk USSR I vyp.* **7**, 11–13 (1965).
111. M. M. Sidlyar, Influence of time-variable heat transfer coefficient upon the temperature field of a plate, *Vestn. Kiev. Univers. Ser. Mat. i Mekh.* No. 7, 52–57 (1965).
112. B. M. Avkhimovich, Calculation of transient temperature fields in constructions by the method of moments, *in* "Metody Raschetov Temper. Polei i Teploizol. Letateln. Apparatov," pp. 3–34. Mashinostr, Moscow, 1966.
113. A. R. Voropaev, "Theory of Heat Transfer of Mine Air and Rocks in Deep Pits." Nedra, Moscow, 1966.
114. M. M. Ivashchenko and M. S. Zolotogorov, Solution of transient heat conduction problems on electric models, *Inzh. Fiz. Zh.* **11**, No. 2, 224–229 (1966).
115. M. M. Ivashchenko and M. S. Zolotogorov, Prescription of boundary conditions in solution of transient heat conduction problem for gas turbines, Sb. "Issled. i raschet vozdushn. okhazhd. gaz. turbin," Trudy Ts. KTI, vyp. 68, 30–38. Leningrad. 1966.
116. Y. V. Vidin, Heating of a cylindric body with inner heat generation at variable heat transfer problem, *Inzh. Fiz. Zh.* **11**, No. 2, 166–170 (1966).
117. H. V. Norden and I. Seppa, On transient heat conduction in the walls of a rectangular gas channel, *Nord tidskr. Inform. Bemande* **6**, No. 2, 144–154 (1966).
118. E. B. Karpin, A. G. Kostyuk, G. K. Zueva, L. V. Pirueva, and V. S. Sokolov, Calculation of transient temperature fields in states and envelopes on computers, *Teploenerget.* No. 3, 53–57 (1966).
119. B. N. Seliverstov, Selection of a mathematical model for transient heat transfer with a one-phase incompressible heat transfer agent, *Inzh. Fiz. Zh.* **11**, No. 4, 545–551 (1966).

120. Y. A. Levin and I. M. Prichodko, Temperature of a flat body with periodic variation of the heat transfer coefficient at the surrounding temperature, *Inzh. Fiz. Zh.* **10**, No. 2, 225–227 (1966).
121. V. V. Ivanov and V. V. Salomatov, Calculation of a temperature field in solids with variable heat transfer coefficient, *Inzh. Fiz. Zh.* **9**, No. 1, 83–84 (1965).
122. G. N. Dulnev, V. I. Cherkasov, and I. A. Yarishev, Thermal conditions of a thin plate heated by a pulse locallized power source, *Inzh. Fiz. Zh.* **11**, No. 3, 382–386 (1966).
123. L. I. Kudryashev, L. I. Zhemkov, V. S. Vekshin, and B. R. Belostotskii, An appromixate solution to the problem of transient heat transfer of an active element in an optic quantum generator, *Zh. Prikl. Spektroskop.* **4**, *vyp.* 1, 12–19 (1966).
124. Y. L. Rozenshtok, Temperature field of bodies under conditions of surrounding temperature and heat transfer coefficient variation with time, Sb. "Teplo i massoperenos," Vol. 6, 165–177. Izd. Nauka i Tekhnika, Minsk, 1966.
125. V. P. Kharitonov, Heat transfer in a gas flow through a fixed packing at variable gas temperature at the inlet, *Inzh. Fiz. Zh.* **12**, No. 2, 205–211 (1967).
126. I. T. Elperin and I. F. Pikus, Heating of particles participating in transient high-intense processes in the absence of temperature gradient, *Inzh. Fiz. Zh.* **12**, No. 1 (1967).
127. B. M. Avkhimovich, Calculation of transient heat conduction in multi-layer plates, *Izv. Vysshikh. Uchebn. Zavedenii Aviats. Tekhn.* **9**, No. 2, 3–8 (1966).
128. G. F. Kohlmayr, Exact maximum slopes for transient matrix heat transfer testing, *Intern. J. Heat Mass Transfer* **9**, No. 7, 671–680 (1966).
129. O. T. Ilchenko and V. E. Prokofiev, Accuracy of boundary drag prescription in grid solution of some transient heat conduction problems with the boundary condition of the 3rd kind, *Inzh. Fiz. Zh.* **12**, No. 6, 758–764 (1967).
130. A. Rivas, L'utilisation des ordinateurs pour la risolution des problèmes de thermique en régime transitoire, *Rev. Gin. Tharm.* **5**, No. 56, 761–769, 757, 759 (1966).
131. D. A. Pereverzev, Transient heat conduction in a belayer hollow infinite cylinder at boundary conditions of the 3rd kind, Sb. "Issled. po teploprovodn," pp. 364–371, Nauka i Tekhn., Minsk, 1961.
132. A. K. Galichskii and V. A. Enazen, Transient heat transfer between heat transfer agent and one-phase flow, *in* "Issled. po teploprovodn.," pp. 492–497. Izd. Nauka i Tekhnika, Minsk, 1961.
133. V. V. Salomatov and E. I. Goncharov, Thermal conditions of heat transfer agent at variable values of heat transfer coefficient and surrounding temperature, *Izv. Akad. Nauk SSSR Energet. i Transp.* No. 4, 99–103, 1967.
134. V. B. Fedorov, V. S. Egorov, and L. E. Fedorova, Calculation of thermal conditions of a periodic quantum generator, *Teplofiz. Vysokikh Temperatur* **5**, No. 5, 884–888 (1967).
135. A. M. Avizov, The simplest heat receivers under the conditions of arbitrary variation of surrounding temperature and heat transfer coefficient with time, *Teplofiz. vysokikh Temperatur* **5**, No. 4, 655–622 (1967).
136. O. N. Suetin, O. T. Ilchenko, and V. E. Prokofiev, Solution of a transient heat conduction problem on electric models at time-variable boundary conditions of the 3d kind, *Inzh. Fiz. Zh.* **14**, No. 6, 1038–1047 (1968).
137. V. K. Li-Orlov and V. N. Volkov, Theory of transient methods of measuring thermal parameters, *in* "Teplo-i massoperenos," vol. 7, pp. 332–339. Izd. Nauka i i tekhnika, Minsk, 1968.
138. N. I. Nazarov and V. K. Kuznetsov, Application of the Biot variational principle

to calculation of temperature fields in uniform and composite slabs at boundary conditions of the 3d kind, *Izv. Vysshikh Uchebn. Zavedenii Aviats. Tekhn.* No. 1, 124–129 (1968).
139. N. I. Nazarov, Calculation of temperature fields in thin wall constructions, *in* "Teplo-i massoperenos," Vol. 8, pp. 302–312. Izd. Nauka i Tekhnika, Minsk, 1968.
140. R. A. Ivanov, Transient temperature fields for heat transfer agent in a channel, *Inzh. Fiz. Zh.* **14**, No. 5, 832–838 (1968).
141. Y. V. Vidin, Transient heat conduction in laminated medium, *Inzh. Fiz. Zh.* **14**, No. 6, 1048–1055 (1968).
142. A. A. Vasiliev, Determination of the temperature field of a finned radiator tube, *Teplofiz. Vysokikh temperatur* **6**, No. 3, 493–497 (1968).
143. Y. G. Yaroshenko and F. R. Sklyar, Approximate solution to the problem of spherical particles heating by a heat source in a counter flow, *Teplofiz. Vysokikh Temperatur* **6**, No. 3, 474–481 (1968).
144. L. A. Brovkin, Method of solution of a heat conduction equation with coefficients depending on temperature, *in* "Teplo-i massoperenos," vol. 8, pp. 177–185. Izd. Nauka i Tekhnika, Minsk, 1968.
145. V. V. Salomatov and E. I. Goncharov, Temperature field of an infinite plate at variable values of the heat transfer coefficient and surrounding temperature, *Inzh. Fiz. Zh.* **14**, No. 4, 743–745 (1968).
146. V. V. Ivanov and V. V. Salomatov, Transient temperature field in solids at variable heat transfer coefficients, *Inzh. Fiz. Zh.* **11**, No. 2, 266–268 (1966).
147. V. V. Ivanov, Study of heat transfer under the conditions of non-linear heat conduction, *Izv. Akad Nauk SSSR Energet. i Trans.* No. 4, 138–139, 1966.
148. V. V. Ivanov, Thermal regime of a solid body at variable heat transfer conditions on its surface, *Izv. Vysshikh. Uchebn. Zavedenii. Aviats. Tekhn.* No. 1, (1967).
149. V. V. Ivanov, Calculation of non-linear heat conduction at a variable heat transfer coefficient, *Izv. Vysshikh Uchebn. Zavedenii Aviats. Tekhn.* No. 2, 1967.
150. V. V. Ivanov and V. V. Salomatov, Relations between temperatures in the processes of non-linear heat conduction, *Inzh. Fiz. Zh.* **13**, No. 2, 232–235 (1967).
151. A. V. Luikov, Heat Conduction Theory, "Vyssh. Shkola." Moscow, 1967.
152. G. T. Aldoshin, V. I. Zhuk and K. I. Shlyakntina, Conjugated problem of heat conduction for fluid flows in a channel, Sb. "Teplo-i massoperenos," Vol. 1, pp. 577–589. Energia, Moscow, 1968.
153. E. K. Kalinin, Transient convective heat transfer and hydrodynamics in channels, *Izv. Akad. Nauk BSSR Ser. Fiz-Tekhn. Nauk* No. 2, 77–86 (1967).
154. R. Siegel, Heat transfer for laminar flow in ducts with arbitrary time variation in wall temperature, *Trans. ASME Ser. E J. Appl. Mech.* **82**, No. 2, 241–249 (1960).
155. R. Siegel, Transient heat transfer for laminar slug in ducts, *Trans. ASME Ser. E. J. Appl. Mech.* **81**, No. 1, (1959).
156. E. M. Dubrovich, Transient heat transfer for laminar fluid flow in tubes *in* "Teoriya, Konstr. Raschet i Ispyt, Dvigat Vuytr. Sgoraniya," (Trudy Inst. Dvigatelei AN SSSR) Vol. 6, 51–60. Izd. Akad. Nauk SSSR, Moscow, 1962.
157. J. Bataill and M. Giat, Transfert de chaleur on de masse instationnaire dans une conduite cylindrique circulaire en régime laminaire, *Compt. Rend. Acad. Sci. Paris*, **t-261**, groupe 2, 1595–1598 (1965).
158. P. V. Tsoi, One method of studying transient heat transfer for a forced fluid flow in tubes of arbitrary section, *in* "Teplo-i massoperenos," Vol. 1, pp. 589–595. Izd. Energia, Moscow, 1968.

159. M. Perlmutter and R. Siegel, Two-dimensional unsteady incompressible laminar duct flow with a step change in wall temperature *Intern. J. Heat Mass Transfer* **3**, No. 2, 94–107 (1961).
160. M. Perlmutter and R. Siegel, Unsteady laminar flow in a duct with unsteady heat addition, *Trans. ASME, Sec. C. J. Heat Transfer* **83**, No. 4, 432–440 (1961).
161. E. M. Sparrow and R. Siegel, Thermal entrance region of a circular tube under transient heating-conditions, *Proc. US Natl. Congr. Appl. Mech. 3rd.* pp. 817–826. Brown University Providence, Rhode Island, (1958).
162. E. M. Sparrow and R. Siegel, Transient heat transfer for laminar forced convection in the thermal entrance region of flat ducts, *Trans. ASME Ser. C. J. Heat Transfer* **81**, No. 1, 29–36 (1959).
163. R. Siegel and M. Perlmutter, Laminar heat transfer in a channel with unsteady flow and wall heating varying with position and time, Paper ASME No. 62-WA-113 (1962); *Trans. Asme Ser. C J. Heat Transfer* **85**, No. 4, 358–365 (1963).
164. J. Delgado Domingos, Transmissao do calor num tubo em escoamento laminar e regime nao estacianario, *Technica* **40**, No. 352, 65–74 (1965).
165. N. I. Nikitenko, Numerical solution of the problem on a temperature field in moving media, *Izv. Vysshikh. Uchebn. Zavedenii Aviats. Tekhn.* No. 1, 26–32 (1963).
166. N. I. Nikitenko, Investigation of heat transfer phenomena in moving medium on digital automatic devices, Sb. "Konvektivn. Teploobmen," pp. 94–104. Naukova Dumka, Kiev, 1965.
167. V. M. Malkin, Transient heat transfer in laminar fluid flow, Sb. "Nauchn. Trudov VNII Metallurg. Teplotekhn.," No. 8, 106–118. Metallurgiz., Sverdlovsk, 1962.
168. K. Millsaps and K. Pohlhausen, Heat Transfer to Hagen-Poiseuile flows, *Proc. Conf. Differential Eq.* pp. 271–294. University of Maryland, (March 1955).
169. J. L. Hudson and S. G. Bankoff, An exact solution of unsteady heat transfer to a shear flow, *Chem. Eng. Sci.* **19**, No. 9, 591–598 (1964).
170. S. C. Chu and S. G. Bankoff, Unsteady transfer to slug flows. Effect of axial conduction, *A.I.Ch.E. J.* **11**, No. 4, 607–612 (1965).
171. Satya Prakash, An exact solution for the problem of unsteady temperature distribution in a viscous flow, *Proc. Natl. Inst. Sci. India* **A32**, No. 4, 360–367 (1966, 1967).
172. E. M. Sparrow and F. N. De Farias, Unsteady heat transfer to duct with time varying inlet temperature and participating walls, *Intern. J. Heat Mass. Transfer* **11**, No. 5, 837–853 (1968).
173. J. L. Hudson and S. G. Bankoff, Asymptotic solutions for the unsteady Graetz problem, *Intern. J. Heat Mass Transfer* **7**, No. 11, 1303–1307 (1964).
174. E. A. Sidorov, Account for unsteady-state conditions in convective heat transfer, *Teploenerget.* No. 4, 79–80 (1958).
175. E. A. Sidorov, Convective heat transfer at unsteady-state conditions, *Izv. Akad. Nauk SSSR Otd. Tekhn. Nauk* No. 9, 116–117 (1958).
176. E. A. Sidorov, Calculation of resistance and convective heat transfer at turbulent unsteady-state conditions, *Inzh. Fiz. Zh.* **2**, No. 11 86–91 (1959).
177. E. M. Sparrow and R. Siegel, Unsteady turbulent heat transfer in tubes, *Trans. ASME Ser. C J. Heat Transfer* **82**, No. 3, 170–180 (1960).
178. W. N. Gill, A note on unsteady forced convection heat transfer, *A.I.Ch.E. J.* **8**, No. 2, 284, 286–287 (1962).
179. V. V. Danenberg and V. I. Plyutinskii, An approximate method of solution of a two-dimensional transient heat transfer problem, "Trudy Vtsnii Kompleks. Avtomatiz.," vyp. **16**, pp. 229–236. Energia, Moscow, 1967.
180. V. V. Danenberg and V. I. Plyutinskii, Solution of transient heat transfer problem

in a turbulent flow, *in* "Trudy VTsNII Kompleks. Avtomatiz," Vol. 16, pp. 279–285. Izd. Energia, Moscow, 1967.
181. V. K. Koshkin, Y. I. Danilov, E. K. Kalinin, G. A. Dreitser, B. M. Galitseysky, and V. G. Izosimov, Unsteady heat transfer in tubes resulting from changes in heat flow, gas flow rate and acoustic resonance, *Proc. Intern. Heat Transfer Conf. 3rd Chicago, 1966*, Vol. 3, paper 87, pp. 57–70. AIChE, New York, 1966.
182. B. M. Galitseysky, G. A. Dreitser, V. G. Izosimov, E. K. Kalinin, and V. K. Koshkin, Experimental study of unsteady heat transfer in tubes with changing heat flux, *Izv. Akad. Nauk BSSR, Ser. Fiz-Tekh., Nauk* No. 2, 65–76 (1967).
183. B. M. Galitseysky, G. A. Dreitser, V. G. Izosimov, E. K. Kalinin, and V. K. Koshkin, Experimental study of unsteady heat transfer in tubes with changing gas flow rate, *Izv. Akad. Nauk BSSR Ser. Fiz-tekhn. Nauk* No. 2, 56–64 (1967).
184. B. M. Galitseysky, G. A. Dreitser, V. G. Izosimov, E. K. Kalinin, and V. K. Koshkin, Unsteady-state heat transfer with changing heat flux and gas flow rate, *Teplofiz. Vysokikh Temperatur* **5**, No. 5, 867–876 (1967).
185. I. S. Kochenov, U. N. Kuznetsov, and V. N. Panaev, Experimental study of unsteady-state convective heat transfer, *in* "Trudy 2 Vsesoyuzn. Sov. Teplo-i Massoobm.," pp. 115. Izd. Nauka i Tekhnika, Minsk, 1964.
186. R. Siegel, Forced convection in a channel with wall heat capacity and with wall heating variable with axial position and time, *Intern. J. Heat Mass Transfer* **6**, No. 7, 607–620 (1963).
187. D. E. Adams and B. Gebhart, Transient forced convection from a flat plate subjected to a step energy input, Paper ASME, NHT-27 (1963); *Trans. ASME Ser. C, J. Heat Transfer* **86**, No. 2, (1964).
188. P. L. Chambre, Theoretical analysis of the transient heat transfer into a fluid flowing over a flat plate containing internal heat sources, "L. M. K. Boelter Anniversary Volume." McGraw-Hill, New York, 1964.
189. M. Soliman and H. A. Johnson, Transient heat transfer for turbulent flow over flat plate of appreciable thermal capacity and containing time-dependent heat source, *Trans. ASME Ser. C J. Heat Transfer* **89**, No. 4, (1967).
190. H. A. Johnson and P. L. Chambre, Transient heat transfer for steady slug flow over a flat plate with uniform exponential heat generation, "L. M. K. Boelter Anniversary Volume." McGraw-Hill, New York, 1964.
191. M. Soliman, Analytical and experimental study of transient heat transfer for external forced convection, PhD Dissertation in Mechanical Engineering, Univ. of California, Berkeley (Jan. 1966); also SAN-1011 Reactor Technology, TID, 4500, 44th ed. (Jan. 1966).
192. M. M. Baida, N. G. Kondrashov, T. L. Perelman, and V. B. Ryvkin, Realization of numerical solution to one heat transfer problem, *Inzh. Fiz. Zh.* **15**, No. 6, 1047–1058 (1968).
193. G. A. Dreitser, V. G. Izosimov, and E. K. Kalinin, Generalization of experimental results of unsteady convective heat transfer with changing heat flux, *Teplofiz. Vysokikh Temperatur* **7**, No. 6 (1969).
194. E. M. Sparrow and J. L. Gregg, Non-steady surface temperature effects on forced convection heat transfer, *J. Aeron. Sci.* **24**, 776–777 (1957).
195. E. M. Sparrow, Combined effects of unsteady flight velocity and surface temperature on heat transfer, *Jet Propulsion* 403–405 (1958).
196. R. D. Cess, Heat transfer to laminar flow across a flat plate with a non-steady surface temperature, *Trans. ASME* **c. 83**, No. 3, 274–280 (1961).
197. T. R. Goodman, Effect of arbitrary non-steady wall temperature on incompressible

heat transfer, *Trans. ASME Sec. C J. Heat Transfer* **84**, No. 4, 347–351 (1962).
198. T. R. Goodman, Application of integral methods to transient non-linear heat transfer, "Advances in Heat Transfer," Vol. I, pp. 51–122. Academic Press, London and New York, 1964.
199. Y. L. Rozenshtok, Unsteady boundary layer on a semi-infinite plate in a viscous fluid flow, *in* "Teplo-i massoperenos," Vol. 1, pp. 277–287. Izd. Nauka i Tekhnika, Minsk, 1965.
200. M. Soliman and P. L. Chambre, On the transient Leveque problem, *Intern. J. Heat Mass Transfer* **10**. No. 2, 164–180 (1967).
201. G. N. Sarma, Unified theory for the solution of the unsteady thermal boundary layer equation, *Proc. Cambridge Phil. Soc.* **61**, No. 3, 809–825 (1965).
202. N. Riley, Unsteady heat transfer for flow over a flat plate, *J. Fluid Mech.* **17**, 97–104 (1963).
203. M. Soliman and H. A. Johnson, Transient heat transfer for forced convection flow over a flat plate of appreciable thermal capacity and containing an exponential time dependent heat source, *Intern. J. Heat Mass Transfer* **11**, No. 1, 27–38 (1968).
204. B. T. Chao and D. R. Jeng, Unsteady stagnation point heat transfer, *Trans. ASME Ser. C J. Heat Transfer* **87**, No. 2, 221–230 (1965); Paper ASME, NHT-2 (1964).
205. Y. M. Grigoriev, Unsteady conduction heat transfer of bodies in infinite medium, *Inzh. Fiz. Zh.* **10**, No. 4, 491–494 (1966).
206. J. A. Gusev, Velocity unsteadiness effect on heat transfer between cylinder and sphere, *Tr. KuAI* **23**, 18–19 (1966).
207. M. Bentwich, G. Szwarcbaum, and S. Sigeman, Time dependent temperature distribution in flow past a sphere, Paper ASME, NHT-38 (1965).
208. D. G. Drake and P. C. Rhodes, Fluctuating heat transfer from a sphere at small Reynolds numbers, *Z. Angew. Math. Phys.* **18**, 342–353 (1967).
209. A. Friedman, Generalized heat transfer between solids and gases under nonlinear boundary conditions, *J. Math. Mech.* **8**, No. 2, 161–183 (1959).
210. R. D. Cess and E. M. Sparrow, Unsteady heat transfer from rotating disk and at a stagnation point. *Intern. Develop. Heat Transfer* Part II, Sec. B, Boulder, Aug. 18-Sep. 1 (1961); Part II Sec. B, New York, ASME, pp. 468–474 (1961).
211. B. D. Katsnelson and F. A. Timofeeva, Study of convective heat transfer between particles and flow at transient conditions, *in* "Teploperedacha i Aerodinamika," 3th ed., Tr. TsKTI, Vol. 12, pp. 119–157, 1949.
212. M. G. Kryukova, Some problems of heat transfer between gas and solid particles, *Inzh. Fiz. Zh.* **1**, No. 4, 10–16 (1958).
213. J. Ya. Bitsyutko, V. K. Shshitnikov, G. V. Sadovnikov, and L. A. Sergeeva. Investigation of the cooling of spherical bodies in an air flow with low values of Bi., *in* "Issled. nestats Teplo-i Massoobmena," pp. 143–160. Izd. Nauka i Tekhnika, Minsk, 1966.
214. J. Ya. Bitsyutko, B. M. Smolsky, and V. K. Shchitnikov, The experimental study of convective cooling of a sphere in a turbulent air flow. *in* "Teplo-i massoperenos," Vol. 1, pp. 173–181. Izd. Energia, Moscow, 1968.
215. L. I. Kudryashev and A. A. Smirnov, The effect of thermal non-steadiness on the heat transfer coefficient in case of external flow around bodies. *Inzh. Fiz. Zh.* **4**, 10, 21–29 (1961).
216. L. I. Kudryashev and A. A. Smirnov. The effect of thermal non-steadiness on the heat output on case of external flow around bodies. *Tr. Ku AI* **12**, 59–76 (1961).
217. L. I. Kudryshev, A. A. Smirnov. Method of taking into account of the thermal

non-steadiness on the convective heat transfer coefficient in case of external flow around spherical bodies in the range of low Reynolds numbers, *in* "Teplo-i massoperenos," Vol. 1, pp. 298–305. Izd. Nauka i Tekhnika, Minsk, 1965.
218. J. N. Rutsky, On the non-stationary heat transfer of thermistor in a forced flow, *in* "Teploenergetichisk. izmeren. i Control," pp. 74–85. Izd. Nauka i Tekhnika, Minsk, 1967.
219. V. P. Popov, Study of local heat transfer on flat surface behind the ledge in supersonic air flow, *in* "Issled. Nestats. Teplo-i Massoobmena," 126–135. Izd. Nauka i Tekhnika, Minsk, 1966.
220. H. K. Ihrig, H. H. Korst, Quasi-steady aspects of the adjustment of separated flow regions to transient external flows. *AIAA J.* 1, 4 (1963).
221. A. L. Parnas, Experimental study of transient heat transfer for a cylinder in a transverse air flow, *in* "Voprosy Nestats. Perenosa Tepla i Massy," pp. 30–35. Izd. Nauka i Tekhnika, Minsk, 1965.
222. I. P. Pelepeichenko and D. F. Simbirsky, Instantaneous heat transfer of thin wires in transient gas flow, "Samoletostr. i Tekhnika vozdushn, Flota, Mezhvedomstv. Nauchno-tekhn. Sb." 5th ed., pp. 82–85. Kharkov, 1966.
223. B. M. Smolsky, V. P. Popov and N. A. Pokryvailo, Experimental study of transient mass transfer of flat plate in an incompressible fluid flow at various angles of attack, *in* "Teplo-i Massoperenos," Vol. 1, pp. 240–249. Izd. Energia, Moscow, 1968.
224. B. T. Chao and L. S. Cheema, Unsteady heat transfer in laminar boundary layer over a flat plate., *Intern. J. Heat Mass Transfer* 11, 1311–1324 (1968).
225. G. A. Ostroumov, Transient heat convection near horizontal cylinder, *Zh. Tekh. Fiz.* 26, vyp. 12, 2720–2730 (1956).
226. R. Siegel, Transient free convection from a vertical flat plate, *Trans. ASME* 80, 347–359 (1958).
227. R. J. Goldstein and E. R. G. Eckert, The steady and transient free convection boundary layer on a uniformly heated vertical plate, *Intern. J. Heat Mass Transfer* 1, 208–218 (1960).
228. E. M. Sparrow and J. L. Gregg, Nearly quasi-steady free convection heat transfer in gases, *Trans. ASME Sec. C J. Heat Transfer* 82, No. 3, 258–260 (1960).
229. K. T. Yang, Possible similarity solutions for laminar free convection on vertical plates and cylinders, *Trans. ASME Ser. E J. Appl. Mech.* 82, No. 2, 230–236 (1960).
230. R. C. L. Bosworth and C. M. Groden, Thermal transients associated with natural convection, *Australian J. Phys.* 13, No. 1, 73–83 (1960).
231. B. Gebhart, Transient natural convection from vertical elements, *Trans. ASME Ser. C J. Heat Transfer* 83, No. 1, 79–91 (1961).
232. P. M. Chung and A. D. Anderson, The transient laminar boundary layer with natural convection, *Trans. ASME Ser. C J. Heat Transfer* 83, No. 4 (1961).
233. E. R. Menold and K. T. Yang *Trans. ASME Ser. E J. Appl. Mech.* 84, No. 1 (1962).
234. J. D. Hellums and S. W. Churchill, Transient and steady state, free and natural convection, numerical solutions, *AIChE J.* 8, No. 5, 690–695, 719 (1962).
235. J. A. Schetz and K. Eichhorm, Unsteady natural convection in the vicinity of a doubly infinite vertical plate, Paper ASME, NWA-162 (1961); *Trans. ASME Sec. C J. Heat Transfer* 84, No. 4, 334–338 (1962).
236. S. L. Zeiberg and W. K. Mueller, Transient laminar flow in a duct with unsteady heat addition, *Trans. ASME Ser. C J. Heat Transfer* 84, No. 2, 141–148 (1962).
237. L. N. Tao, On unsteady heat transfer of combined free and forced convection in circular tubes, *Trans. ASME Ser. E J. Appl. Mech.* 85, No. 2 (1963).

238. J. N. Mason, Study of free convective heat transfer at extremely high Prandtl numbers from horizontal heated cylinders, Thesis, presented to the Fac. School Air Force Inst. Technol. Air Univ. NAD-41960, IV, p. 74. (Clearing House Fed. Sci. Techn. Inform.), 1963.
239. B. Gebhart, Transient natural convection for vertical elements for time dependent internal energy generation-appreciable thermal capacity, *Intern. J. Heat Mass Transfer* **6**, No. 11, 951–957 (1963).
240. B. Gebhart, Transient natural convection from vertical elements — appreciable thermal capacity, *Trans. ASME* **C 85**, No. 1, 10–14 (1963).
241. J. A. Schetz and R. Eichhorn, Natural convection with discontinuous wall temperature variations, *J. Fluid Mech.* **18**, No. 2, 167–176 (1964).
242. R. J. Goldstein and D. G. Briggs, Transient free convection about vertical plates and circular cylinders, *Trans. ASME Ser. C J. Heat Transfer* **86**, No. 4, 490–500 (1964).
243. W. T. Lawrence and J. C. Chato, Heat Transfer effects on the developing laminar flow inside vertical tubes, Paper ASME, NWA/HT-11, 99 (1965).
244. D. Forster, Onset of convection in a layer of fluid cooled from above, *Phys. Fluids* **8**, No. 10, 1770–1774 (1965).
245. B. C. Raychaudhari, Transient thermal response of enclosures: the integrated thermal time-constant, *Intern. J. Heat Mass Transfer* **8**, No. 11, 1439–1449 (1965).
246. G. A. Ostroumov, Free convection in closed cavities, A review of work carried out at Perm. USSR, *Intern. J. Heat Mass Transfer* **8**, 253–268 (1965).
247. E. V. Kudryavtsev and I. A. Turchin, Dependence of transient heat transfer on heat flux density, *Inzh. Fiz. Zh.* **10**, No. 5 (1966).
248. K. T. Yang, Remarks on transient laminar free convection along a vertical plate, *Intern. J. Heat Mass Transfer* **9**, No. 5, 511–513 (1966).
249. H. Z. Bakarat, Transient laminar free-convection heat and mass transfer in two-dimensional closed containers containing distributed heat source, Paper ASME, NWA/HT-28 (1965).
250. G. S. H. Lock and G. A. Glatz, Unsteady free convection from a large horizontal surface, *Proc. Intern. Heat Transfer Conf. 3rd Chicago*, Vol. 2 (1966); *Am. Inst. Chem. Eng.* **N. 4**, 205–213 (1966).
251. H. Z. Barakat and J. A. Clark, Analytical and experimental study of the transient laminar natural convection flows in partially filled liquid containers, *Proc. Intern. Heat Transfer Conf. 3rd Chicago* Vol. 2 (1966); *Am. Inst. Chem. Eng.* **N. 4**, 152–162 (1966).
252. J. W. Tatom and W. D. Carlson, Transient turbulent free convection in closed containers, *Proc. Intern. Heat Transfer Conf. 3rd Chicago* Vol. 2 (1966); *Am. Inst. Chem. Eng.* **N. 4**, 163–171 (1966).
253. F. Landis and H. Yanowitz, Transient natural convection in a narrow vertical cell, *Proc. Intern. Heat Transfer Conf. 3rd Chicago*, Vol. 2 (1966); *Am. Inst. Chem. Eng.* 139–151 (1966).
254. L. B. Evans and N. E. Stefany, An experimental study of transient heat transfer to liquids in cylindrical enclosures, *Chem. Eng. Progr. Symp. Ser.* **62**, No. 64. 209–215 (1966).
255. G. S. Ambrok, Initial period of natural convection development, in "Issled. nestats, teplo-i massoobmena," pp. 82–91. Izd. Nauka i tekhnika, Minsk, 1966.
256. B. C. Raychandhuri, S. P. Jain, C. L. Gupta, and M. L. Gupta, Heat transmission in insulated masonry structures under unsteady heat flow conditions, *Indian J. Technol.* **4**, No. 3, 86–93 (1966).

257. K. Aziz and J. D. Hellums, Numerical solution of the three-dimensional equations of motion for laminar natural convection. *Phys. Fluids* **10**, No. 2, 314–324 (1967).
258. B. Gebhart, R. P. Dring, and C. E. Polymeropoulus, Natural convection from vertical surfaces. The convection transient regime, *Trans. ASME Ser. C J. Heat Transfer* **89**, No. 1 (1967).
259. V. I. Yudovich, Stability of convection fluxes, *PMM* **31**, No. 2, 272–281 (1967).
260. V. I. Polezhaev, Numerical solution of the system of two-dimensional unsteady-state Navier-Stokes equations for compressible gas in closed envelope, *Izv. Akad. Nauk SSSR, Mech. Zhidk. i Gasa* No. 2, 103–111 (1967).
261. C. Y. King and W. W. Webb, Photoelastic observation of transient heat transfer across a solid-fluid boundary, *Trans. ASME Sec. C. J. Heat Transfer* **89**, No. 1, 65–68 (1967).
262. J. Madejski, Non-stationary natural convection heat transfer. Special solutions for vertical flat plates, *Arch. Mech. Stosowanej* **19**, No. 3, 421–431 (1967).
263. D. D. Joseph and L. N. Tao, Unsteady free and forced convection in vertical-annular and annular sector tubes, "Development Mechanics," Vol. 2, Part 1 pp. 403–420. Pergamon Press, Oxford, 1965.
264. B. Gebhart and R. P. Dring, The leading edge effect in transient natural convection from a vertical plate, *Trans. ASME* **B89**, No. 3, 274–275 (1967).
265. J. A. Clark, V. S. Arpaci, and K. M. Treadwell, Dynamic response of heat exchangers having internal heat sources. Part I -*Trans. ASME* **80**, No. 3, 612–623 (1958); Part II-*ibid* **80**, No. 3, 625–634 (1958); Part III-*Trans. ASME Ser. C J. Heat Transfer* **81**, No. 4, 253–266 (1959).
266. W. J. Yang, J. A. Clark, and V. S. Arpaci, Dynamic response of heat echangers having internal heat sources (part IV), *Trans. ASME Sec. C J. Heat Transfer* **83**, No. 3, 1961.
267. Masami Masubuchi, Dynamic response and control of malti-pass heat exchangers, *Trans. ASME Ser. D J. Basic Eng.* **82**, 51–65 (1960).
268. W. J. Yang, Dynamic response and resonance phenomenon of a single-solid single. fluid heat exchangers (Part II, Influence of the wall fluid heat-capacity ratio), *Trans. ASME* **27**, No. 183, 1892–1907 (1961).
269. W. J. Yang, Dynamic response and resonance phenomenon of a single-solid single-fluid heat exchangers (Part III, Transient response), *Trans. ASME* **28**, 188, 551–558 (1962).
270. W. J. Yang, Dynamic response of heat exchangers to the disturbance of wall temperature, Paper ASME No. 62-WA-27 (1962).
271. W. J. Yang, Frequency response of heat exchangers having sinusoidally space-dependent-internal heat generation, Paper ASME, NHT-21 (1962).
272. A. A. Armand, Calculation of transient processes in heat exchangers, *in* "Teploobmen pri vys. tepl. nagruzkakh i dr. spets. usl.," pp. 113-136. Izd. GEI, Moscow, 1959.
273. J. S. Turton, A method of evaluating transient response of gasto-gas heat exchangers, *J. Mech. Eng. Sci.* **2**, No. 4, 349–358 (1960).
274. H. G. Elrod, Improved lumped parameter method for transient heat conductions, *Trans. ASME Sec. C J. Heat Transfer* **82**, No. 3, 181–188 (1960).
275. G. D. Rabonovich, Problem on unsteady cooling of limited liquid volume, *Inzh. Fiz. Zh.* **4**, No. 3, 58–63 (1961).
276. G. D. Rabinovich, Transient heat transfer in a counter-flow recuperative exchangers, *Inzh. Fiz. Zh.* **4**, No. 2, 58–62 (1961).
277. A. Hempel, On the dynamics of steam-liquid heat exchangers, *Trans. ASME* **D83**, No. 2, 244–252 (1961).

278. A. A. Armand, Calculation of transient processes in heat exchangers with variable parameters of heat transfer agent, *in* "Povysh. Parametr. Para i Moshchn, Agregat. v Teploenerg.," pp. 479–493, Izd. GEI, Moscow, 1961.
279. H. H. Rosenbrock, The transient behaviour of distillation columns and heat exchangers, *Trans. Inst. Chem. Engrs. (London)* **40**, No. 6, 376–384 (1962).
280. A. L. Iskra, Calculation of heaters with forced motion of fluid, *Inzh. Zh.* **2**, vyp. 1, 17–28 (1962).
281. J. R. Hume and V. J. Skoglund, Theoretical solution of a transient fluid-cooled heat generator, *J. Aerospace Sci.* 1156–1163 (1962).
282. F. J. Stermole and M. A. Larsen, Dynamic response of heat exchangers to flow rate changes, *Ind. Eng. Chem. Fundament* **2**, No. 1, 62–67 (1963).
283. G. D. Rabinovich, Theory of thermal calculation of recuperative heat changers, *Izd. Akad. Nauk BSSR* Minsk, 1963.
284. F. J. Stermole and M. A. Larsen, The dynamics of flow-forced distributed parameter, *AIChE J.* **10**, No. 5, 688–694 (1964).
285. V. V. Danenberg, V. I. Plyutinskii, K. I. Proskuryakov, and Y. V. Kharitonov, Experimental device for the study of heat transfer dynamics, *Tr. VTsNII Kompleks avtomat*, **8**, 145–156 (1964).
286. P. Foraboschi, Su un problema di scambiotermics tra fluidi in controcorrente, *Nota I-Calore* **35**, No. 4, 158–166 (1964); *Nota II-Calore* **37**, No. 2, 57–62 (1966).
287. A. I. Mitskevich, Transient thermal process in a pipeline, *Tr. TsKTI Kotloturbostr.* **45**, 52–67 (1964).
288. V. I. Senkin, Transient process in a pipeline with heat generation in pipe wall, *Tr. TsKTI Kotloturbostr.* **45**, 68–84 (1964).
289. Cz. Graczyk, Zagadnienia dynamiki procesow termoenergetycznych, *Zeoz. nauk Politechn, slaskiej* No. 123, 141 (1964).
290. B. N. Devyatov, "Theory of Transient Processes in Technological Apparatuses." *Izd. Sib. Otd. Akad. Nauk SSSR*, 1964.
291. R. K. Thomasson, Frequency response of linear counter-flow heat echangers, *J. Mech. Eng. Sci.* **6**, No. 1, 13–24 (1964).
292. W. J. Yang, Transient heat transfer in a vapor heated heat exchanger with arbitrary timewise-variant flow perturbation, *Trans. ASME C***86**, No. 2, 133–142 (1964); Paper ASME, NHT-21 (1963).
293. R. Isermann, Das regeldynamische verhalten von uberhitzern, *Fortschr. Ber. VDI Z. Reihe* **6**, No. 4, 45 (1965).
294. I. C. Finlay and N. Dalgleish, Response of single-pass heat exchangers to disturbances in flow rate, *J. Mech. Eng. Sci.* **7**, No. 3, 318–327 (1965).
295. J. R. Gartner and H. L. Harrison, Dynamic characteristics of water to air crossflow heat exchangers, *Trans. Am. Soc. Heat Refrig. Air Conditioning Engr. Part I* **71**, 212–224 (1965).
296. B. G. Volik, Dynamic characteristics of pipeline, included into the contour controlling thermal processes, *Avtomat. i telemekhan.* **26**, No. 3, 539–544 (1965).
297. R. Y. Ladiev and Z. Y. Kozanevich, Dynamic characteristics of vapour liquid heat exchanger for account of unconstant parameters of liquid, *Kn. Khim. Mashinostr. Nauch. Tekhn. Sb.* No. 1, 120–123 (1965).
298. Z. Y. Kozanevich and R. Y. Ladiev, Transfer functions of parallel-counter-flow heat exchanger for account of unconstant parameters of liquid, *Avtomat. Khim. Proizvodstva, Kiev* **2**, 45–46 (1965).
299. Z. Y. Kozanevich, An approximation of heat exchanger dynamics on models with concentrated parameters, *Kn. Khim. Mashinostr., Kiev* **2**, 170–176 (1965).

Heat Transfer and Hydrodynamics in Channels 501

300. B. P. Korolkov, Transient characteristics of heat exchangers with independent heating, *Izv. Akad. Nauk SSSR Energet. i Transport* No. 3, 121–130 (1965).
301. P. S. Lall and R. I. Schoenhals, Dynamic analysis and experimental measurements for a single fluid heat exchanger, Paper ASME, NWA/HT-9, 3 (1965); *Trans. ASME Ser. C J. Heat Transfer* **88**, No. 1 (1966).
302. Sakai Kunio and Umeda Tosioki, Dynamic characteristics of heat generating energy units, *Toshiba Rev. Intern. Ed.* **20**, No. 6, 542–548 (1965).
303. A. A. Armand and V. V. Krasheninnikov, Dynamic characteristics of heat exchangers operating within near-critical region, *Teploenerget.* No. 1, 64–70 (1966).
304. I. C. Finlay, Determining the dynamic response of heat exchangers, in "Heat Transfer Survey" *Chem. Process. Engrs.* **47** 142–150 (1966).
305. H. Thal-Larsen and W. V. Loscutoff, Fluid center-line velocity effect on fluid temperature transient in a heat exchanger, *Trans. ASME Ser. D J. Basic Eng.* **88**, No. 2, 408–414 (1966).
306. T. J. Csermerly and L. E. Ostrander, Identification of the frequency response function of a heat exchanger by statistical correlation and spectral analysis using discrete interval binary noise input signal, *Proc. Intern. Heat Transfer Conf. 3rd Chicago*, AIChE, *1966* Vol. 1, pp. 149–158. 1966.
307. P. A. Andriyanov and M. M. Maslennikov, Method of analytical determination and analysis of dynamic parameters of full-time operating apparatuses, "Teplo-i massoperenos," Vol. 6, pp. 421–433. Naika i Tekhnika, Minsk, 1966.
308. B. P. Korolkov, Approximation of transcendented functions of heat exchangers with independent heating, *Teploenerget.* No. 7, 79–81 (1966).
309. B. P. Korolkov, One function often met in the studies of the dynamics of thermal units, *Izv. Akad. Nauk SSSR Energet. i transport* No. 3, 132–138 (1966).
310. E. D. Shestov and L. L. Poltav'seva, Calculation of dynamic characteristics of parallel -and counter-flow heat exchangers, *Tr. TsNIIKA* **16**, 201–228 (1967).
311. V. M. Rushchinsky, J. Y. Khvostova, and V. N. Tsyurik, Dynamic equations for boiler sections with one-phase medium, *Tr. TsNIIKA vyp* **16**, 140–200 (1967).
312. T. W. Larsen, Rapid calculation of temperature in a regenerative heat exchanger having arbitrary initial solid and entering fluid temperatures, *Intern. J. Heat Mass Transfer* **10**, No. 2, 149–168 (1967).
313. V. V. Krasheninnikov, Transient processes in boiling heat exchangers with arbitrary small disturbances, *Sb. Konferents. Molod. Spetsialistov VTI Doklady* 68–80 (1968).
314. G. E. Myers, J. W. Mitshell and R. F. Norman, The transient response of crossflow heat exchangers evaporators and condensers, *Trans. ASME Sec. C J. Heat Transfer* **89**, No. 1, 75–80 (1967); *Paper ASME NWA/HT* **34** (1967).
315. M. Y. Khait, Engineering method of calculating dynamic characteristics of casing-tube heaters, *Inzh. Fiz. Zh.* **12**, No. 6, 722–730 (1967).
316. E. P. Serov and B. P. Korolkov, "Dynamics of Processes in Heat and Mass Exchangers." Energia, Moscow, 1967.
317. V. V. Lanenberg and V. I. Plyutinsky, Calculation of the dynamics of a heat exchanger with one-phase incompressible heat transfer agent, in "Priborostr. sredstva avtomatiz. i sistemy upravl," pp. 275–291. Nauka, Moscow, 1967.
318. A. A. Shevyakov and R. V. Yakovleva, "Engineering Calculations of Heat Exchanger Dynamics." Mashinostr., Moscow, 1968.
319. I. V. Kotlyar and V. N. Ermolchik, Calculation of changes in regenerator parameters at transient operational conditions in gasturbine devices, *Teploenerget.* No. 8, 65–67 (1968).
320. E. Tyan-Tsi, Calculation of transient temperature fields in constructions with

account for thermal radiation at unsteady fligh conditions, *Izv. Vysshikh. Uchebn. Zavedenii Aviats. Tekhn.* No. 2 (1962).

321. A. L. Burka, Transient radiation-convection heat transfer on a rectangular, *Zh.P. MTF.* No. 5, 162–166 (1964).
322. A. L. Burka and N. A. Rubtsov, Transient heat transfer by radiation of two opaque finite bodies, *Inzh. Fiz. Zh.* 3, No. 5, 773–778 (1965).
323. L. P. Batov, Y. I. Karker, and M. G. Kogan, Transient radiation heat transfer in the system of coaxial cylinders, *Inzh. Fiz. Zh.* 11, No. 3, 289–295 (1966).
324. L. A. Glenn and S. Desoto. An analog method for the solution of unsteady radiant heat transfer problems with combined conduction and convection. *J. Spacecraft Rockets* 3, No. 2, 224–230 (1966).
325. A. L. Grosbie, R. Viscanta. Transient heating or cooling of one-dimentional solids by thermal radiation. *"Proc. Intern. Heat Transfer Conf. 3rd Chicago, 1966* Vol. 5, pp. 146–153. AIChE, New York, 1966.
326. Yu. V. Vidin. Transient temperature field in a plate under conditions of the combined action of thermal radiation and convection. *Inzh. Fiz. Zh.* 12, No. 5 (1967).
327. A. L. Burka, Transient radiation heating of a cylindrical body, *Zh. PMTF* No. 2, 76–80 (1968).
328. G. P. Boikov, Laws of thermal radiation regime in solids for radiation heat transfer, *Teplofiz. vysokikh. Temperatur* 6, No. 3, 482–486 (1968).
329. A. L. Grosbie and R. Viscanta, Transient heating or cooling of a plate by combined convection and radiation, *Intern. J. Heat Mass Transfer* 11, No. 2, 305–317 (1968).
330. R. Viscanta and P. S. Bathla, Unsteady energy transfer in a layer of gray gas by thermal radiation, *Z. Angew. Math. Phys.* 18, No. 3, 353–367 (1967).
331. K. K. Vasilevsky, Transient heat and mass transfer in a semi-infinite capillary-porous body at boundary conditions of the first kind, *Inzh. Fiz. Zh.* 7, No. 4 (1964).
332. F. T. Hung and R. G. Navins, Unsteady-state heat transfer with a flowing fluid through porous solids, ASME, NHT-10, 1965.
333. A. A. Aleksashenko and V. A. Aleksashenko, Approximate method of solution of heat and mass transfer problems, *in* "Teplo-i massoperenos," vol. 8, pp. 108–114. Izd. Nauka i Tekhn., Minsk, 1968.
334. V. N. Kharchenko, Heat transfer inside porous material under transient conditions, *Inzh. Fiz. Zh.* 15, No. 1, 149–152 (1962).
335. J. W. Elder, Transient convection in a porous medium, *J. Fluid Mech.* 27, No. 3, 609–623, 1967.
336. U. N. Kuznetsov, Unsteady heat Transfer investigation, presented at All Union *Heat Transfer Confer. Young Scientists Minsk, January 1969.*
337. U. N. Kuznetsov, Solution of unsteady Grats problem, *Teplofiz. Vysokikh Temperatur* 7, No. 4, 697–705 (1969).

Heat Transfer and Friction in Turbulent Pipe Flow with Variable Physical Properties

B. S. PETUKHOV

High Temperature Institute, Academy of Science of the USSR, Moscow, USSR

I.	Introduction	504
II.	Analytical Method	507
	A. Basic Equations	507
	B. Eddy Diffusivities of Heat and Momentum	510
	C. Analytical Expressions for Temperature and Velocity Profiles, Heat Transfer, and Skin Friction.	516
III.	Heat Transfer with Constant Physical Properties.	521
	A. Analytical Results	521
	B. Experimental Data	525
IV.	Heat Transfer and Skin Friction for Liquids with Variable Viscosity.	528
	A. Theoretical Results.	528
	B. Experimental Data and Empirical Equations	530
V.	Heat Transfer and Skin Friction for Gases with Variable Physical Properties	533
	A. Analytical Results	533
	B. Experimental Data and Empirical Equations	540
VI.	Heat Transfer and Skin Friction for Single-Phase Fluids at Subcritical States	543
	A. Analytical Results	543
	B. Experimental Data and Empirical Equations for Normal Heat Transfer Regimes	550
	C. Experimental Data for Regimes with Diminished and Enhanced Heat Transfer	555
VII.	Conclusion	560
	Nomenclature	561
	References	561

I. Introduction

Heat transfer in turbulent pipe flow has been investigated for almost 60 years. Nusselt's paper published in 1910 was probably the first one analyzing this problem on a scientific basis (*1*). In this paper devoted to the heat transfer of turbulent gas flow in tubes, the similarity method was originally used for the correlation of experimental data on heat transfer. This is the reason for the continued interest in Nusselt's paper.

During subsequent years different investigators performed numerous experimental studies on heat transfer processes in turbulent pipe flow for various fluids including liquid metals. As a result they formulated relations for the Nusselt number versus the Reynolds and Prandtl numbers for a wide range of Re and Pr. Reynolds (*2*) was the first who theoretically studied heat transfer in turbulent pipe flow. The relationship obtained between heat flux and wall shear stress, known as the Reynolds analogy, is valid only for $Pr = 1$. Some investigators have improved upon Reynolds analysis. For example, Taylor (*3*) and Prandtl (*4, 5*) took into account approximately the influence of fluid flow peculiarities at the wall on heat transfer, assuming the flow to consist of a turbulent core and viscous (laminar) sublayer. Karman (*6*) improved this model by the introduction of an intermediate layer between a laminar sublayer and a turbulent core. The expressions for heat transfer obtained by Karman and Prandtl are true for constant physical properties over the range 0.7 to 10–20 for Pr. The last restriction concerns the fact that they neglected turbulent heat transfer in a viscous sublayer (this leads to essential errors when Pr is large) and heat transfer by conduction in the turbulent core (this is not true for low Pr numbers). Further development of analytical methods for heat transfer in a turbulent pipe flow with constant physical properties was achieved when investigators digressed from the above assumptions and began to use more accurate relationships for the distributions of velocity and eddy diffusivities of heat and momentum along the pipe cross section (*7–14*). For example, Lyon (*8*) obtained an expression for the Nusselt number in the case of constant heat flux at the wall. This expression predicts the heat transfer rate, if the distributions of velocity and turbulent diffusivity of heat are known. The use of more accurate relationships for distributions of velocity and eddy diffusivities of heat and momentum require the application of numerical methods. Numerical calculations of heat transfer in turbulent pipe flow for constant physical properties were carried out in (*8–11, 13, 14*). The results of these predictions covering a wide range of Re and Pr numbers as a rule are in good agreement with experimental data.

Thus, nowadays the problem of heat transfer in a turbulent quasi-steady fluid flow with constant physical properties in circular tubes has been rather fully investigated.

In reality, fluid physical properties depend on temperature. That is why heat transfer relations obtained with the assumption of constant physical properties can only be used in practice either at small temperature differences in a flow or with physical properties changing slightly in the temperature range considered. In this case the effect of changing physical properties can be approximately accounted for by choosing the properties at a certain average fluid temperature.

In heat transfer systems used in different fields of engineering, large temperature drops and high heat fluxes are often realized. In this case the large temperature gradients occur in a fluid flow. For example, in nuclear reactors the heat flux (K cal./m² hr) may be as high as several millions. In cooling systems for jet propulsion engines it may rise to several score of millions while a heat flux of hundreds of millions may occur in some special kinds of apparatus. Liquids and gases whose physical properties are very responsive to temperature changes are often used as heat transfer fluids. Gases flowing at large temperature differences or some liquids (single-phase fluid) at subcritical states serve as examples. In these cases it is impossible to consider physical properties constant, because great errors would otherwise result. Under such conditions the analysis of the flow and heat transfer should include the dependence of physical properties on temperature.

For various types of fluids and for a given fluid the variation of the physical properties with temperature and pressure is not the same over different ranges of the state parameters. For such a fluid under these varying conditions it is presently impossible to describe the fluid flow and heat transfer by a single relationship valid for all conditions. As a consequence, the problem of a fluid flow and heat transfer with variable physical properties divides into several problems, and each problem corresponds to a certain type of dependence of physical properties on temperature and pressure. Therefore, the analytical expressions for fluids with constant physical properties are not universal in the case of variable physical properties. Theoretical studies of flow and heat transfer in fluids with variable physical properties are hindered by different mathematical and physical difficulties. The mathematical difficulties can be explained by the fact that the momentum and energy equations in the case of variable physical properties are coupled and nonlinear. However, these difficulties can be overcome, e.g., by using numerical methods and with the help of computers. The difficulties of a physical nature are more serious. They may be attributed to the inability to prescribe analytical expressions for the turbulent diffusivities of heat and momentum for fluids possessing variable physical properties. These expressions have been more or less studied for fluids with constant physical properties only; consequently, in the case of variable properties we have only a few theories which have not been verified experimentally. That is why analytical solutions of heat transfer in a turbulent flow with variable physical properties are not so accurate, e.g., in the case of a laminar flow. They must be verified by comparison with experimental data.

Experimental study of fluid flow and heat transfer for variable physical properties is also a very difficult problem because experiments at high temperatures, large heat fluxes, and high pressures are not easy to perform. Other aspects of the problem are the difficulties of interpretation and correlation of experimental data, because heat transfer and skin friction at variable physical properties depend on many parameters. That is why the dynamics and the heat transfer of turbulent flow with variable physical properties have not been studied in full. Nevertheless during the last 10–15 years striking progress has been made in this field.

Several theoretical papers are devoted to the fluid mechanics and heat transfer of turbulent pipe flow with variable physical properties.

Kutateladze (15–17) studied a gas flow at large temperature differences and small subsonic velocities. He obtained a correlation between the friction factor and the temperature ratio parameter (the wall temperature to the bulk gas temperature ratio) for the limiting case of Re $\to \infty$. And for a gas Pr ≈ 1, it was assumed that the same relationship was valid for the heat transfer coefficient. These relationships are also approximately valid for the finite values of Re.

Deissler (10, 11, 18) and Goldmann (19) have developed methods for the calculation of heat transfer and skin friction for an incompressible fluid with an arbitrary temperature dependence of its properties. The essence of their methods is in the simultaneous numerical integration of the energy and momentum equations formulated on the assumption that heat flux and shear stress are constant (or changing linearly) along the pipe radius. The methods of Deissler and Goldmann differ with respect to the calculation of the eddy diffusivities of heat and momentum for constant and variable fluid properties. Deissler performed calculations of heat transfer and skin friction for some gases and a liquid; Deissler and Goldmann performed these calculations for water above the critical point.

Petukhov and Popov (14) have developed another method of calculating the heat transfer and skin friction for an incompressible fluid with an arbitrary temperature dependence of its properties. Analytical expressions for both heat transfer and skin friction and also for velocity and temperature profiles obtained from the energy and momentum equations are the basis of this method. With the help of these expressions we can calculate heat transfer and friction using the method of successive approximations. Later on this method was used for calculation of heat transfer and friction of some gases with and without dissociation, and also for carbon dioxide at supercritical state parameters.

The fluid mechanics and heat transfer of turbulent pipe flow with variable physical properties have been studied both theoretically and experimentally. Published papers are available which contain many experimental results. In particular they present data obtained on heat transfer and skin friction for liquids under the conditions of substantial changes in viscosity, for some gases at large temperature differences, and for water, carbon dioxide, and some other substances at supercritical states. Some of the papers contain empirical equations correlating the experimental results.

In this paper we shall consider heat transfer and skin friction in turbulent pipe flow with variable physical properties. The constant properties solution will be considered only so far as is necessary for the flow and heat transfer analysis with variable physical properties.

II. Analytical Method

A. Basic Equations

Turbulent flow is, of its nature, transient. Velocity, temperature, and other properties change continuously in time at every point of a turbulent flow. These changes are irregular fluctuations with respect to some temporal mean. This behavior allows us to represent different turbulent flow properties as the sum of the mean value, in time, and a pulsation of this value. So we can describe the field of real (instantaneous) velocities as a field of averaged (in time) velocities and the superimposed field of velocity fluctuations. We can do the same with temperature, pressure, and density fields and with other dependent variables. With this approach, transfer processes in a turbulent flow are controlled by two mechanisms: molecular and convective (turbulent). The first mechanism results in the appearance of viscous stresses proportional to the gradients of the averaged velocity and heat fluxes due to heat conduction which are proportional to the averaged temperature gradients. The second mechanism gives rise to turbulent stresses caused by momentum transfer due to velocity fluctuations and turbulent heat fluxes caused by heat transfer resulting from velocity and temperature fluctuations. This approach suggested by Reynolds allows us to pass from energy, momentum, and continuity equations for instantaneous values to the corresponding equations for the averaged values. Hence, the solution of turbulent flows is reduced to the analysis of the averaged equations in combination with analytical expressions for turbulent diffusivities arising from some physically motivated assumptions in accordance with experimental data.

Furthermore, we shall consider a quasi-steady[1] axisymmetric turbulent flow of an incompressible fluid with variable physical properties. We shall restrict our problem to the analysis of the fluid flow and heat transfer in circular pipes far from the entrance, i.e., in that region where thermal and velocity boundary layers coincide. At present it is possible to analyze this problem only approximately. Therefore we shall make the following assumptions:

1. Flow velocities are not large, so the energy dissipation can be neglected.
2. The effect of body forces is small in comparison with that of viscosity and inertia forces.
3. Physical properties change weakly over the range of temperature fluctuations (i.e., from T to $T + T'$, where T is the averaged temperature value, T' is the temperature fluctuation); therefore, the physical properties at a given point can be considered constant and equal to the physical properties at the averaged temperature for this point.
4. Change of heat flux along the axis caused by thermal conductivity and turbulent diffusivity is small compared with its change along the radius.
5. Change of normal stresses (viscous and turbulent) along the coordinate axes is small in comparison with the change in the shear stresses.

The averaged energy, momentum, and continuity equations for the conditions we have formulated can be written in a cylindrical coordinate system:

$$\rho \left(W_x \frac{\partial h}{\partial x} + W_r \frac{\partial h}{\partial r} \right) = \frac{1}{r} \frac{\partial}{\partial r} \left[r \left(\lambda \frac{\partial T}{\partial r} - \rho \overline{W_r' h'} \right) \right] \tag{1}$$

$$\rho \left(W_x \frac{\partial W_x}{\partial x} + W_r \frac{\partial W_x}{\partial r} \right) = -\frac{\partial P}{\partial x} + \frac{1}{r} \frac{\partial}{\partial r} \left\{ r \left[\mu \left(\frac{\partial W_x}{\partial r} + \frac{\partial W_r}{\partial x} \right) - \rho \overline{W_x' W_r'} \right] \right\} \tag{2}$$

$$\left(W_x \frac{\partial W_r}{\partial x} + W_r \frac{\partial W_r}{\partial r} \right) = -\frac{\partial P}{\partial r} + \frac{\partial}{\partial x} \left[\mu \left(\frac{\partial W_r}{\partial x} + \frac{\partial W_x}{\partial r} \right) - \rho \overline{W_x' W_r'} \right] \tag{3}$$

$$\frac{\partial (\rho W_x)}{\partial x} + \frac{\partial (\rho W_r)}{\partial r} + \frac{\rho W_r}{r} = 0 \tag{4}$$

[1] A turbulent flow in which averaged properties do not change with time.

where x and r are the axial and radial coordinates (x coincides with the pipe axis); W_x and W_r are the averaged (in time) values of the axial and radial components of the velocity vector, respectively; ρ, h, T, and P are the averaged density, enthalpy, temperature, and pressure; W_x', W_r', h', and T' are the fluctuations of velocity (in the axial and radial directions), enthalpy, and temperature, respectively; λ and μ are the thermal conductivity and dynamic viscosity at temperature T and pressure P.

Since fully developed flow is being considered, it can be assumed that the change in the axial component of the mass velocity along the axis x is small, i.e.,

$$\partial(\rho W_x)/\partial x \approx 0$$

Therefore, from Eq. (4), the radial velocity component $W_r = 0$. In addition, if we assume that the change of viscous and turbulent shear stresses along the x axis is small, i.e.,

$$\frac{\partial}{\partial x}\left(\mu \frac{\partial W_x}{\partial r} - \rho \overline{W_x' W_r'}\right) \approx 0$$

then it is seen from Eq. (3) that $\partial P/\partial r = 0$, i.e., pressure P is constant over the cross section.

With these assumptions the set of Eqs. (1)–(4) is reduced to the following two equations:

$$\rho W_x \frac{\partial h}{\partial x} = \frac{1}{r}\frac{\partial}{\partial r}\left[r\left(\lambda \frac{\partial T}{\partial r} - \rho \overline{W_r' h'}\right)\right] \tag{5}$$

$$\rho W_x \frac{\partial W_x}{\partial x} = -\frac{dP}{dx} + \frac{1}{r}\frac{\partial}{\partial r}\left[r\left(\mu \frac{\partial W_x}{\partial r} - \rho \overline{W_x' W_r'}\right)\right] \tag{6}$$

On the right-hand side of Eq. (5), in parentheses, the expression for the heat flux is given by

$$q = \lambda(\partial T/\partial r) - \rho \overline{W_r' h'} \tag{7}$$

The first term of this sum is the heat flux due to conductivity, while the second item is accounted for by the eddy diffusivity of heat.

On the right-hand side of Eq. (6), the expression for shear stress is given by

$$\sigma = -[\mu(\partial W_x/\partial r) - \rho \overline{W_x' W_r'}] \tag{8}$$

Here the first term is the viscous stress and the second term is the turbulent stress.

In theoretical investiagations the assumption is often made that the heat flux and shear stress vary linearly with respect to r, i.e.,

$$q = q_w R \quad \text{and} \quad \sigma = \sigma_w R \tag{9}$$

or it is supposed that these quantities are constant along the radius

$$q = q_w \quad \text{and} \quad \sigma = \sigma_w \tag{10}$$

Here q_w and σ_w are the heat flux and the shear stress at the pipe wall, respectively; $R = r/r_0$ is the dimensionless radius; r_0 is the pipe radius.

We can see that the first assumption from (9) is fulfilled only in the case of slug[2] flow, and the second assumption only for fully developed flow with constant physical properties. As for assumption (10), it is not fulfilled for pipe flow. However, in a number of cases assumption (9) and even (10) do not introduce great errors into our analysis. This may be attributed to the fact that in calculations of heat transfer and skin friction a correct description of the flow near the wall is the most essential, and in this region assumptions (9) and even (10) are fulfilled approximately. Quantitative error estimations are given in Section III, p. 521 (see Table I, p. 522).

Assumptions (9) and (10) allow us to replace Eqs. (5) and (6) by simpler ones which can be obtained from Eqs. (7) and (8) as a result of the substitution of q and σ from (9) and (10).

For calculations with the help of Eqs. (5) and (6) it is necessary to express the turbulent heat flux $\rho \overline{W_r' h'}$ and the turbulent shear stress $\rho \overline{W_x' W_r'}$ as functions of the independent variables and the averaged flow properties. Then it is convenient to introduce the coefficients of the eddy diffusivities of heat and momentum.

B. Eddy Diffusivities of Heat and Momentum

By the definition, the eddy diffusivity of heat is

$$\epsilon_q = \frac{-\overline{W_r' h'}}{\partial h / \partial r} \tag{11}$$

and the eddy diffusivity of momentum is

$$\epsilon_\sigma = \frac{-\overline{W_x' W_r'}}{\partial W_x / \partial r} \tag{12}$$

---

[2] A slug flow is a flow with a uniform (over the pipe cross section) velocity profile.

If in Eqs. (7) and (8) we express $\overline{W_r'h'}$ and $\overline{W_x'W_r'}$ in terms of ϵ_q and ϵ_σ and consider that $\partial h/\partial r = C_p\,\partial T/\partial r$ (as pressure is constant over the cross section), then the expressions for heat flux and shear stress become

$$q = (\lambda + \rho C_p \epsilon_q)\, \partial T/\partial r \tag{13}$$

$$\sigma = -(\mu + \rho \epsilon_\sigma)\, \partial W_x/\partial r \tag{14}$$

Though we have introduced eddy diffusivities of heat and momentum it is still very difficult to determine $\rho\overline{W_r'h'}$ and $\rho\overline{W_x'W_r'}$. These difficulties remain in the determination of ϵ_q and ϵ_σ which should be considered as unknown functions of the independent variables and the averaged flow properties.

We might determine the eddy diffusivities of heat and momentum from semi-empirical turbulence theories such as Prandtl's mixing length theory or Karman's local similarity theory. But the relations arising from the semi-empirical turbulence theory do not give the correct description of the eddy diffusivities of heat and momentum near the wall and near the pipe axis. More reliable information on the eddy diffusivities of heat and momentum may be obtained on the basis of experimental data correlations in light of the semiempirical turbulence theory. In the case of fully developed flow with constant physical properties, as is seen from Eq. (6), the shear stress changes linearly along the radius, i.e., $\sigma = \sigma_w R$. Substituting this relation into (14) and transforming to universal coordinates while nondimensionalizing, we obtain

$$\frac{\epsilon_\sigma}{\nu} = \frac{1 - \eta/\eta_0}{d\varphi/d\eta} - 1 \tag{15}$$

where $\varphi = W_x/v^*$ is the dimensionless velocity, $\eta = v^*y/\nu$ is the universal independent variable, $v^* = (\sigma_w/\rho)^{1/2}$ is the so-called friction velocity, $y = (r_0 - r)$ is the coordinate reference point at the wall, and $\eta_0 = v^*r_0/\nu$.

Thus, for the experimental determination of the eddy diffusivity of momentum ϵ_σ we need to measure only the velocity distribution along the tube section.

Many investigators have measured velocity profiles in circular and plane tubes in a turbulent isothermal flow. The surveys of these investigations can be found in a number of books and articles (20–23). The results of measurement show that the dimensionless velocity φ and the dimensionless eddy diffusivity of momentum ϵ_σ/ν are continuous functions of η, and near the axis they are also functions of the radius R (or of the Reynolds number)[3]:

$$\varphi = \varphi(\eta, R) = \varphi(\eta, \text{Re})$$
$$\epsilon_\sigma/\nu = (\epsilon_\sigma/\nu)(\eta, R) = (\epsilon_\sigma/\nu)(\eta, \text{Re})$$

For $\eta \to 0$, $\varphi = \eta$ and $\epsilon_\sigma \to 0$. At small η, $\epsilon_\sigma \sim \eta^m$, where $m \geqslant 3$. Within a range of $\eta > 30$ and $R > 0.85$, Prandtl's logarithmic law of velocity distribution is valid:

$$\varphi = (1/\kappa) \ln \eta + A \tag{16}$$

(κ and A are constant) and consequently $\epsilon_\sigma \sim \eta$. Finally, when $\eta > 30$ and $R < 0.85$, φ and ϵ_σ depend both on η and R; when $R \approx 0.5$, ϵ_σ passes through a maximum and tends to some constant quantity while approaching the axis.

Some authors suggested empirical and semi-empirical relations for the eddy diffusivity of momentum for an isothermal pipe flow. These relationships take into account the above-mentioned aspects of the variation in ϵ_σ. We shall discuss only those which will be of use in further investigation. In the range $0 \leqslant \eta \leqslant 26$ Deissler (*11, 24*) suggested the equation

$$\epsilon_\sigma/\nu = n^2 \varphi \eta [1 - \exp(-n^2 \varphi \eta)] \tag{17}$$

where $n = 0.124$. For $\eta > 26$ he recommends using Karman's equation obtained with the assumption of local similarity of the velocity field:

$$\frac{\epsilon_\sigma}{\nu} = \kappa^2 \frac{(d\varphi/d\eta)^3}{(d^2\varphi/d\eta^2)^2} \tag{18}$$

where $\kappa = 0.36$

Deissler has noted that the velocity profile in a turbulent core calculated from (18) for ϵ_σ/ν is described by logarithmic relationships (16) when $\kappa = 0.36$. Therefore we can approximate ξ_σ/ν with the following equation instead of (18):

$$\epsilon_\sigma/\nu = 0.36(1 - \eta/\eta_0)\eta - 1 \tag{19}$$

which is obtained from the substitution of (16) into (15).

Reichardt (*7*) has suggested two equations:

---

[3] We should notice that $R = r/r_0 = 1 - \eta/\eta_0$, and $\eta_0 = v^* r_0/\nu = \frac{1}{2} \text{Re}(\xi/8)^{1/2}$ where $\xi = \xi(\text{Re})$ is the friction factor.

For $0 \leq \eta \leq 50$:
$$\epsilon_\sigma/\nu = \kappa[\eta - \eta_n \tanh(\eta/\eta_n)] \tag{20}$$

For $\eta > 50$:
$$\epsilon_\sigma/\nu = (\kappa/3)\,\eta(0.5 + R^2)(1 + R) \tag{21}$$

where $\kappa = 0.4$ and $\eta_n = 11$.

With large Prandtl numbers the main temperature change occurs directly in the vicinity of the wall. In this case, for the calculation of heat transfer, it is very important to describe turbulent transfer processes near the wall correctly. We shall analyze Eqs. (17) and (20) from this point of view. At small η Eqs. (17) and (20) can be simplified by series expansions of the exponential function and hyperbolic tangent, and by considering only the first two terms of the series. In addition, we can take $\varphi = \eta$ in (17) which results in

$$\epsilon_\sigma/\nu = c_1 \eta^4 \tag{17a}$$

and

$$\epsilon_\sigma/\nu = c_2 \eta^3 \tag{20a}$$

where $c_1 = n^4 = 2.365 \times 10^{-4}$ and $c_2 = \kappa/3\eta_n^2 = 1.102 \times 10^{-3}$.

Thus, at small η, according to Deissler $\epsilon_\sigma \sim \eta^4$, and according to Reichardt $\epsilon_\sigma \sim \eta^3$; also Eq. (17a) gives lower ϵ_σ values than those predicted by Eq. (20a).

In the literature there is no consensus on the value of the exponent m in the equation for the region close to the wall $\tilde{\epsilon}_\sigma = c\eta^m$ (here $\tilde{\epsilon}_\sigma = \epsilon_\sigma/\nu$). We can only say from theoretical considerations that $m \geq 3$.[4] Being inaccurate, measurements near the wall do not produce a reliable value for m. We can indirectly infer m values from comparisons of the predicted values of heat and mass transfer with experimental data at large Pr or Sc. Such a comparison was made in (25) on the basis of a statistical analysis of the experimental data on heat and mass transfer which shows that m ranges from 3 to 3.2. Apparently, this is close to reality.

Thus, Eq. (20) which gives $m = 3$ is more likely to describe the mechanisms of turbulent transfer near the wall than Eq. (17). Equations (20) and (21) have some other advantages over Eqs. (17)–(19). The relationship for ϵ_σ/ν, suggested by Reichardt, has no discontinuities in the range $\eta < 50$ which is important for the calculation of heat

---

[4] We can show by the continuity equation that the function $\tilde{\epsilon}_\sigma(\eta)$ and its first two derivatives go to zero when $\eta = 0$ and thus $m \geq 3$ [see (21)].

transfer and skin friction. Besides, Eq. (21) takes into account changes in ϵ_σ/ν with R (and with Re) in the central part of the tube, and on the tube axis it gives a nonzero value of ϵ_σ which varies with Re that is in full agreement with experimental data. This means that the velocity gradient calculated from Eq. (21) is zero along the tube axis. Therefore, we shall use mainly Eqs. (20) and (21). In Fig. 1, ϵ_σ/ν versus η is plotted for

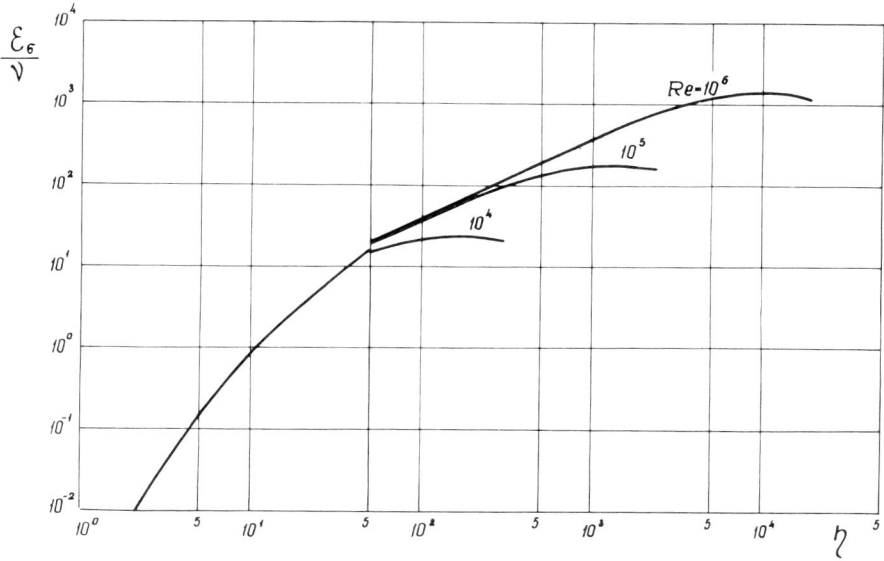

FIG. 1. ϵ_σ/ν vs η and Re according to Reichardt's data.

various Re numbers as calculated from Eqs. (20) and (21). Nevertheless, it should be noted that, with small Pr numbers, calculations of heat transfer (in the case of constant physical properties) using Eqs. (17) and (18) and (20) and (21) give similar results (see Section III). The problem of heat diffusivity in a turbulent flow has been investigated less than that of momentum diffusivity. If we proceed from Reynolds' concept that the turbulent diffusivities of heat and momentum are identical, we should take the ratio of the eddy diffusivities $\beta \equiv \epsilon_q/\epsilon_\sigma = 1$. As a matter of fact, Prandtl's mixing length theory gives the same result. Reichardt (7) made an assumption that the β value near the wall equals unity, and it increases as the distance from the wall increases.

Having measured velocity and temperature profiles in a flow, and using Eq. (5) and relation (11), we can determine the eddy diffusivity of heat ϵ_q from this experimental data. Unfortunately such measurements are not numerous. Besides, the accuracy of experimental values of ϵ_q is rather poor (especially in the wall region). The majority of measurements was carried out with liquid metals and seldom with air. In this paper we shall not discuss the problems of eddy diffusivity and heat transfer for liquid metals. The measurements with air (26–29) show that at some small distance from the wall $\beta = 1.2$–1.5, but it decreases when the distance from the wall and Re increase. However, some other workers (30) report that β is approximately unity near the wall and it increases with the distance from the wall. Lack of experimental data has as yet prevented us for determining the general relations for ϵ_q or β over a wide range of Re and Pr.

Presently, in predicting heat transfer, β is usually taken as one (at least for Pr $\gtrsim 1$) due to the ambiguity of estimating β. This is justified since the results at constant properties predicted with such an assumption are in good agreement with experimental data.

If the physical properties change with temperature, it becomes necessary to take into account the influence of variable physical properties on the turbulent diffusivity expressions. As is known, this problem has not yet been systematically investigated. Therefore, a solution is usually based on some assumptions, and their validity can be confirmed only indirectly by comparing predicted values of heat transfer and skin friction with experimental data.

Deissler (11, 24) assumed that, to calculate the eddy diffusivity of momentum with variable physical properties, relations (17) and (18) may be used although they were obtained at constant physical properties. In these relations the kinematic viscosity is considered a variable. In accordance with such an assumption the equation for ϵ_σ/ν takes the form

$$\frac{\epsilon_\sigma}{\nu} = n^2 \varphi_w \eta_w \frac{\nu_w}{\nu} \left[1 - \exp\left(-n^2 \varphi_w \eta_w \frac{\nu_w}{\nu}\right) \right] \qquad (22)$$

$$\frac{\epsilon_\sigma}{\nu} = \kappa^2 \frac{\nu_w}{\nu} \frac{(d\varphi_w/d\eta_w)^3}{(d^2\varphi_w/d\eta_w^2)^2} \qquad (23)$$

where

$$\varphi_w = W_x/v_w^*, \qquad \eta_w = v_w^* y/\nu_w, \qquad v_w^* = (\sigma_w/\rho_w)^{1/2}$$

Constants n and κ have the same values as in the case of constant physical properties. Equation (22) is recommended when $\eta_w < 26$, and Eq. (23) is used when $\eta_w > 26$.

Goldmann (19) suggested a method of calculating ϵ_σ with variable physical properties using the hypothesis that local turbulence characteristics at a given point depend on physical properties at that point and do not depend on physical properties changing in the vicinity of that point. In light of this hypothesis, Goldmann has come to the conclusion that the velocity distribution with variable physical properties is described by the same relations between the generalized variables φ^+ and η^+ as those between the variables φ and η with constant physical properties. The generalized variables are described by

$$\varphi^+ = \int_0^{W_x} \frac{dW_x}{(\sigma_w/\rho)^{1/2}} \quad \text{and} \quad \eta^+ = \int_0^y \frac{1}{\nu}\left(\frac{\sigma_w}{\rho}\right)^{1/2} dy \tag{24}$$

If ρ and ν are constants, then $\varphi^+ = \varphi$ and $\eta^+ = \eta$.

Transforming in Eqs. (15), from variables φ and η to φ^+ and η^+ we can see that for the determination of ϵ_σ/ν with variable properties we can use the same relations as in the case of constant properties provided φ and η are substituted for φ^+ and η^+, respectively.

The comparison of predicted heat transfer results for a gas at high T_w/T_β values with experimental data shows that the calculation method for ϵ_σ with variable properties suggested by Goldmann produces better agreement with experimental heat transfer values than Deissler's method.

The ratio of the diffusivities of heat and momentum at variable properties, as in the case of constant properties, is usually taken to be one ($\beta = 1$).

C. Analytical Expressions for Temperature and Velocity Profiles, Heat Transfer, and Skin Friction

Consider the problem of turbulent fully developed quasi-steady flow and heat transfer in a circular tube, assuming that the fluid is incompressible and its physical properties display some arbitrary temperature dependence. The problem is analyzed for the case of constant heat flux which is prescribed at the wall ($q_w = $ const) (14).

If the assumptions of Section II, A are applied, this problem may be described by the energy and momentum equations (5) and (6). The left-hand side of Eq. (6) can be written as

$$\rho W_x \frac{\partial W_x}{\partial x} = \frac{\partial(\rho W_x^2)}{\partial x} - W_x \frac{\partial(\rho W_x)}{\partial x} \approx \frac{\partial(\rho W_x^2)}{\partial x}$$

As earlier, $\partial(\rho W_x)/\partial x \approx 0$ was assumed (see Section II, A). Taking all this into account and also using relationships (13) and (14) we obtain the equations

$$\rho W_x \frac{\partial h}{\partial x} = \frac{1}{r}\frac{\partial}{\partial r}(rq) \tag{25}$$

$$\frac{\partial}{\partial x}(P + \rho W_x^2) = -\frac{1}{r}\frac{\partial}{\partial r}(r\sigma) \tag{26}$$

where

$$q = (\lambda + \rho C_p \epsilon_q) \, \partial T/\partial r \tag{13}$$

$$\sigma = -(\mu + \rho \epsilon_\sigma) \, \partial W_x/\partial r \tag{14}$$

In order to transform Eqs. (25) and (26) to ordinary differential equations, we shall make the additional assumption that the derivatives with respect to x on the left-hand side of Eqs. (25) and (26) are constant across the tube cross section, i.e.,

$$\partial h/\partial x = f_1(x) \tag{27}$$

$$\partial (P + \rho W_x^2) \, \partial x = f_2(x) \tag{28}$$

With variable physical properties and especially with variable heat capacity and constant heat flux at the wall, assumption (27) holds to a greater degree than the usual assumptions of linear change of q along the radius or of uniform (over the section) longitudinal temperature gradient $[\partial T/\partial x = f(x)]$.

For liquids and gases flowing at small subsonic velocities, the pressure gradients due to longitudinal density changes $\partial/\partial x(\rho W_x^2)$ are, by far, smaller than the total pressure gradient dP/dx. Since P does not change along the tube section assumption (28) is also well founded under these conditions.

First, analytical expressions must be found for the enthalpy and temperature fields and the Nusselt number.

Multiplying Eq. (25) by $r \, dr$, taking into account assumption (27), and integrating with respect to the radius from 0 to r_0, we obtain

$$\partial h/\partial x = 2q_w/\overline{\rho W_x} r_0 \tag{29}$$

where $\overline{\rho W_x}$ is the (over the section) bulk velocity. Substituting (29) into (25) and integrating from 0 to r we obtain the expression for the heat flux distribution along the radius

$$\frac{q}{q_w} = \frac{2}{R} \int_0^R \frac{\rho W_x}{\overline{\rho W_x}} R \, dR \tag{30}$$

where $R = r/r_0$.

Since the pressure P is uniform over the tube section, we have

$$\partial h/\partial r = C_p \, \partial T/\partial r \tag{31}$$

Using this relationship, we can write Eq. (13) in the form

$$q = (\lambda/C_p)(1 + \epsilon_q/a) \, \partial h/\partial r$$

or

$$q = (\lambda/C_p)(1 + \beta \Pr \epsilon_\sigma/\nu)\, \partial h/\partial r \tag{32}$$

where $\beta = \epsilon_q/\epsilon_\sigma$. Simultaneous solution of (30) and (32) gives

$$\frac{\partial h}{\partial r} = q_w\, d \int_0^R \frac{\rho W_x}{\overline{\rho W_x}} R\, dR \bigg/ \frac{\lambda}{C_p}\left(1 + \beta \Pr \frac{\epsilon_\sigma}{\nu}\right) R \tag{33}$$

where $d = 2r_0$ is the tube diameter.

Having integrated (33) from R to 1, we obtain an enthalpy distribution equation

$$h_w - h = q_w\, d \int_R^1 \left\{ \int_0^R \frac{\rho W_x}{\overline{\rho W_x}} R\, dR \bigg/ \frac{\lambda}{C_p}\left(1 + \beta \Pr \frac{\epsilon_\sigma}{\nu}\right) R \right\} dR \tag{34}$$

where h_w is the enthalpy at the wall.

Solving (30) and (13) simultaneously and integrating from R to 1, we find the analogous temperature distribution equation

$$T_w - T = \frac{q_w\, d}{\lambda_w} \int_R^1 \left\{ \int_0^R \frac{\rho W_x}{\overline{\rho W_x}} R\, dR \bigg/ \frac{\lambda}{\lambda_w}\left(1 + \beta \Pr \frac{\epsilon_\sigma}{\nu}\right) R \right\} dR \tag{35}$$

Now we shall calculate the bulk enthalpy (h_b), or, to be more exact, the enthalpy difference $h_w - h_b$. By definition,

$$h_w - h_b = 2 \int_0^1 (h_w - h) \frac{\rho W_x}{\overline{\rho W_x}} R\, dR \tag{36}$$

Substituting $h_w - h$ from (34) into (36) and integrating by parts, we obtain

$$h_w - h_b = 2 q_w\, d \int_0^1 \left\{ \left(\int_0^R \frac{\rho W_x}{\overline{\rho W_x}} R\, dR \right)^2 \bigg/ \frac{\lambda}{C_p}\left(1 + \beta \Pr \frac{\epsilon_\sigma}{\nu}\right) \right\} R\, dR \tag{37}$$

Now let us introduce the heat transfer coefficient from the definition

$$\alpha = \frac{q_w}{T_w - T_b} = \frac{q_w \bar{C}_p}{h_w - h_b} \tag{38}$$

and the Nusselt number

$$\mathrm{Nu}_w = \frac{\alpha\, d}{\lambda_w} = \frac{q_w \bar{C}_p\, d}{(h_w - h_b)\, \lambda_w} \tag{39}$$

where T_w and T_b are the wall temperature and the bulk temperature, respectively; λ_w is the value at temperature T_w; $\bar{C}_p$ is the average specific heat of the fluid within the temperature range T_b to T_w,

$$\bar{C}_p = \frac{h_w - h_b}{T_w - T_b} = \frac{1}{T_w - T_b} \int_{T_b}^{T_w} C_p \, dT \tag{40}$$

Substituting $h_w - h_b$ from (37) into (39), we obtain the expression for the Nusselt number

$$\frac{1}{\mathrm{Nu}_w} = 2 \frac{C_{pw}}{\bar{C}_p} \int_0^1 \left\{ \left(\int_0^R \frac{\rho W_x}{\overline{\rho W_x}} R \, dR \right)^2 \Big/ \frac{\lambda}{\lambda_w} \frac{C_{pw}}{\bar{C}_p} \left(1 + \beta \Pr \frac{\epsilon_\sigma}{\nu} \right) R \right\} dR \tag{41}$$

For constant physical properties, we can reduce Eq. (41) to the well-known Lyon integral (8)

$$\frac{1}{\mathrm{Nu}} = 2 \int_0^1 \left\{ \left(\int_0^R \frac{W_x}{\overline{W_x}} R \, dR \right)^2 \Big/ \left(1 + \beta \Pr \frac{\epsilon_\sigma}{\nu} \right) R \right\} dR \tag{42}$$

Now we shall deduce analytical expressions for the velocity profile and the friction factor.

Having multiplied Eq. (26) by $r \, dr$ and taking into account assumption (28), we integrate first with respect to r from 0 to r, and second from 0 to r_0. Dividing the first of the obtained expressions by the second we get

$$\sigma = \sigma_w R$$

This result is the consequence of assumption (28) and it essentially means that we approximated the real shear stress distribution (along the radius) by a linear one. Having substituted σ into (14) and integrated from R to 1 we get the velocity distribution equation

$$W_x = \frac{\sigma_w r_0}{\mu_w} \int_R^1 \frac{R}{(\mu/\mu_w)(1 + \epsilon_\sigma/\nu)} dR \tag{43}$$

where μ_w is the dynamic viscosity at temperature T_w.

The bulk velocity over the section is

$$\overline{\rho W_x} = 2 \int_0^1 \rho W_x R \, dR$$

After having substituted W_x from (43) we obtain

$$\overline{\rho W_x} = \frac{\sigma_w d\rho_w}{\mu_w} \int_0^1 \frac{\rho}{\rho_w} \left(\int_R^1 \frac{R}{(\mu/\mu_w)(1 + \epsilon_\sigma/\nu)} dR \right) R \, dR \tag{44}$$

where ρ_w is the density at temperature T_w.

By definition, the friction factor is

$$\xi_w = 8\sigma_w \rho_w / (\overline{\rho W_x})^2 \tag{45}$$

Substituting $\overline{\rho W_x}$ from (44) into (45), we obtain the expression

$$\xi_w = \frac{8}{\text{Re}_w} \left\{ \int_0^1 \frac{\rho}{\rho_w} \left(\int_R^1 \frac{R}{(\mu/\mu_w)(1 + \{\epsilon_\sigma/\nu\})} dR \right) R\, dR \right\}^{-1} \tag{46}$$

where

$$\text{Re}_w = \overline{\rho W_x}\, d/\mu_w$$

If the physical properties of the fluid are constant, expression (46) takes the form

$$\xi = \frac{8}{\text{Re}} \left\{ \int_0^1 \left(\int_R^1 \frac{R}{1 + \epsilon_\sigma/\nu} dR \right) R\, dR \right\}^{-1} \tag{47}$$

where

$$\text{Re} = \overline{\rho W}\, d/\mu$$

Using the equations derived and the successive approximation method, we can calculate the Nusselt number and friction factor for a fluid with variable physical properties. For convenience of calculation in Eqs. (35), (41), (43), and (46) we should transform from the dimensionless radius R to the universal coordinate η_w. The terms R and η_w are connected by the following relation:

$$R = 1 - \eta_w/\eta_{0w}$$

where

$$\eta_w = v_w^* y/\nu_w, \qquad \eta_{0w} = v_w^* r_0/\nu_w, \qquad v_w^* = (\sigma_w/\rho_w)^{1/2}$$

The procedure used to calculate is as follows:

1. Values of η_{0w}, $q_w d$, T_w, and P are prescribed.

2. Having chosen one of the relationships for ϵ_σ/ν, the first approximation of the velocity and temperature profiles is calculated from Eqs. (43) and (35). Physical properties of the fluid are assumed constant and equal to their values at the wall temperature.

3. The distribution of the physical properties over the pipe section is calculated from the obtained (in the first approximation) temperature profile. Then the same equations determine ϵ_σ/ν and the velocity and temperature profiles for the second approximation, physical properties variation being taken into account. Then, in the same way, the temperature profile is again calculated for the third approximation and all

higher approximations. The calculation is performed till the difference in temperature distribution of the $(n+1)$th and nth approximations becomes smaller than some prescribed value within the range of which we can neglect the change of the physical properties.

4. ϵ_σ/ν and the velocity distribution are calculated from the physical properties distribution obtained in the last approximation. Then from Eq. (41) we can find the value

$$\mathrm{Nu_w}/\bar{C}_p = q_w\,d/(h_w - h_b)$$

Since $q_w d$ and T_w are prescribed, then from the last relation we can find the average bulk enthalpy h_b and the appropriate bulk temperature T_b. After that the number $\mathrm{Nu_w} = q_w d/\lambda(T_w - T_b)$ is determined. Using Eq. (46) we can calculate the friction factor ξ_w.

The procedure is simplified in the case of constant physical properties, where the distribution of ϵ_σ/ν and a velocity profile are calculated, and Nu and ξ obtained from Eqs. (42) and (47).

Naturally, numerical calculation demands the use of computers.

III. Heat Transfer with Constant Physical Properties

A. Analytical Results

Consider the heat transfer solution for the case of a fully developed flow with constant properties in a circular tube with constant heat flux at the wall. The calculation has been done by Petukhov and Popov (14) by the method discussed in Section II, C. The eddy diffusivity of momentum was calculated from Reichardt's Eqs. (20) and (21); $\beta = \epsilon_q/\epsilon_\sigma$ was taken to be one. The calculation was done for $\mathrm{Re} \sim 10^4$–5×10^6 and $\mathrm{Pr} \sim 0$–2000. The method of calculation takes into account the variation of the heat flux q and shear stress σ along the radius. In order to estimate the errors which could appear with the assumption of uniform q and σ along the radius, several values of Nu and ξ were calculated for $q = q_w$ and $\sigma = \sigma_w$. The ratios of the corresponding values of Nu and ξ for varying q and σ to these values for $q = q_w$ and $\sigma = \sigma_w$ are tabulated and presented in Table I. As can be seen from the table, the assumption of uniform q and σ produces noticeable errors in Nu and ξ values, especially for low Re and Pr.

Nu vs Pr for various Re, according to the predicted values, is plotted in Fig. 2 (for $\mathrm{Pr} \geqslant 0.5$). The calculations over the range Re and

TABLE I

Ratio of Nu and ξ Calculated Taking into Account Changes of q and σ along the Radius to Their Values when $q = q_w$ and $\sigma = \sigma_w$.

Pr \ Re	10^4	10^5	10^6
	\multicolumn{3}{c}{$Nu/Nu_{q=q_w,\sigma=\sigma_w}$}		
1	1.14	1.11	1.10
10	1.09	1.08	1.07
100	1.08	1.06	1.05
1000	1.07	1.06	1.04
	\multicolumn{3}{c}{$\xi/\xi_{\sigma=\sigma_w}$}		
	1.17	1.12	1.10

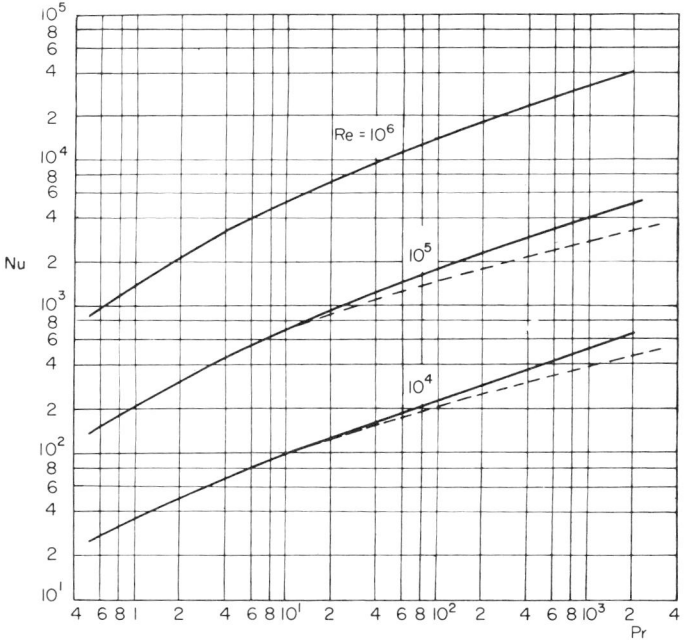

Fig. 2. Nu vs Re and Pr from predictions by Petukhov and Popov (solid lines) and Deissler (dotted lines).

Pr ~ 10^4–5×10^6 and 0.5–2000, respectively, are described by the interpolation equation

$$\mathrm{Nu} = \frac{(\xi/8)\,\mathrm{Re}\,\mathrm{Pr}}{K_1(\xi) + K_2(\mathrm{Pr})(\xi/8)^{1/2}\,(\mathrm{Pr}^{2/3} - 1)} \tag{48}$$

where

$$\xi = (1.82 \log \mathrm{Re} - 1.64)^{-2} \tag{49}$$

$$K_1(\xi) = 1 + 3.4\xi, \qquad K_2(\mathrm{Pr}) = 11.7 + 1.8\,\mathrm{Pr}^{-1/3}$$

The disagreement of the predicted Nu with Eq. (48) is within 1% except for the ranges $5 \times 10^5 < \mathrm{Re} < 5 \times 10^6$ and $200 < \mathrm{Pr} < 2000$ where it is 1–2%.

If in Eq. (48) K_1 and K_2 are taken constant and equal to 1.07 and 12.7, respectively, the equation becomes simpler:

$$\mathrm{Nu} = \frac{(\xi/8)\,\mathrm{Re}\,\mathrm{Pr}}{1.07 + 12.7(\xi/8)^{1/2}\,(\mathrm{Pr}^{2/3} - 1)} \tag{50}$$

This equation suggested in paper (13) describes the predicted results with an accuracy of 5–6% over a range of 10^4–5×10^6 for Re and 0.5–200 for Pr, and with 10% accuracy for $0.5 < \mathrm{Pr} < 2000$ and the same range of Re.

With very large Pr number an analytical expression for Nu (or Sh)[5] may be obtained. In this case the integrand in Eq. (42) decreases rapidly when the distance from the wall increases and it becomes negligibly small at low $\eta = v^*y/\nu$. Expression (20a) for ϵ_o/ν is true for low η. Substituting ϵ_o/ν from (20a) into Eq. (42), assuming $R \approx 1$ in the integrand expression, and making a change of variables from R to η, we obtain

$$\mathrm{Nu} = \frac{1}{2\eta_0} \int_0^1 \frac{1}{1 + c_2\,\mathrm{Pr}\,\eta^3}\,d\eta$$

Performing the indicated integration yields

$$\mathrm{Nu} = \frac{3\sqrt{3}}{\pi}\,(c_2\,\mathrm{Pr})^{1/3}\,\eta_0$$

or

$$\mathrm{Nu} = K(\xi/8)^{1/2}\,\mathrm{Re}\,\mathrm{Pr}^{1/3} \tag{51}$$

---

[5] In calculation of mass transfer instead of Nu and Pr numbers, their diffusional analogies Sherwood (Sh) and Schmidt (Sc) numbers are used.

where

$$K = \frac{3\sqrt{3}}{2\pi} c_2^{1/3} = 0.0855$$

At $\Pr \geqslant 100$, Eq. (51) agrees with the results of more accurate numerical calculations to within an error of 10% and within 2% for $\Pr \geqslant 1000$. For small Prandtl numbers Eq. (51) produces overestimated Nusselt numbers which increase with decreasing Pr and increasing Re.

It is of interest to compare the results of this analysis with similar calculations of other authors. Deissler (11, 24) calculated the heat transfer for constant properties over the ranges $4 \times 10^3 < \text{Re} < 2 \times 10^5$ and $0.73 < \Pr < 3000$ by using Eqs. (13) and (14) and assuming $q = q_w$ and $\sigma = \sigma_w$. From Eqs. (17) and (18) he calculated an eddy diffusivity assuming $\beta = 1$. Sparrow et al. (31) calculated the heat transfer over the range $10^4 < \text{Re} < 5 \times 10^5$ and $0.7 < \Pr < 150$ by directly solving the energy equation. Here they used Deissler's expression for the velocity profile and eddy diffusivities, Eqs. (17) and (19), assuming $\beta = 1$. Deissler and Sparrow et al. produced similar results (it is natural since actually they used the same expressions for ϵ_a/ν). Therefore in Fig. 2 the results of the analysis discussed earlier are compared only with the results of Deissler obtained over a wider range of Prandtl numbers.

With $\Pr \leqslant 10\text{--}20$ the results of the earlier analysis are in good agreement with the results of Deissler. However, for higher Prandl number the results disagree. When $\Pr = 100$ the Nusselt number according to Deissler is 15% lower and when $\Pr = 1000$ it is 25% lower its value calculated by Eq. (48). For very large Pr (or Sc) [Pr (or Sc) ~ 200] Deissler recommends the equation

$$\text{Nu} = K_D (\xi/8)^{1/2} \text{Re} \Pr^{1/4} \tag{52}$$

where

$$K_D = 2\sqrt{2}\,n/\pi = 0.112$$

was obtained with the assumption that the eddy diffusivity of momentum ϵ_a/ν is described by Eq. (17a).

Unlike Eq. (51), the expression (52) gives a weaker dependence of Nu on Pr. That is why at very large Prandtl numbers (or Sc) Nu (or Sh) calculated from (52) appears to be lower than that calculated from Eq. (51) (at $\text{Sc} = 10^5$ it is lower by approximately a factor of two). Only by comparing the predicted results with the experimental data can we solve the problem to the extent that the predicted results correspond to reality. Before the discussion of the experimental data we shall note a very important fact.

The predicted results listed in this paragraph refer to the case of heat transfer with constant heat flux at the wall (q_w = const, Nu = Nu_q). If heat transfer occurs at a constant wall temperature (t_w = const, Nu = Nu_t) the relation of Nu with Re and Pr differs from that for the case when q_w = const. However, theoretical analysis carried out in (9, 32, 33) reveals the following: the difference in Nu_t and Nu_q (when Re and Pr are the same) takes place only when Pr numbers are small, i.e., mainly for liquid metals. When both Pr and Re increase this difference decreases rapidly. When Pr = 0.7 and Re = 10^4, Nu_q is already only 4% greater than Nu_t. When Re increases to 10^5 this difference decreases to $\sim$2%, and at Pr = 10 and $R \geqslant 10^5$ the difference is less than 1%.[6] Thus, when Pr $\geqslant 0.7$ and Re $\geqslant 10^4$ the results of the heat transfer calculations both at q_w = const and at t_w = const are valid.

B. Experimental Data

A great number of experimental papers on heat transfer in turbulent pipe flow have been published. Unfortunately, in many cases measurement accuracy was not high; therefore, heat transfer coefficients obtained experimentally often contain substantial errors which are difficult to estimate. Little experimental data of rather high accuracy have been reported in recent years (34–37). Mainly heat transfer for air and water flow has been measured, i.e., approximately over a range of 0.7–10 for Prandtl numbers. Only a few authors have obtained heat transfer data at Pr from 10 to 100–150 and a little higher. Heat transfer measurements were not performed for Pr $\sim$ 1000 because of the great experimental difficulties. Therefore mass transfer experimental data were used over the range of Pr (Sc to be more exact) for which the exact analogy between heat and mass transfer processes was valid.

For comparison with the predicted results the most reliable experimental data have been chosen. The main characteristics of the data are given in Table II. The heat transfer experimental data were extrapolated to the zero wall and flow temperature difference to avoid the affect of the dependence of the fluid physical properties upon the temperature. In some cases such extrapolation was performed rather accurately, while in other cases only approximately. Naturally the mass transfer experimental data need not be extrapolated because all the measurements were performed under isothermal conditions.

In Fig. 3 the predicted results described by Eq. (48) are compared with

---

[6] All the listed results are for the case of fully developed Nu number. The difference between Nu_q and Nu_t can be larger at the thermal entrance region.

TABLE II

THE MAIN CHARACTERISTICS OF HEAT AND MASS TRANSFER EXPERIMENTAL DATA

Ref.	Symbols	Fluid	l_0/d [a]	l/d [a]	$Re \cdot 10^{-3}$	Pr or Sc
Volkov and Ivanova (38)	⊖	Air	0	48–370	12.5–350	~0.7
Petukhov and Roizen (39)	⊕	Air	40	39	15–280	~0.7
Sukomel and Tsvetkov	◐ ◑	Air, helium	50	80	9–40	0.67–0.71
Allen and Eckert (34)	●	Water	96	30	13–110	8
Dipprey and Sabersky (37)	▽	Water	48.5	46.2	150	1.2–5.9
Yakovlev (36)	△	Water	~4	70–80	19–140	2–12
Malina and Sparrow (35)	○	Water, oil	96	30	12–100	3, 48, 75
Sterman and Petukhov (40)	□	Monoisopropyldiphenyl	0, 30	89–125	22–260	12–35
Hamilton (41)[b]	⊕	Water and water solutions of glycerine and metaxyl	75	13	10–100	430–10⁵

[a] l_0/d and l/d are relative lengths of calming nonheated and heated sections.

[b] In Hamilton (41) an experimental tube wall made of benzoic acid served as a solid phase.

the experimental data. The ordinate represents the ratio of the experimental Nusselt numbers to Nu_0 predicted by Eq. (48) both at the same Re and Pr. The abscissa represents Re or Pr. As seen in the figure, experimental and predicted data are in good agreement. The divergence

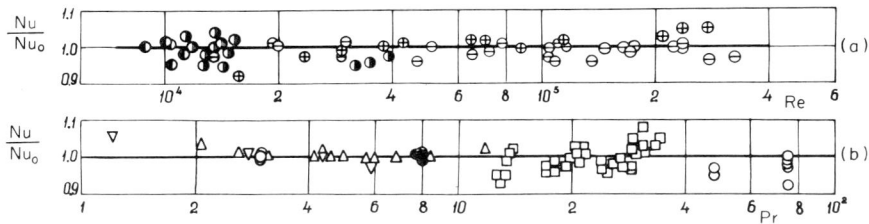

FIG. 3. Experimental Nusselt numbers (Nu), predicted Nusselt numbers (Nu_0): (a) air, (b) liquids (for symbols see Table II).

of the experimental data from predictions does not exceed 5–6% (except for a few points), this being within the range of accuracy for both predicted and experimental data.

In Fig. 4 the predicted results are compared with the mass transfer

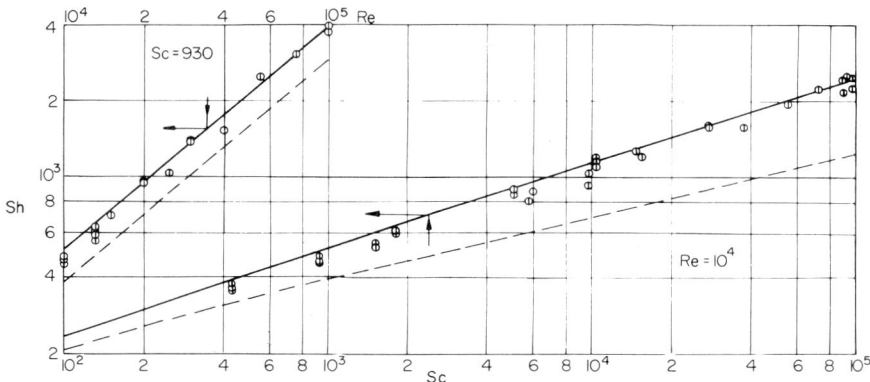

FIG. 4. Comparison of mass transfer experimental data (circles) with the analytical results of Petukhov and Popov (solid lines) and Deissler (dotted lines).

experimental data at large Sc. The solid curve in the lower plot corresponds to Eq. (48) (when Sc $< 10^3$) and to Eq. (51) (when Sc $> 10^3$). In the upper plot the solid curve corresponds to Eq. (51). The dotted lines show Deissler's calculations (when Sc is approximately 10^3 and higher, the curves being drawn according to Eq. (52)). From this figure, Eqs. (48) and (51) are in quite good agreement with the experimental data over a wide range of Sc (up to 10^5). In comparison with the experimental data Deissler's calculations produce lower Sh numbers. Therefore, when Sc $= 10^3$, the difference is approximately 20%, and when Sc $= 10^5$, the difference increases to 50%.

The analysis shows that, for practical heat and mass transfer calculations over a range of Pr or Sc ~ 0.5–10^3, Eq. (48) should be used, but for calculations in the range of Pr or Sc $\sim 10^3$–10^5, Eq. (51) is valid. Equation (50) may also be used for Pr or Sc ranging from 0.5 to 200. As mentioned above, these equations describe the range of Reynolds numbers from 10^4 to 5×10^6.[7]

---

[7] With high Re and simultaneously high Pr or Sc, the validity of Eqs. (48), (50), and (51) has not yet been verified experimentally due to the absence of experimental data. However, from the theoretical considerations we can assume that with high Re they are also in good agreement with experimental data.

Empirical equations of the following type are widely used in practice:

$$\text{Nu} = c\,\text{Re}^m\,\text{Pr}^n \tag{53}$$

Comparing Eq. (53) with Eqs. (48) and (51) it is easy to see that with Eq. (53) at constant c, m, and n it is impossible to describe to a reasonable accuracy the change of Nu number with Re and Pr over a wide range of these parameters. A direct comparison of Eq. (53) with experimental data leads to the same conclusion. Allen and Eckert (34) have shown that for Re $\sim 1.3 \times 10^4$–11×10^4 and Pr = 8, Eq. (53) (when $c = 0.023$, $m = 0.8$, and $n = 0.4$) produces an error of up to 20%. An equation of the type (53) can be used for Nu = Nu (Re, Pr) only assuming that c, m, and n are functions of Re and Pr. For Re $\sim 10^4$–5.10^6 and Pr (or Sc) ~ 0.5–10^5, m changes from 0.79 to 0.92, while n varies from 0.33 to 0.6.

IV. Heat Transfer and Skin Friction for Liquids with Variable Viscosity

A. Theoretical Results

For liquids (condensed medium) far from their critical point only dynamic viscosity varies greatly with temperature; all the other physical properties (ρ, C_p, λ) depend on temperature rather weakly. Therefore while investigating nonisothermal liquid flow, a model with variable viscosity may be used as a good approximation, other physical properties being assumed constant.

Deissler's paper (11) should be noted as one of the papers devoted to the analysis of flow and heat transfer for liquids with variable viscosity. His analysis reduces to the simultaneous solution of Eqs. (13) and (14) in dimensionless form by means of the successive approximations method. As in the case of constant properties, $q = q_w$ and $\sigma = \sigma_w$ are assumed. The eddy diffusivity of momentum is calculated from Eqs. (22) and (23), and β is taken to be one. The variation of viscosity with temperature is taken into account only near the wall ($\eta < 26$). The temperature dependence of viscosity is formulated as

$$\mu/\mu_w = (t/t_w)^K$$

where K is a constant; K varies from -1 to -4. As for the turbulent core viscosity, it is considered constant here.

Deissler has calculated heat transfer and skin friction over the range of 1–10^3 for Pr and 4×10^3–2×10^5 for Re. His results are given as Nu = Nu (Re$_x$ Pr$_x$) [where the subscript x means that the physical properties are evaluated at the characteristic temperature as defined below] and $\xi = \xi(\mathrm{Re}_x)$ for constant physical properties, if in calculating Re_x and Pr_x the values of dynamic viscosity are taken at the reference temperature:

$$t_x = x(t_w - t_b) + t_b \tag{54}$$

Pr vs x is plotted in Fig. 5 for both heating and cooling of the fluid. The upper plot may be used for the calculation of heat transfer, the plot

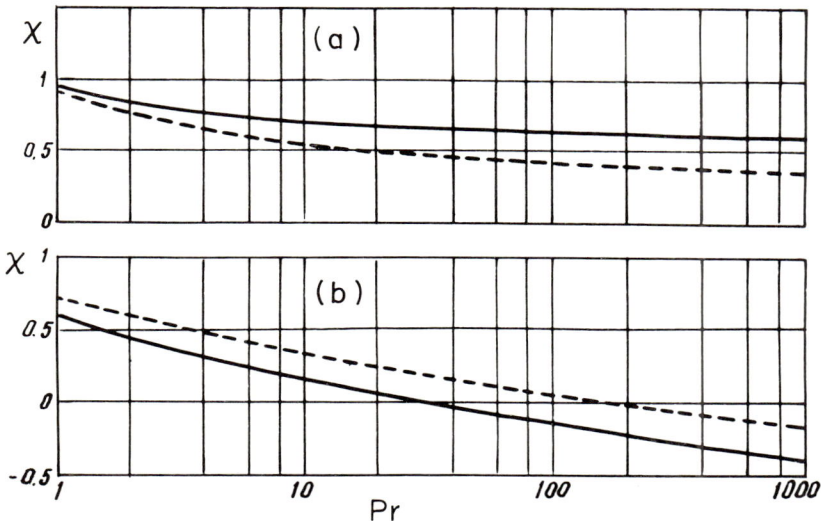

FIG. 5. Reference temperature for heat transfer (a) and friction (b) calculations by Deissler's method for fluids with variable viscosity. Key: solid line, heating of fluid; dashed line, cooling of fluid.

below is for the calculation of skin friction. The values of x in Fig. 5 correspond to $K \sim -1.0$ to -4.0 and $\mu_w/\mu_B \sim 0.5$–2.0 (approximately). The exponent K does not affect strongly the shape of the curves $x(\mathrm{Pr})$; nevertheless, the curves appear to be quite different in the cases of liquid heating and cooling.

Thus, according to Deissler's theory the effect of variable viscosity on heat transfer and skin friction does not depend on Re and changes only with Pr. However, when $\mathrm{Pr} \geqslant 10$, the reference temperature for the calculation of heat transfer varies weakly with Pr and is close to the arithmetic mean of t_w and t_b.

In the following paragraph we shall discuss the agreement between the predicted results and the experimental data.

B. Experimental Data and Empirical Equations

Heat transfer and skin friction experimental data obtained under the conditions of essentially varying viscosity are not numerous; with rare exception they are not accurate and often do not agree well with each other. Therefore we shall use only that small amount of data which may be considered the most reliable.

Analyzing the experimental data we assume that the relation of Nu with Re and Pr, and the relation between ξ and Re at variable physical properties (in this case at variable viscosity) is the same as in the case of constant properties. This assumption is confirmed by the experimental data for liquids with variable viscosity, gases with variable physical properties, and certain other cases (see the following sections).

Of course, assuming a similar variation of Nu with Re and Pr, and ξ with Re for both constant and variable physical properties is only approximate. For example, Allen and Eckert discovered experimentally (34) (see Fig. 6) that variable viscosity affected heat transfer to a greater

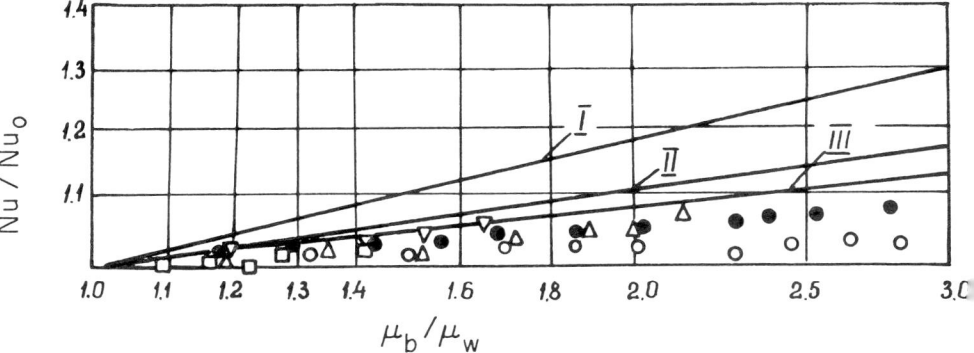

Fig. 6. Variable viscosity influence on the heat transfer in heated water: ○, ●, △, ▽, and □ are Allen and Eckert's experimental data for Re = 13,000, 20,500, 35,500, 62,500, and 110,000 at Pr = 8. I, Deissler's predictions; II, $Nu/Nu_0 = (\mu_b/\mu_w)^{0.14}$; III, $Nu/Nu_0 = (\mu_b/\mu_w)^{0.11}$.

or lesser degree depending on Re. This result is of great interest and deserves to be studied in more detail. However, the error in Nu arising from not including the influence of Re on Nu and μ_b/μ_w is several percent. Because of the absence of systematic data it is impossible at present to take into consideration such effects.

If we proceed from the given assumption, the effect of variable viscosity on heat transfer and skin friction can be estimated from the following relationships:

$$\mathrm{Nu}/\mathrm{Nu}_0 = f_{\mathrm{Nu}}(\mu_w/\mu_b), \qquad \xi/\xi_0 = f_\xi(\mu_w/\mu_b) \qquad (55)$$

where Nu and ξ are the Nusselt number and the friction factor at variable viscosity obtained experimentally; Nu_0 and ξ_0 are the same numbers calculated by assuming constant physical properties with Re and Pr the same as for the corresponding Nu and ξ; μ_w and μ_b are the dynamic viscosities at T_w and T_b, respectively. All the physical properties (except μ_w) in the expressions for the dimensionless numbers are calculated at the bulk temperature t_b for the given tube section.

In Fig. 6, Allen and Eckert's experimental data (34) on heat transfer for the case of water heating are compared with Deissler's predicted results and some empirical relationships. As is readily seen, the predicted results describe qualitatively the effect of variable viscosity on heat transfer but produce quantitatively overestimated values.

The relative change of heat transfer due to viscosity dependence on temperature can be expressed by the equation

$$\mathrm{Nu}/\mathrm{Nu}_0 = (\mu_b/\mu_w)^n \qquad (56)$$

where Nu_0 is calculated from Eq. (48) or (50) and the n expression is determined from the experimental data.

As we can see from Fig. 6, $n = 0.14$ as suggested by Sieder and Tate (42) is overstated; $n = 0.11$ as suggested in (13, 43) corresponds to experimental data for liquid heating better than $n = 0.14$.

To choose the correct value of n in Eq. (56) the heat transfer experimental data corresponding to heating and cooling for several liquids over a wide range of values μ_w/μ_b (the main characteristics of these data are presented in Table III) where treated. The results of the treatment are

TABLE III

THE MAIN CHARACTERISTICS OF HEAT TRANSFER EXPERIMENTAL DATA FOR FLUIDS WITH VARIABLE VISCOSITY

Ref.	Symbols	Fluid	l/d	Re · 10⁻³	Pr	μ_w/μ_b
Yakovlev (36)	○	Water	70–80	19–123	2–12	0.19–0.77
Kreith and Summerfield (45)	◇	Butyl alcohol	38	42–78	23–30	0.08–0.45
Petukhov (13)	●	Transformer oil	88	5–44	39–61	1.2–8.6
Petukhov (13)	□	Oil MS	88	5–14	134–140	1.6–38

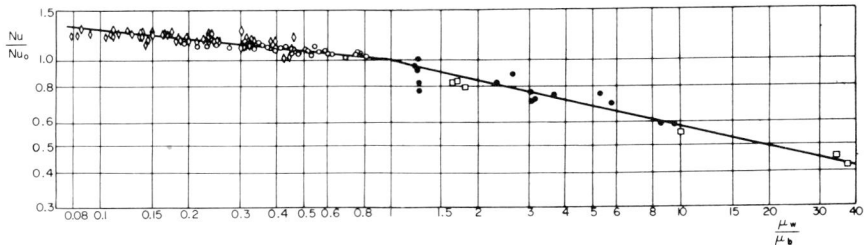

FIG. 7. Variable viscosity influence on the heat transfer in different fluids for heating and cooling (for symbols see Table III).

given in Fig. 7. The averaging curves drawn through the experimental points correspond to $n = 0.11$ when the fluid is heated ($\mu_w/\mu_b < 1$) and $n = 0.25$ when the fluid is cooled ($\mu_w/\mu_b > 1$). The value $n = 0.25$ is in agreement with Mikheev's recommendation (44). He suggested that one should take into account the effect of variable physical properties by means of $(Pr_b/Pr_w)^{0.25}$ which takes the form $(\mu_b/\mu_w)^{0.25}$ in the case of varaible viscosity and constant C_p and λ.

Thus, to calculate heat transfer in a turbulent flow with variable viscosity we can use Eq. (56) with $n = 0.11$ for the case of heating and $n = 0.25$ for the case of cooling. Equation (56) is valid over a range of 0.08–40 for μ_w/μ_b, 10^4–1.25×10^5 for Re, and 2–140 for Pr.

Figure 8 illustrates the effect of variable viscosity on the friction factor where the measured values of Allen and Eckert (34) and Rohonczy (46) are presented. The former are obtained with heated water and the latter with cooled water. Deissler's predicted results and some empirical relationships are given in this figure. Both the predicted results and the empirical equation suggested by Sieder and Tate (42)

$$\xi/\xi_0 = (\mu_w/\mu_b)^{0.14}$$

in comparison with experimental data produce a weaker dependence of the friction factor on μ_w/μ_b.

The experimental data plotted in Fig. 8 are well described by the following simple equations:

Under heating ($\mu_w/\mu_b < 1$):

$$\xi/\xi_0 = \tfrac{1}{6}(7 - \mu_b/\mu_w) \tag{57}$$

Under cooling ($\mu_w/\mu_b > 1$):

$$\xi/\xi_0 = (\mu_w/\mu_b)^{0.24} \tag{58}$$

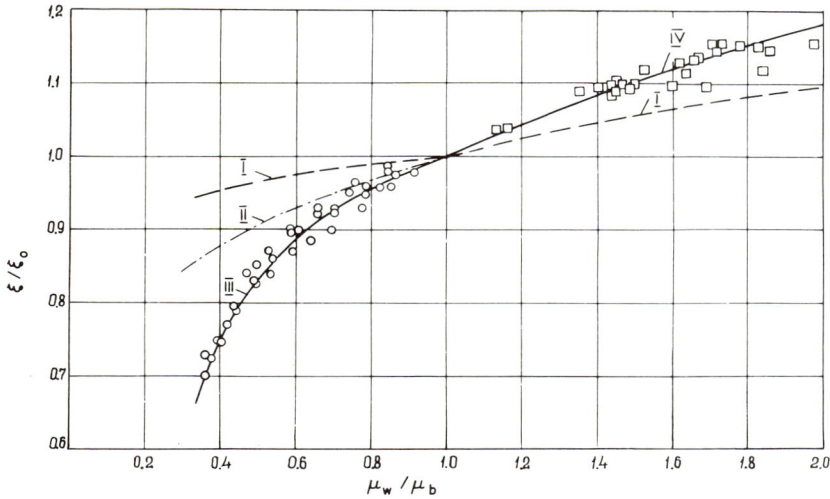

Fig. 8. Variable viscosity influence on friction in water for both heating and cooling: (○) Allen and Eckert experiments (Re = 13,000–110,000, Pr = 8); (□) Rohonczy's experiments (Re = 33·10³–225·10³, Pr = 1.3–5.8). I, Deissler's calculation (when Pr = 8 for heating and Pr = 2.3 for cooling); II, $\xi/\xi_0 = (\mu_w/\mu_b)^{0.14}$; III, $\xi/\xi_0 = 1/6(7 - \mu_b/\mu_w)$; IV, $\xi/\xi_0 = (\mu_w/\mu_b)^{0.24}$.

The friction factor in an isothermal flow ξ_0 is calculated from Eq. (49).

Equations (57) and (58) are true over a range of 0.35–2 for μ_w/μ_b, 10^4–23×10^4 for Re, and 1.3–10 for Pr. They are probably true over an even wider range of these parameters. However, this should be verified experimentally.

V. Heat Transfer and Skin Friction for Gases with Variable Physical Properties

A. Analytical Results

Consider the analysis of the heat transfer and skin friction for a turbulent gas flow in a circular tube, far from the entrance with constant heat flux at the wall. The solution was obtained by Petukhov and Popov (14) using the method described in Section II, C. The physical gas properties ρ, C_p, λ, μ were considered as given functions of temperature. The variation of density with pressure and the energy dissipation in the flow was neglected. Therefore the analysis is valid only for gas flows with small subsonic velocities. The eddy diffusivity of momentum was determined according to Eqs. (20) and (21) and was extended to the case of variable properties by introducing Goldmann's variable (24). The eddy diffusivity of heat ϵ_q was taken equal to ϵ_σ (i.e., $\beta = 1$).

The calculations were carried out for air and hydrogen over the following ranges of the characteristic parameters: 10^4–4.3×10^6 for Re_b and 0.37–3.1 for T_w/T_b in air, 10^4–5.8×10^6 for Re_b and 0.37–3.7 for T_w/T_b in hydrogen (here $\text{Re}_b = \overline{\rho W} d/\mu_b$).

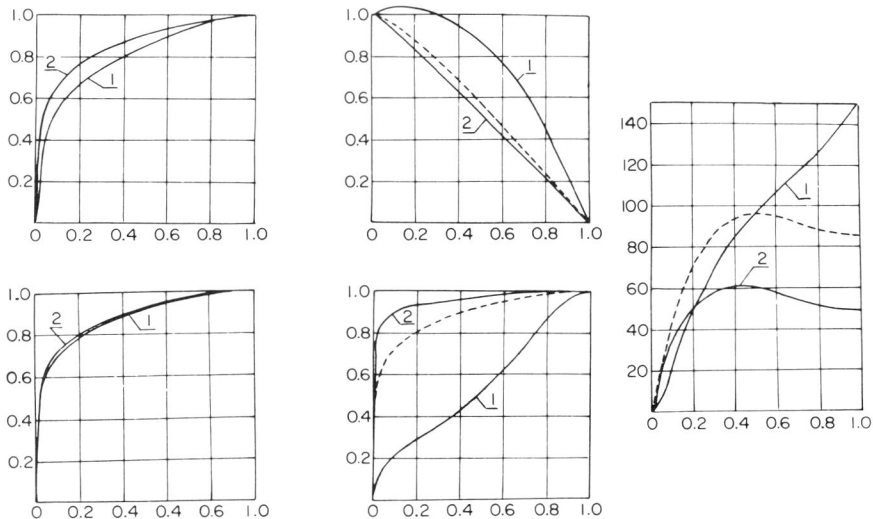

FIG. 9. Distribution (along radius) of dimensionless flow variables at variable (solid lines) and constant (dotted lines) air properties for $\text{Re}_w \simeq 43 \cdot 10^3$ and $\text{Pr} \simeq 0.70$–0.71: (1) $T_w = 1000°\text{K}$, $T_0 = 154°\text{K}$, $T_w/T_b = 3.11$; (2) $T_w = 300°\text{K}$, $T_0 = 902°\text{K}$, $T_w/T_b = 0.383$.

Figure 9 illustrates how variable physical properties affect the distribution of the dimensionless flow parameters along the radius: temperature, velocity, mass velocity, heat flux, and eddy diffusivity of momentum. All the curves correspond to Pr_w and $\text{Re}_w = \overline{\rho W} d/\mu_w$ (these numbers have approximately the same values). Due to the variation of physical properties the temperature profile appears more concave for cooling than for heating. The property variation doesn't affect the velocity profile so strongly as it does the mass velocity profile. For the case of the fluid being heated, the mass velocity profile is flatter than for the cooled fluid. As is seen from Eq. (30) a change in the mass velocity profile results in a redistribution of the relative heat flux along the radius. For heating, when ρW_x decreases at the wall and, consequently, convective heat transfer along the axis decreases too, a maximum occurs in the q/q_w distribution. For cooling, when the mass velocity profile is more full, the q/q_w distribution becomes nearly linear. The property variation also affects greatly the distribution (along the radius) of the relative eddy diffusivity of momentum.

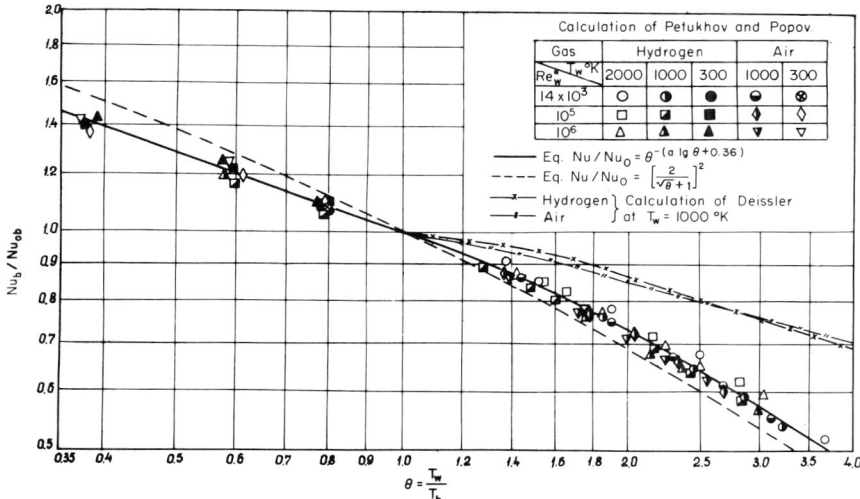

FIG. 10. Heat transfer versus temperature ratio parameter according to the analysis.

Figure 10 represents the results of the heat transfer calculations as Nu_b/Nu_{0b} vs θ, where Nu_b and Nu_{0b} are the Nusselt numbers for variable and constant gas properties, respectively and the same Re_b and Pr_b [8]; $\theta = T_w/T_b$ is the temperature ratio parameter. From Fig. 10, the predicted points diverge to a degree depending on the type of gas, wall temperature, and Reynolds number $Re_w{}^* = Re_w \rho_w/\rho_b$. This divergence is not surprising because we cannot take into account the influence of variable physical properties on heat transfer by means of only one temperature ratio parameter. Even so, the divergence of the points is not great. Other things being equal, the error in Nu_b for air and hydrogen does not exceed 1%; Nu_b for hydrogen at $T_w = 2000°K$ is only 3–5% greater than at $T_w = 1000°K$; the error in Nu_b for air and hydrogen when $Re_w{}^* = 14 \times 10^3$ and 10^6 is 3%.

---

[8] From here on, the subscripts b and w mean that the physical gas properties are evaluated at temperatures T_b and T_w when calculating the corresponding dimensionless numbers.

If the above-mentioned small errors in $\mathrm{Nu_b}$ are neglected, the analytical results can be correlated by the equation

$$\mathrm{Nu_b}/\mathrm{Nu_{0b}} = \theta^n \tag{59}$$

where

$$n = -(a \log \theta + 0.36)$$

For cooling, $a = 0$. For heating, $a = 0.3$, and consequently n decreases with increasing θ. With these values for n, Eq. (59) describes the solution for air and hydrogen with an accuracy of $\pm 4\%$. For simplicity we can take n to be constant for heating also. Then, when $n = -0.47$, Eq. (59) describes the analytical results within $\pm 6\%$. In the case of heated air, $n = -0.5$ produces slightly better results.

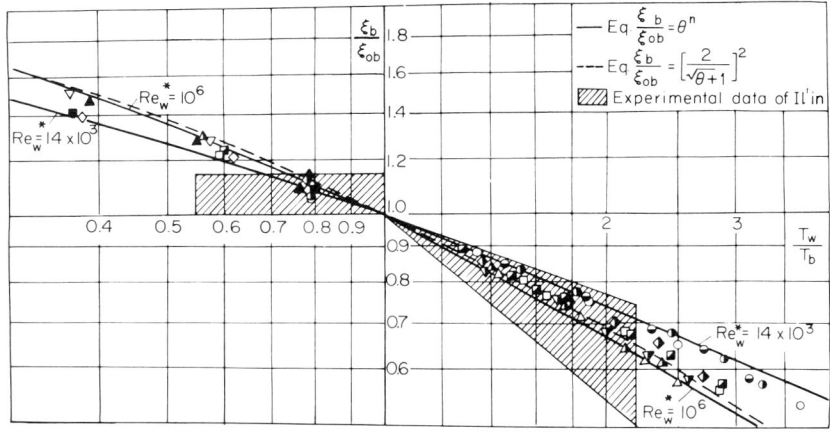

FIG. 11. Friction versus temperature ratio parameter (dots denote Petukhov and Popov's predictions; for symbols see Fig. 10).

In Fig. 11 the predicted friction factor is plotted as ξ_b/ξ_{0b} vs θ. Here $\xi_b = 8\sigma_w\rho_b/(\overline{\rho W_x})^2$ and ξ_{0b} are the friction factors at variable and constant physical gas properties and at the same $\mathrm{Re_b}$, respectively. It should be noted that, contrary to the case of heat transfer, $\mathrm{Re_w}^*$ greatly affects the shape of the curve ξ_b/ξ_{0b} vs θ, but as for the wall temperature and the type of gas, their influence is not very significant. The solutions for air and hydrogen are presented by the equation

$$\xi_b/\xi_{0b} = \theta^n \tag{60}$$

where

$$n = -0.6 + 5.6(\mathrm{Re_w}^*)^{-0.38}$$

for heating and

$$n = -0.6 + 0.79(\text{Re}_w^*)^{-0.11}$$

for cooling.

Equation (60) describes the calculated results within 2–3% over the range 0.37–3.7 for θ and 14×10^3–10^6 for Re_w^*. As Re_w^* varies over the indicated range, n goes from -0.44 to -0.58 for the case of heating and from -0.32 to -0.42 for the case of cooling. If in Eq. (60) n is taken as -0.52 for heating and -0.38 for cooling, this equation describes the calculated data to within 7% accuracy in the first case and 4% in the second.

Kutateladze and Leontiev (17) have obtained an analytical expression for the functions describing the influence of variable physical properties on heat transfer and skin friction in a turbulent gas flow when $\text{Re} \to \infty$. By making some assumptions (σ and q, vary identically with respect to the radius for both constant and variable properties, the velocity and temperature fields are similar, $C_p = \text{const}$) they obtained the following relationship:

$$\frac{\text{Nu}_b}{\text{Nu}_{ob}} = \frac{\xi_b}{\xi_{ob}} = \left(\frac{2}{\sqrt{\theta}+1}\right)^2 \qquad (61)$$

The authors also recommend this expression for finite Re values, based on the empirical fact that the influence of varying physical properties on heat transfer and skin friction depends weakly on Re.

Equation (61) produces stronger dependence of heat transfer and skin friction on the temperature ratio parameter than Petukhov and Popov's analysis, but the difference is not more than 10% (see Figs. 10 and 11). At large Re Eq. (61) is in good agreement with Eq. (60).

Deissler and Presler (47) analyzed the heat transfer for a number of gases (argon, helium, air, hydrogen, and carbon dioxide), taking into account the temperature dependence of their physical properties. They used the same method (Deissler's method) as for fluids with variable viscosity (see Section IV, A), the only difference being in considering all the physical properties of the fluid (ρ, C_p, μ, λ) as functions of temperature. The results are given by the equation

$$\text{Nu}_x = \text{Re}_x^{3/4}/31 \qquad (62)$$

The subscript x denotes that the physical properties in Nu_x and Re_x are evaluated at the temperature

$$T_x = x(T_w - T_b) + T_b$$

TABLE IV

Main Results of Heat Transfer Experimental Investigations for Gas with Variable Physical Properties

Ref.	Gas	l/d	$\mathrm{Re}_b \cdot 10^{-3}$	θ	Method of accounting variable properties influence
(a) Average heat transfer:					
Il'in (48)	Air	59, 62	7–60	0.56–2.3	$\mathrm{Nu}_b = c\,\mathrm{Re}_b^{0.8}\,\theta^n$ θ 0.5–0.9 0.9–1.2 1.2–2.3 c 0.0218 0.0212 0.0223 n 0 −0.27 −0.58
Humble, Lowdermilk, and Desmon (49)	Air	30–120	7–300	0.46–3.5	$\mathrm{Nu}_b = 0.023\,\mathrm{Re}_b^{0.8}\,\mathrm{Pr}_b^{0.4}\,\theta$ (for $x/d > 60$) $n = 0$ at $\theta < 1$ $n = -0.55$ at $\theta > 1$
Bialokoz and Saunders (50)	Air	29–72	124–435	1.1–1.73	$\mathrm{Nu}_b = 0.022\,\mathrm{Re}_b^{0.8}\,\mathrm{Pr}_b^{0.4}\,\theta^{-0.5}$
Weight and Walters (51)	Hydrogen			1–4	$\mathrm{Nu}_b = 0.021\,\mathrm{Re}_b^{0.8}\,\mathrm{Pr}_b^{1/3}\,\theta^{-0.575}$ at great l/d
Taylor and Kirchgessner (52)	Helium	60, 92	3.2–60	1.6–3.9	$\mathrm{Nu}_f = 0.021\,\mathrm{Re}_f^{0.8}\,\mathrm{Pr}_f^{0.4}$ at great l/d
McCarthy and Wolf (53)	Hydrogen	43, 67	7–1500	1.5–2.8	$\mathrm{Nu}_b = 0.023\,\mathrm{Re}_b^{0.8}\,\mathrm{Pr}_b^{0.4}\,\theta^{-0.3}$
McCarthy and Wolf (54)	Hydrogen, helium	21–67	5–1500	1.5–9.9	$\mathrm{Nu}_b = 0.045\,\mathrm{Re}_b^{0.8}\,\mathrm{Pr}_b^{0.4}\left(\dfrac{l}{d}\right)^{-0.15}\theta^{-0.7}$

(b) Local heat transfer:

Author	Fluid				Equation
Wieland (55)	Helium, hydrogen	250		<2.8	$Nu_f = 0.021\ Re_f^{0.8}\ Pr_f^{0.4}$ far from the entry
Taylor (56)	Hydrogen, helium	77		1.5–5.6	$Nu_f = 0.021\ Re_f^{0.8}\ Pr_f^{0.4}$
McEligot, Magee, and Leppert (57)	Air, helium, nitrogen	160		1.1–2.5	$Nu_b = 0.021\ Re_b^{0.8}\ Pr_b^{0.4}\ \theta^{-0.5}$ far from the entry
Kirillov and Malugin (58)	Nitrogen	138	7–160	1.1–2.3	$Nu_b = 0.021\ Re_b^{0.8}\ Pr_b^{0.4}\ \theta^{-0.5}$ at $x/d > 30$
Lelchuk, Elphimov, and Fedotov (59)	Air, carbon dioxide, argon	77–206	14–600	1.1–2.7	$Nu_b = 0.021\ Re_b^{0.8}\ Pr_b^{0.4}\ \theta^{-0.5}$ at $x/d > 50$ and $M \ll 1$
Perkins and Worsoe-Schmidt (60)	Nitrogen	160	18–280	1.3–7.5	$Nu_b = 0.024\ Re_b^{0.8}\ Pr_b^{0.4}\ \theta^{-0.7}$ at $x/d > 40$
Volkov and Ivanov (38)	Air	48–370	14–400	1.1–2.1	$Nu_b = 0.0193\ Re_b^{0.8}\ Pr_b^{0.4}\ \theta^{-0.55}$ at $x/d > 100$
Petukhov, Kirillov, and Maidanic (61)	Nitrogen	80–100	13–300	1–6	$Nu_b = 0.021\ Re_b^{0.8}\ Pr_b^{0.4}\ \theta^n$ at $x/d > 80$ $n = -(0.9 \log \theta + 0.205)$

where the parameter x depends on the type of gas, Re_x, and T_w. At $Re_x > 10^5$, $x \simeq 0.4$ for all the gases.

From Fig. 10, Deissler's results for air and hydrogen are very close to each other. However, from these results we can see that the change in heat transfer with the temperature ratio parameter is less pronounced in comparison with the results of (14) and (17).

B. Experimental Data and Empirical Equations

A great number of experimental papers are devoted to the investigation of the heat transfer between the tube wall and the gas flow at large temperature differences when physical properties cannot be considered constant. Table IV presents a schematic summary of the main results. Heat transfer in the case of gas heating for constant heat flux at the wall (and sometimes for variable q_w) is the object of the majority of the papers. In the case of gas cooling heat transfer, experimental data are neither numerous nor complete. In the earlier papers investigators measured the average (along tube) heat transfer coefficients. The data obtained from these investigations, especially in short tubes, do not reveal the real relation between the local Nu and θ. Therefore, in recent papers the local heat transfer is the prime subject of the investigation.[9] Heat transfer measurements were performed mainly with diatomic gases (air, N_2, H_2) and to a lesser degree with monatomic (He, Ar) and triatomic (CO_2) gases. The experiments cover the temperature ratio parameter range approximately from 0.4 to 4. In some papers the values of θ approached 6 and even 10. But such high values of θ were as a rule found in the entrance region of the tube.

The results of heat transfer measurements at large temperature differences between the wall and the gas flow are usually presented as

$$Nu_b = c \, Re_b^{0.8} \, Pr_b^{0.4} \, \theta^n \tag{63}$$

Sometimes θ is not included in an equation of the type (63) because the influence of variable physical properties is sometimes accounted for by the appropriate choice of a reference temperature. For example, with physical property values chosen at the temperature $T_f = \frac{1}{2}(T_b + T_w)$, the experimental data can be correlated satisfactorily.

In the entrance region of the tube the coefficient c and also the exponent n [from (61) and (62)] in Eq. (63) change with x/d. Far from the entrance, i.e., when $x/d > 40$–100, c becomes constant and n becomes independent

---

[9] Table IV does not provide heat transfer data in relatively short tubes, as heat transfer expressions in the entrance region have not been discussed in this paper.

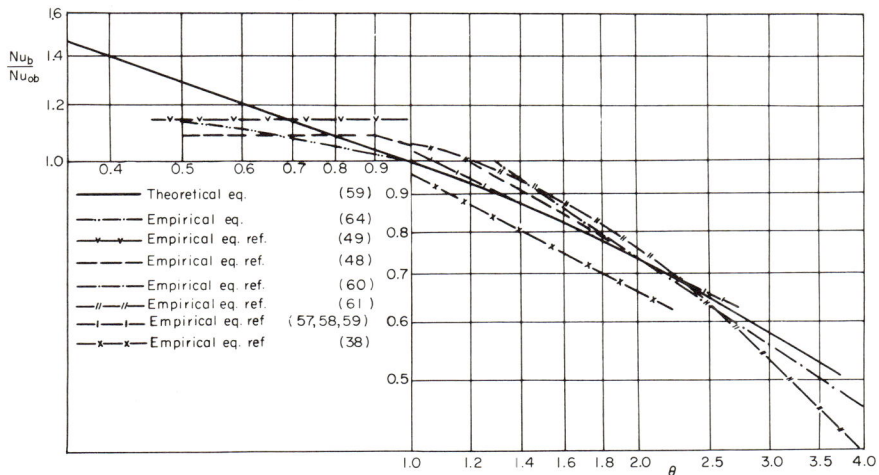

FIG. 12. Heat transfer versus temperature ratio parameter according to predictions and empirical equations.

of x/d. Table IV contains empirical equations obtained by different investigators for distances far from the entrance. In Fig. 12 some of these equations are compared with each other and with Eq. (59) which represents the analytical results. In Fig. 12 the abscissa is the ratio of Nu_b at variable properties to Nu_{0b} at constant properties, calculated from Eq. (48).

For the case of gas heating Table IV and Fig. 12 show that c varies from 0.019 to 0.024, and n ranges from -0.3 to -0.7. This differences is most probably attributed to the fact that n is actually not constant. Its value decreases while θ increases. Both theoretical analysis and experimental data (48, 61) confirm this. Therefore, using Eq. (63) as the mathematical representation of the experimental data when $n = $ const, different investigators obtain different values for n depending on the range of θ. Extrapolation of these data (obtained when $\theta \geqslant 1$) to $\theta = 1$ leads naturally to different values of c.

For the case of gas heating Fig. 12 shows good agreement of the analytical results Eq. (59) with the experimental data only if the empirical equations are not extrapolated past the limits imposed upon θ by the experiments. Also, within this restriction a good degree of consistency is observed for all the experimental data considered. Neither the experimental nor the analytical results reveal a noticeable influence of the Reynolds number and the type of gas on the shape of the curve Nu_b/Nu_{0b} vs θ. Apparently this influence, at least for the gases investigated (see Table IV), is within the range of experimental error. Thus, for practical heat transfer calculations in the case of gas heating when $1 \leqslant \theta \leqslant 4$, Eq. (59) may be used.

As was mentioned above, heat transfer for gas cooling has not been investigated extensively. In particular, there is no experimental data on local heat transfer. Therefore, the analytical results can only be compared with experimental data on average heat transfer. Empirical equations of Fig. 12 show that for gas cooling the heat transfer is larger than when $\theta = 1$. Over the range of the experiments the ratio $\text{Nu}_b/\text{Nu}_{0b}$ was found to be independent of θ. Perhaps the scatter of points, found in the experiments, prevented the actual relation between Nu_b and θ from being observed. Nevertheless, in the case of gas cooling, even under the restriction stated, the relation between $\text{Nu}_b/\text{Nu}_{0b}$ and θ, from experimental data, appears to be significantly weaker than that obtained analytically. Further experiments and theoretical studies are necessary to explain these differences. To obtain new data for practical calculations of heat transfer under the conditions of gas cooling, the simple equation suggested by Ivashchenko (63) from the treatment of Ilyin's experimental data (48) may be used:

$$\text{Nu}_b/\text{Nu}_{0b} = 1.27 - 0.27\theta \tag{64}$$

This equation is valid when $0.5 \leqslant \theta \leqslant 1$.

The skin friction in the pipe flow of a gas at large differences between the wall and flow temperatures has been studied (48, 57, 60, 62). Unfortunately, the data obtained in these papers are contradictory. This probably stems from the inaccuracy of the measurements (the experimental data differs from the recommended relations by 20–25%). Ilyin's experimental data (48) are given in Fig. 11. From these data the value of n in Eq. (60) is -0.58 for the case of gas heating and $n = 0$ for gas cooling, though here $\xi_b/\xi_{0b} > 1$. Thus, in the case of gas heating Ilyin's data are in good agreement with the analytical results, but they are lower than the analytically predicted data in the case of gas cooling.

Perkins and Worsoe-Schmidt (60) found that far from the tube entrance $(x/d > 55)$ when[10] $1 \leqslant \theta \leqslant 4$ the following equation is valid:

$$\xi_b/\xi_{0w} = \theta^{-0.6} \tag{65}$$

where ξ_{0w} is the friction factor in an isothermal flow. In calculating $\text{Re}_w{}^*$, the physical gas properties in the expression are evaluated at the wall temperature, i.e., $\text{Re}_w{}^* = (\overline{\rho W} d/\mu_w)(\rho_w/\rho_b)$. If we assume that

---

[10] At the entrance region θ values approached 7.5.

$\xi_{0w} \sim \text{Re}_w^{*-0.2}$ and $\mu \sim T^{0.68}$, it can easily be shown that Eq. (65) takes the form

$$\xi_b/\xi_{0b} = \theta^{-0.264} \tag{66}$$

But, when the experimental data are plotted as ξ_b/ξ_{0b} vs θ as indicated by (60), there is considerable scatter in the results. According to the data of Lel'chuk and Dyadyakin, the exponent in Eq. (66) should be -0.16 (instead of -0.264), while McEligot's work (57) indicates a value of -0.1 for $\theta \sim 1.0$–2.4. Thus the experimental data from Refs. (57, 60, 62) differ significantly. The general trend of these papers is to suggest a weaker relationship between ξ_b/ξ_{0b} and θ than predicted analytically, thus indicating the need for further investigation.

VI. Heat Transfer and Skin Friction for Single-Phase Fluids at Subcritical States

A. Analytical Results

By heat transfer in single-phase media at subcritical states (or, shortly, in the supercritical range) is meant that the heat transfer takes place at supercritical pressures and at subcritical or pseudocritical temperatures (i.e., at temperatures corresponding to the maximum heat capacity at constant pressure).

We can explain heat transfer anomalies in the supercritical region by the fact that physical properties in this region change considerably and in a special way.

Figure 13 shows the behavior of physical properties with temperature. This figure presents the data on carbon dioxide at $P = 100$ bar. The majority of the physical properties do not change monotonically: heat capacity at constant pressure has characteristic maxima; heat conductivity and viscosity coefficients usually pass through a minimum.

Density changes considerably but the volumetric coefficient of thermal expansion attains a maximum. Therefore, even though the temperature difference in the flow is small ($T_w - T_b \approx 10$–20 °C), the physical properties change considerably across the tube. For instance, if $T_b < T_m < T_w$ (here T_m is a pseudocritical temperature) as the distance from the wall increases, C_p increases rapidly, goes through a maximum, and then decreases. In the case of heating density increases rapidly from the wall to the tube axis. The variation of physical properties along the radius is especially significant at high heat fluxes and at large temperature differences between the wall and the fluid, respectively.

If heat transfer takes place at rather small temperature differences then

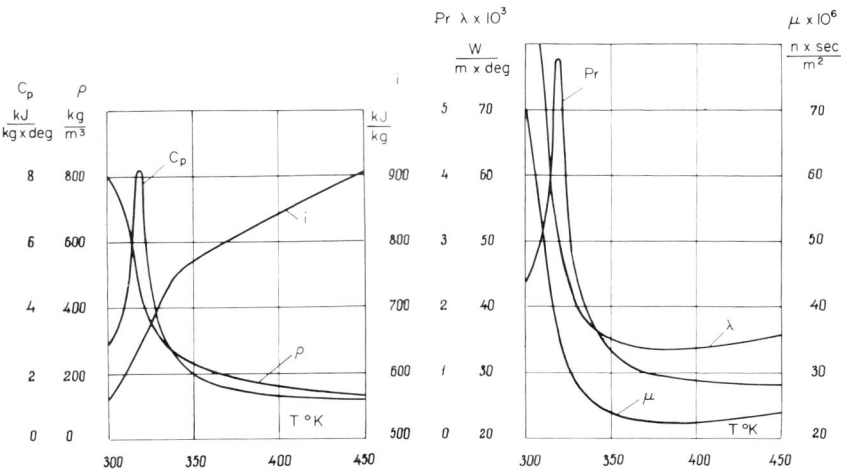

FIG. 13. Carbon dioxide physical properties at $P = 100$ bar.

irrespective of the change of physical properties with temperature, the calculation of such a process may be carried out by assuming that the physical properties are constant. Therefore, Eq. (48) is also valid for turbulent liquid pipe flow when $T_w - T_b \to 0$ in the supercritical region. Figure 14 depicts the heat transfer coefficient α calculated from Eq. (48) for carbon dioxide vs T_b at different pressures $P > P_{cr}$.

Of course, the behavior of α essentially depends on the specific variation of the carbon dioxide physical properties with T and P. American investigators Deissler and Goldmann (19, 64) were the first who theoretically analyzed heat transfer and skin friction at supercritical states, taking into account the variation of the physical properties of the fluid with temperature. In both papers the authors studied heat transfer in a fully developed turbulent liquid flow in a circular tube, the heat flux at the wall being constant ($q_w = $ const). Having written Eqs. (13) and (14) in a dimensionless form, they solved these equations simultaneously by the successive approximation method. The distribution of q and σ along the radius was taken either constant (Deissler) or linear (Goldmann). Deissler and Goldmann's methods differ from one another because they used different methods of calculating the eddy diffusivity of momentum at variable fluid properties. Deissler used Eqs. (22) and (23) and Goldmann determined ϵ_σ from the universal velocity profile (for constant properties the profile is calculated by Deissler's method) in terms of his own variables φ^+ and η^+ [see relationships (24)]. In both cases β was taken to be one.

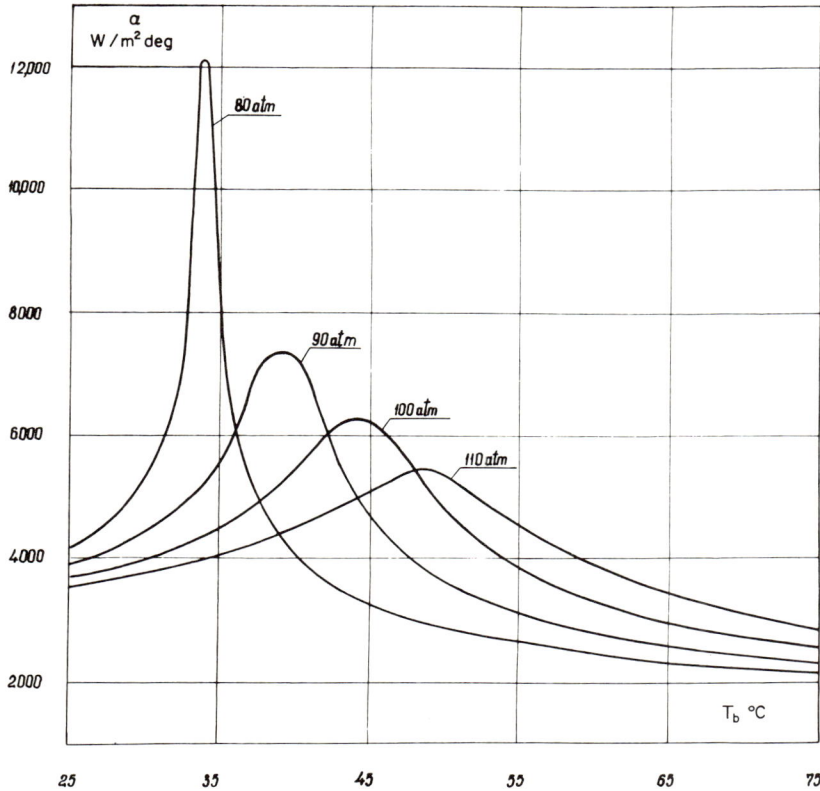

Fig. 14. Heat transfer coefficient for carbon dioxide in the supercritical region calculated with the assumption of constant physical properties ($d = 6.7$ mm, $G = 100$ kg/hr).

Deissler predicted heat transfer and skin friction for the case of heated water over a temperature range of 204–650°C, pressure being equal to 344 bar. The results were correlated by using the reference temperature which proved to be a complicated and nonmonotonic function of T_w/T_b and T_w. This function is plotted in Fig. 18. Goldmann also carried out heat transfer and skin friction calculations for the case of heated water at a pressure of 344 bar, but over a wider temperature range of 260–840 C. The analytical solution of the heat transfer is presented by the equation

$$\frac{q_w\, d^{0.2}}{(\rho W)^{0.8}} = f(T_w, T_b) \tag{67}$$

which is based on the assumption that with both constant and variable

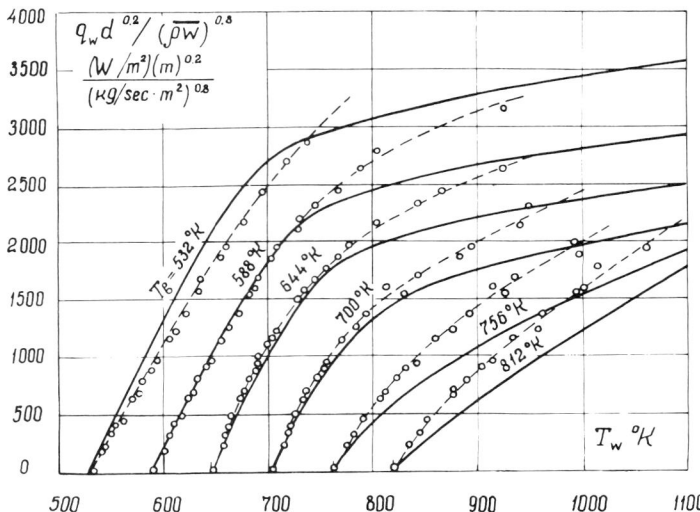

FIG. 15. Goldmann's analytical results for water at $P = 344$ bar (solid lines) in comparison with experimental data (dotted lines; circles represent averaged experimental data). Ranges of parameters: $d = 1.27$–1.9 mm, $q_w = (0.32$–$9.5) \cdot 10^6$ W/m²; $\overline{\rho W} = (2$–$4) \cdot 10^3$ kg/sec-m².

properties the heat transfer varies with $\overline{\rho W}$ and d in the same way. The function $f(T_w, T_b)$ is plotted in Fig. 15. If the physical properties were constant f would be a linear function of T_w. Thus, divergence from a linear relation characterizes the influence of variable physical properties.

Using Goldmann's method Tanaka et al. (65) performed heat transfer calculations for carbon dioxide at $P = 78.5$ bar for both a heated and a cooled fluid. Their results are given in Fig. 16. As is seen in the figure, the graph of the heat transfer coefficient α vs t_b at $t_b \approx t_m$ passes through a maximum, which may be attributed to the presence of the appropriate maxima of $C_p(T)$ and $\Pr(T)$ (see Fig. 13). The maximum value of α in the case of cooling ($q_w < 0$) is higher than that in the case of heating ($q_w > 0$) and decreases when q_w increases. This is because the conductivity in the viscous sublayer for cooling is almost twice that for heating, and it decreases when q_w and T_w increase. For cooling α decreases with increasing q_w at a smaller rate than for heating.

For the calculation of heat transfer and skin friction in a supercritical region, Popov (66) used the method given in Section II, C. The eddy diffusivity of momentum was determined from Eqs. (20) and (21) extended to the case of variable physical properties by introducing

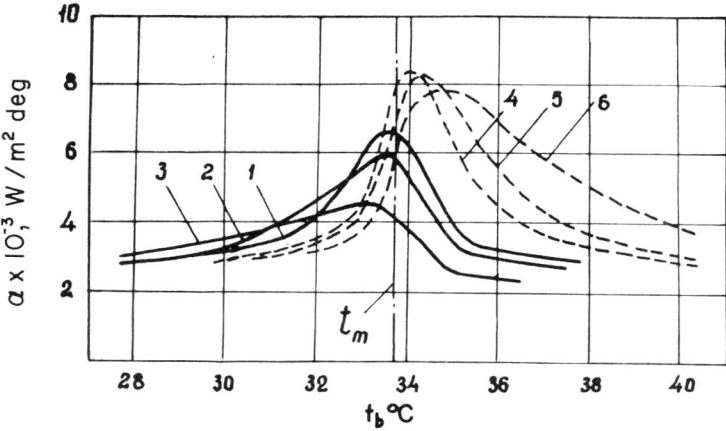

FIG. 16. α vs t_b for carbon dioxide both heated and cooled ($P = 78.5$ bar, $G = 3.9 \times 10^{-2}$ kg/sec, $d = 10$ mm) at $q_w \times 10^{-4}$ W/m²: (1) 0.70; (2) 1.4; (3) 2.8; (4) -0.70; (5) -1.4; (6) -2.8.

Goldmann's variable (24). The value of β was taken to be one. The calculation was done for the heating of carbon dioxide in a circular pipe far from the inlet at constant q_w. Calculations were made for $P/P_{\mathrm{cr}} = 1.33$ ($P = 98$ bar).

$$0.94 \leqslant T_b/T_m \leqslant 1.24, \quad \text{and} \quad 0.97 \leqslant T_w/T_m \leqslant 1.24.$$

The results are presented as interpolation equations for heat transfer and skin friction.

Comparison of the analytical results with the experimental data shows that the relation of the Nu number to Re and Pr and to ξ and Re at variable properties is approximately the same as at constant physcial properties. Then the ratios $\mathrm{Nu}_b/\mathrm{Nu}_{0b}$ and ξ_b/ξ_{0b} depend only on the behavior of the physical properties with temperature. For the given liquid $\mathrm{Nu}_b/\mathrm{Nu}_{0b}$ is a function of T_b, T_w, and P or T_b/T_m, T_w/T_m, and P/P_{cr}, which is more convenient for the comparison of the data obtained for different pressures and liquids.

In Figs. 15, 17, and 18, some predicted values are compared with experimental data and empirical equations. The behavior of $\mathrm{Nu}_b/\mathrm{Nu}_{0b}$ with T_w/T_m is clearly seen in Fig. 18. When $T_b/T_m < 1$ heat transfer increases with T_w/T_m increasing; when $T_w/T_m \approx 1$, it reaches the maximum and then decreases. If $T_b/T_m \geqslant 1$ heat transfer decreases when T_w/T_m increases (see Fig. 17).

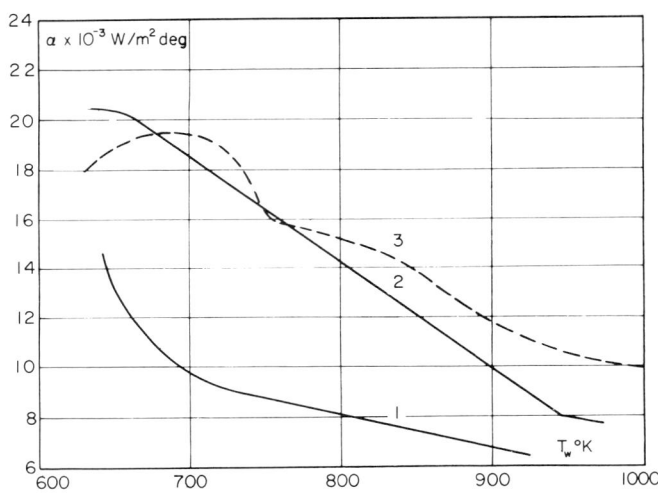

FIG. 17. α vs T_w for water at $P = 345$ bar, $T_w/T_b = 1.25$, $\overline{\rho W} = 2150$ kg/m²-sec, and $d = 9.4$ mm: (1) Deissler's predictions; (2) Goldmann's predictions; (3) Svenson et al.'s empirical equation.

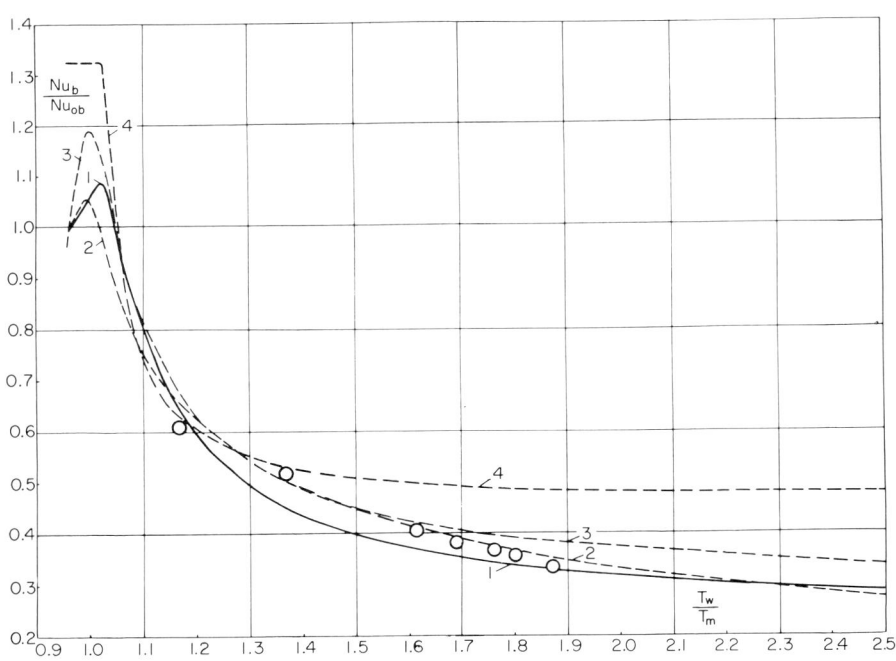

FIG. 18. Nu_b/Nu_{0b} vs T_w/T_m for carbon dioxide at $P = 98$ bar, $T_b = 303°$K, $G = 100$ kg/hr, and $d = 4.08$ mm: (1) Popov's predictions; (2, 3, 4) empirical equations (69), (71), and (68), respectively; circles represent experimental data.

It can be seen from Fig. 17, taken from (67), that Deissler's theoretical results produce lower heat transfer values in comparison with experimental data (approximately two times lower). Bringer and Smith (68), who calculated the heat transfer for heated carbon dioxide by Deissler's method, have found a disagreement of 25% with their measured values. Having calculated the heat transfer by using Deissler's method but Prandtl's expression for the mixing length in a turbulent core and approximating the influence of the density fluctuations, Melik-Pashaev (69) have obtained better agreement between calculated and experimental data. Goldmann's predicted results (see Figs. 15 and 17) are in good agreement with the experimental data when $T_w - T_b \lesssim 100°C$. Disagreement increases with increasing $T_w - T_b$, reaching 25% when $T_w - T_b = 250$–$300°C$. Povov's theoretical results (66) are in agreement with the experimental data for carbon dioxide within 20%. Figure 18 illustrates this comparison.

Hsu and Smith (70) made an attempt to take into account (approximately) the influence of density fluctuations on the turbulent momentum diffusivity. The calculations for carbon dioxide have shown that this correction improves the agreement between the predicted and experimental heat transfer data. However, the predictions for hydrogen (71) with a correction for density fluctuations displayed less favorable agreement with experimental data than the same predictions without the correction. Thus, the problem of accounting for fluctuations due to density and other physical properties has not as yet been satisfactorily solved.

One might be tempted to expect a substantial influence of free convection on heat transfer in a supercritical forced flow, because density changes considerably with temperature. Hsu and Smith (70) analyzed this problem theoretically for the case of a turbulent flow of heated carbon dioxide in a vertical tube. As should be expected, the effect of free convection on heat transfer becomes more pronounced with higher Gr and lower Re. For example, when $Gr = 10^8$ and $Re = 10^4$ the heat transfer rate increases approximately two times due to free convection; when $Gr = 10^8$ and $Re = 10^5$ free convection does not affect the heat transfer.

The analysis of the predicted results reveals that when physical properties do not change considerably over the flow cross section it is possible to satisfactorily describe the heat transfer mechanism in the supercritical region (but not for all possible regimes of flow and heat transfer). The disagreement found between the predicted values and the experimental data can probably be attributed to the inexact analytical methods and to some uncertainty in the estimation of the physical properties at subcritical states.[11] In case of significant variations in physical properties over the tube cross section and for some specific flow and heat transfer regimes (the so-called regimes with diminished and enhanced heat transfer, see below) the present theoretical methods do not permit the satisfactory description of heat transfer at supercritical states. In this case considerable disagreement between the predicted results and the experimental data is observed. There is every reason to believe that this disagreement may be attributed to the unsatisfactory estimation of the variable physical properties effect (mainly of density) on turbulent diffusivity. So, for example, in the work of Hall *et al.* (*72*) who studied heat transfer in the carbon dioxide flow in a plane tube with one wall heated and the other cooled, it is shown that for large density gradients in the layer adjacent to the wall the theoretical results obtained by Goldmann's method and the experimental data can differ by a factor of two from one another. Unfortunately, the problem of turbulent diffusivity at variable physical properties has not as yet been completely studied. Therefore, relevant experimental data and theoretical results would be of great interest.

B. Experimental Data and Empirical Equations for Normal Heat Transfer Regimes

At present there is a considerable number of experimental works on heat transfer in the supercritical region. In the vast majority of these works investigators analyzed the heat transfer for turbulent flows of water and carbon dioxide in circular tubes for heating when $q_w =$ const. Table V gives an incomplete summary of these works. Some papers are devoted to the heat transfer for heated hydrogen (*71, 88*), oxygen (*89*) and Freon-12 (*90*). Experimental data on the heat transfer of cooled fluids are not available.

Judging from the experimental data obtained for small temperature differences in a flow when the physical properties do not change considerably, heat transfer in the supercritical region displays no striking differences and is described well by the known relationships for constant properties.

For rather large temperature differences in a flow, i.e., under the conditions when the physical properties change significantly, all the possible flow and heat transfer regimes can be classified in the following

---

[11] For example, according to some data, thermal conductivity at the pseudocritical temperature has a pronounced maximum; according to other data, it changes monotonically. In this paper the latter is adopted. But in Ref. (*72*) it is shown that the consideration of the thermal conductivity maximum makes no difference for theoretical predictions.

TABLE V

EXPERIMENTAL HEAT TRANSFER DATA OF WATER AND CARBON DIOXIDE AT SUPERCRITICAL PARAMETERS

Author	P/P_{cr}	T/T_m	T_b/T_m	$Re_b \cdot 10^{-3}$	Pr_b	$q_w \cdot 10^{-6}\,W/m^2$
(a) Water							
Armand, Tarasova, and Kon'kov (73)	1.01–1.16	0.91–1.12	0.87–1.09	35–180	0.89–25	0.18–0.73
Vikhrev, Barulin, and Kon'kov (74)	1.1–1.2	0.6–1.4	0.5–1.05	18–800	0.9–8.7	0.23–1.25
Miropol'sky and Shitsman (75, 76)	1.02–1.25	0.81–1.17	0.97–1.12	13–570	0.84–55	0.23–2.5
Shitsman (77)	1.02–1.11	1.04–1.57	0.85–1.25			0.27–1.1
Alad'ev, Vel'tishchev, and Kondratiev (78)	1.11–1.38	0.81–1.44	0.74–1.30	21–101	0.8–14	0.08–0.13
Dickinson and Welch (79)	1.09–1.4	0.69–1.35	0.59–1.22	58–860	0.75–16	0.87–1.7
Chalfant and Randall (80, 81)	1.56	0.74–1.6	0.74–1.25	64–1020	0.82–3.3	3.1–9.5
Swenson, Carver, and Kakarala (67)	1.03–1.87	0.56–2.1	0.54–1.9	27–680	0.8–9.4	0.2–1.8
Bischop and Sandberg (67)	1.03–1.25	0.89–1.35	0.87–1.22	36–530	0.89–16	0.32–3.6
Yamagata, Nishikawa, Hasegawa, and Fujii (82)	1.11	0.84–1.33	0.6–1.25	12–400	0.9–15.4	0.12–0.93
(b) Carbon dioxide							
Krasnoshchekov and Protopopov (83–85)	1.06–1.46	0.98–2.6	0.96–1.7	65–500	0.85–40	0.02–2.5
Melik-Pashaev (69)	1.2–5.7	1.4–2.9	0.91–1.2	150–650	1.0–8.1	up to 9.5
Bringer and Smith (68)	1.13	0.99–1.06	0.97–1.02	38–270	2.6–7.4	0.02–0.31
Koppel and Smith (86)	1.00–1.03		0.95–1.06	30–300	0.9–11	0.06–0.63
Wood and Smith (87)	1.004	0.97–1.05	0.97	910–950	0.85–10.5	0.01–0.07
Hall, Jackson, and Khan (72)	1.51	0.84–1.32	0.84–1.32	52–400	2.5–10.5	0.002–0.25

way: (a) normal regimes, (b) regimes with diminished heat transfer, (c) regimes with enhanced heat transfer.

Normal regimes are those for which the relations observed in experiments can be explained and analyzed by the existing concepts of turbulent flow and heat transfer formulations for variable physical properties.

Regimes with diminished and enhanced heat transfer are those with lower or higher heat transfer coefficients, respectively, in comparison with those which should have been expected if the regime had been normal. Such a classification of flow and heat transfer regimes is, of course, rather conventional and most likely it is not a matter of principle differences between these regimes but has appeared as a result of insufficient knowledge of the actual mechanism of turbulent diffusivity with variable physical properties. A rather elementary model of turbulent diffusivity used at present allows us to describe normal regimes satisfactorily, but it appears insufficient and inadequate for description of more complicated phenomena involved in regimes with diminished and enhanced heat transfer.

In heat exchanger systems at supercritical states, normal regimes occur mostly. Heat transfer formulations for these regimes are essentially in good agreement with predicted results. Several empirical equations are known for the calculation of heat transfer in the normal regime. One of the first equations was suggested by Miropolskii and Shitsman (75, 91). This equation has the form

$$\text{Nu}_b = 0.023 \, \text{Re}_b^{0.8} \, \text{Pr}_{\min}^{0.8} \tag{68}$$

where $\text{Pr}_{\min}$ is the smaller of the two Pr numbers calculated from the bulk temperature (Pr_b) and the wall temperature (Pr_w).

Krasnoshchekov and Protopopov (85), using their experimental data for CO_2 and the experimental data of other workers for CO_2 and H_2O, suggested the equation

$$\text{Nu}_b = \text{Nu}_{0b}(\rho_w/\rho_b)^{0.3} \, (\bar{C}_p/C_{pb})^n \tag{69}$$

where Nu_{0b} is the Nusselt number for constant physical properties determined from Eq. (50); $\bar{C}_p$ is the average heat capacity for the temperature range T_b to T_w,

$$\bar{C}_p = \frac{h_w - h_b}{T_w - T_b} \tag{70}$$

h_w and h_b being the enthalpy at T_w and T_b, respectively; n is the exponent depending on T_w/T_m and T_b/T_m. This relationship for n is

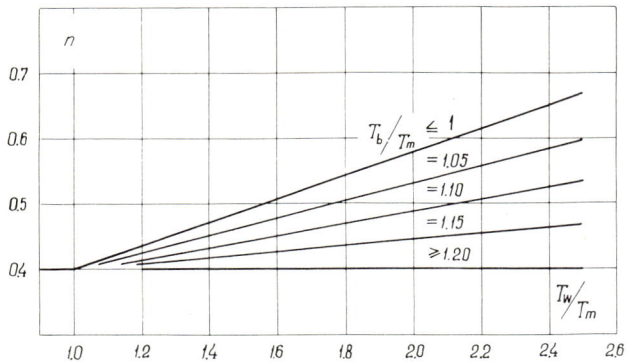

FIG. 19. n vs T_w/T_m and T_b/T_m.

plotted in Fig. 19. Equation (69) describes the experimental data with an accuracy of $\pm 15\%$ and is valid for the following range of

$1.01 \leqslant P/P_{cr} \leqslant 1.33$; $0.6 \leqslant T_b/T_m \leqslant 1.2$; $0.6 \leqslant T_w/T_m \leqslant 2.6$

$2 \cdot 10^4 \leqslant \text{Re}_b \leqslant 8 \cdot 10^5$; $0.85 \leqslant \text{Pr}_b \leqslant 55$; $0.09 \leqslant \rho_w/\rho_b \leqslant 1.0$

$0.02 \leqslant \bar{C}_p/C_{pb} \leqslant 4.0$; $2.3 \cdot 10^4 \leqslant q_w \leqslant 2.6 \cdot 10^6 \ \text{W/m}^2$; $l/d > 15$

Swenson et al. (67) have correlated their data on the heat transfer for water by the equation

$$\text{Nu}_w = 0.00459 \, \text{Re}_w^{0.923} \, \bar{\text{Pr}}_w^{0.613} \, (\rho_w/\rho_b)^{0.231} \tag{71}$$

where $\bar{\text{Pr}}_w = \bar{C}_p \mu_w/\lambda_w$ and $\bar{C}_p$ is determined from Eq. (70).

Equation (71) describes the experimental data of the authors with a mean square error of $\pm 10\%$ and covers the range of the characteristic parameters shown in Table IV.

Besides the above mentioned, other emirical equations are known; for example, Bischop's (67) equation for heat transfer in water and the equation of Hess and Kunts (93) for heat transfer in hydrogen.

In Figs. 17, 18, and 20 empirical equations are compared with each other, with theoretically obtained equations, and directly with experimental data (Fig. 18). Equation (68) procedures more or less satisfactory results only when $T_w/T_m < 1.05$ for water and when $T_w/T_m < 1.4$ for carbon dioxide. This equation does not describe the range of high T_w/T_m. Another disadvantage of Eq. (68) is that it does not give the correct results in the constant properties limit. As is mentioned above, Eqs. (69) and (71) describe the experimental data

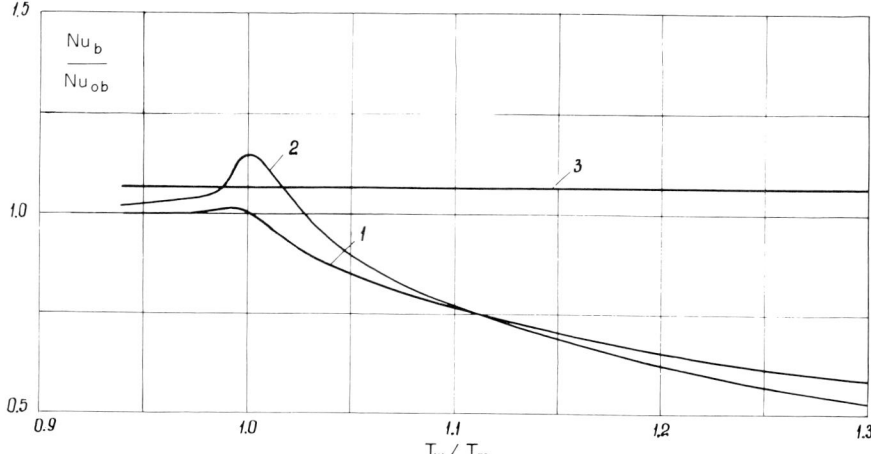

FIG. 20. Nu_b/Nu_{0b} for water at $P = 235.4$ bar, $T_b = 573°K$, $\overline{\rho W} = 2150$ kg/m² sec, and $d = 9.4$ mm: 1, 2, and 3 are empirical equations (69), (71), and (68), respectively.

quite satisfactorily, are self-consistent, and are in good agreement with the analytical results. Therefore, Eqs. (69) and (71) can be used to calculate the normal heat transfer regimes in the case of heated water and carbon dioxide when there is no free convection influence.

The problem of free convection influence on turbulent forced convection has as yet been investigated very little. Nevertheless, the available data show that under certain conditions this influence may be considerable. For example, according to (92), in the flow of water through a uniformly heated horizontal tube at $P = 245$ bar, $\overline{\rho W} = 374$ kg/m²-sec, and $q_w = 0.45 \cdot 10^6$ W/m², the nonuniformity of the circumferential wall temperature distribution reached 220°C. In the case of a vertical or inclined pipe flow with small Re values we should also expect a substantial effect of free convection on the heat transfer.

Tarasova and Leontiev (94) studied skin friction for the case of water flowing through smooth tubes in the supercritical region. The experiments were carried out in downward and upward flows in heated tubes over the pressure range of 226–265 bar, with heat fluxes from $0.6 \cdot 10^6$ to $2.3 \cdot 10^6$ W/m² and in the range of Re_b from $5 \cdot 10^4$ to 63.10^4. The averaged friction factor $\bar{\xi}_b$ was determined in the section of length $l = 50d$ and $75d$, following a heated or nonheated section of length $50d$. The experimental results given in Fig. 21 are described by the equation

$$\bar{\xi}_b/\bar{\xi}_{0b} = (\bar{\mu}_w/\bar{\mu}_b)^{0.22} \tag{72}$$

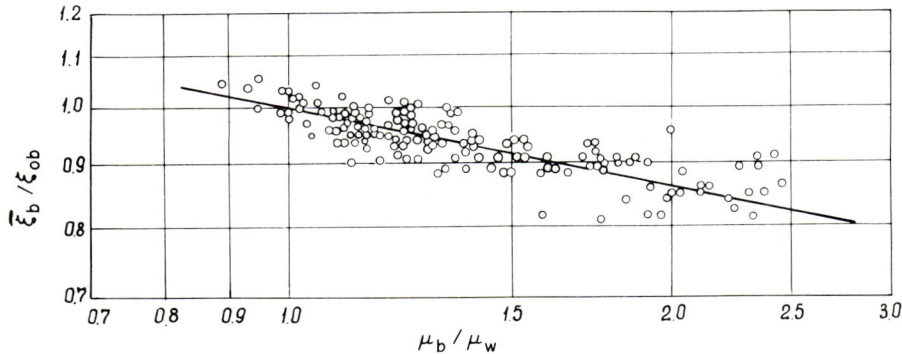

FIG. 21. Average friction factor for water when $P = 226$–265 bar and $q_w = 0.6 \times 10^6$–2.3×10^6 W/m².

where $\bar{\xi}_b = 8\bar{\sigma}_w \bar{\rho}_b / (\overline{\rho W})^2$; ξ_{0b} is calculated from Eq. (49) when $\text{Re} = \text{Re}_b = \overline{\rho W} d / \bar{\mu}_b$; $\bar{\rho}_b$ and $\bar{\mu}_b$ are averaged values of ρ_b and μ_b over the tube section under consideration (they are calculated from the prescribed distribution of bulk enthalpy along the tube length); $\bar{\mu}_w$ is the averaged value of μ_w in this section.

C. Experimental Data for Regimes with Diminished and Enhanced Heat Transfer

A number of experimental studies (74, 77, 82, 89, 95) have led to the discovery of diminished heat transfer regimes. Such regimes are seen very clearly in Shitsman's experiments (77) with water at $P = 226$–245 bar. For fluid heating at $T_b < T_m$ and a certain combination of mass velocity and wall heat flux, heat transfer sharply decreases and the wall temperature increases rapidly (in the case of $q_w = \text{const}$).[12] The heat transfer and wall temperature found in this case are not described by the relations valid for the normal heat transfer regime.

In Fig. 22 the t_w and t_b variations along the tube length are plotted for upward liquid flow for several regimes of normal and diminished heat transfer (77). When $q_w = 221 \times 10^3$ and 281×10^3 W/m² (Curves 1 and 2), normal regimes are observed. But when $q_w = 300 \cdot 10^3$ W/m² (Curve 3), the wall temperature in the tube section length $10d$–$15d$ increases sharply and becomes approximately 100°C higher than the expected one. With further increase in heat flux (Curves 4 and 5), the

---

[12] Usually (but not always) diminished regimes occur in the cases when T_w is higher than T_m.

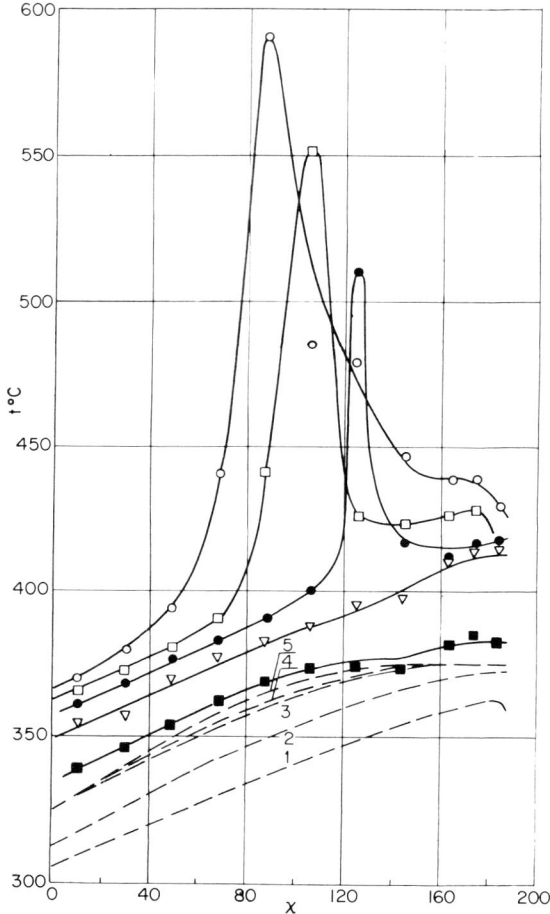

FIG. 22. Distribution of t_w (solid lines) and t_b (dotted lines) along the tube length for water at $P = 226$ bar, $\overline{\rho W} = 430$ kg/m²-sec, and $q_w \times 10^{-3}$ W/m²: 1 (■), 221; 2 (▽), 281; 3 (●), 300; 4 (□), 337; 5 (○), 386.

wall temperature at the maximum point increases, the maximum t_w displaces towards the tube entry, and the region with high wall temperature expands covering a substantial part of the tube.

The data on decreased heat transfer regimes have been obtained only for water and they are far from being complete. In Fig. 23 limit heat fluxes q_w^{lim} are plotted (i.e., the values at which heat transfer begins to diminish) versus the bulk enthalpy in that flow cross section from which

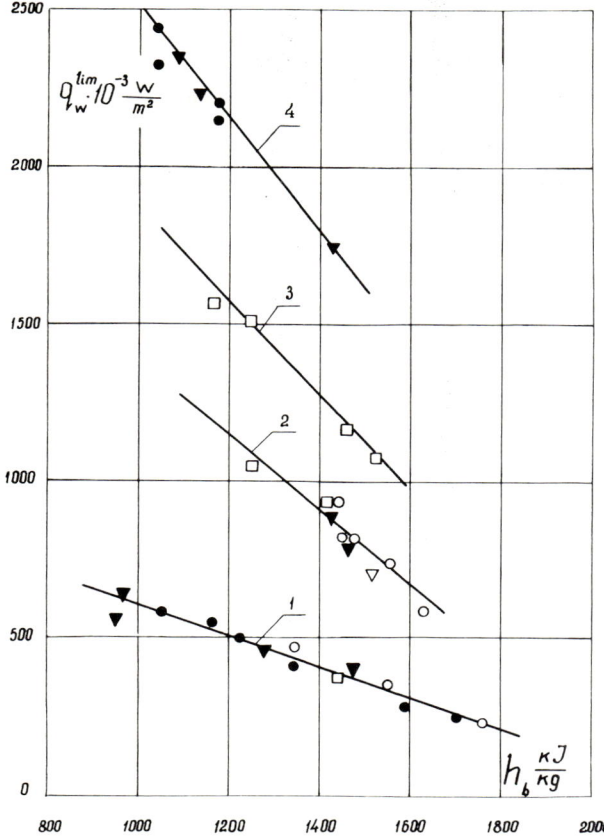

Fig. 23. Limiting values of heat fluxes for the downward flow of water in a tube d ($d = 8$ mm) with $\overline{\rho W}$ measured in kg/m² sec: (1) 430; (2) 700; (3) 950; (4) 1500. P is equal to: (○) 226, (●) 235, (▽) 245, (▼) 294, and (□) 343 bar.

the wall temperature begins increasing rapidly.[13] A straight line corresponds to each value of bulk velocity; in other words, the relation of q_w^{lim} vs h_b can approximately be considered linear. It is of interest to note that q_w^{lim} at prescribed values of h_b and $\overline{\rho W}$ does not depend on the pressure, at least over the pressure range of 226–343 bar. The area below the appropriate curves in Fig. 23 corresponds to normal heat transfer regimes (from the start of heating to the section under consideration

[13] Silin's treatment of experimental data (77) is shown in Fig. 23.

which naturally can also be the outlet); the area above these curves corresponds to decreased heat transfer regimes.

When decreased heat transfer regimes appear in a certain range of heat fluxes and other parameters, considerable pressure fluctuations as well as fluctuations in T_w and T_b at the entrance are observed. For example, according to (77), when $P = 245$ bar the amplitude of the pressure fluctuations reached 25 bar.

The appearance of diminished heat transfer regimes can probably be attributed to the change in flow mechanisms and, in particular, turbulent diffusivity processes caused by large changes in the physical properties over the flow cross section. Free convection appearing as a consequence of large density gradients can affect considerably the flow regime. This is confirmed by the fact that diminished heat transfer regimes are observed in an upward flow in heated tubes and are not found in a downward flow. In the latter case normal regimes occur under the same conditions which may be apparently connected with more intensive fluid mixing when the directions of forced and free convection at the wall are opposite. The small amount of available data (96) shows that, in the case of a horizontal flow, diminished heat transfer regimes can occur too, though not in a very pronounced manner.

Improved heat transfer regimes were discovered by Goldmann (64) and again later on in other works; for example, in (97). The experiments described by Goldmann were performed on a water flow in an electrically heated tube 1.58 mm in diameter and 203 mm in length. Figure 24 is a typical plot of liquid and wall temperature distribution along the tube length when $P = 345$ bar and $\overline{\rho W} = 2 \cdot 10^3$ kg/sec-m².

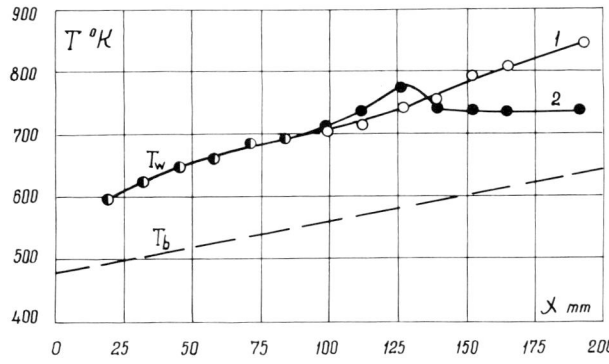

FIG. 24. Distribution of T_w (solid lines) and T_b (dotted lines) along the tube length at the onset of the improved heat transfer regime: (1) $q_w = 3.54 \times 10^6$ W/m²; (2) $q_w = 3.67 \times 10^6$ W/m². Experiments with water at $P = 345$ bar, $\overline{\rho W} = 2 \times 10^3$ kg/sec-m², and $d = 1.58$ mm.

When $q_w = 3.54 \times 10^6$ W/m² (Curve 1), the wall temperature increases along the whole tube and is described by the relationships valid for the normal heat transfer regime. But with a small increase in heat flux to $q_w = 3.67 \times 10^6$ W/m² (when P, $\overline{\rho W}$, and T_b are the same as at the entrance) the wall temperature at a distance $x \simeq 125$ mm from the entrance (after a small local increase) decreases rapidly and over the rest of the tube, which is $50d$ in length, has a constant value. In this case the constant value $T_w = 733°$K is a little bit higher than $T_m \simeq 691°$K. Heat transfer in the region with constant T_w is much higher and T_w is much lower in comparison with the values which should have been expected for the normal regime.

Improved heat transfer regimes are always accompanied by a loud sound similar to a sharp whistle with a frequency of approximately 1400–2200 Hz.

Improved heat transfer regimes arise only with high heat fluxes (usually higher than 3×10^6 W/m²) and when $T_b < T_m$ after the wall temperature reaches a value somewhat lower or, more often, somewhat higher than T_m. Heat fluxes corresponding to the onset of improved heat transfer, all other conditions being equal, increase with increasing P and $\overline{\rho W}$ and decrease when the fluid temperature at the entrance increases. When improved heat transfer regimes occur, a further increase of heat flux causes the wall temperature as a whole to increase in the region adjacent to the exit, while for a given heat flux it remains constant along the length of this region.

Apparently heat transfer improvement is connected with a rather considerable change in physical properties (mainly ρ and C_p) over the flow cross section at high heat fluxes. Under the conditions when improved heat transfer regimes are observed, fluid density at the wall is several times smaller and heat capacity is several times higher than in the core. Fluid particles coming from the core to the hot wall due to turbulent diffusivity possess relatively high thermal conductivity and small heat capacity. With high temperature differences between the layer adjacent to the wall and the particles which come from the core, the particles (64) are heated rapidly and explosively dilated. This process yields intensive fluid mixing in the wall layer and, consequently, higher heat transfer. The existing theory neglects the above-mentioned heat transfer mechanisms with variable physical properties. Therefore, the theory is not yet able to describe improved heat transfer regimes satisfactorily.

For an interpretation of heat transfer diminution and enhancement some investigators use the analogy with surface liquid boiling. They identify diminuation with the heat transfer crisis of subcooled liquid boiling and enhancement with the nucleate boiling of a subcooled liquid. This analogy is based on the following considerations. The supercritical region can be conventionally subdivided into two regions with a line of maximum heat capacity on isobars serving as their border. The region of $T < T_\mathrm{m}$ is conventionally considered a liquid phase and the region $T > T_\mathrm{m}$ is taken to be a vapor phase. The fact that both of the abovementioned phenomena occur when $T_\mathrm{b} < T_\mathrm{m}$ and $T_\mathrm{w} \gtrsim T_\mathrm{m}$ at sufficiently high heat fluxes[14] (i.e., when densities in the core and at the wall are substantially different) was the reason for an attempt to explain these phenomena by the processes similar to those which take place during subcooled liquid boiling. However, these assumptions demand careful experimental verification and theoretical consideration.

VII. Conclusion

All that has been said regarding heat transfer in turbulent pipe flow can be summarized in the following manner. The analytical methods allow us to describe heat transfer mechanisms for constant liquid properties quite satisfactorily and to take into account the influence of the variation of physical properties with temperature versus heat transfer and skin friction in a number of important cases. Disagreement between theoretical and experimental results observed in other cases, in particular, with a considerable change in physical properties over the flow cross section, may be attributed to imperfect methods of estimating the effect of the variation of physical properties on turbulent diffusivity. Therefore, further refinement of the analytical methods demands enhanced study of turbulent diffusivity with respect to variable physical properties.

Important experimental material has been accumulated on heat transfer and skin friction for variable physical properties. However, certain portions of this material possess relatively low accuracy that prevents its successful use. For a number of important cases there has been no systematic data collection, or that which is available is scanty and contradictory. Therefore, the need for further experimental investigations, with a high degree of accuracy, into the fluid mechanics and heat transfer for variable physical properties is quite urgent.

---

[14] The former takes place at relatively lower heat fluxes; the latter occurs at much higher heat fluxes.

Heat Transfer and Friction in Turbulent Pipe Flow

Nomenclature

- α heat transfer coefficient
- $\beta = \epsilon_q/\epsilon_\sigma$, the ratio of the eddy diffusivity of heat to the eddy diffusivity of momentum
- C_p specific heat at constant pressure
- $d = 2r_0$, tube diameter
- ϵ_q eddy diffusivity of heat
- ϵ_σ eddy diffusivity of momentum
- η $v^* y/\nu$, universal coordinate
- φ W_x/v^*, dimensionless velocity
- φ^+, η^+ Goldmann's variables; see Eqs. (24)
- G mass flow rate
- Gr   Grashof number
- h enthalpy
- κ const
- l tube length
- λ thermal conductivity
- μ dynamic viscosity
- Nu   Nusselt number
- ν kinematic viscosity
- ξ friction factor, $\sim \xi = 8\sigma\rho/(\overline{\rho W_x})^2$
- Pr   Prandtl number
- P pressure
- q heat flux
- Re   Reynolds number
- $R = r/r_0$, dimensionless radial coordinate
- r coordinate along radius
- ρ density
- $\overline{\rho W_x}$ average mass velocity over a given cross section
- Sc   Schmidt number
- Sh   Sherwood number
- σ shear stress
- T temperature, °K
- t temperature, °C
- $\theta = T_w/T_b$, temperature ratio parameter
- v^* friction velocity
- W_x, W_r velocity vector components
- x coordinate along tube axis
- $x = (T_x - T_b)/(T_w - T_b)$, dimensionless characteristic temperature
- y coordinate along the normal to the wall measured from the wall

Subscripts

- b   properties at bulk temperature
- cr   critical
- m   pseudocritical
- 0   predicted results or constant properties
- w   values at the wall or properties at the wall temperature
- x   characteristic temperature

References

1. W. Nusselt, *Mitt. Forsch.-Arb. Ing.-Wes. (VDI-Forsch.-Heft)* No. 89, 1–38 (1910).
2. O. Reynolds, "Scientific Papers," Vol. 1. Cambridge Univ. Press, London and New York, 1901.
3. G. Y. Taylor, Tech. Rept. Adv. Comm. Aer, Vol. II, pp. 423–429. Rep. Mem. No. 272 (May 1916).
4. L. Prandtl, *Phys. Zs. Bd.* **11**, 1072–1078 (1910).
5. L. Prandtl, *Phys. Zs, Bd.* **29**, 487–489 (1928).
6. T. Karman, *Trans. ASME* **61**, 705–710 (1939).
7. H. Reichardt, *Arch. Ges. Warmetechnik*, **6/7**, 129–143 (1951).
8. R. N. Lyon, *Chem. Eng. Progr*, **47**, No. 2 (1951).
9. R. A. Seban and T. T. Shimazaki, *Trans. ASME* **73**. No. 6, 803–809 (1951).
10. R. G. Deissler, *Trans. ASME* **73**. No. 2, 101–107 (1951).

11. R. G. Deissler, NACA Tech. Rept. 1210 (1955).
12. G. P. Piterskikh, *Khim. Prom.* No. 8, 480–485 (1954).
13. B. S. Petukhov and V. V. Kirillov, *Teploenerget.* No. 4 (1958).
14. B. S. Petukhov and V. N. Popov, *Teplofiz. Vysok. Temperatur (High Temperature Heat Physics).* **1**. No. 1 (1963).
15. S. S. Kutateladze *Teploenerget.* No. 7940 (1956).
16. S. S. Kutateladze, *Zh. Prik. Mekhan i Tekhn. Fiz.* No. 1, 129–132 (1960).
17. S. S. Kutateladze and A. I. Leontiev, "Turbulentnyi pogranichnyi sloi szhimaemogo gaza" (Turbulent Boundary Layer of a Compressible Gas). Izv. Sibirsk. Otd. Akad. Nauk SSSR, 1962.
18. R. G. Deissler, *Trans. ASME* **76**. No. 1, 73–86 (1954).
19. K. Goldmann, *Chem. Eng. Progr. Symp. Ser., Nucl. Eng. Part I* **50**. No. II, 105–113 (1954).
20. J. O. Hinze, "Turbulence. An Introduction to its Mechanism and Theory," 4th ed. McGraw-Hill, New York, 1959.
21. A. S. Monin and A. M. Yaglom, Statisticheskaya gidromekhanika Mekhanika turbulentnosti (Statistical fluid mechanics and turbulence mechanisms). *Izv. "Nauka"* 1 (1965).
22. L. G. Loitsyansky, Trudy Vsesouznogo S'ezda po teoreticheskoi i prikladnoi mekhanike 1960 (Proceedings of the All-Union Congress on Theoretical and Applied Mechanics), *Izv. Akad. Nauk SSSR* (1962).
23. J. Kestin, P. D. Richardson, *Intern. J. Heat Mass Transfer* **6**. No. 2, 147–189 (1963).
24. R. G. Deissler, *in* "Turbulent Flows and Heat Transfer" (C. C. Lin, ed.). Princeton Univ. Press, 1959.
25. B. A. Kader, *Mekhan. zhidkosti i gaza.* No. 6, 157–163 (1966).
26. R. A. Seban and T. T. Shimazaki, *Proc. Gen. Disc. Heat Transfer.* Inst. Mech. Engineers, London, 1951.
27. W. Corcoran, F. Page, W. G. Schlinger, and B. H. Sage. *Inv. Eng. Chem.* **44**, (1952).
28. S. D. Cavers, N. T. Hsu, W. G. Schlinger, and B. H. Sage, *Ind. Eng. Chem.* **45**, No. 10, 2139–2145 (1953).
29. C. A. Sleicher, *Trans. ASME* **80**. No. 3 (1958).
30. H. Ludwieg, *Z. fur Flugwissenschaften* **4** (1956).
31. E. M. Sparrow, T. M. Hollman, and R. Siegel, *Appl. Sci. Res. Sect. A* **7**, 37–52 (1957).
32. C. A. Sleicher and M. Tribus, *Trans. ASME* **79**, No. 4, 789–797 (1957).
33. R. Siegel and E. M. Sparrow, *J. Heat Transfer (Trans. ASME Ser. C)* p. **82**, No. 2, 152–153 (1960).
34. R. W. Allen and E. R. G. Eckert, *J. Heat Transfer (Trans. ASME Ser. C)* **86**, No. 3 (1964).
35. J. A. Malina and E. M. Sparrow, *Chem. Eng. Sci.* **19**, 953–962 (1964).
36. V. V. Yakovlev, *At. Energ.* No. 3 (1960).
37. D. F. Dipprey and R. H. Sabersky, *Intern. J. Heat Mass Transfer* **6**, No. 5, 329–353 (1963).
38. P. M. Volkov and A. V. Ivanova, Heat transfer and hydrodynamics in elements of power equipment, *Tr. TsKTI* **73**, (1966).
39. B. S. Petukhov and L. I. Roisen, *Teplofiz. Vysoc. Temperatur* **1**, No. 3 (1963).
40. L. S. Sterman and V. V. Petukhov, *in* "Teplo i Massoperenos" (Heat and Mass Transfer), Vol. I. Izv. Nauka itekhnika, Minsk, 1965.
41. R. M. Hamilton, Ph.D. Thesis, Cornell Univ. (June, 1963).
42. E. N. Sieder and G. N. Tate, *Ind. Eng. Chem.* **28**. No. 12 (1936).
43. F. Kreith and M. Summerfield, *Trans. ASME* **71**. No. 7 (1949).

44. M. A. Mikheev, *Izv. Akad. Nauk SSSR Otd. Tekhn. Nauk.* No. 10 (1952).
45. F. Kreith and M. Summerfield, *Trans. ASME* **72**. No. 6 (1950).
46. G. Rohonczy, *Schweizer Arch.* No. 5 (1939).
47. R. G. Deissler and A. F. Presler, *Intern. Heat Transfer Conf. Colorado-London* Part III, papers 68 (1961–1962).
48. L. N. Il'in, *Kotloturbostroenie* No. 1 (1951).
49. L. V. Humble, W. H. Lowdermilk, and L. G. Desmon, NACA Rept. 1020 (1951).
50. J. Bialokoz and O. Saunders, *Combustion and Boilerhouse Engineering* (November, 1956).
51. F. C. Wright and H. Walters, WADC Techn. Rept. 59–423 (August 1959).
52. M. F. Taylor and T. A. Kirchgessner, *Am. Rocket Soc. J.* **30**, No. 4 (1960).
53. J. R. McCarthy and H. Wolf, *Am. Rocket Soc. J.* **30**. No. 4 (1960).
54. J. R. McCarthy and H. Wolf, Rept. No. RR-60-12, Rocketdyne, Canoga Park, California (1960).
55. W. F. Wieland, *AICHE Symp. Nucl. Eng. Heat Transfer, Chicago, 1962*.
56. M. F. Taylor, "Heat Transfer and Fluid Mechanics Institute," p. 251. Stanford Univ. Press, Stanford, California, 1963.
57. D. M. McEligot, P. M. Magee, and G. Leppert, *J. Heat Transfer (Trans. ASME Ser. C)* **87**, No. 1 (1965).
58. V. V. Kirillov and Yu. S. Malugin, *Teplofiz. Vysok. Temperatur (High Temp. Heat Phys.)* **1**. No. 2 (1963).
59. V. L. Lelchuk, G. I. Elphimov, and Yu. P. Fedotov, *in* "Teplo i Massoperenos" (Heat and Mass Transfer), v. I. Minsk, 1965.
60. H. C. Perkins and P. Worsoe-Schmidt, *Intern. J. Heat Mass Transfer*, **8**, No. 7, 1011–1031 (1965).
61. B. S. Petukhov, V. V. Kirillov, and V. N. Maidanik, *Intern. Heat Transfer Conf. 3rd, Chicago* **1**, paper 28 (1966).
62. V. L. Lelchuk and B. V. Dyadyakin, *in* "Voprosy Teploobmena" (Heat and Mass Transfer Problems). Izd. Akad. Nauk SSSR, 1959.
63. N. I. Ivashchenko, *Teploenerg.* No. 2 (1958).
64. K. Goldmann, International developments in heat transfer, Rep. 67. *Intern. Heat Transfer Conf. Colorado*, Part III, Re p. 67 (1961).
65. H. Tanaka, N. Nishiwaki, and M. Hirata, *Semi-Intern. Symp. Heat Mass Transfer Thermal Stress, Tokyo, 1967. Abstracts. J. SME* (1967).
66. V. N. Popov, *in* "Teplo i Massoperenos" (Heat and Mass Transfer), Vol. 1. Izd. Nauka i Tekhnika, Minsk, 1965.
67. H. S. Swenson, J. R. Carver, and C. R. Kakarala, *J. Heat Transfer (Trans. ASME Ser. C.* **87**, No. 4 (1965).
68. R. P. Bringer and J. M. Smith, *AIChE J.* **3**. No. 1 (1957).
69. N. I. Melik-Pashaev, *Teplofiz. Vysok. Temperatur* (High Temp. Heat Phys.) **4**, No. 6 (1966).
70. I-Yu Hsu and D. Smith, *Heat Transfer (Trans. ASME Ser. C)* **83**, No. 2, 94–104 (1961).
71. E. J. Szetela, *Am. Rocket Soc. J.* **32**, 1289–1292 (1962).
72. W. B. Hall, J. D. Jackson, and S. A. Khan, *Proc. Intern. Heat Transfer Conf. 3rd.* **1**, 257–266 (1966).
73. A. A. Armand, N. V. Tarasova, and A. S. Kon'kov, *in* "Teploobmen pri Vysokikh Teplovykh Nagruzkakh i Drugikh Spetsialnykh Usloviyakh" ("Heat Transfer Under High Heat Fluxes and Other Special Conditions"). Gosenergoizdat, Moscow, 1959.
74. Yu. V. Vikhrev, Yu. D. Barulin and A. S. Kon'kov, *Teploenerg.* No. 9, 80–82 (1967).

75. Z. L. Miropol'sky and M. E. Shitsman, *Zh. Tekhn. Fiz.* **27**, 10 (1957).
76. Z. L. Miropol'sky and M. E. Shitsman, *in* "Issledovaniya Teplootdachi k Pary i Vode Kipyashchei v Trubakh Pri Vysokikh Davleniyakh" (Investigations of Heat Transfer to Vapor and Water Boiling in Tubes under High Pressure). Atomizdat, Moscow, 1958.
77. M. E. Shitsman, *Teplofiz. Vysok. Temperatur* **1**. No. 2, 267–275 (1963).
78. I.T. Alad'ev, N. A. Vel'tishchev and N. S. Kondratiev, *in* "Teploperedacha i Teplovoe Modelirovanie" (Heat Transfer and Heat Simulation), p. 158. Izd. Akad. Nauk SSSR, 1959.
79. N. L. Dickinson and C. P. Welch, *Trans. ASME* **80**, 746–752 (1958).
80. A. I. Chalfant, PWAC-109 (June, 1954).
81. D. G. Randall, NDA 2–51 (Nov. 1956), TID–7529, Pt. 3 (Nov. 1957).
82. K. Yamagata, K. Nishikawa, S. Hasegawa, and F. Fujii, *Semi-Intern. Symp. Papers Heat Mass Transfer Thermal Stress, Tokyo*, 1967. *J. SME* **2** (1967).
83. B. S. Petukhov, E. A. Krasnoshchekov, and V. S. Protopopov, Part III, Rept. 67, 1961. *Intern. Heat Transfer Conf. Intern. Develop. Heat Transfer, Colorado,* 1961.
84. E. A. Krasnoshchekov, V. S. Protopopov, Van-Fan, and I. V. Kuraeva, *in* "Teplo i massoperenos" (Heat and Mass Transfer), Vol. Izd. Nauka i tekhnika, Minsk, 1965.
85. E. A. Krasnoshchekov and V. S. Protopopov, *Teplofiz. Vysok. Temperatur* **4**, No. 3, 389–398 (1966).
86. L. B. Koppel and J. M. Smith, Transfer, Part III, Rept. 69. Intern. Heat Transfer Conf., *Intern. Develop. Heat Transfer Colorado* 1961. Part III, Rept. 69 (1961).
87. R. D. Wood and J. M. Smith, *AIChE J.* **10**, No. 2 (1964).
88. R. C. Hendricks, R. W. Graham, Y. Y. Hsu, and A. A. Meckrios, *Am. Rocket Soc. J.* **32**, 244–252 (1962).
89. W. B. Powell, *Jet Propulsion* **27**, No. 7, 776–783 (1957).
90. J. P. Holman, S. N. Rea, and C. E. Howard, *Intern. J. Heat Mass Transfer* **8**, No. 8, 1095–1102 (1965).
91. M. E. Shitsman, *Teploenerget.* No. 1 (1959).
92. M. E. Shitsman, *Teploenerget.* No. 7 (1966).
93. Kh. L. Hess and Kh. R. Kunts, *J. Heat Transfer (Trans. ASME Ser. C)*, **87**, No. 1 (1965).
94. N. V. Tarasova and A. I. Leontiev, *Teplofiz. Vysok. Temperatur* **4**. No. 4 (1968).
95. K. R. Schmidt, Mitteilungen der Vereinigung der Grosskesselbesitzer, Heft 63 (Dezember, 1959).
96. Yu. V. Vikhrev and V. A. Lokshin, *Teploenerget.* No. 12 (1964).
97. N. L. Kaphengauz, *Dokl. Akad. Nauk SSSR* **173**. No. 3, 577–559 (1967).

Author Index

Numbers in parentheses are reference numbers and indicate that an author's work is referred to although his name is not cited in the text. Numbers in italics show the page on which the complete reference is listed.

A

Abarbanel, S. S., 95(156), *130*
Abbott, D. E., 109, *130*
Acrivos, A., 61, *127*
Adams, D. E., 460, 469(187), 475, *495*
Adams, J. A., 312(101), *366*
Agnone, A., 41(81), 65(81), *126*
Ai, D. K., 53(100), *127*
Aitsam, A. M., 407, 408(62), 409(62), *489*
Akman, R. G., 418(101), *491*
Alad'ev, I. T., 551, *564*
Alber, I. E., 29(38), 119, *124*
Alber, J. E., 107(187), *132*
Aleksashenko, A. A., 474(333), *502*
Aleksashenko, V. A., 474(333), *502*
Aldoshin, G. T., 416(25), 418(152), *487*, *493*
Allen, R. W., 525(34), 526, 528, 530(34), 531, 532, *562*
Ambrok, G. S., 474(255), 483(255), *498*
Amitonov, R. V., 416(90), *490*
Anderson, A. D., 474(232), 480(232), 482, 483(232), *497*
Andriyanov, P. A., 474(307), *501*
Armand, A. A., 474(272, 278, 303), *499*, *500*, *501*, 551, *563*
Arpaci, V. S., 474(265, 266), *499*
Attridge, J. L., 94(145), *129*
Auxer, W. L., 41(89), 42(89), 69(89), 75(89), 82(89), 94(89), 95(89), *126*
Avizov, A. M., 418(135), *492*
Avkhimovich, B. M., 418(112, 127), *491*, *492*
Aziz, K., 474(257), 483(257), *499*

B

Bach, J., 334, 336(91), *366*
Badrinarayan, M. A., 29(48), *124*

Baida, M. M., 465, *495*
Ball, E. F., 418(103), *491*
Bankoff, S. G., 435(169, 170, 173), 437(169, 170, 173), *494*
Barakat, H. Z., 474(249, 251), 483(249, 251), 484, *498*
Barnes, J. W., 17(27), 18(27), 25(27), 27(27), *123*
Barulin, Yu. D., 551, 555(74), *563*
Bataill, J., 424(157), 430(157), *493*
Batchelor, G. K., 5, 59, 60, *122*, *127*
Bates, W. J., 197, *364*
Bathla, P. S., 474(330), *502*
Batov, L. P., 474(323), *502*
Batt, R. G., 36(66), 37(66), 51(66), 52(66), *125*
Bauer, R. C., 25, *131*, *132*
Baum, E., 20(31, 32, 33), 22, 23(33), 24(32), 36, 38, 39(32), 40(32), 101(161, 162, 163, 164), 102, 104, 105(164), 106(164), 107(163), *123*, *130*
Beastall, D., 29(42), *124*
Beck, J. V., 416(81, 83, 84), *490*
Belostotskii, B. R., 418(123), *492*
Bennett, F. D., 205, 220, 222(66), *364*, *365*
Bennett, G. S., 211(61), *365*
Benoit, J. R., 196(30), *363*
Bentwich, M., 474(207), *496*
Berger, S. A., 6, 38, 98(12), *122*, *125*
Berkovich, S. Y., 418(102), *491*
Berry, C. J., 98(159), *130*
Bershader, D., 244(75), *365*
Bertram, M. H., 79(134), *128*
Bialokoz, J., 538, *563*
Biondo, P. P., 57(101), 58(101), 62(101), 64(101), 65(101), 107(101), *127*
Bird, G. A., 2(3), *122*

565

Bitsyutko, J. Ya., 474(213, 214), 478, *496*
Blue, R. E., 244(72), *365*
Boehler, J. P., 38(72), *125*
Bogdonoff, S. M., 35(61), 80(135), 95(155), *125*, *128*, *130*
Boikov, G. P., 474(328), *502*
Boldmen, D. R., 404, *488*
Born, M., 135(1), 169, 179, 195, 219, 220, 292, *362*
Bornage, H., 36(62), *125*
Bosworth, R. C. L., *497*
Bradshaw, P., 407, *488*
Brag, K. N. C., 95(157), *130*
Brdlik, P. M., 312(100), *366*
Briggs, D. G., 474(242), 480(242), 482, *498*
Bringer, R. P., 549, 551, *563*
Brodowicz, K., 312(99), *366*
Brovkin, L. A., 418(144), *493*
Brun, E. A., 38(72), *125*
Burbank, P. D., 94(147), *129*
Burch, J. M., 201, *364*
Burggraf, O. R., 60, 62(104, 105), *127*, 416(82), *490*
Burghart, G. H., 11(21), 30(21), 34(21), *123*
Burka, A. L., 474(321, 322, 327), *502*
Buss, H., 14(25), *123*
Buynyachenko, G. P., 416(86), *490*

C

Carichner, G. E., 69(120), 71, *128*
Carlson, W. D., 474(252), 483(252), 484, *498*
Carslon, W. O., 312(97), *366*
Carriere, P., 87, *129*
Carslow, H. S., *489*
Carstens, M. K., 405, 406(60), 407(60), 408(60), 409(60), 410(60), *489*
Carver, J. R., 549(67), 551, 553(67), *563*
Cavers, S. D., 515(28), *562*
Centolanzi, F. J., 77(133), *128*
Cess, R. D., 474(196, 210), 475, 476, *495*, *496*
Chakalev, K. N., 376(6), 414(6), 415(6), 463(6), 464(6), 467(6), *486*
Chalfant, A. I., 551, *564*
Chambre, P. L., 460, 468, 474(200), 477, *495*, *496*
Channapragada, R. S., 17, *123*, 416(93), *490*

Chao, B. T., 474(204, 224), 476, 477, *496*, *497*
Chapman, D. R., 5, 7, 13, 33, 39(22), 84, 89(7), *122*, *123*
Charwat, A. F., 11(21), 23, 30(21), 33, 34(21), 35(53), 41(83), 44, 51(53), 68(83, 115), 69(115), 70(115), 73, 75(130), 76(83), 77(130), 78, 79(83), 80(130), 81(130), 82, *123*, *124*, *126*, *127*, *128*
Chato, J. C., 474(243), 480(243), 483(243), *498*
Cheema, L. S., 474(224), 477, *497*
Cheng, R., 38(73), *125*
Chentsov, A. K., 418(102), *491*
Cherkasov, V. I., 418(122), *492*
Childs, M. E., 7(17, 18), 15(17), *123*
Chu, S. C., 435(170), 437(170), *494*
Chung, P. M., 474(232), 480(232), 482, 483(232), *497*
Churchill, S. W., 345, 346, *366*, 474(234), 480(234), 483(234), *497*
Clark, J. A., 474(251, 265, 266), 483(251), 484, *498*, *499*
Clark, R. J., 211, *365*
Clay, W. G., 36(63), *125*
Coats, D. E., 107(187), *132*
Cohen, C. B., 111, *131*
Collins, I. K., 94(147), *129*
Cope, W. F., 94(145), *129*
Corcoran, W., 515(27), *562*
Crawford, D. R., 116(181c, 181d), 117 (181c, 181d), *131*
Creekmore, H. S., 94(146), *129*
Cresci, R. J., 41(79, 82), 65(79, 82), *126*
Crocco, L., 5, 36, 85(6), 108, 109, 112, *122*
Cross, E. J., 114, *131*
Csermerly, T. J., 474(306), *501*
Curle, N., 89, 108(143), *129*

D

Dahm, W. K., 40(75), 41(75), 65(75), 66(75), 75(75), 77(75), *125*
Daily, I. W., 406, 407, 409(61), 410, *489*
Dalgleish, N., 474(294), *500*
Danenberg, V. V., 447(179), 448(180), 474(285, 317), *494*, *500*, *501*
Danilov, Y. I., 449(181), 453(181), 456 (181), 457(181), *495*
Davis, B. M., 98(159), *130*

AUTHOR INDEX

De Farias, F. N., 435(172), 467, *494*
Deissler, R. G., 404, *488*, 504(10, 11), 506, 512, 515, 524, 528, 534(24), 537, 544(24), *561*, *562*, *563*
Delery, J., 30(50), 31(51), 51(96), 65(96), 66(96), 113(51), *124*, *127*
Delgado Domingos, J., 434(164), *494*
Demetriades, A., 36(64, 65), *125*
Denison, M. R., 20(31, 32), 22, 24(32), 36(32), 38(32), 39(32), 40(32), 101(163), 104, 107(163), *123*, *130*
Desmon, L. G., 538, *563*
Desoto, S., 474(324), *502*
Deverall, L. I., 416(93), *490*
Devoino, A. N., 417(97), *490*
Devyatov, B. N., 474(290), *500*
Dewey, C. F., Jr., 22, 36(67), 37(67), 41(67, 83), 68(83, 115), 69(115), 70(115), 73(83), 76(83), 78(83), 79(83), 82(83), *123*, *125*, *126*, *127*
Dhanak, A. M., 61(106), *127*
Dickinson, N. L., 551, *564*
Dipprey, D. F., 525(37), 526, *562*
Donaldson, I. S., 51(95), 52(95), *126*
Dorodnitsyn, A. A., 116, *131*
Drake, D. G., 474(208), *496*
Drake, R. M., 211(57), *364*
Dreitser, G. A., 449(181, 182, 183, 184, 193), 453(181, 182, 183, 184, 193), 456(181, 182, 183, 184), 457(181, 182, 183, 184), *495*
Dring, R. P., 312 (103, 104), 366 474(258, 264), 480(258,264), 483(258), *499*
Drougge, G., 95(150), *129*
Dubrovich, E. M., 424(156), *493*
Dulnev, G. N., 418(122), *492*
Dunham, W. H., 73(125), *128*
Duric, M. D., 396(33), *487*
Dyadyakin, B. V., 540(62), 542(62), 543(62), *563*

E

East, L. F., 73(116, 128), *127*, *128*
Eckert, E. R. G., 40(76), 41(76), 77(76), 83(136a), *126*, *128*, 211, 312(95-98), *364*, *366*, 474(227), 480(227), 483, *497*, 525 (34), 526, 528, 530(34), 531, 532, *562*
Eggink, H., 29(42), *124*
Egorov, V. S., 418(134), *492*

Ehlers, R. C., 404(58), *488*
Eichhorm, K., 474(235), 480(235), 482, *497*
Eichhorn, R., 474(241), 480(241), 482(241), *498*
Elder, J. W., 474(335), *502*
Elperin, I. T., 418(126), *492*
Elphimov, G. I., 539, *563*
Elrod, H. G., 418(100), 474(274), *491*, *499*
Emery, A., 42(90), *126*
Emery, J. C., 95(149), *129*
Enazen, V. A., 418(132), *492*
Erdmann, S. F., 202, *364*
Erdos, J., 87, 89(140), 90, 91(140), *129*
Eremenko, E. V., 399, 401, *488*
Ermolchik, V. N., 474(319), *501*
Evans, L. B., 474(254), 483(254), 484, *498*

F

Favre, A., 36(62), *125*
Fedorov, V. B., 418(134), *492*
Fedorova, L. E., 418(134), *492*
Fedotov, Yu. P., 539, *563*
Fedyaevsky, K. K., 410(72), 411(71, 72), *489*
Ferris, D. H., 404, *488*
Finlay, I. C., 474(294, 304), *500*, *501*
Foraboschi, P., 474(286), *500*
Forster, D., 474(244), 483(244), *498*
Fox, J., 76, *128*
Françon, M., 174, 195, 197(37), 199, *363*, *364*
Fridlender, N. A., 418(109), *491*
Friedman, A., 474(209), *496*
Fromm, J., 2, *122*
Fujii, F., 551, 555(82), *564*

G

Gabor, D., 200, *364*
Gadd, G. E., 29(39), 35(57), 36, 94(145), 95(151, 154, 157), 108(154, 167), *124*, *129*, *130*
Galichskii, A. K., 418(132), *492*
Galitseysky, B. M., 449(181, 182, 183, 184), 453(181, 182, 183, 184), 456(181, 182, 183, 184), 457(181, 182, 183, 184), *495*
Gandelsman, A. F., 390(15, 19), 405(15, 19), *463*
Gardiner, C. P., 17(27), 18(27), 25(27), 27(27), *123*

Garrison, C. J., 411(74), *489*
Gartner, J. R., 474(295), *500*
Gates, J. W., 201(46), *364*
Gaviglio, J., 36(62), *125*
Gayhart, E. L., 202, *364*
Gebhart, B., 211, 312(103, 104), *365, 366,* 460, 469(187), 474(231, 239, 240, 258, 264), 475, 480(231, 239, 240, 258, 264), 481, 482, 483(258), *495, 497, 498, 499*
Giat, M., 424(157), 430(157), *493*
Gibson, T. S., 73(123), *128*
Gill, W. N., 447, *494*
Ginevskii, A. S., 410(72), 411(71, 72), *489*
Ginoux, J. J., 33, 67, 73, 80, 87(185), 92(144), 93(144), 95(144), 96(144), *124, 127, 128, 129, 131*
Glatz, G. A., 474(250), 483(250), *498*
Glenn, L. A., 474(324), *502*
Glick, H. S., 109, *130*
Glushko, G. S., 398, 399, *488*
Goin, K. L., 29(45), *124*
Goldmann, K., 506, 515, 544, 558, 559(64), *562, 563*
Goldstein, R. J., 221, *365,* 474(227, 242), 480(227, 242), 483, *497, 498*
Goldstein, S., 23, *123,* 312(96), 397(41), *366, 488*
Golik, R. J., 109(174, 176), 110(174), 112(174), 113(174), *131*
Golosov, A. S., 416(25), *487*
Golovastikov, P. P., 418(102), *491*
Gomez, A. V., 23, *123*
Goncharov, E. I., 418(133, 145), *492, 493*
Goodman, T. R., 474(197, 198), 475, 476, 480, *495, 496*
Goodwin, F. K., 116(183), 117(183), 118(183), *131*
Gordov, A. N., 418(98, 99), *490*
Graczyk, Cz., 474(289), *500*
Graham, R. W., 404, *488,* 550(88), *564*
Grange, J. M., 109(175), 119, *131*
Greber, I., 95(156), *130*
Green, A., 197(32), *364*
Greenberg, R. A., 57(101), 58(101), 62(101), 64(101), 65(101), 107(101), *127*
Gregg, J. L., 474, 476, 480(228), 481, *495, 497*
Grey, J. D., 95(152), *129*
Grigoriev, Y. M., 474(205), 479, *496*

Grigull, U., 150(11), 221, 312(87), 316(87), 331(90), 340(92), *363, 365, 366*
Groden, C. M., *497*
Gromeka, I. S., 396, *487*
Gromyko, A. G., 416(90), *490*
Grosbie, A. L., 474(325, 329), *502*
Gukhman, A. A., 390(15, 19), 405, *486*
Gupta, C. L., 474(256), 483(256), *498*
Gupta, M. L., 474(256), 483(256), *498*
Gusev, J. A., 474(206), *496*

H

Haber, F., 197(35), *364*
Hahn, G., 265(79), *365*
Haines, K. A., 200(45), *364*
Haji-Sheikh, A., 415(79), *489*
Hakkinen, R. J., 95(156), *130*
Hall, R. G. N., 201(46), *364*
Hall, W. B., 550, 551, *563*
Hama, F. R., 35(59a, 59b, 59c), 44, 50(59a, 59b, 59c), 51(59a, 59b, 59c), 52(59a, 59b, 59c), 59, 105(59a, 59b, 59c), *125*
Hamilton, R. M., 526, *562*
Hankey, W. L., 114, *131*
Hannes, H., 205, 211, 275, 320, 322(89), *364, 365, 366*
Hanrew, W. L., 406(61), 407(61), 409(61), 410(61), *489*
Hansen, G., 211, *365*
Hanson, A. R., 40(75), 41(75), 65(75), 66(75), 75(75), 77(75), *125*
Harrison, H. L., 474(295), *500*
Hasegawa, S., 551, 555(82), *564*
Hastings, R. C., 29(41), *124*
Hauf, W., 244(76), 312(87), 316(87), 331(90), *365, 366*
Haugen, R. L., 61(106), *127*
Hause, C. D., 211(61), *365*
Hayasi, N., 418(107), *491*
Hayes, W. D., 105(165), *130*
Hellums, J. D., 474(234, 257), 480(234), 483(234, 257), *497, 499*
Hempel, A., 474(277), *499*
Hendricks, R. C., 550(88), *564*
Hertofilis, S. A., 95(153), *129*
Hess, Kh. L., 553, *564*
Hildebrand, B. P., 200(45), *364*
Hill, P. R., 414, 418, *489*

Hinze, J. O., 393(22), 394(22), 400(22), 487, 511(20), 562
Hirata, M., 546(65), 563
Hitchcock, J. E., 312(105), 366
Hitz, J. A., 41(83), 68(83, 115), 69(115), 70(115), 73(83), 76(83), 78(83), 79(83), 82(83), 126, 127
Holder, G. W., 29(39), 35(57), 36, 124, 195, 363
Hollman, T. M., 524(31), 562
Holloway, P. F., 94(146), 129
Holt, M., 109(171), 116, 117(181a, 181b, 181c), 118, 130, 131
Honda, M., 109, 130
Hsu, I-Yu, 549, 563
Hsu, N. T., 515(28), 562
Hsu, Y. Y., 550(88), 564
Hudson, J. L., 435(169, 173), 437(169, 173), 494
Humble, L. V., 538, 563
Hume, J. R., 474(281), 500
Hung, F. T., 474(332), 502
Hurlbut, F. C., 38(73), 125

I

Ihrig, H. K., 474(220), 479, 497
Ilchenko, O. T., 418(129, 136), 492
Il'in, L. N., 538, 541(48), 542, 563
Inouye, K., 418(107), 491
Isermann, R., 474(293), 500
Ishirawa, S., 396(34), 487
Iskra, A. L., 474(280), 500
Iuchi, M., 69(119), 128
Ivakin, V. P., 379(11), 380(11), 486
Ivanov, R. A., 418(140), 493
Ivanov, V. V., 418(121, 146, 147, 148, 149, 150), 492, 493
Ivanova, A. V., 526, 539, 562
Ivashchenko, M. M., 418(114, 115), 491
Ivashchenko, N. I., 542, 563
Izosimov, V. G., 449(181, 182, 183, 184, 193), 453(181, 182, 183, 184, 193), 456(181, 182, 183, 184), 457(181, 182, 183, 184), 495

J

Jackson, J. D., 550(72), 551, 563
Jaeger, J. C., 489

Jain, P., 396(39), 488
Jain, S. P., 474(256), 483(256), 498
Jamin, J., 195, 363
Jeng, D. R., 474(204), 476, 496
Johnson, H. A., 460, 461(189), 467, 474(203), 477, 495, 496
Johnstone, R. K. M., 211, 365
Jordan, J. M., 406(61), 407(61), 409(61), 410(61), 489
Joseph, D. D., 474(263), 499

K

Kabelitz, H. P., 71(121, 122), 76(121, 122), 79(122), 80(122), 128
Kader, B. A., 390(19), 405(19), 486, 513(25), 562
Kahl, G. D., 205, 222(66), 364, 365
Kakarala, C. R., 549(67), 551, 553(67), 563
Kalinin, E. K., 376(1), 387(14), 412(14), 416(96), 419(153), 449(181, 182, 183, 184, 193), 453(181, 182, 183, 184, 193), 456(181, 182, 183, 184), 457(181, 182, 183, 184), 486, 490, 493, 495
Kalishevsky, L. L., 409, 410, 489
Kaphengauz, N. L., 558(97), 564
Karker, Y. I., 474(323), 502
Karlsson, S. K. F., 411(73), 489
Karman, T., 504, 561
Karpin, E. B., 418(118), 491
Katsnelson, B. D., 474(211), 477, 496
Katsnelson, G. G., 390(19), 405(19), 486
Katz, E., 29(44), 124
Kavanau, L. L., 35(60), 125
Kawaguti, M., 396(39), 488
Kays, W. M., 404, 488
Keller, H. B., 2, 122
Kennard, R. B., 278, 365
Kennedy, E. D., 14, 123
Kestin, J., 511(23), 562
Khait, M. Y., 474(315), 501
Khan, S. A., 550(72), 551, 563
Kharchenko, V. N., 474(334), 502
Kharitonov, V. P., 418(125), 492
Kharitonov, Y. V., 474(285), 500
Khristichenko, P. I., 418(105), 491
Khvostova, J. Y., 474(311), 501
Kierkus, W. T., 312(99), 366
Killermann, F., 190, 363

Kinder, W., 195(27), 206, 211, 218, 224, *363*, *364*
King, C. Y., 474(261), 483(261), *499*
King, H. H., 20(32), 22, 24(32), 36(32), 38(32), 39(32), 40(32), *123*
Kirchgessner, T. A., 538, *563*
Kirchhoff, G., 2, *122*
Kirillov, V. V., 504(13), 523(13), 531(13), 539, 540(61), 541(61), *562*, *563*
Kirk, F. N., 18, *123*
Kistler, A. L., 67, *127*
Klineberg, J. M., 109(175), 119(175), *131*
Kmonicek, V., 416(94), *490*
Knowles, C. P., 211, *365*
Knuth, E. L., 383(12), 474(12), 476, *486*
Kochenov, I. S., 408, 409(66), 457, *489*, *495*
Kogan, M. G., 474(323), *502*
Kohlmayr, G. F., 418(128), *492*
Komoda, H., 69(119), *128*
Kondrashov, N. G., 465(192), *495*
Kondratiev, N. S., 551, *564*
Kon'kov, A. S., 551, 555(74), *563*
Koppel, L. B., 551, *564*
Korolkov, B. P., 474, *501*
Korst, H. H., 5, 7, 15, *122*, *123*, 474(220), 479, *497*
Koshkin, V. K., 376(1), 449(181, 182, 183, 184), 453(181, 182, 183, 184), 456(181, 182, 183, 184), 457(181, 182, 183, 184), *486*, *495*
Kostyuk, A. G., 418(118), *491*
Kotlyar, I. V., 474(319), *501*
Kovalev, S. D., 416(95), 417(97), *490*
Koverianov, V. A., 416(89), *490*
Kozanevich, Z. Y., 474(297, 298, 299), *500*
Krasheninnikov, V. V., 474(303, 313), *501*
Krasnoshchekov, E. A., 551, 552, *564*
Krause, F. R., 40(75), 41(75), 65(75), 66(75), 75(75), 77(75), *125*
Kraushaar, R., 199, *364*
Kreith, F., 531, *562*, *563*
Krishnamurty, K., 69(126), 73(126), *128*
Krug, W., 135(7), 195, 322, *363*, *365*
Kryukova, M. G., 474(212), 478, *496*
Kubota, T., 14, 22, 33(52), 89(52), 91(52), 92(52), *123*, *124*
Kudryashev, L. I., 376(8), 418(123), 463(8), 464, 466, 474(215, 216, 217), 478, 479, 480, *486*, *492*, *496*

Kudryavtsev, E. V., 376(6, 7), 414, 463, 464(6, 7), 467, 474(247), 480(247), 483 (247), *486*, *498*
Kuehn, D. M., 5(7), 7(7), 33(7), 84(7), 89(7), *122*
Kuhn, H., 197(36), *364*
Kunts, Kh. R., 553, *564*
Kuraeva, I. V., 551(84), *564*
Kutateladze, S. S., 506, 537, *562*
Kuznetsov, I. A., 418(106), *491*
Kuznetsov, V. K., 418(138), *492*
Kuznetsov, U. N., 408, 409(66), 437, 457, 474(336, 337), *489*, *495*, *502*
Kvon, V. I., 402, *488*

L

Ladenburg, R., 195, 268(78), 274(78), *363*, *365*
Ladiev, R. Y., 474(297, 298, 299), *500*
Lall, P. S., 474(301), *501*
Lamla, E., 205, 211, *364*
Landis, F., 474(253), 483(253), *498*
Landolt-Börnstein, Ç., 284(83), *366*
Larsen, M. A., 474(282, 284), *500*
Larsen, T. W., 474(312), *501*
Larson, H. K., 5(7), 7(7), 33(7), 39(74), 77, 78(74), 82(74), 84(7), 89(7), *122*, *125*
Larson, R. E., 40(75), 41(75), 65, 66, 75(75), 77(75), *125*
Lassiter, T. W., 73(123), *128*
Lau, E., 322, *365*
Lawrence, W. T., 474(243), 480(243), 483(243), *498*
Lees, L., 5, 29(38), 33(52), 36, 85(6), 89(52), 91(52), 92(52), 108, 109, 110, 112, 113(172, 174), 119, *122*, *124*, *130*, *131*
Leith, E. N., 200, *364*
Lelchuk, V. L., 539, 540(62), 542(62), 543(62), *563*
Leontiev, A. I., 379, 380(11), *486*, 506, 537, 554, *562*, *564*
Leppert, G., 539, 542(57), 543(57), *563*
Levin, Y. A., 418(120), *492*
Levy, A., 42(90), *126*
Lewis, J. E., 33, 89, 91(52), 92(52), *124*, *129*
Libby, P. A., 108(166), *130*
Li-Orlov, V. K., 418(137), *492*

AUTHOR INDEX 571

Liyv, U. R., 407, 408(62), 409(62, 63, 64, 65), 410(65), *489*
Lock, G. S. H., 474(250), 483(250), *498*
Löwe, F., 197(35), *364*
Loitsyansky, L. G., 511(22), *562*
Lokshin, V. A., 558(96), *564*
Lommel, E., 228, *365*
Lorenz, G. C., 41(84), *126*
Loscutoff, W. V., 474(305), *501*
Love, E. S., 44, *126*
Lowdermilk, W. H., 538, *563*
Ludwieg, H., 515(30), *562*
Luikov, A. V., 376, 377, 418(151), 464, 465(3, 4), *486*, *493*
Lundgren, T. S., 415(79), *489*
Lykoudis, P. S., 6, *122*
Lynes, L. L., 116(183), 117(183), 118(183), *131*
Lyon, R. N., 504, 519, *561*

M

McCarthy, J. R., 538, *563*
McEligot, D. M., 539, 542(57), 543, *563*
McFadden, P. W., 312(101), *366*
Mach, L., 195, 268, *363*
Madejski, J., 474(262), *499*
Magee, P. M., 539, 542(57), 543(57), *563*
Maidanik, V. N., 539, 540(61), 541(61), *563*
Malina, J. A., 525(35), 526, *562*
Malkin, V. M., 435(167), 437(167), *494*
Malugin, Yu. S., 539, *563*
Markov, S. B., 390(18), 402, *486*
Markovin, M. V., 396(28a), 405, *487*
Martellucci, A., 41(81), 65(81), 108(166), *126*, *130*
Masami Masubuchi, 474(267), *499*
Maslennikov, M. M., 474(307), *501*
Mason, J. N., 474(238), 483(238), *498*
Maull, D. J., 73(116), *127*
Meckrios, A. A., 550(88), *564*
Melik-Pashaev, N. I., 549, 551, *563*
Meng, J. C. S., 116(181b), 117(181b), 118, *131*
Menold, E. R., 474(233), 480(233), 482, *497*
Mercer, W. E., 312(105), *366*
Michaelson, A. A., 196, 197, *363*

Mikheev, M. A., 532, *563*
Millsaps, K., 435(168), 437(168), *494*
Minkwitz, G., 225, 226(70), 227(70), 228 (70), *365*
Mirande, J., 30(50), 31(51), 113(51), *124*
Mironov, B. P., 379(11), 380(11), *486*
Miropol'sky, Z. L., 551, 552, *564*
Mitalas, G. P., 416(85), *490*
Mitshell, J. W., 474(314), *501*
Mitskevich, A. I., 474(287), *500*
Mix, T., 82(138), *129*
Močalov, V. A., 312(100), *366*
Monin, A. S., 395(23), 398, *487*, *488*, 511(21), 513(21), *562*
Moore, F. K., 396(28), *487*
Moretti, P. M., 404, *488*
Morley, E. W., 197(31), *363*
Morrow, J. D., 29(44), *124*
Mozorov, M. G., 69(117), 73(117), *127*
Mueller, W. K., 474(236), 484(236), *497*
Mukhin, V. A., 379(11), 380(11), *486*
Muntz, E. P., 41, 65(80), *126*
Myers, G. E., 474(314), *501*

N

Nash, J. F., 18, 28, 29(30), *123*
Naurits, L. P., 390(19), 405(19), *486*
Navins, R. G., 474(332), *502*
Naysmith, A., 41(86, 87), *126*
Nazarov, N. I., 418(138, 139), *492*, *493*
Neiland, V. Ya., 396(32), *487*
Nelson, W., 98(160), 100(160), 101, *130*
Nestler, D. E., 41(89), 42, 69(89), 75(89), 82(89, 139), 94(89), 95(89), *126*, *129*
Newlander, R. A., 94(147), *129*
Nicoll, K. M., 75(131), 76(131), 80, *128*
Nielsen, J. N., 109(171), 116, 117, 118, *130*, *131*
Nikitenko, N. I., 435(165, 166), 436(165, 166), *494*
Nikitin, I. K., 400, *488*
Nishikawa, K., 551, 555(82), *564*
Nishiwaki, N., 546(65), *563*
Norden, H. V., 418(117), *491*
Norman, R. F., 474(314), *501*
North, R. J., 195(23), *363*
Nurick, W. H., 11(21), 30(21), 34(21), *123*
Nusselt, W., 504, *561*

O

O'Brien, R. J., 312(102), *366*
Oleinik, O. A., 396(38), *487*
Olive, K. W., 406(61), 407(61), 409(61), 410(61), *489*
Osipova, V. A., 416(87, 88), *490*
Ostrach, S., 396(28), *487*
Ostrander, L. E., 474(306), *501*
Ostroumov, G. A., 474(225, 246), 483(225, 246), *497, 498*
Oswatitch, K., 3(5), *122*
Ou, J. W., 396, 397(40), 431, *488*

P

Paal, L. L., 407(62), 408(62), 409(62), *489*
Page, F., 515(27), *562*
Page, R. H., 7(17, 18, 19), 15(17, 19), *123*
Pai, Shi-I., *488*
Pak, M. I., 416(87, 88), *490*
Pallone, A., 41(77, 78), 65(77, 78), 87, 89(140), 90, 91(140), *126, 129*
Pan, F., 61, *127*
Panaev, V. N., 457(185), *495*
Panchurin, N. A., 405, 408, 410, *489*
Parnas, A. L., 474(221), 479, *497*
Pavlovskii, Iu. N., 116, *131*
Pavlovsky, G. I., 418(101), *491*
Pearce, W. M., 312(105), *366*
Pearcey, H. H., 95(158), *130*
Pelepeichenko, I. P., 474(222), 480, *497*
Perelman, T. L., 376(2, 3, 4, 5), 377, 464, 465(3, 4, 192), *486, 495*
Pereverzev, D. A., 418(131), *492*
Perkins, H. C., 539, 542, 543(60), *563*
Perlmutter, M., 425(159), 428, 430(159, 160), 433(160, 163), 434(163), 458, 459 (159), *494*
Pery, R., 38(72), *125*
Petukhov, B. S., 378(9), 396, 397(9), 419, 466(9), *486*, 504(13, 14), 506, 521, 523(13), 526, 531, 533(14), 539, 540(61), 541(61), *562, 563, 564*
Petukhov, V. V., 526, *562*
Philpot, J. ST. L., 162, *363*
Pikus, I. F., 418(126), *492*
Pirueva, L. V., 418(118), *491*
Piterskikh, G. P., 504(12), *562*
Pleshchenkov, G. A., 416(95), 417(97), *490*

Plumblee, H. E., 73(123), *128*
Plyutinskii, V. I., 447(179), 448(180), 474(285, 317), *494, 500, 501*
Pohlhausen, K., 435(168), 437(168), *494*
Pokryvailo, N. A., 474(223), 480(223), *497*
Polezhaev, V. I., 474(260), 483(260), *499*
Poltav'seva, L. L., 474(310), *501*
Polyakov, Y. A., 416(91), *490*
Polymeropoulus, C. E., 474(258), 480(258), 483(258), *499*
Popov, V. N., 504(14), 506, 521, 531, 533, 546, 549, *562, 563*
Popov, V. P., 474(219, 223), 479, 480(223), *497*
Prakash, S., 435(171), *494*
Prandtl, L., 390, *487*, 504, *561*
Pratt, A. W., 418(103), *491*
Prescott, R., 202, *364*
Presler, A. F., 537, *563*
Price, E. W., 211, 212, *365*
Prichodko, I. M., 418(120), *492*
Probstein, R. F., 105(165), *130*
Prokofiev, V. E., 418(129, 136), *492*
Prokopets, S. I., 418(105), *491*
Proskuryakov, K. I., 474(285), *500*
Protopopov, V. S., 551, 552, *564*

R

Rabonovich, G. D., 474(275, 276, 283), *499, 500*
Randall, D. G., 551, *564*
Raychandhuri, B. C., 474(245, 256), 483(245, 256), *498*
Rayleigh, J. W. S., 197, *364*
Rebuffet, R., 29(49), *124*
Redekopp, L. G., 73, 75(130), 77(130), 80(130), 81(130), 82, *128*
Reeves, B. L., 14(25), *123*
Regan, J. D., 29(39), *124*
Reichardt, H., 504(7), 512, 514, *561*
Reimann, M., 317(88), *366*
Reshotko, E., 111, *131*
Reynolds, O., 504, *561*
Rhodes, P. C., 474(208), *496*
Richardson, P. D., 511(23), *562*
Rienitz, J., 135(7), 195(7), *363*
Riley, N., 474(202), 475, *496*
Rivas, A., 418(130), *492*
Rogers, E. W. E., 98, *130*

AUTHOR INDEX 573

Rohonczy, G., 532, *563*
Roisen, L. I., 526, *562*
Roller, J. E., 405, 406(60), 407(60), 408(60), 409(60), 410(60), *489*
Rom, J., 38, 41(88), 109(69), *125, 126*
Ronchi, V., 199, *364*
Roos, J. N., 41(83), 68(83, 115), 69(115), 70(115), 73(83), 76(83), 78(83), 79(83), 82(83), *126, 127*
Rosenbrock, H. H., 474(279), *500*
Rosenhead, L., 396(29), *487*
Roshko, A., 6, 9(20), 29(20), 31(20), 33, 34, 48(56), 52(56), 69(118), 72(118), 77(56), 96, 109, 110, 112(172), 113(172), *122, 123, 124, 128, 130, 131*
Rossiter, J. E., 73(124), *128*
Rotta, J., 398, 399, *488*
Rottenkolber, H., 202, 221, *364, 365*
Rozenshtok, Y. L., 418(104, 124), 474(199), 475, *491, 492, 496*
Rubtsov, N. A., 474(322), *502*
Runge, J., 136(10), *363*
Rushchinsky, V. M., 474(311), *501*
Rutsky, J. N., 474(218), 479, *497*
Ryvkin, V. B., 376(5), 377(5), 465(192), *486, 495*

S

Sabersky, R. H., 525(37), 526, *562*
Sadovnikov, G. V., 474(213), 478(213), *496*
Sage, B. H., 515(27, 28), *562*
Sakai Kunio, 474(302), *501*
Salomatov, V. V., 418(145, 146, 150, 121, 133), *492, 493*
Saltzman, E. J., 29(46), *124*
Sarma, G. N., 396(36, 37), 474(201), *487, 496*
Sarpkaya, T., 411(74), *489*
Saunders, O., 538, *563*
Saydah, A. R., 41(89), 42(89), 69(89), 75(89), 82(89), 94(89), 95(89), *126*
Schaaf, S. A., 38(73), *125*
Schardin, H., 135(2), 162, 166, 187(2), 189, 268, 278, *362*
Schepers, H. J., 71(121, 122), 76(121, 122), 79(122), 80(122), *128*
Schetz, J. A., 474(235, 241), 480(235, 241), 482, 483(241), *498*
Schlinger, W. G., 515(27, 28), *562*
Schmidt, E., 189, *363*

Schmidt, E. M., 41(82), 65(82), *126*
Schmidt, J. F., 404(58), *488*
Schmidt, K. R., 555(95), *564*
Schneider, L., 23(37), *124*
Schödel, G., 284(85), 286, 340(85, 92), 343(85), 344(85), *366*
Schoenhals, R. I., 474(301), *501*
Schulz, G., 135(7), 195(7), 225, 226(70), 227(70), 228(70), *363, 365*
Scott, C. J., 40(76), 41(76), 77(76), *126*
Seban, R. A., 42, *126*, 504(9), 515(26), 525(9), *561, 562*
Seginer, A., 41(88), *126*
Selikhovkin, S. V., 409, 410, *489*
Seliverstov, B. N., 418(119), *491*
Senkin, V. I., 474(288), *500*
Seppa, I., 418(117), *491*
Sergeeva, L. A., 474(213), 478(213), *296*
Serov, E. P., 474, *501*
Shabrin, A. N., 402, *488*
Sherberg, M. G., 38, *125*
Shestov, E. D., 474(310), *501*
Rushchinsky, V. M., 474(311), *501*
Shevyakov, A. A., 474, *501*
Shimazaki, T. T., 504(9), 515(26), 525(9), *561, 562*
Shine, A. J., 312(102), *366*
Shitsman, M. E., 551, 552, 555, 557(77), 558(77), *564*
Shlyaktina, K. I., 418(152), *493*
Shorin, G. N., 390(15), 405(15), *486*
Shshitnikov, V. K., 474(213, 214), 478(213, 214), *496*
Shumakov, N. V., 376(6, 7), 414(6), 415(6), 463(6, 7), 464(6, 7), 467(6), *486*
Sidlyar, M. M., 418(110, 111), *491*
Sidorov, E. A., 440, 441, 442(174, 176), *494*
Sieder, E. N., 531, 532, *562*
Siegel, R., 420, 421(155), 422(154), 424(154), 425(159), 428, 430(154, 159, 160), 431, 433(160, 163), 434(163), 442, 444(177), 445(177), 446(177), 447, 458, 459(159), 469(177), 474(226), 480, *493, 494, 495, 497*, 524(31), 525(33), *562*
Sigeman, S., 474(207), *496*
Silberstein, L., 205, *364*
Simbirsky, D. F., 474(222), 480, *497*
Simon, H. A., 312(98), *366*
Sirieix, M., 17, 27, 29(40), 30(50), 31(51), 34, 51(96), 65(96), 66(96), 113(51), *123, 124, 127*

Sklyar, F. R., 418(143), *493*
Skoglund, V. J., 474(281), *500*
Slattery, R. E., 36(63), *125*
Sleicher, C. A., 515(29), 525(32), *562*
Smeets, G., 194, 199(21), *363*
Smirnov, A. A., 376(8), 463(8), 464, 466, 474(215, 216, 217), 478, 479, 480, *486, 496*
Smith, D., 549, *563*
Smith, F. H., 197, *364*
Smith, J. M., 549, 551, *563, 564*
Smith, M. E., 38, *125*
Smith, W., 211, *365*
Smolsky, B. M., 474(214, 223), 478(214), 480(223), *496, 497*
Soehngen, E. E., 211(57), 312(94), *364, 366*
Softley, E. J., 41, 65(80), *126*
Sokolov, V. S., 418(118), *491*
Solignac, J. L., 17, 27, 34, *123*
Soliman, M., 460, 461, 467, 474(200, 203), 477, *495, 496*
Sommerfeld, A., 136, 170(9), *363*
Sparrow, E. M., 415, 431, 435(172), 442, 444(177), 445(177), 446(177), 447, 467, 469(177), 474, 476, 480(228), 481, *489, 494, 495, 496, 497*, 524, 525(33, 35), 526, *562*
Spee, B. M., 73(127), *128*
Sperling, J., 176, 188(17), *363*
Stamm, H. G., 224, *365*
Steegmaier, K., 265(80), *365*
Stefany, N. E., 474(254), 483(254), 484, *498*
Stephenson, D. G., 416(85), *490*
Sterman, L. S., 526, *562*
Stermole, F. J., 474(282), *500*
Sterret, J. R., 94(146), 95(149), *129*
Stewartson, K., 13, *123*, 396, *487*
Stolz, G., 416(80), *489*
Strack, S. L., 41(84, 85), *126*
Struminskii, V. V., 396, 397(27), *487*
Suetin, O. N., 418(136), *492*
Summerfield, M., 531, *562, 563*
Svensson, H., 162, *363*
Swenson, H. S., 549(67), 551, 553, *563*
Szetela, E. J., 549(71), 550(71), *563*
Szwarcbaum, G., 474(207), *496*

T

Takami, H., 2, *122*
Talmor, E., 390(16), 405, *486*
Tanaka, H., 546, *563*
Tang, H. H., 17, 18, 25(27), 27(27), *123*
Tani, I., 69(119), 109, *128, 130*
Tanner, L. H., 201(46), *364*
Tao, L. N., 474(237, 263), 484(237), *497, 499*
Tarasova, N. V., 551, 554, *563, 564*
Tate, G. N., 531, 532, *562*
Tatom, J. W., 474(252), 483(252), 484, *498*
Taylor, G. Y., 504, *561*
Taylor, J. K., 293, *366*
Taylor, M. F., 538, 539, *563*
Temkin, A. G., 412, 416(86, 90), *489, 490*
Temple, E. B., 202, *364*
Ter-Mkrtchyan, A. A., 376(1), *486*
Thal-Larsen, H., 474(305), *501*
Thiry, F., 73, 80(129), *128*
Thomann, H., 82(137), *128*
Thomasson, R. K., 474(291), *500*
Thomke, G. J., 9(20), 29(20), 31(20), 33, 34, 48(56), 52(56), 77(56), 96, *123, 131*
Tilton, L. W., 293, *366*
Timofeeva, F. A., 474(211), 477, *496*
Todisco, A., 41(77, 78), 65(77, 78), *126*
Tohuda, N., 396(30), 474(30), 476(30), *487*
Tolansky, S., 135(6), 195, *363*
Treadwell, K. M., 474(265), *499*
Trepaud, P., 38(72), *125*
Tribus, M., 525(32), *562*
Trilling, L., 95(156), *130*
Trucco, A., 41(81), 65(81), *126*
Tsetserin, V. A., 408, *489*
Tsoi, P. V., 425, 435(158), *493*
Tsyurik, V. N., 474(311), *501*
Turchin, I. A., 474(247), 480(247), 483(247), *498*
Turton, J. S., 474(273), *499*
Twyman, F., 197(32), *364*
Tyan-Tsi, E., 474(320), *501*
Tyson, T. J., 107, *132*

U

Uberoi, M. S., 390, *487*
Umeda Tosioki, 474(302), *501*
Upatnieks, J., 200, *364*
Usanov, V. V., 390(15, 19), 405(15, 19), *486, 487*

V

Vanden Eykel, E. E., 82(139), *129*
Van-Fan, 551(84), *564*
Van Hise, V., 35(58), *125*
VanVoorhis, C. C., 268, 274, *365*
Vas, I. E., 95(155), *130*
Vasilevsky, K. K., 474(331), *502*
Vasiliev, A. A., 418(142), *493*
Vasiliev, O. F., 402, *488*
Vekshin, V. S., 418(123), *492*
Vel'tishchev, N. A., 551, *564*
Vidin, Y. V., 418(108, 116, 141), *491*, *493*
Vidin, Yu. V., 474(326), *502*
Vikhrev, Yu. V., 551, 555(74), 558(96), *563*, *564*
Vilenskii, V. D., 378, 384, 434, 438, *486*
Viscanta, R., 474(325, 329, 330), *502*
Viviand, H., 38(72), *125*
Vogenitz, F. W., 109(174), 110(174), 112(174), 113(174), *131*
Volik, B. G., 474(296), *500*
Volkov, P. M., 526, 539, *562*
Volkov, V. N., 418(137), *492*
Voropaev, A. R., 418(113), *491*

W

Wachtell, G. P., 244(73), *365*
Walters, H., 538, *563*
Webb, W. H., 109(174, 176), 110, 112, 113, *131*
Webb, W. W., 474(261), 483(261), *499*
Weinbaum, S., 45, 46(93), 47, 48, 49, 50, 52(93), 53, 55(98), 82(98), *126*, *127*
Weinberg, F. J., 135(3), 162, 166, 195, *362*
Weise, A., 265(79), *365*
Weiss, R. F., 45, 48, 50, 53, 57, 58, 61, 62, 64(101), 65(101), 98(160), 100(160), 101, 107, *126*, *127*, *130*
Welch, C. P., 551, *564*
Weyl, E. J., 135(5), *362*
White, R. A., 29(47), *124*
Wieland, W. F., 539, *563*
Wiener, O., 162, *363*
Wiggs, M. M., 79(134), *128*
Winbrow, W. R., 29(43), *124*
Winckler, J., 244(74), 268(78), 274(78), *365*
Wolf, E., 135(1), 169, 179(1), 195, 219(1), 220(1), 292(1), *362*
Wolf, H., 416(81), *490*, 538, *563*
Wolter, H., 135(4), 162, 175, 178, 187(4), 201(4), *362*, *363*
Wood, R. D., 551, *564*
Woodgen, A., 95(157), *130*
Worsoe-Schmidt, P., 539, 542, 543(60), *563*
Wright, F. C., 538, *563*
Wyborny, H. P., 71, 76(121, 122), 79, 80, *128*

Y

Yaglom, A. M., 395(23), *487*, 511(21), 513(21), *562*
Yakovlev, V. V., 525(36), 526, 531, *562*
Yakovleva, R. V., 474, *501*
Yakura, J. K., 33, 35(53), 44, 51(53), *124*
Yamagata, K., 551, 555(82), *564*
Yang, K. T., 396(35), 474(229, 233, 248), 480(233, 248), 482, *487*, *497*, *498*
Yang, W. J., 396, 397(40), 431, 474(30, 1,3 266, 268, 269, 270, 271, 292), 476(30,31), *487*, *488*, *499*, *500*
Yanowitz, H., 474(253), 483(253), *498*
Yarishev, I. A., 418(122), *492*
Yarkho, S. A., 376(1), *486*
Yaroshenko, Y. G., 418(143), *493*
Yudovich, V. I., 474(259), 483(259), *499*
Yung, A. D., 398(42), 411(42), *488*

Z

Zakkay, V., 41(79), 65(79), *126*
Zalutskii, E. V., 390(17), *486*
Zaric, Z., 403, *488*
Zehnder, L., 195, *363*
Zeiberg, S. L., 474(236), 484(236), *497*
Zernike, F., 201, *364*
Zhemkov, L. I., 416(92), 418(123), *490*, *492*
Zhuk, V. I., 416(25), 418(152), *487*, *493*
Zolotogorov, M. S., 418(114, 115), *491*
Zueva, G. K., 418(118), *491*
Zukoski, E. E., 92, 94(148), 108(184), *129*

Subject Index

A

Abbe's theory, 186
Annulus, 312

C

Channel flow
  chemical reactions, 418
  mass transfer
  one-dimensional model, 374
  thermal conductivity variable, 533
  viscosity variable, 528, 533
Conjugated problems, 368, 373, 411, 457
Concentration boundary layer, 182
Converging-diverging flows, 403
Cornu's spiral, 172

D

Diffusion coefficient, 164

E

Eddy diffusivities, 510, 560
Eikonal equation, 136
Entrance effects, 378, 396, 410, 431, 525, 540

F

Fermat's principle, 145
Fraunhofer-Fresnel diffraction, 184
Fresnel biprism

G

Gladstone-Dale equation, 276, 287

H

Holography, 200
Huygens' principle, 146, 169
Hydrogen flame, 326

I

Illumination intensity, 151
Interferometer adjustment, 211
Interferometer interference contrast, 237

J

Jamin interferometer, 197

L

Liquid metal, 437
Lorentz–Lorenz equation, 284
Lyon integral, 519
Luminous intensity

O

Osculation plane, 141

Q

Quasi-stationary flows, 388

R

Rotating cylinder, 317

S

Schliere, 135, 144, 150, 274, 320
Schlieren enthalpy, 275
Schlieren interference, 202
Schlieren lens, 155, 184, 187

SUBJECT INDEX

Separated flows
  acoustic radiation, 73
  boundary layer solutions, 101, 107, 110, 116
  breakaway separation, 6
  Chapman-Korst model, 7, 17, 22
  critical point, 34, 56, 112
  downstream facing step, 4, 33, 41, 114
  eddy viscosity, 16, 120
  free interaction, 84
  invisid solutions, 2
  lip shocks, 19, 50, 55
  low density flows, 38
  notch induced separation, 5, 60, 68, 75
  ramp induced separation, 4, 73
  recompression, 27, 34
  recovery temperature, 94
  reviews articles, 6
  shock-induced separation, 4, 5
  slow flows, 2, 61
  stability, 36
  supersonic flows, 19
  transition, 35, 38, 77, 87, 92
  transpiration, 80
  unsteady heat transfer, 479
  upstream facing step, 4, 5, 6, 67, 83, 84, 89, 95
  wedge flows, 37, 51, 52, 65
Snell's law, 144

T

Thermal boundary layer, 180
Thermal diffusion, 344
Thermal diffusivity measurements, 332

W

Wandbindung, 156, 159, 258

Z

Zhukovsky number, 383

QC
320
A1A3
v.6
1970

SEP 21 1970